Product Design and Manufacturing

A Product Life Cycle Approach

Product Design and Manufacturing

A Product Life Cycle Approach

Editors

Srinivasa Prakash Regalla
Professor
Mechanical Engineering
and
Coordinator
Centre for Product Design and Realization (CPDR)
BITS Pilani, Hyderabad

Kuldip Singh Sangwan
Associate Professor and Head
Department of Mechanical Engineering
BITS Pilani, Hyderabad

Publishing Globally

NEW AGE INTERNATIONAL (P) LIMITED, PUBLISHERS
New Delhi • Bangalore • Chennai • Cochin • Guwahati
Hyderabad • Kolkata • Lucknow • Mumbai
Visit us at www.newagepublishers.com

Published by New Age International (P) Ltd., Publishers
First Edition: 2014

BRANCHES

- **Bangalore** 37/10, 8th Cross (Near Hanuman Temple), Azad Nagar, Chamarajpet, Bangalore-560 018
Tel.: (080) 26756823, Telefax: 26756820, **E-mail: bangalore@newagepublishers.com**
- **Chennai** 26, Damodaran Street, T. Nagar, Chennai-600 017
Tel.: (044) 24353401, Telefax: 24351463, **E-mail: chennai@newagepublishers.com**
- **Cochin** CC-39/1016, Carrier Station Road, Ernakulam South, Cochin-682 016
Tel.: (0484) 2377004, Telefax: 4051303, **E-mail: cochin@newagepublishers.com**
- **Guwahati** Hemsen Complex, Mohd. Shah Road, Paltan Bazar, Near Starline Hotel, Guwahati-781 008
Tel.: (0361) 2513881, Telefax: 2543669, **E-mail: guwahati@newagepublishers.com**
- **Hyderabad** 105, 1st Floor, Madhiray Kaveri Tower, 3-2-19, Azam Jahi Road, Nimboliadda, Hyderabad-500 027
Tel.: (040) 24652456, Telefax: 24652457, **E-mail: hyderabad@newagepublishers.com**
- **Kolkata** RDB Chambers (Formerly Lotus Cinema) 106A, 1st Floor, S.N. Banerjee Road, Kolkata-700 014
Tel.: (033) 22273773, Telefax: 22275247, **E-mail: kolkata@newagepublishers.com**
- **Lucknow** 16-A, Jopling Road, Lucknow-226 001
Tel.: (0522) 2209578, 4045297, Telefax: 2204098, **E-mail:lucknow@newagepublishers.com**
- **Mumbai** 142C, Victor House, Ground Floor, N.M. Joshi Marg, Lower Parel, Mumbai-400 013
Tel.: (022) 24927869, Telefax: 24915415, **E-mail: mumbai@newagepublishers.com**
- **New Delhi** 22, Golden House, Daryaganj, New Delhi-110 002
Tel.: (011) 23262368, 23262370, Telefax: 43551305, **E-mail: sales@newagepublishers.com**

ISBN: 978-81-224-3608-2

₹ 399.00

C-13-08-7100

Printed in India at Ajit Printers, Delhi.
Typeset at In House, Delhi.

PUBLISHING FOR ONE WORLD
NEW AGE INTERNATIONAL (P) LIMITED, PUBLISHERS
7/30 A, Daryaganj, New Delhi-110002
Visit us at **www.newagepublishers.com**

PREFACE

The objective of this book is to put together the subject knowledge pertaining to the current philosophy, techniques and latest knowledge on the subject of product design and manufacturing with a particular emphasis on the product lifecycle approach. Without the consciousness of the product's lifecycle the outlook is likely to miss the bounds, may progress unilaterally in only one direction without a balance among the different important issues being treated. Thus the framework of product lifecycle was used to provide as a binding constraint to drive the discussions towards a converging idea of reducing lead time, improving product quality and reducing useless redesign efforts that might result due to lack of coordination among the people belonging to various stages of the product lifecycle. This framework is depicted on the cover page of this book.

The editors derived the motivation to bring out this book from observations made and lessons learnt from the deliberations of a national conference on design for product life cycle, organized by them at BITS, Pilani. Secondly, there appeared the lack of a book treating this subject at senior undergraduate level and graduate level. This book is a collective effort of not only the authors but also several eminent engineer scientists. Their names are mentioned at topic wise. The editors are immensely thankful to them for their valuable contributions. The editors are thankful to Prof. L. K. Maheswhari, former Vice-Chancellor and Prof. B. N. Jain, Vice-Chancellor of BITS Pilani for their encouragement and support.

No book is either complete or perfect in itself. Nevertheless, there is always some scope for improvement that may miss the editors' vision due to the magnitude of the work to be managed while writing a book of this size. We will be grateful to the readers who are most welcome to send suggestions for further improvement by email. Every effort will be made to make corrections and add improvements in the next edition.

Editors

Srinivasa Prakash Regalla

regalla@hyderabad.bits-pilani.ac.in

Kuldip Singh Sangwan

kss@pilani.bits-pilani.ac.in

LIST OF CONTRIBUTORS

AMARESH CHAKRABARTI
Professor, Mechanical Engineering
IDeaS Lab, CPDM, Indian Institute of Science, Bangalore

A. KANDA
Professor, Mechanical Engineering
Indian Institute of Technology, Delhi
Hauz Khas, New Delhi – 110016

A. MAKADIA
V.V.P. Engineering College, Rajkot

A. PATEL
V.V.P. Engineering College, Rajkot

AJAY RATHORE
Assistant Professor, Mechanical Engineering
Malaviya National Institute of Technology
Jaipur, Rajasthan

A. K. SEN
Terna Engineering College, Navi Mumbai

ANAND G.
Assistant Professor, Mechanical Engineering
Birla Institute of Technology and Science, Pilani
Vidya Vihar, Pilani – 333031, Rajasthan

ANEES Y. KHAN
Birla Institute of Technology and Science, Pilani
Vidya Vihar, Pilani – 333031, Rajasthan

ATUL THAKUR
Indian Institute of Technology, Bombay, Powai
Mumbai

B. RAGHUNATH
A.D. Patel Institute of Technology, Gujarat

B. RAVI
Professor, Department of Mechanical Engineering, Indian Institute of Technology
Bombay, Powai Mumbai

B. K. ROUT
Birla Institute of Technology and Science Pilani
Vidya Vihar, Pilani – 333031, Rajasthan

BADARINATH AMBATI
Altair Engineering India Pvt. Ltd., Bangalore

C. JEGADHEESAN
Kongu Engineering College, Erode

C. S. P. RAO
National Institute of Technology, Warangal

DARSONO TJOKROAMIDJOJO
Professor, Department of Mechanical Engineering, University of North Dakota USA

DEBABRATA MAJI
NERIST, Nirjuli, Arunachal Pradesh

A CHENNAKESAVA REDDY
JNTU College of Engineering, Hyderabad

GOKULA VIJAYKUMAR A. V.
IDeaS Lab, CPDM, Indian Institute of Science Bangalore

J. MADAN
Indian Institute of Technology, Delhi
Hauz Khas, New Delhi – 110016

J. THAKKAR
A. D. Patel Institute of Technology, Gujarat

K. P. RAO
GRASIM, Jodhpur

K.V. CHANDRATRE
N.D.M.V.P.S'S College of Engineering, Nashik

KAJAL KUMAR MANDAL
NERIST, Nirjuli, Arunachal Pradesh

KODALI RAMBABU
Professor & Group Leader
Mechanical Engineering
Birla Institute of Technology and Science, Pilani
Vidya Vihar, Pilani – 333031, Rajasthan

KULDIP SINGH SANGWAN
Associate Professor & Head
Department of Mechanical Engineering
Birla Institute of Technology and Science
Pilani, Vidya Vihar, Pilani – 333031, Rajasthan

L. VEDHAVALLI
Carborundum Universal Limited, Chennai

M. D. BETGERI
Department of Mechanical Engineering
Budhgaon

M. Y. KHIRE
Department of Mechanical Engineering
Budhgaon

MADHUMITA DUTTA
VCET, Thane

MONICA SHARMA
Birla Institute of Technology and Science, Pilani
Vidya Vihar, Pilani – 333031, Rajasthan

RAGHAVENDRA DHARMARAO
Altair Engineering India Pvt. Ltd., Bangalore

OM PRAKASH YADAV
Professor, Department of Mechanical Engineering, University of North Dakota USA

P. MUTHUKUMAR
Kongu Engineering College, Erode, Tamil Nadu

P. V. M. RAO
Associate Professor, Mechanical Engineering,
Indian Institute of Technology, Delhi
Hauz Khas, New Delhi – 110016

PRABIR SARKAR
IDeaS Lab, CPDM, Indian Institute of Science, Bangalore.

PRADEEP YAMMIYAVAR
Professor, Mechanical Engineering
Indian Institute of Technology, Guwahati
Assam

PRASHANT DIXIT
Indian Institute of Technology, Guwahati
Assam

R. K. MITTAL
Professor & Director
Birla Institute of Technology and Science
Pilani, Dubai Campus, Dubai, UAE.

R.P. TERKAR

Terna Engineering College, Navi Mumbai

S. A. PUJARI

Textile & Engineering Institute
Ichalkaranji, Pune

S. G. DESHMUKH

Professor, Mechanical Engineering
Indian Institute of Technology, Delhi
Hauz Khas, New Delhi – 110016

S. JEROME

Kongu Engineering College, Erode,
Tamil Nadu

S. K. SAHA

Professor, Mechanical Engineering
Indian Institute of Technology, Delhi
Hauz Khas, New Delhi – 110016.

S. P. SINGH

Indian Institute of Technology, Delhi
Hauz Khas, New Delhi – 110016

S. PATI

Indian Institute of Technology, Delhi
Hauz Khas, New Delhi – 110016

S. S. MANTHA

Terna Engineering College, Navi Mumbai

S. S. PANDE

Professor, Department of Mechanical Engineering, Indian Institute of Technology, Bombay, Powai, Mumbai

S. LAKSHMINARASIMHAN

TVS Needles, Madurai

S.M. KANNAN

Thiagarayar College, Madurai

SAMPATRAO D. MANJARE

Birla Institute of Technology and Science
Pilani – Goa Campus, Goa

SANDEEP MUNDE

Department of Mechanical Engineering
Indian Institute of Technology, Bombay
Powai, Mumbai

SANDEEP S. PATIL

Aurora Engineering College, Hyderabad

SRIKANTA ROUTROY

Assistant Professor, Mechanical Engineering
Birla Institute of Technology and Science, Pilani
Vidya Vihar, Pilani – 333031, Rajasthan

SRINIVAS JAMPANI

Birla Institute of Technology and Science, Pilani
Vidya Vihar, Pilani – 333031, Rajasthan

SRINIVAS KOTA

IDeaS Lab, CPDM, Indian Institute of Science, Bangalore

SRINIVASA PRAKASH REGALLA

Professor, Department of Mechanical Engineering, Birla Institute of Technology and Science, Pilani – Hyderabad Campus, Jawahar Nagar Shameerpet Mandal, Ranga Reddy District, Hyderabad – 500078
regalla@hyderabad.bits-pilani.ac.in

SUNIL PADUCHURU

Altair Engineering India Pvt. Ltd., Bangalore

SUSHRUT PAVANASKAR

Indian Institute of Technology, Bombay, Powai Mumbai

T. G. RAO

Indian Institute of Technology Delhi
Hauz Khas, New Delhi – 110016

T. K. KUNDRA

Professor, Mechanical Engineering
Indian Institute of Technology, Delhi
Hauz Khas, New Delhi – 110016

T.A. JANARDHAN REDDY

Osmania University, Hyderabad

UDAY M. CHAUDHARI
Aurora Engineering College, Hyderabad

V. B. SUNIL
Indian Institute of Technology, Bombay
Powai, Mumbai

V. D. SHINDE
Textile & Engineering Institute
Ichalkaranji, Pune

V. E. ANNAMALAI
Carborundum Universal Limited, Chennai

V. R. NAIK
Textile & Engineering Institute,
Ichalkaranji, Pune

V. R. V. R. KRISHNAN
Birla Institute of Technology and Science
Pilani, Vidya Vihar, Pilani – 333031
Rajasthan

V. V. THIGLE
N.D.M.V.P.S'S College of Engineering
Nashik

VIKRAM BATRA
Indian Institute of Technology
Guwahati, Assam

Y. RAVI KUMAR
National Institute of Technology, Warangal

CONTENTS

SECTION - I

COLLABORATIVE DESIGN, PRODUCT DESIGN AND PRODUCT LIFE CYCLE MANAGEMENT

COLLABORATIVE DESIGN

1.1. INTRODUCTION

Collaborative design enable designers located geographically in various parts of the world through Internet tools. The designers may be of different knowledge domains and located at disparate and geographically far and distant places, anywhere around the world. The participation in product design must take place in real time as well as off line simultaneously or concomitantly. The need for collaborative design has arisen only in the last decade and due to two main reasons. First reason is to adopt the modern concept of *design for product life cycle* in order to reduce the lead times of new or redesigned products. This is of great importance and has a valid connection to the current scenario of globalized economies. Lead time is the total duration of time from the ideation of the product to the actual introduction of the product into the market. In the olden days, customers were much more conservative, brand conscious and were complacent to use the same *Cinni* or *Usha* fan repeatedly over the years. Even if the first fan failed, the customers would again purchase the product of the same brand. This was because there were not many manufactures in the market to drive the competition and quality forward. In contrast to that, in the current markets, customers have been offered with an array of options of high quality products to choose, in any product category. The companies have already begun to think about themselves as placed in a global market and are enthusiastically vying to produce better quality and less costly products to stay in the market. This has lead to what is called *dynamic markets*, in which the customers' tastes keep changing fast because the customers are now habituated to upgrade or change their needs at a much more rapid phase. This has necessitated, in general, a need for lesser lead time for any product. If the lead time is high, the idea will die away and cease to be relevant to the customers' tastes even before it can see the light of the market.

In collaborative design technology, the designers need not physically meet at a meeting place to participate in the design of the product. They can give their contribution by sitting in their office room, using their desktop computer both in on-line and off-line modes. This has enabled to design by use of computer software that works on client-server communication model. A simple collaborative design environment (CDE) may employ the paradigm of one server and multiple clients whereas fully developed and efficient CDE may employ multiple distributed servers and multiple clients' paradigm. Both the CDEs have extensively benefited from the recent developments in networking technology and the techniques of designing multi-user software. The idea of "dotcom" business may have failed in the late twentieth century but the efforts put by various companies and consortiums have taken the web based technology ahead by leaps and bounds. All those developments are now paying dividends in an indirect way.

1.2. SIMPLE COLLABORATIVE DESIGN ENVIRONMENTS

The main challenge in building any CDE is to capture the design semantics, design decisions and record them in the most authentic way. When two designers sit face to face, to design a product, the communication between them constitutes the so called "design semantics". It could be speech, written or drawing based communication. A well-built CDE should be able to capture that communication in most realistic form, as truer to the face-to-face communication.

1.3. DESIGN SEMANTICS

With respect to the above discussion, on how effective is the collaborative design methodology vis-à-vis the face-to-face brainstorm, it has become necessary to distinguish between two different types of information exchange. Those parts of total communication that serve to describe actual design decisions are termed as *design semantics*. These design semantics have justifications of design decisions, and properties associated with performance and behaviour of designs. Undocumented design decisions and vague design descriptions, on the other hand can lead to misunderstanding and confusion in the collaborative development of a design. Therefore, all other parts of the total information exchange are termed as non-semantic information. Having defined these terms, it is now clear that an ideal CDE should maximize the design semantics and minimize the other types of communication.

Whenever a new collaborative design workbench (CDW) is designed, the efficacy of which can be described in terms of the percentage of design semantics it finally permits. Such capturing of design semantics, in different CDEs, has been demonstrated with different data models by various researchers.

1.4. OBJECT ORIENTED DATA MODELS

In the objective oriented model of CDE, the information is integrated by means of certain structured framework named as design prototype, which represent the information relevant to design. In this approach, each class of design objects are characterized by its function, behaviour and structure. The project data can be stored centrally and accessed by separate programs. For example, the part being designed may be modeled by the constructive solid geometry (CSG) approach of solid modeling where the relevant Boolean operations constitute the structured framework. As long as the CAD system is used, any designer is capable of exceling CSG modeling paradigm and the information exchange can have no awkward transition virtually.

1.5. COLLABORATIVE DESIGN WORKBENCHES (CDW)

In this methodology, a client-server paradigm is implemented in which the client software resides on the designer's computer and either simply allows snapshot images of the part in the required orientation or may allow full list viewing options for the solid model. The client software also supports a toolbar that can enable the designer to embed textual, graphical, speech and video information to be embedded on the top of such snapshot views by proper linking thereby capturing multidisciplinary interpretations of the design. All the data in this model also is again stored in the central server to be shared by the designers globally located. Certain workbenches employed third party CAD model viewers, which permit rotation of the object also. Normally, one of the neutral formats such as IGES, WRL, STEP etc. are used to store and play the part information.

In CDW, a computer-based design environment, the shared workspace has an electronic and distributed space. Some of the issues that are taken for granted in a shared physical workspace, such as access to the person who produced the design idea or the familiarity of informal or formal group meetings, have to be carefully considered when the workspace is a distributed, electronic representation. Initially, when the designers begin to collaboratively design the part on the workbench, each has own understanding of the part design and possible solution methodologies. However, the shared or overlapping quantity of such understanding among all the designers forming the design team for that part is small in the beginning. As the time passes and collaborative design continues, the portion of this shared information should grow rapidly. The faster it grows, the more efficient is CDW. The components of a CDW address the concept of shared understanding are information sharing, communication media, process management, and exploration space as shown in the Fig. 1.1. These can also be described as functionalities of the CDW which we will discuss one by one.

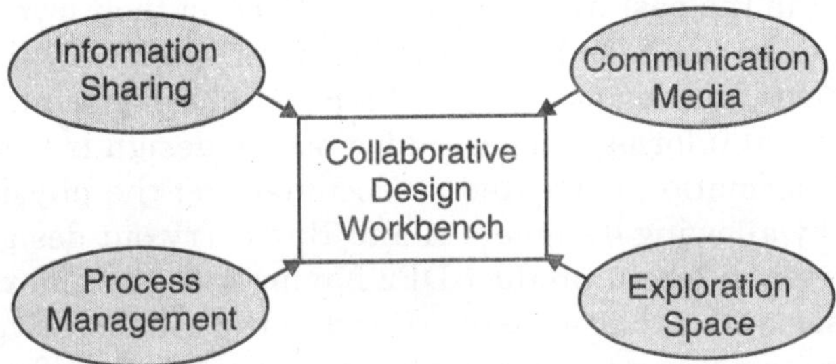

Fig. 1.1: The basic framework of CDW showing the four components or functions

Information sharing requires that the CDW should support representation and access of shared design information to each of the participant and their respective computer applications. *Communication media* represents shared workspace that can be evaluated by number of modes in which the communication among designers is possible. One mode of communication is the exchange of design information generated among designers. In order to ensure consistency, designers have to refer to this information, to update their own solution states. A second mode of communication is the informal exchange of ideas through conversation, gesturing and rough sketches. The focus in this view of the shared workspace is on the visual interaction of the designers and the graphical representation of the design solution. The process of planning, managing and controlling design activities has been identified as a key requirement for the development of computer-integrated design applications. *Process management* function of the CDW includes management of information into various versions or the software coding thereof, management of the server, installation and maintenance of the client software on designers desktops, management of the accent control system, networking and security issues. The management and administration of the CDW should be carried out in a way that the designer faces little problem in the process of his full participation. One of the most inefficient process management is denial of access to certain portions of the design information to the designer. This may result in undue delay in design decision making or poor quality design semantics, which makes a debated designer to hardly focus on his work.

The CDW should also be so designed as to serve as an *exploration space* for new ideas than merely making available a record of various versions of the design as it proceeds through the various stages of collaborative design. The function of exploration space is to expand possibilities of new design solutions and even misunderstanding of others opinion as catalyst for nucleation of new ideas. For example, any suggestion on "revolving" of a new chair being designed can lead to multiple ideas for creative chair as either oscillating about a horizontal axis or revolving about a vertical axis. Similarly, for another example, suggestion of a "remote" nature of function of a robot being designed for welding in toxic environments can lead to multiple ideas such as like the robot being controlled by a remote controller or virtual reality controller or the operation being conducted in a remote environment or in an airborne vehicle.

1.6. HOW DESIGNERS COMMUNICATE AND DOCUMENT THEIR DESIGNS

Protocol studies, aimed at finding out how much more and what type of design semantics are documented in a CDE as compared to computer aided design carried out alone by a design engineer, were carried out in the past by Maher [1] in a CDW of their own named as Computer Mediated Collaborative Design (CMCD). In order to bring out specifically the active documented justification for a decision, or function or purpose of a physical component, which are named as design semantics, the textual information embedded on the design is treated as documented semantics and graphic information is treated as documenting the physical component. The experiments conducted, by allowing designers A and B to carry out designing a product DP1 individually first and then a different product DP2 having same complexity as DP1 in CDW, showed that the design semantics that were expressed in the collaborative design were not as much more as expected in comparison to individually carried out design. Two reasons were found to be responsible for this.

Firstly, current CAD systems with all of their emphasis on spatial geometry take up all the time and attention of the designer in dealing with the spatial geometry of the product. That results in much lesser opportunity and fewer tools for the designer for recording design semantics. In a CAD system, the designer associates attributes with a CAD element or object and in a whiteboard environment the designer uses the text tool. These tools do not provide any structure to the design semantics and they do not encourage or facilitate the documentation of design semantics. Secondly, designers tend to document less information in a collaborative session than when they are working alone because they can describe the design semantics verbally or orally to the other designers. From this point of view restricting those tools that permit direct verbal conversation on the CDW can give better results because then the designers would be forced to type the same in textual form or embed them as a audio link, thus resulting in more design semantics. The intensive information exchange via audio-video conferencing between the parties during a CDW session would rather result in a valuable amount of the semantic information that is properly documented. Video conferencing only allows meetings to take place while the participants are in remote locations, but does not provide any additional facilities for recording or formalizing the meeting and hence is not suitable for collaborative design. This is because the designers describe their design semantics verbally, through video and audio channels, and this information is not included in the final design document. To be able to record and extract this information, a CDW tool should be able to provide means for keeping track of designers' activities, indirectly forcing them to record the semantics that stand behind their design solutions. If properly put to use through appropriately designed CDW software, the existing highly powerful and fast computer media not only can allow us to document more design information, but also can give the opportunity to keep the whole design documentation in electronic form, available for further use and processing.

1.7. A DATABASE CENTRIC COLLABORATIVE DESIGN WORKBENCH

Designing a new product that is found out to be in compliance with the latest needs of the customer in the twenty first century is more of a task of linking the design knowledge of the existing products to the newer needs of the customers rather than designing from level zero or scratch. This is because design engineers have at their disposal a huge amount of design knowledge of various kinds of products among which the features of one to another do not vary too much. The sheer magnitude of this knowledge base demands, for easy retrieval and storage, a systematically conceived database management system that is specifically designed for design knowledge. In addition to that, this database must be integrated with a well designed collaborative environment wherein designers placed at globally far and distant places can communicate with each other to undertake the design of the products. The whole system must be woven around a widely accepted engineering design methodology to be technically and practically sound. The entire design, manufacturing and measurement processes can be integrated into a single workbench. This section has focus on the design database and its structure for typical mechanical engineering products and the collaborative design environment. The database was built in the MS Access and the collaborative design workbench was built in the Visual Basic and Visual C++.

Engineering design methodology has attracted widespread attention in the recent years. The primary reason for this increased attention is the ability of the design decisions to affect the quality of the product. Moreover, the cost of redesign to maximize quality can be prohibitively high in the traditional life cycle of a product in which, the market survey to know the requirement of customers is followed by initial conception of the shape and size of the product, then by the material selection, then by stress analysis to arrive at the safe dimensions, then by prototype manufacture, then by testing. During the first stages of the design, often the product is designed from scratch with features that do not vastly vary from most other products. Most product features are similar and hence this old method needs change.

In the past few decades as the theory of engineering design has developed, variations of the above methodology were proposed. In the methodology proposed by the classical work by Hubka [2], the Spiral of Progress improves the completeness of information of the design as the design process progresses from problem assignment to finally the assembly drawings. The possibility of the Spiral of Progress retracting back due to some design decisions being found erroneous or unviable at a later stage is not taken into account. Nevertheless the last stage of the analysis-synthesis-evaluation approach by Hubka [2] was to suggest any redesign to be effected. Such possibility was first hinted in an earlier work [3] in which a two-dimensional view of design progress was enunciated. The vertical series of activities indicate the phases in the life cycle of the product whereas the horizontal set of activities indicate the problem-solving process that may take place at many stages of vertical dimension. This concept was much developed in the model of Pahl and Beitz [4], where the possibility of upgrading and improving the design at all stages was emphasized. A complete description and discussion on various engineering design methodologies proposed in past is given in [5]. The importance and necessity of a design workbench that can enable design of a product collaboratively by a design team whose members are globally dispersed was already brought out [6, 7]. In an industrial environment where the manufacturing lead time has to be minimized, each new product cannot be designed from scratch. Moreover, it is very sensible to utilize the knowledge information of previously designed similar products. This is practicable if the information is properly stored and made possible to be easily retrieved. A systematic procedure should be used for storage and retrieval process of design information.

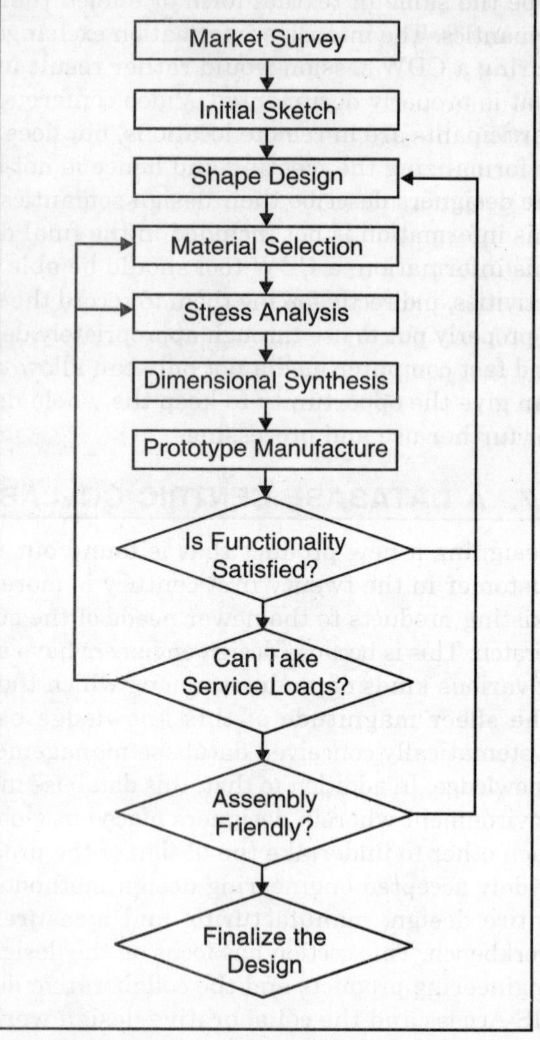

Fig. 1.2: The current design procedure

Thus, design of each new product or new version of the product involves a lot of repetitive and redundant work. During the manufacturing of the prototype and testing, many errors are found in the product, either from the point of view of manufacturability or assembly. It may even be with respect to lack of ultimate conformity with the service requirements. All these kinds of errors necessitate redesign. For redesign, the part drawing has to be redirected back to the design team that tries to understand the viewpoint of the shop floor engineers and tries to correct the mistakes in the design. This kind of flow of the part drawing between the design department and the shop floor may occur many times before a satisfactory product is finalized and concluded to be produced in the necessary quantity. The traditional design methodology is what is shown in Fig. 1.2 sans the redirecting arrows. In a manufacturing environment where the lead time is to be minimized, the traditional design methodology is hardly in place.

1.8. DESIGN REPOSITORY

The design repository is a very important component of the **Co**llaborative **D**esign and **Ma**nufacturing **W**ork**b**ench (CoDMaWb), which is a model CDW developed by the first author. Unlike the usual database where information pertaining to a given entity is stored in fields and records which can be of different types (integer, character, boolean etc.), a design repository has to be constructed in a different way. Here we need to store designs. A CAD model is the representation of the design. It is essential to identify the distinguishing features of the product and develop a code in the name of which each feature of the product's design is stored. There are many methods of coding parts including those suggested by group technology. The design team can develop its own coding method. In any case, the coding system should ensure that no duplication occurs and that the code can be computerized. The designs, in the form of CAD models, reside in the design repository on the computer memory disk and the interface between the database system and the repository is through the code. The general layout of this system is given in Fig. 1.3. In what follows, the design repository is exemplified through screen snapshots of CoDMaWb.

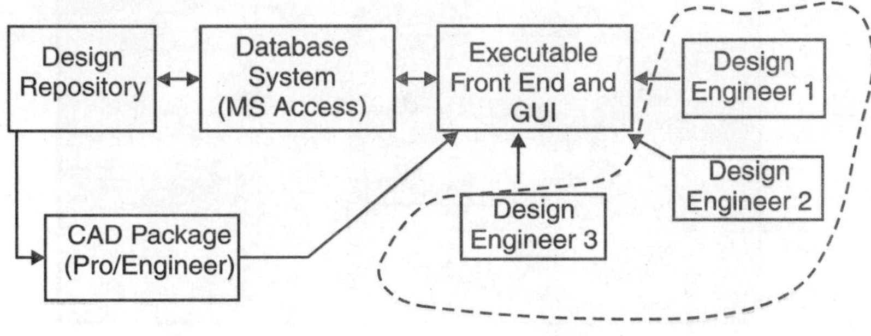

Fig. 1.3: The general layout of the CoDMaWb

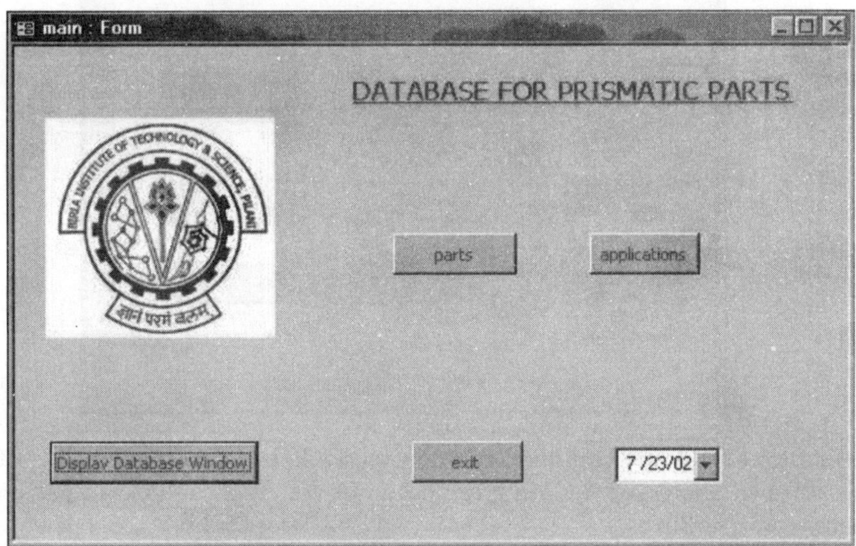

The front end's main menu, which was created in MS Access shows two options. The designer can choose any category or ancillary applications. The *parts* button takes to the list of the various products whose designs are available in the repository.

The different parts of the designs which are stored in the repository can be the screw, bearing, nut, bolt etc. if the design being carried out is of a mechanical system.

What is important is to identify the distinguishing features of each part that can uniquely yet completely define the part. For example, the pitch diameter, pitch, material, type and length are features for screws. The code developed for screws should contain information of the values for all these features. If the type of screw is not changing, only pitch will differentiate between various screws in the parametric geometric design. This aspect must be properly taken into account while building the code.

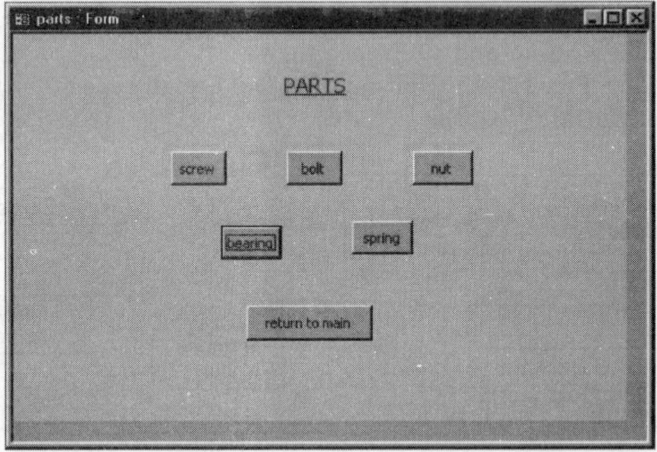

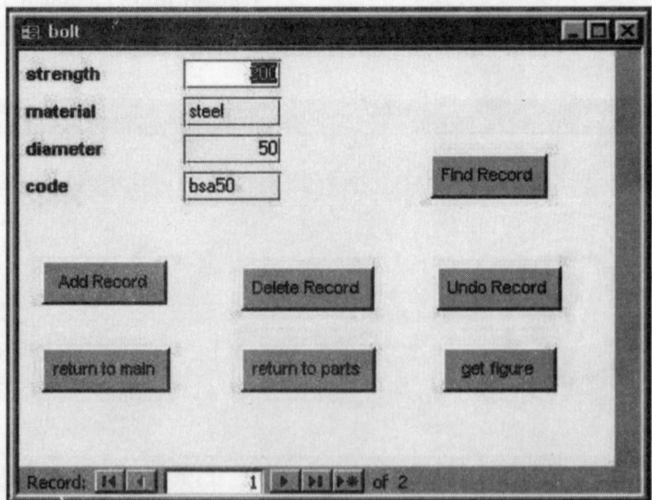

The Access forms for some sample parts are shown below. They are bolt, nut and screw. The present workbench allows for updating the database for each category of parts with respect to their respective part features.

Thus, in addition to building the initial database of parts, the database can be updated whenever a new design that is not similar and not derivable from the existing drawings is encountered. The CAD model of the part should also be stored in the repository with the same code. The *get figure* button enables retrieval of the CAD model in a CAD package depending on the format it is stored in. Pro/Engineer and IGES were used in the present workbench.

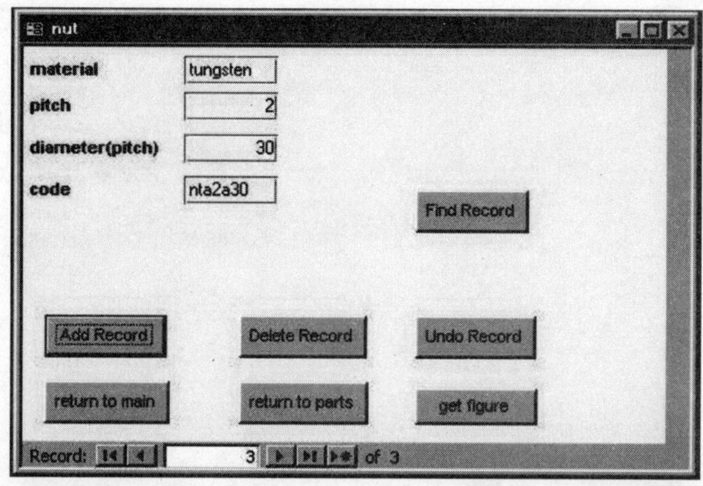

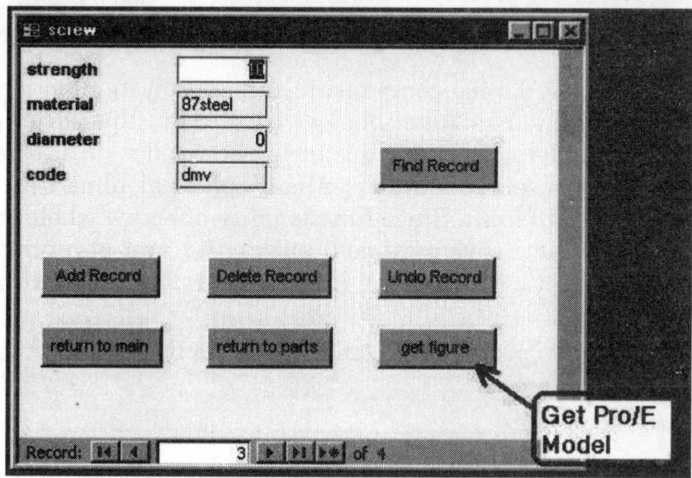

Whenever a new product is to be designed, retrieving the design that is nearest to it from the repository is not any easy task. This is because there is rarely an exact match in the database. Therefore, some kind of grouping and coding are required.

1.8.1. Grouping the Machine Elements

The component coding for the purpose repository maintenance for future design requirements can be done in two methods. The first method is named as 'static method' and the other can be named as 'functional method'. The static method of extracting the characteristic features has long been practiced even since the fundamental principles of group technology and computerized cataloguing have been introduced. In static method, the components are grouped according to their static features such as shape, size and weight considerations. For example, one simple classification of components is rotational and prismatic parts, which is based on the general shape and more minute features such as hole, slot and keyway etc. are not taken into consideration.

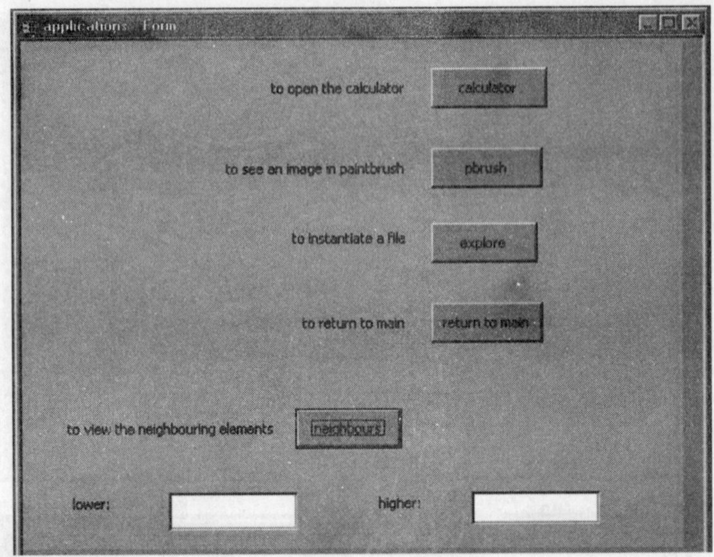

The second method of grouping components is based on its function in addition to static features and hence, it is named as 'functional method'. The superiority of functional method over static method can be understood by distinguishing between a super-elastic ball and a lollipop. Leaving out the stick in the latter, both are essentially spherical, almost equal in size and weight. But the functionalities are different. Since functionality affects a whole lot of design decisions, including material, manufacturing methods, packaging and shipping methods, it is very important to observe that static features are insufficient information with respect to which the components can be stored.

The functional method of grouping of components can be based on the following component features:

(1) Dependent functionality
(2) Dependent mobility degrees of freedom
(3) Material related characteristics – density
(4) Prime mover
(5) Size, weight and shape constraints.

1.9. JAVA BASED EXPERIMENTAL MODEL OF COLLABORATIVE DESIGN WORKBENCH

As experimentation towards a versatile CDW that incorporates modern ideas and some new ideas we generated, collaborative design of a product and the corresponding mediation among the designers involved has been systematized through a web-based tool that we called as collaborative design workbench (CoDMaWb). Instead of using existing INTERNET based tools (such as *ThinkCycle* etc.) or commercial collaboration tools (such as Co | Create, AIMSNET etc.), a custom built workbench has been devised using Java so that independently the related aspects could be studied. The primary attribute of this workbench is its future portability across platforms (because Java has been used). The current implementation is limited to the Windows environment in connection with the fact that no explicit code has been included to connect to applications such as Microsoft office tools, MATLAB, Pro/Engineer etc. because Windows has an in-built linking facility. When, on the other hand, this workbench is ported to another OS such as Linux, additional code may have to be added to link files to their programs. In the workbench, separate user and administrator controls have been implemented to facilitate administration of access and design data at the same time allowing free and open communication among the designers. The implementation and testing of the workbench has been done by converting a computer into a web server using Apache Tomcat and the user interface is provided through HTML code. As an example, the workbench was tested for the case of projects of the course ME C312: Design of Machine Elements at BITS, Pilani. The initial tests brought out many lessons to be learnt to improve the workbench and consequently many modifications have been included.

Collaborative design is fairly a new term and the closely focused research began only in late 1990s. That is not surprising because the most important technological ingredients of this philosophy are the computer networks, INTERNET, internetworking technologies and web-based data and graphics editing tools. One other very important ingredient, that is not new but comparatively much older, is the science and tools of computer aided design (CAD). It all began in the US with the initiative of industries to develop tools that will enable interaction among designers who might be physically separated at distant places but still are needed to make effective contributions in the development of a new product. Physical meetings and the necessary logistics obviously entail much higher costs of product development and high lead times. Otherwise, the company has to be complacent with the engineers available at hand at the given point of time. With this as the primary motivation, many industries requested software companies to develop tools that would enable collaboration. Another parallel motivation was developed to convert the old techniques of logistics management into modern supply chain management tools with full integration achieved among various parties through the INTERNET based tools.

Collaborative design principles were widely tested and it was established that with such collaboration in the conceptual design stage as well as in the final documentation stage, the effectiveness was shown to improve by means of collaborative design workbench [6, 7, 8]. In the collaborative design and manufacturing workbench (CoDMaWb) was proposed in which manufacturing aspects were also integrated.

The development of computer tools for collaborative design would start with a clear understanding of the nature of the design, review, and communication activities encountered in traditional environments, where teams work together in the same room with clay models, drawings, and CAD systems. Therefore, there is a need to focus on establishing or validating precise task and performance models of these traditional environments. These will provide guidelines for future research and yardsticks for validating and justifying the results.

Computerized design collaboration tools should strive to provide a similar, or even more effective, design environment to teams distributed in different locations, possibly in different time zones, probably involving people with no CAD or graphics expertise, connected via private high-bandwidth networks or plain phone lines. Progress toward productivity will involve advances in a variety of disciplines.

There is an extensive research work going on in the field of collaborative CAD from areas like: shape manipulation, interactive graphics, user-interfaces and virtual reality environments, 3D modeling and compression, shared databases and versioning, interface standards, decision support and documentation, multimedia communication and mark-up tools, and product data management. Finally, we must ensure that these research results will impact our industries.

Therefore, contributions describing the role and importance of collaborative design in the overall product cycle and the interdependencies between the design activity and other business processes should be encouraged.

1.9.1. Tools and Methodology of Java-based CDW

In this section, the various Java technologies and features used in the development of the collaborative design workbench are presented. We first look at the relevant software tools, namely, servlets, Apache Tomcat, Java server web development kit.

Collaborative work bench tool is developed on Java platform as it can provide platform independency. Apache Tomcat server has been configured with a port number of 8080 on a computer with ip address "172.16.3.54" as an example. This server can be accessed by all other computers on local area network (LAN). They act as clients. There are few steps to be followed in order to check this application running.

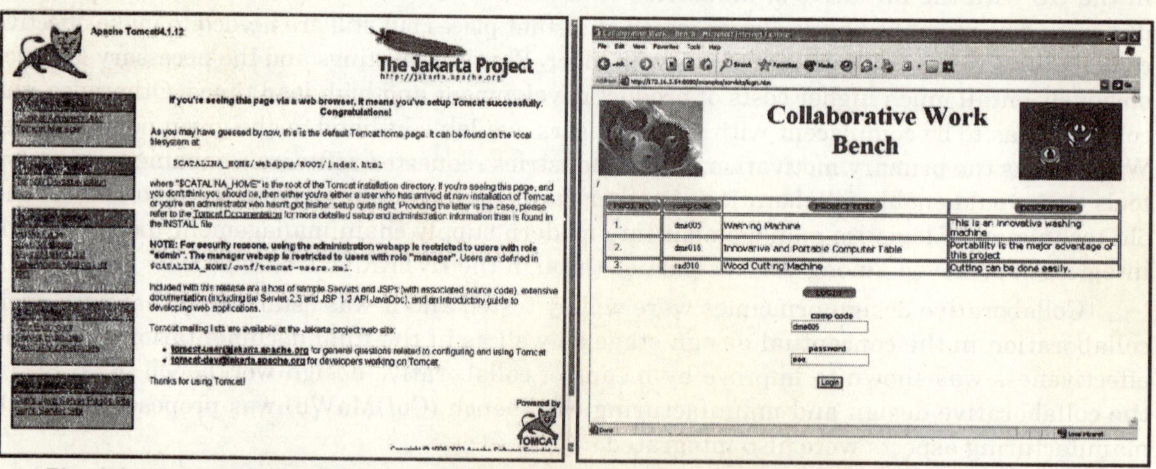

Fig. 1.4: Illustrates the official home page of Tomcat web server

Fig. 1.5: Main page: project list

1. Type http://172.16.3.54:8080/index.jsp URL in a web browser. The following page in Fig. 1.4 appears. Otherwise, it means that client is not able to reach the server.
2. Type http://172.16.3.54:8080/examples/servlets/login.htm URL in the web browser. The home page of the application gets opened. This page can be seen in Fig. 1.5.

The above mentioned URLs are as implemented at BITS, Pilani and the reader may appropriately substitute the URL relevant to his LAN while building his own CDW.

Home page of the application, lists out the projects that are available and their respective computer codes. Every member of a project will be given the password of the particular project before hand. Now, the member would be able to login with the computer code and password.

Every project is associated with a corresponding project page. Here the members of the particular project are listed out along with their ID Nos. This page has two features, uploading and downloading of files. Uploading feature has a size restriction of 1 MB (Fig. 1.6).

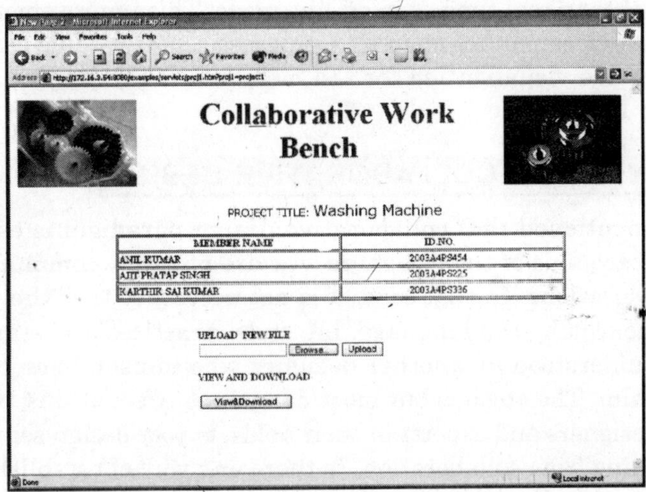

Fig. 1.6: Project page

Files can be downloaded with any size depending on the space allocated by the server. The illustration of the project page can be visualized in Fig. 1.7.

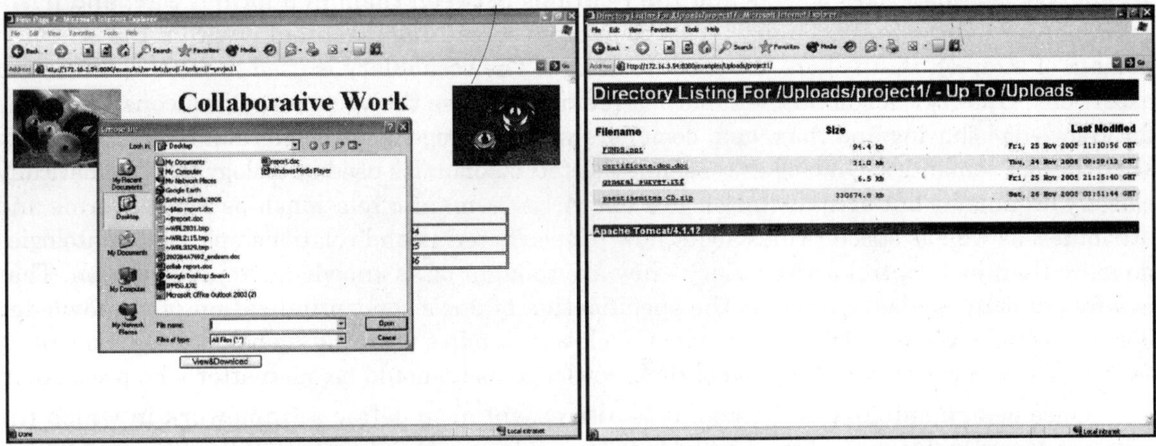

Fig. 1.7: Downloading the CAD files or other designer comments from server via the workbench and listing of the various design semantics available for the given product

Various observations were made by testing this Java based CDW on which certain conclusions can be drawn as follows. The user authentication module is presently project-wise. That is, users have to login with their project computer codes and the respective project passwords. This can be improved by adding member-wise authentication also. Each member can be provided with a login id and password for each project area as per the discretion of the administrator, meaning that there will be a dual authentication.

Session tracking in the servlets can be used to create a session for each user and maintain his records throughout the session against the project in question. A discussion forum module can be introduced for the convenience of the team members so that they can post messages on different threads either in real time or in off-line mode. This improvement is particularly helpful in increasing the design semantics shared. A progress report module can be introduced for each project so as to provide visualization for the members of the project about the deadlines for completion on time line.

1.10. ONTOLOGY BASED COLLABORATIVE DESIGN

We have already mentioned that collaborative design paradigm is essentially to exploit the heterogeneous backgrounds of the designers who are probably communicating from physically far off and diverse locations. In that case, it is not necessary that the intents of one designer, which is fully influenced by the language, life style, practices and standards that are local to him, can be fully understood by another designer who himself/herself is influenced by those conditions local to him. The obvious but most dangerous effect of this, despite all of them being highly competent designers and experts in their fields, is poor design semantics and consequently a poor design resulting from collaboration. Is there any way of establishing a 'code of language' that is a sort of a new design language dictionary using the words/phases within which these designers can communicate so that there is no 'communication gap'?

Yes, the solution to this problem is offered by the so called *design ontology*. Ontology is the specification of a conceptualization of a design knowledge domain. *Ontology* is a controlled vocabulary that describes objects and the relations between them in a formal way, and has a grammar for using the vocabulary terms to express something meaningful within the specified domain of interest, that is here the product design. The vocabulary is used to make queries and assertions. Ontological commitments are agreements to use the vocabulary in a consistent way for knowledge sharing and thus, help designers with heterogeneous backgrounds to successfully share design semantics. Ontologies resemble faceted taxonomies used in biology to systematically classify organisms but actually ontologies use richer semantic relationships among terms and attributes, as well as stricter rules about how to specify terms and relationships. Since ontologies do more than just control a vocabulary, they are thought of as knowledge representation. That is why ontology is also defined as the specification of one's conceptualization of a knowledge domain. Ontology is derived from the two Greek words, *ontos* meaning "to be" and *logos* meaning "word." Thus, it is wordy description of the knowledge as it should be, no matter who presents it.

Once design ontology is defined, it is also essential to define a framework in which the terms of design ontology need to be embedded to enable proper use and connectivity with different applications [9]. This framework is provided by the *Semantic Web*, an initiative coined in 2004 by the World Wide Web Consortium (W3C). W3C defined Semantic Web as follows. The Semantic Web provides a common framework that allows *data* to be shared and reused across application, enterprise, and community boundaries. It is based on the Resource Description Framework (RDF) and ontology. RDF is used to represent information and to exchange knowledge in the

Web. In order to enable publishing and sharing sets of terms called ontologies, and supporting advanced Web search, software agents and knowledge management, the W3C has created the Web Ontology Language (OWL).

REVIEW QUESTIONS

1. How collaborative design is different from the traditional design of mechanical components?
2. What is meant by design semantics? Recall a situation when you designed a mechanical product in a team environment. List down the design semantics you and your member(s) shared during that activity.
3. Download Cosmoplayer plug-in and visit many available sites that show the potential of the virtual reality modeling language. Argue whether you think VRML can be a potential medium to show virtual assembly to collaborating engineers.
4. What is ontology? How it is relevant to product design?
5. Build a Java based CDW by yourself and using it, conduct collaborative design with your friends. Document your experiences.

REFERENCES

1. Maher M.L., Cicognani A. and Simoff S., *"An Experimental Study of Computer Mediated Collaborative Design"*, IEEE Fifth Workshops on Enabling Technologies: Infrastructure for Collaborative Enterprises, June 19–21, 2006, Stanford University, California, USA, pp. 268–273.
2. Hubka V., 1980, *"Principles of Engineering Design"*, Butterworth Scientific Ltd., UK.
3. Hall A.D., 1962, *"A Methodology for Systems Engineering,"* Van Nostrand, Princeton, New Jersey.
4. Pahl G., Beitz W., Feldhusen J. and Grote K.H., 2007, *"Engineering Design : A Systematic Approach"*, 3rd Edition, Springer-Verlag, London.
5. Birmingham R., Cleland G., Driver R. and Maffin D., 1997, *"Understanding Engineering Design: Context, Theory and Practice"*, Prentice Hall Europe, England.
6. Allen R.H., Nidamarthi S., Srinivasa Prakash Regalla and Ram Sriram, *"Enhancing Collaboration Using an Internet Integrated Workbench,"* In Proceedings of ASME DETC/DEC Conference (Edited by M. Gartner et al.) Paper No. DAC8573, Las Vegas, Nevada, USA, Sept. 1999, pp. 12.
7. Nidamarthi S., Allen R.H., Srinivasa Prakash Regalla and Ram Sriram, *"Observations from multidisciplinary, Internet Based Collaboration on a Practical Design Project,"* International Conference on Engineering Design, ICED 99, Munich, Germany, Vol.2, Aug. 1999, pp. 709–714.
8. Srinivasa Prakash Regalla, *"Database Centric Collaborative Design and Manufacturing Workbench"*, Proceedings of International Conference on Multimedia and Design (ICMD), September 2002, Mumbai.
9. Szykman S., Sriram R.D. and Regli W.C., *"The Role of Knowledge in Next Generation Product Development Systems"*, Journal of Computing and Information Science in Engineering, 2001, pp. 3–11.

PRODUCT LIFE CYCLE ENGINEERING AND MANAGEMENT

2.1. INTRODUCTION

Design for excellence implies considering all aspects of product life cycle early during its development. It is a natural evolution and integration of DFX (design for manufacture, assembly, service, environment, etc.) technologies and methodologies. Development of an excellent product involves creating, exchanging and managing a huge amount of data during design and optimization among the team-members, who may be located all over the world. This can be achieved by integrating CAX (CAD, CAE, CAM), product data management (PDM) and collaborative engineering tools, referred to as product life cycle management (PLM). This is a rapidly evolving field, and provides a fruitful area of collaborative research and development between academia and industry.

Continuous improvement in product features, as well as price, quality and response competitiveness, has become essential to survive and succeed in the global market. Innovative features impart for additional functionality, better user interface, higher efficiency or smaller size. Price competitiveness relies on manufacturing a product within a target cost given by target market price minus desired profit. Quality competitiveness involves designing robust products plus production systems to get the right quality first time and every time. Response competitiveness implies faster development and introduction of a new product, by working proactively with respect to customer expectations. These however, will not suffice in years to come, when customers start expecting excellence in products.

Excellent products are those which are designed considering all aspects of their lifecycle: not only functionality, but also ease of manufacture, assembly, delivery, installation, use, safety, reliability, service and eventually reuse, recycling or disposal (Fig. 2.1).

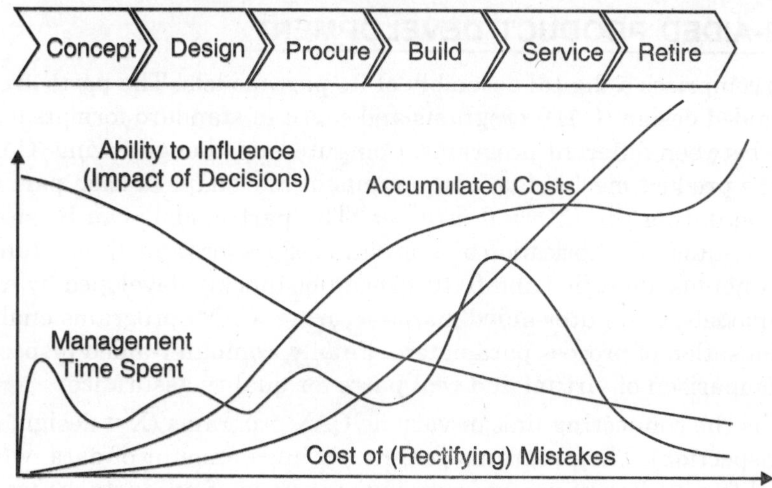

Fig. 2.1: Need for early consideration of product life cycle issues

Designing excellent products is a challenge. Products are already very complex, incorporating a host of continuously emerging technologies. This in turn requires specialist teams to handle different activities in product development life cycle: concept design, engineering analysis, tooling development, manufacturing planning, part manufacture, product assembly, and delivery. The teams need to work with each other concurrently, to optimise the product with respect to various life cycle issues and to launch the product early in the market. Concurrent engineering is difficult to practice when the specialist teams are located in different parts of the world, an increasing trend with globalization.

The solution lies in connecting the team members and product life cycle experts through a digital communication network and providing them appropriate software programs to create, analyse and modify a virtual model of the product. The model and results are stored in digital form in a central or distributed server and accessible to all team members over a local area network or Internet. This approach to product development can be referred to as collaborative product life cycle engineering (Fig. 2.2). The building blocks are briefly described next.

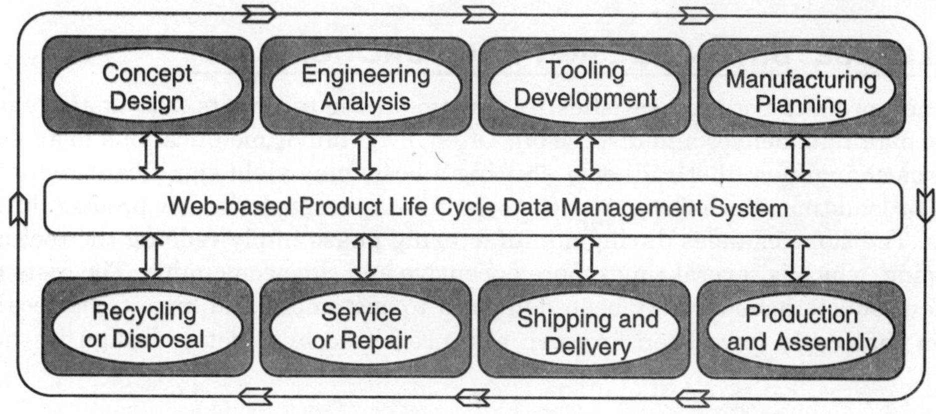

Fig. 2.2: Collaborative product life cycle engineering framework

2.2. COMPUTER-AIDED PRODUCT DEVELOPMENT

The virtual product comprises a digital assembly of its part models. The parts are modelled in 3D using computer-aided design (CAD) programs and saved in standard formats (e.g. IGES and STEP) for exchange between different programs. Computer-aided engineering (CAE) programs enable simulating the product mechanism and optimising the shape of each part under static/dynamic loads by simulating the internal stresses. The part models can be sent to a rapid prototyping system for automatic fabrication of a physical replica for form, fit and function testing. The tooling models (moulds, dies, jigs and fixtures) can be quickly developed by modifying the corresponding part models. Computer-aided manufacturing (CAM) programs enable planning, simulation and optimisation of process parameters. Finally, computer-aided inspection systems enable automatic comparison of virtual and real parts for quality assurance.

The 3D model is the connecting link in various CAX programs (X = design, engineering, manufacture and inspection). The programs generate a huge amount of data, which includes the solid models of different iterations and previous versions of products, as well as tooling, materials, process plans and results of analysis. This necessitates a systematic approach to data storage, verification and retrieval, which is achieved by a product data management (PDM) system.

The PDM systems have rapidly evolved over the last decade and now allow distributed storage and remote access over Internet. They provide better data translation for exchange among various programs. Security is handled by data encryption during exchange and facilities for setting limits of access by non-core team members. Some useful utilities for collaboration are also provided. For example, comments or suggestions can be attached to product features. Two or more team members can synchronise their computer displays, point out product features (by an arrow visible to all team members) and discuss various improvements using a messaging or videoconferencing facility.

A complete set of CAX, PDM and collaboration utilities is now referred to as a product life cycle management (PLM) solution. The set may be offered entirely by a single vendor (such as Dassault Systems, France; Unigraphics, USA or Parametric Technology Corp, (USA), or can be put together by integrating the most suitable program for each application (e.g. Solid Works for CAD, ANSYS for CAE, Delcam for mould design and CAM, MoldFlow for plastic flow simulation and Matrix One for PDM).

2.3. EXCELLENCE: BEYOND DESIGN FOR FUNCTIONALITY

A well-designed product not only satisfies its functional requirements, but is also easier to manufacture, maintain, deliver and dispose off. Often, even minor modifications to an existing design (such as changing a fillet radius or shifting a boss) may yield significant reduction in costs, defects or lead-time. Such changes are easy and inexpensive for a new product during its design phase. The same changes during manufacturing phase imply redoing the tooling and process planning, which is several times more expensive and time-consuming. The costs mount further if the products have already been delivered to customers and have to be recalled to fix a problem that could have been foreseen and prevented by a better design in the first place.

Design for manufacture (DFM) is influenced by product geometry, material and process. For example, injection moulded plastic parts need to be designed with a uniform wall thickness to minimize the production cycle time and avoid sink marks. Metal castings must be designed for directional solidification for ease of feeding. Forged parts must have generous radii of curvature to facilitate flow of metal. Sheet metal parts made by blanking operations must be designed for efficient nesting and thereby stock utilization.

Similar guidelines are also available for improving a product design for ease of assembly, service, transportation and environment. Design for assembly (DFA) guidelines includes reduction of number of parts and number of assembly directions. Design for service guidelines includes modular design and ease of access. Design for delivery guidelines includes provision of suitable hooks and handles for lifting the part. Design for environment (DFE) guidelines include use of recyclable materials (metals are better than plastics), reduced packaging, and improving the energy efficiency of the product.

The guidelines for DFX (X = manufacture, service, environment, etc.) are distilled from practical experience on previous products, and may not be applicable to new materials or processes. They may also conflict with each other, but provide no quantitative feedback to resolve such conflicts. Moreover, most designers have little time, interest or knowledge in handling product life cycle issues beyond functionality. One way to overcome these problems is to invite external experts to evaluate the design for various DFX considerations and suggest suitable improvements. Since the experts are likely to be inaccessible (in terms of time or distance), a web-based PLM system greatly facilitates such collaborations.

The collaborative engineering framework also yields an important by-product: a vast amount of data and information captured in digital form during development of several generations of products. Special data mining tools can be used to process the accumulated data and produce new knowledge. All such information and knowledge can be made accessible over the network, and used in developing future generations of products. Knowledge management (KM) is in fact, rapidly becoming an important part of PLM.

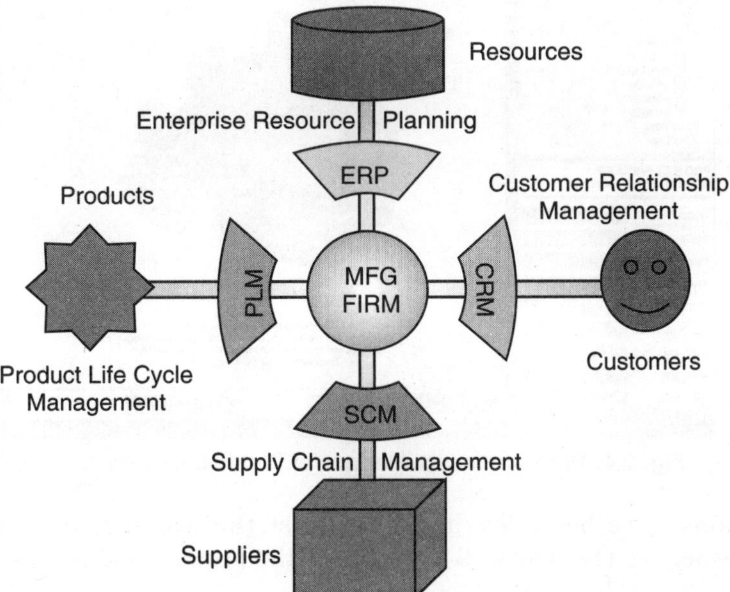

Fig. 2.3: Product life cycle management in relation to other enterprise IT applications

The focus of PLM on engineering data and knowledge contrasts with enterprise resource planning (ERP), supply chain management (SCM) and customer relationship management (CRM), which largely handle business process data (Fig. 2.3). There is however, an increasing overlap between the systems and the distinctions are gradually blurring. In future, it is likely that the systems will have seamless connections and data exchanges with each other, and will provide access to any relevant data through a standard (but customisable) user interface.

For example, at IIT Bombay, India, research and development focus is on developing a systematic approach and software for collaborative product life cycle engineering of cast components, used in automobile, aerospace, farming, mining, machine tools, electrical, consumer and many other products. Over the last one decade, several elements of a framework code-named WebICE (Web-based Intelligent Collaborative Engineering) have been developed and used for solving industrial problems (Fig. 2.4). The framework allows product, tooling and manufacturing engineers to design, analyse and improve a casting for lower cost and higher quality assurance using software tools and Internet.

The backbone of the system is an XML-based casting markup language (CastML), which captures over 1000 different items of information about a casting project. The items range from part geometric parameters to material properties, tooling element designs, process plans and costing details. Other files containing 2D images, 3D models, library options and knowledge-bits (if-then rules) related to the casting are linked to various nodes of the CastML tree. The data is stored in a web server. Team members can access and view it using a standard browser. The large size of solid models prevents their real-time exchange over standard network connections. To overcome this problem, a special program was developed to compress, upload, download, decompress and display the solid models.

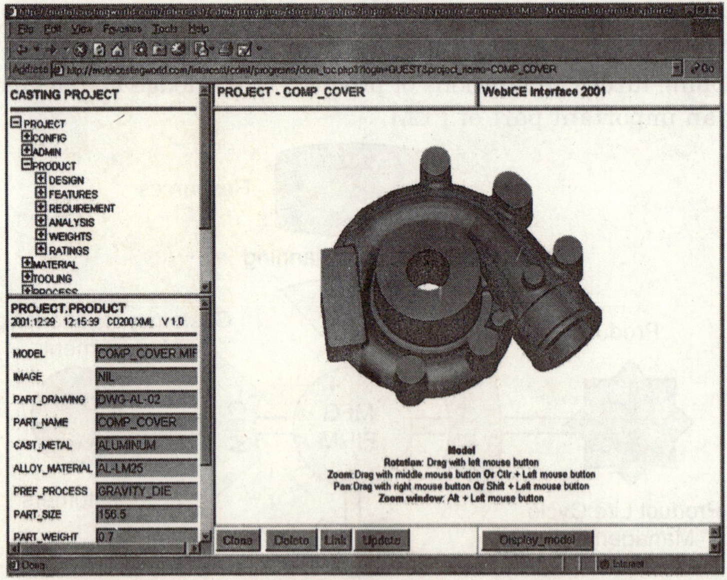

Fig. 2.4: Browser-based interface of WebICE system

Several programs have been developed to assist the product, tooling and foundry engineers in their respective tasks as well as evaluate the product and process. This includes casting alloy selection, process selection, process planning, methoding (feeding and gating design), casting defect prediction, product-process-producer compatibility analysis, shape complexity estimation, tooling process selection, cost estimation, feature recognition and DFX checks. Simpler programs are implemented in PHP/Java and executed at the server itself, whereas computation-intensive programs are implemented in C++ and run at client computers, followed by exchange of results over the network.

The system has been used to troubleshoot and optimize many castings ranging in weight from a few kg (e.g. compressor casing, medical equipment housing in aluminium-alloys) to several tons (e.g. hydraulic pump, press tool parts in ferrous metals). The projects are carried out by active involvement of product, tooling and manufacturing engineers, and coupled with continuing education to impart the necessary theoretical background.

Different methodologies and software tools for collaborative engineering, including intelligent CAX, DFX, PLM and KM are hot areas of research, since they have a direct and immediate impact on the competitiveness of manufacturing firms. Today, many research labs in leading universities and several journals and conferences are dedicated to various aspects of collaborative engineering.

India is now known worldwide for its IT prowess. We are however, lagging behind in applying IT to our manufacturing industry. Product life cycle engineering and management implies leveraging IT (software and communications) to develop excellent products and achieve price, quality and response competitiveness. Early adaptation of PLM in our industry can help India leapfrog and catch up with manufacturing giants like China, Japan and Germany. Though the PLM philosophy is not new, the software tools and methodologies for successfully implementing it in an organisation are still evolving and are a fruitful area of research and development. Our focus must be on development of low-cost and easy-to-use PLM software to benefit even small and medium size firms, providing them a level-playing field to compete in the global arena. This requires collaborative R and D between engineering institutes, IT firms and manufacturing industry.

REVIEW QUESTIONS

1. Why the cost of rectifying mistakes becomes prohibitively high towards the end stages of the product? What should be the next step?
2. How the Internet or the World Wide Web (WWW) helps in integrating the various modules of the product life cycle system? What are the inherent advantages it offers?
3. If the valuable product data is being shared through WWW, how the security of information is ensured? What are the techniques?
4. CAD data of products can be often huge, running into several gigabytes sometimes. What strategy can be adopted to share such information through WWW in real time?
5. Do research and find out the current status of PLM application in metal forming industries. Prepare a summary.

3

LEAN NEW PRODUCT DEVELOPMENT

3.1. INTRODUCTION

Recent years have seen great strides in the reorganization of manufacturing around the lean manufacturing (LM) philosophy. The prime reason being that in most of the markets, demand has slowed or declined and the cost competitiveness of an organization has considerably increased due to competition. LM has paved the way for improving the competitive position of an organization by drastically reducing the wastes in manufacturing. Similar condition prevails in the field of product development and manufacturers have become more concerned about the efficiency in developing new products. They have been facing profitability problems due to the high costs of developing and manufacturing many new products or product variations. These organizations have recognized that effective product realization requires flexibility and leanness across the whole process of design to manufacture. The advocates of lean philosophy have already emphasized on the application of lean principles organization wide, but due importance was not given till recent years. Now, researchers have started applying the principles of LM to product development process and are inundating the literature with topics like wastes in product development process, tools and techniques, barriers etc. In this chapter, a conceptual framework, which will provide a comprehensive structure or a guideline for the organizations to make their product development process leaner is explained.

The new product development (NPD) process has a great influence on the outcome of product, both technically and economically. It is precisely in this process where the customer's ideas and speculations are conceptualized into a physical model, defining his/her needs and requirements into form, function, fit, procedures, drawings, and technical specifications. Developing new products faster than competitors do is a formidable strategic weapon to succeed in increasingly turbulent markets. But it is a daunting task as the nature of product design process is quite complex because of the following reasons:

- It involves thousands of decisions, with numerous interdependencies and a large number of participants are involved, such as product managers, development engineers, service engineers, market consultants and representatives from top management in addition to customers and suppliers. Each category of professionals has a different background, culture, and learning style.

- Trade-offs between multiple, competing design criteria must be made throughout the design process, often with inadequate information and under intense budget and schedule pressure.
- Early design stages are notoriously hard to evaluate and control against progress milestones because of the lack of physical deliverables such as drawings. Further, it is difficult to measure the amount of work completed and remaining on any given task in the project.
- New product development projects extends over a period of years and are increasingly subject to uncertainty because of the pace of technological change, the rapid shifting of market opportunities, and the inability to keep pace with relentless pressure to reduce time and cost.
- The administration and management of design have been barely explored and exemplified, hence resulting in chaos, causing poor communication, lack of adequate documentation, deficient or missing input information, unbalanced resource allocation, lack of coordination between disciplines, and erratic decision making.

Researchers have attempted several methods to solve the above-mentioned problems, through techniques like project management, concurrent engineering, process models, new organizational forms, along with information technology support. Even though these 'state-of-the-art' design management approaches contain many interesting and seemingly effective new features, they are fragmented and applied on a stand-alone basis during the product development, thus not yielding the desired output of faster introduction of new products with the minimal use of resources and cost. There is still a shortfall in the expected or desired improvements of new product development [1]. In recent years, manufacturing has seen great strides in its reorganization around the lean manufacturing (LM) philosophy. The prime reason being that in most of the major markets, demand has slowed or even declined while the cost competitiveness of an organization has considerably increased due to competition and LM has paved the way for improving the position of an organization by drastically reducing the wastes in manufacturing. A similar condition prevails in the field of product development. Further manufacturers are plagued by the above-mentioned problems affecting the efficiency in developing new products. In addition, they are also facing profitability problems that are due to the high costs committed for developing so many new products or product variations. These organizations have recognized that effective product realization requires flexibility and leanness across the whole process of design to manufacture. With the evolution of lean principles and its success in manufacturing side, researchers have started applying these principles to the NPD process. The advocates of lean philosophy, under the concept of lean enterprise have already emphasized on the application of lean principles organization wide, but due importance has not been given till recent years. Now, researchers have started applying the principles of LM to product development process and are inundating the current literature with topics like wastes in product development process, tools and techniques, barriers etc.

3.2. AN OVERVIEW OF LEAN MANUFACTURING

Lean theory originated in the practices of the Toyota Motor Corporation, codified and popularized through publications such as "The machine that changed the world" and "Lean thinking" written by Womack and Jones [2]. These writings focus mainly on the manufacturing aspect of business rather than the engineering and design processes, but maintain that the same principles can be applied to both shop floor and non-shop floor activities. Lean is defined as the search for perfection through the elimination of waste and the insertion of practices that contribute to reduction in cost and schedule while improving performance of products, processes and organization as a whole. As categorized by Womack and Jones [2], the main tenets of lean are:

- *Define value*: Value can be defined as information/material in a form that the customer is willing to pay for. The customer defines value. Value is created by the producer.
- *Identify the value stream*: The value stream consists of the actual tasks required to bring a specific product through three critical processes of design, order and make.
- *Flow the product*: It can be achieved through eliminating activities that are pure waste.
- *Pull*: Pull is letting the customer pull value from enterprise. That means the enterprise should not make anything until it is needed. When it is needed the organization needs to make it as quickly as possible.
- *Strive for perfection*: There is no end to the process of reducing efforts, space, costs and mistakes. Perfection is continuous radical and incremental improvement and is continuous banishment of muda (waste).

3.3. LEAN NEW PRODUCT DEVELOPMENT PROCESS

Product development is the set of activities beginning with the perception of a market opportunity and ending in the production, sale, and delivery of a product. Most organizations follow a structured process to develop new products. A typical NPD process consists of the following stages as shown in Fig. 3.1 in which different activities are carried out in each stage. These activities can be classified as value added activity, non-value added activity and required non-value added activity.

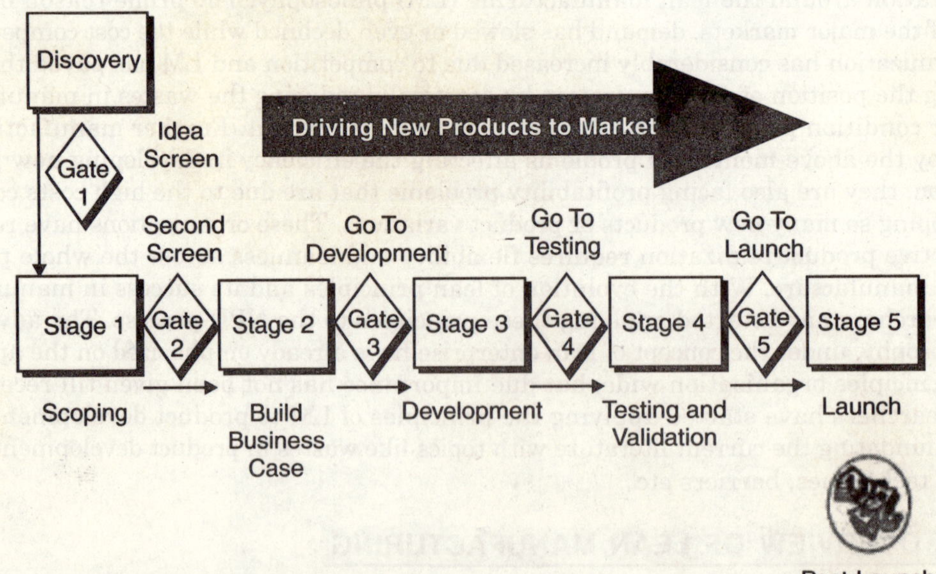

Fig. 3.1: A generic stage-gate new product development process [3]

The objectives of lean new product development (LNPD) are to eliminate those non-value added activities and reduce the required non-value added activities. This concept is not new, rather an extension of lean manufacturing concepts across the enterprise. Smith and Reinertsen [4] identified the application of just-in-time manufacturing philosophy (a component of lean) in the NPD process. They described how 'pull' approach can be applied to information and if it is established in a development team, only the downstream persons can ask for whatever information they need. But the term LNPD was first introduced in the book—'The machine that changed the world' [2]. Since the focus of the book was on manufacturing and assembly processes, LNPD did not deserve much attention among the companies to foresee it as an efficient product development methodology. The basic tenets of lean can be adapted to the NPD process as follows:

- *Define value*: It refers to providing the customer with the right product, with right quality (form, fit and function), for the right price, at the right time.
- *Identify the value stream*: The value stream is the NPD process, which can be defined as "specific activities that involve gathering customer requirement, conceptualizing, designing, testing, manufacturing and providing a specific product, from concept to launch".
- *Flow the product*: It is described as "the progressive achievement of tasks along the value stream of the product development process, so that a product proceeds from requirements to concept to design to launch, order to delivery, and raw materials into the hands of the customer with no stoppages, scrap or backflows."
- *Pull*: It is defined as "a system of cascading customer and design instructions from downstream to upstream activities in which no activity is performed or nothing is conceived by the upstream supplier until the downstream customer signals a need."
- *Strive for perfection*: It refers to continuously and relentlessly improving the value, value stream, flow, and pull in business operations—i.e., NPD process.

Hence LNPD can be defined as the application of lean principles to the product development process, to eliminate waste and non-value adding activities. It helps companies to develop a seamlessly flowing product development value stream with minimal waste, defined and pulled by the customer [5]. There are 7 categories of wastes defined in LM, which are applicable to manufacturing, namely—overproduction (producing ahead of demand), inventory (stocks more than the absolute minimum), transportation of materials, unnecessary movement by employees during their work, waiting for the next process step, defective products production, over processing of parts due to poor tool and product design. These wastes in manufacturing are relatively easy to identify and quantify which can be extrapolated and applied to a product development processes. When extrapolating to the product development process, it must be clearly understood that it is information that flows through the product development process, just as physical material flows through the manufacturing process. The quality of the information flow is characterized in terms of Form, Fit, Function, and Timeliness (FFFT). Further, value within product development develops not only as a function of the FFFT of the information included within the design package, but also as a function of how well that design package, the process, procedures and activities, allows the final product to meet the FFFT desires of the customer. As the information flows and matures through this NPD process, the tasks performed add value to the information, transforming it from its initial state of raw data to the completed build to package. However, not everything that begins as raw data results in useful information.

McManus and Millard [5] and Haque and Moore [6] have listed down various wastes that occur during NPD process. Table 3.1 identifies and classifies such wastes in terms of seven wastes identified in LM.

Table 3.1: 7 Wastes of lean manufacturing in a product development process

7 Wastes in LM	Applied to new product development process
Over production or early production	• Too many products or projects • Over design or providing unnecessary features • Over specifications/tolerance or too much details provided for a part • Redundant development—*i.e.* Reuse not practiced
Transportation or movement (with respect to information and prototype)	• A long hierarchical structure for decision making and approval • Information system and information formats—has a lack of common/compatible standards resulting in repeated communication, manual transfer and conversion • Too many data interfaces or multiple sources • Poor interface control (with respect to departments) or management of design data
Unnecessary inventory	• Too much information • Poor configuration management • Working on designs never used, completed or delivered • Unnecessary documents, physical prototypes • Lack of use of standard parts or lack of commonality;
Waiting or Delays	• Information created too early • Unavailable information • Unclear decision criteria or lack of needed information • Late delivery of information due to approval procedures, meetings etc. • Waiting for test results or process information
Inappropriate processing or poor NPD process design	• Stuck up with too many meetings yielding no results or consensus • Failure to identify and manage design risk • Technology acquired, but not used • Unnecessary serial effort • Too many iterations or re-work due to changing priorities or requirements • Unnecessary data conversions • Excessive verification and testing • Underutilization of design knowledge, for example in costly parts • Working with wrong level of detail or defective information
Unnecessary motion or inefficient performance of design	• Lack of standardization of processes • Information forwarded to wrong people • Poor re-use resulting in unnecessary development activities • Complicated retrieval • Over authorization • Over or inappropriate tolerance • Working with wrong/incomplete information • Out of sequence working (due to poor integration) • Inappropriate changes (changes not customer driven or not of benefit to business) • Data acquired then not used

Contd...

Defects	• Warranty issues • Use of immature and inappropriate technology • Poor make or buy decisions • Poor design for X—manufacture, assembly, cost, reliability and supply • Under specifying • Inadequate use of design tools generating poor design • Poor supplier identification • Failure to understand and capture requirements or poor understanding of customer needs • Conversion errors • Unclear requirements or incomplete, ambiguous and inaccurate information generation

It is necessary that these wastes have to be eliminated to make the product development process leaner. Kearney [7] found that a total of 58% of cost is wasted in the design of railway vehicles including failure to exploit previous design experience, over design and poor design for manufacturing: Table 3.2 lists the opportunities for waste reduction in the NPD process.

Table 3.2: Opportunities for reduction of waste in NPD process [7]

Area of waste reduction	Percentage of design waste
Designs never used, completed or delivered	Unknown
Downtime while finding information, waiting for test results etc.	33–50%
Unnecessary documents and prototypes	18%
Underutilization of design knowledge, for example in costly parts	8%
Validating manufacturing errors early in the design process	17%
Poor designs producing product defects	15%

In addition to waste reduction, LNPD also holds the promise of dramatically improving a company's competitive position. Its implementation offers the following benefits [8]:

- The extensive use of suppliers as expert developers increases technological sophistication of products as well as creates potential for faster product development with fewer engineering hours.
- Emphasis of collaboration between different functional areas within the company helps improved manufacturability of products.
- A conscious effort to use quality as a guiding principle throughout the whole development process results in fewer production start-up problems.
- Faster time to market.

In a nutshell, the differences between a lean product development and traditional product development are shown in Table 3.3.

Table 3.3: Differences between LNPD and NPD [9]

Lean product development	Traditional product development
Lean thinking	Functional management
Rapid model replacement	Slow model replacement
Frequent model-line expansion	Infrequent model line expansion
Heavy weight project managers	Light weight project coordinators
Overlapping compressed phases	Sequential long phases
High levels of supplier engineering	High levels of in-house engineering
Design team and project-manager continuity	Department member continuity
Good communication mechanisms	Walls between departments
Cross-functional teams	Narrow skills in specialized departments

3.4. TOOLS AND TECHNIQUES IN LNPD

Womack *et al.* [2] in their book "The machine that changed the world" have described LNPD in terms of number of techniques. Karlsson and Ahlstrom [8] have discussed about these techniques briefly:

- *Supplier involvement*: Suppliers are involved from the beginning of a new product development project, instead of being involved in one detailed design specifications for the subcontracted parts. Usually these suppliers are responsible for developing complete modules for the product often without detailed specifications.
- *Cross-functional teams*: These teams consist of members from different functional areas in the company. The aim is to integrate all functional aspects in the product from the beginning. Thus, marketing, production and other functions participate with the aim to provide different kinds of input in all phases of the development project.
- *Integrating mechanisms*: These are strategies for effectively coordinating actions across teams and groups within a program. As catalysts, they facilitate information flow across communication barriers such as a company or program's organizational structure, incentive systems, location, leadership styles, cultural differences, and management traditions [10]. Some of the integrative mechanisms are: co-location, improved information and communications technologies, training, town meetings, manager mediation, participant mediation etc.
- *Use of a heavy weight team structure*: It is a team which has a project manager who has direct access to and is responsible for the work of all those involved. This will improve communication, create stronger commitment to the project and bring about a focus for cross-functional problem solving.
- *Strategic management*: Each development project is managed via visions and objectives instead of detailed specifications.

In addition to this, there are several other techniques, which are discussed briefly below:

- Open Book Management (OBM) is defined in the following statement: It is possible to create a work environment in which each employee can make an impact on the company's profitability. Its success stems from creating a partnership with all employees. This partnership is developed through the application of a combination of proven and innovative human resource initiatives that are based on trust and sharing and free from the concept of exploitation [11]. Thus, it becomes clear that the ideals of OBM fit very nicely with the empowerment ideals of lean.

- *Theory of Inventive Problem Solving (TRIZ)*: It is a family of techniques developed originally in Russia for product invention and creativity. The fundamental belief of TRIZ is that invention can be taught. All inventions can be reduced to a set of rules and principles. It consists of about 40 inventive principles. A general methodology comprises three steps: First, determine why the problem exists. Second, "state the contradiction". Third, imagine the ideal solution. Many TRIZ ideas require some technical knowledge, or at least a technical aptitude [12].
- *Concurrent/simultaneous engineering*: It means that different activities in the development effort are performed in parallel. For instance, instead of first developing the product and then tools to manufacture it, the development of product and tools takes place simultaneously. CE has been developed in western product environments into a much broader concept, encompassing all functions in the product life cycle, not just engineering [13]. But there exists a wide difference between CE and LNPD. Haque and James-Moore [6] have discussed in detail about the differences, which are shown in Table 3.4.

Table 3.4: Differences between concurrent engineering and lean new product development

Concurrent engineering	Lean new product development
Lacks an enterprise-wide common strategic direction or statement for implementation. It is geared towards improving the NPD process and thus promotes specialized tools such as the Design for 'X', QFD, etc.	Lean is by definition an enterprise initiative with a common format for all business processes with the single strategic goal of eliminating waste and improving the flow of value. Its basic/original definition does not, however, address adequately the needs of NPD processes
Lacks a life cycle approach; that is, does not answer the question: where does an organization start and how does organization sustain the movement? The focus is on the 'what to do' and not the 'how to'	Provides a life cycle approach with both the 'what' and 'how to', starting and ending with the customer, with a continuous drive for waste elimination. The five principles provides an implementation route map for Lean
Liable to different interpretations and definitions	Universal definition: 'value' and 'waste' as main factors
Does not classify and contextualize waste. Waste elimination is a by-product of CE activities	Wastes identified, classified and contextualized within given value streams, and then eliminated
Promotes customer focus and improvement of information flow, but does not explicitly define a systematic approach	Explicitly promotes (a) creation of value stream maps based on customer demands, (b) flow is only truly possible after elimination of waste, and (c) the value creating process be pulled at a customer defined rate
It promotes overlapping of activities and hence the flow of partial information based on downstream process needs. This greatly improves information flow	It can be seen as the engineering version of a manufacturing-type pull system

3.5. TOOLS FOR LNPD

LNPD encompasses a number of tools [8] and some of the tools are briefly discussed here:

- Failure mode effect analysis (FMEA): It is a structured approach to the identification and evaluation through a risk priority number (RPN) of possible modes of failure in a product or process design. Failure is taken in its broadest sense, not only as a catastrophic breakdown but also as a consequence of not meeting a customer's requirements. The aim is to anticipate and design out all possible failures before they occur thereby removing the cost to manufacture, warranties and customer satisfaction.
- Quality function deployment (QFD): QFD is a tool, which should be used in a product development team. The QFD process involves mapping customer requirements onto specific design features and manufacturing processes through a series of matrices. This matrix has the most general structure and is often called the house of quality (HOQ). After the house of quality matrix, a number of additional matrices may be used to deploy the customer requirements through production planning. The ability of QFD to be deployed in this manner makes it unique among formal methods in its ability to span life cycle processes [14].
- Design for X: Design for X means designing products which are easier to manufacture, assemble, maintain, disassemble, repair, reuse, refurbish and recycle while meeting the desired dimensions, cost, value etc. Design for manufacture and assembly is a key facilitator of design and manufacturing integration. Through the use of some simple rules and additional numerical evaluation, products may be effectively and efficiently examined for their ease of manufacture and assembly. Three well-known design for assembly techniques are those of Boothroyd-Dewhurst and Lucas design for assembly (DFA) and Hitachi assemblability evaluation method (AEM). The process of challenging the existence of each component in a product is the key to efficient assembly. Products that consist of the minimum number of parts are not only enhanced for assembly but also provide benefits through reduced stock holding and inventory, reduced manufacturing or sourcing costs, and increased reliability. In addition to DFA analysis, design for manufacture (DFM) analysis is used to aid in the detailed design of parts for manufacture. DFM tools such as design for machining and design for sheet metalworking have been developed by the Boothroyd-Dewhurst partnership to address specific processes and the design of parts suited to those processes [15]. Similarly, Leaney [16] has discussed in detail about the other design for X concepts.
- Quality engineering: According to Taguchi [17] quality engineering pertains to the evaluation and improvement of the robustness of products, tolerance specifications, the design of engineering management processes, and the evaluation of the economic loss caused by the functional variation of products. Taguchi evaluates quality through a quality loss function. The quality loss function is expressed as the square of the deviation of an objective characteristic from its target, assuming the target to be the desire to meet customer satisfaction; any deviation from that value will mean a level of reduced satisfaction for the customer. In practice the Taguchi approach to quality engineering provides an analytical tool for designers to develop new products that can perform the desired functions, while keeping production costs below those for competitive products.

- Value engineering: Value engineering is a team based evaluative technique, which assigns a value to a product. The process attempts to enhance the value of the product by increasing its functional capability, for the same or lower cost. Or inversely, reducing the cost whilst maintaining the same functional capability. The goal is to eliminate unnecessary features and functions by optimizing the value to cost ratio [18]. The process of value engineering consists of five phases: information, function, speculation, evaluation, and implementation. These phases span the following activities; information gathering and defining the function of the product and its constituent components, assigning a value to each component, generating and assessing alternatives, and finally implementing the proposed solutions.
- Set-based engineering: Set-based engineering (SBE) seeks large sets of options that satisfy the requirements developed in the product definition phase. The first component of SBE is to develop sets of designs, i.e., groups of design alternatives, for a given design problem. Rather than trying to identify one solution, engineers should instead develop a variety of design options, and then gradually eliminate alternatives, until only single option remains. Independent exploration of design sets enables several engineering specialties to consider a design problem from their own perspective (i.e., to allow each specialty to work on a sub-problem) and then to effectively recombine those independent alternatives into an integrated final solution. Set-based engineering can therefore be looked at from a product architecture standpoint as well as from a design standpoint [19].
- The Design Structure Matrix (DSM): The DSM technique is a well developed method for analyzing the sequences of, and information flows between the tasks in a process. An "n-squared" matrix is used to depict the information flow from one task to another. The matrix can be numerically optimized to minimize iterations and maximize the potential for concurrent work. A complete description of the method, with tutorials for its use, can be found at the MIT-UIUC DSM website (2002). There are four different types of DSMs, (1) component based or architecture DSM, useful for modeling system component relationships and facilitating appropriate architectural decomposition strategies; (2) team-based or organization DSM, beneficial for designing organization structures to account for inter team information flow; (3) activity-based or schedule DSM, advantageous for modeling project schedules based on activity information dependencies; and (4) parameter-based or (low level) schedule DSM, effective for planning design decisions and activities based on physical design parameter relationships.
- Modularity: Modularity is often seen purely as a process of decomposition or demarcation of product architecture into subassemblies, but modules have a number of characteristics that provide fundamental differences between them and convenient groups of components in a subassembly:
  - Modules have their main functional interactions within rather than between modules
  - Modules have one or more well defined functions that can be tested in isolation from the system and are a composite of the components of the module
  - Modules are independent and self contained and may be combined and configured with similar units to achieve a different overall outcome.

Modularity is typically utilized for its ability to rationalize variety through the partitioning of product functions [4, 20] and allow for flexibility of application. However, variety is only one aspect of product modularity. One of the key elements of modularity is its fresh approach to meet the requirements of effective new product introduction.

- System dynamics: System dynamics modeling, a simulation technique pioneered at MIT is used to evaluate system characteristics, interactions, and sensitivities. The tool has been found to be of great use in analyzing processes whose process components and characteristics are known and understood. System dynamic modeling has been used for understanding of management practices; rework cycles, sustainment and population/ inventory assessment, and many others [21].
- Information technology: It refers to the use of computers in the product development process. It facilitates better communication through software tools like Email system and product data management (PDM), which can be used in Internet as well as intranet environments. It begins with the use of CAD software tool and has been extended towards modeling and simulation and recently, it has gone beyond CAD/CAM integration. Some of the information technology tools that are most commonly used in NPD process are discussed briefly:
  - Modeling and simulation: Modeling and simulation have become essential tools in the product development process. The fidelity and performance achieved by the complex systems today are due in large part to our enhanced computer capability for design and development, specifically through modeling and simulation. Examples include Pro-Engineer, CATIA, Unigraphics, ANSYS etc.
  - Product data management: These are software, which can integrate with the other modeling and simulation software and maintain a database of models pertaining to different components, sub assemblies, finished product. It may also have Key Characteristic Databases (KCB). Many engineering companies use Key Characteristics (KC), to manage the risk of variation in complex products. Key Characteristics are product features, manufacturing processes, and assembly characteristics that significantly affect a product's performance, function, fit, and form. KCs can be identified at a system level and flowed down to a detailed part level. KCs are also used to identify critical fabrication, sub-assembly, and assembly processes that are used to produce critical parameters. A typical example is Windchill from PTC.

It must be clearly understood that LNPD should not be confused with its tools and techniques. Thus, implementing one or few of the techniques contained in the total concept is not sufficient for achieving LNPD. A successful move towards lean product development requires approaching these interrelated techniques as elements of a coherent whole.

3.6. FRAMEWORK FOR LNPD

A framework is referred as a prescriptive set of things to do, while some authors have chosen to portray a framework through diagrams or graphical representations. Aalbregtse *et al.* [22] defined a framework as being "a clear picture of the leadership goal for the organization and should present key characteristics of the to-be style of business operations". i.e. it means that one should design and develop a framework representing the modus operandi, the systems to be developed, the activities to be carried out and the ultimate vision of the new style of management in the organization. Framework helps in translating a theory into practice through some systematic means. A framework may consist of various elements or blocks, which an organization needs to follow or adopt, when it tries to implement a change in the current way of functioning. Before we discuss the framework for LNPD, it is necessary to understand why a framework is required. A framework is needed:

- To illustrate an overview of a philosophy or change process to be adopted so as to communicate a new vision of the organization.
- It forces management to address a substantial list of key issues which otherwise might not be addressed.
- It gives an insight into the organization's strengths and weaknesses.

A framework for LNPD typically consists of various activities represented in the form of sequence of activities or steps, along with the necessary tools specified. The sequence of activities can be classified into various phases namely—Define, Review, Investigate, Verify, Execute and Repeat. In short it is called as a DRIVER framework for LNPD. The various activities that need to be carried out for an effective lean new product development are described in Table 3.5 and they are discussed briefly below:

Table 3.5: A DRIVER framework for LNPD

S. No.	Phase	Steps/Activities	Tools used
1.	Define	Understand the current product development process	Process mapping tools like flow chart, process chart etc.
2.		Identify a pilot project to which LNPD to be applied	Decision making tools like cost-benefit analysis, risk analysis or multi-attribute models
3.		Understand the waste and value in current development process	Value stream mapping—depicting the current state
4.	Review	Understand the causes of wastes in NPD	
5.		Estimate how the future NPD process should be by eliminating waste	Value stream mapping—depicting the future state
6.	Investigate	Identify potential solutions to eliminate each wastes	• Techniques like co-location, training, cross-functional teams, integrating mechanisms, concurrent engineering etc. • Tools like Design for X, QFD, Modularity, Simulation and modeling, design structure matrix, value engineering etc.
7.	Verify	Implement potential solutions on a pilot project	
8.		Check the status of implemented solutions	Techniques like performance measurement, Audit, checklist etc. are used
9.	Execute	Implement similar solutions across all NPD projects	Advocate effective documentation of project incorporating, failures, corrective actions and novel concepts, tools or techniques used
10.	Repeat	Continuously review the LNPD process and repeat the cycle	

Define Phase

As explained in the define phase, the first step of 'understanding the current product development process' is accomplished through mapping the current NPD process as shown in Fig. 3.1. Similarly, the product, which is critical to the organization, should be selected based on experience or using some decision-making tools. Once the critical product under development/to be developed is identified, the next important step in making a NPD process leaner is to apply value stream mapping (VSM). Freire and Alarcon [23] have demonstrated the application of VSM to the NPD process. VSM is of two types—the current state map and the future state map. The current state map of the design process is shown in Fig. 3.2. Though this current state map is incomplete, as no data has been shown, similar map can be drawn for the real life NPD process. From this current state map, different wastes, as represented in Table 3.1, should be carefully identified. For example, the triangles represent storage or waiting time between each stage. It can also be found that the current process has six stages of processing including Data collection, Design, Review, Correction, Release, and Distribution and may include any of the wastes discussed in Table 3.1.

Review Phase

In this phase the potential causes, which create wastes, are analyzed. A future state map will be drawn by foreseeing how the NPD process will be, when wastes are removed. Freire and Alarcon [23] have also demonstrated the future state (ideal state) map for the design process, which is shown in Fig. 3.2. From this figure, following differences can be found in comparison with Fig. 3.2. The six stages in the current state map have been reduced to four stages, which include design, review/correction, release and distribution. Further, a pull process has been employed before the design stage and the distribution stage by incorporating supermarket system, whereas it was push process in the current state map. These improvements may sometime represent the ideal state, which may not be achievable. But a future state map acts as a blue print and provides direction for improvement activities.

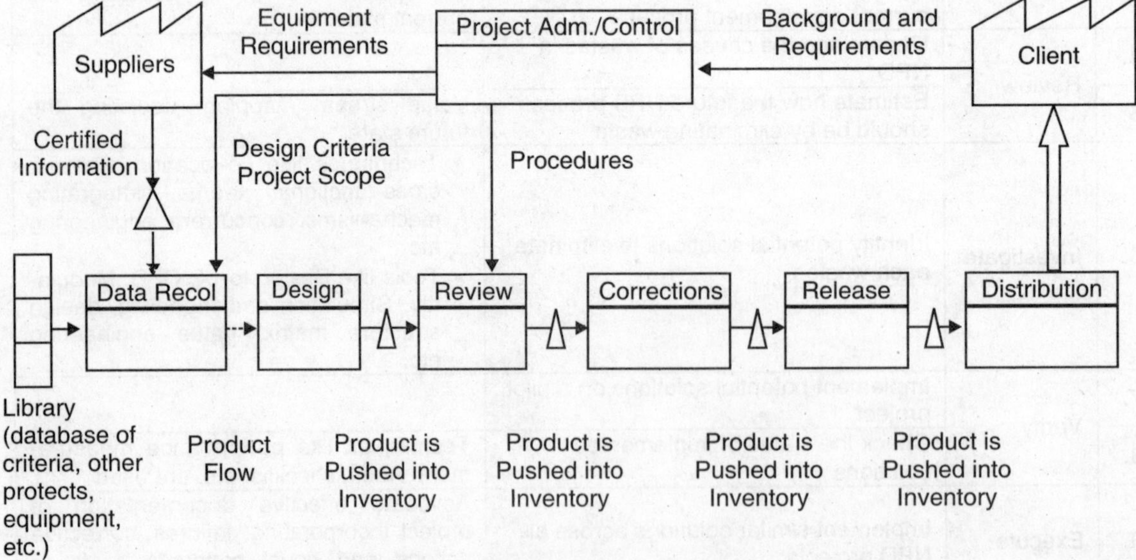

Fig. 3.2: Current state value stream map of a design process [23]

Investigate

During this phase, appropriate tools and techniques discussed above, should be selected and used to identify potential solutions for eliminating the wastes in NPD. For example, if too much time is consumed for generating 3D drawings in the design stage, then the organization can implement sophisticated CAD tools like 3D modeling software, Product life cycle management software etc., to facilitate reuse and recycle of design. Also incorporating appropriate Email software can reduce the waste in information transactions. To facilitate better communication, techniques like co-location; cross-functional teams can be established. Similarly, the different tools and techniques have been mapped to eliminate different types of wastes identified earlier. Table 3.6 provides a description of appropriate tools and techniques, which can be used to eliminate or reduce the wastes involved in a NPD. It should be understood that the Table 3.6 does not provide a complete list of tools and techniques, but it provides only a sample of tools that can be used to eliminate particular waste.

Verify

Once the necessary tools and techniques are applied, it is necessary to verify the output - to check whether the desired goal is achieved or not, whether LNPD has yielded desired benefits, whether the people are using these tools consistently etc. This can be accomplished by set of tools and techniques like audits, check sheets or performance measurement systems focused specially on NPD.

Execute

If improvements are seen in the pilot project, then it should be replicated across all the products that are developed. To enable easy and smooth replication, proper documentation should be carried out and maintained during the pilot project. This can be accomplished by the use of information technology tools and techniques like document standards, process capability database, knowledge management server etc. Such techniques facilitate easy storage and retrieval.

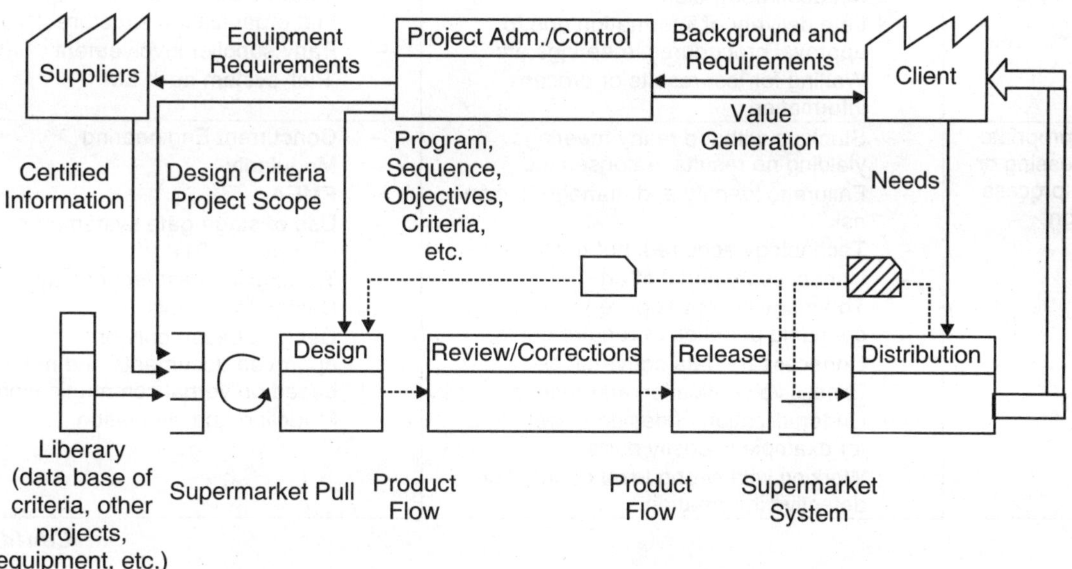

Fig. 3.3: Future state value stream map of design process [23]

Repeat

Since LM is based on the principle of striving for perfection, all the five phases discussed should be continuously carried out in a cycle.

Table 3.6: Tools and techniques used in LNPD for elimination of wastes

7 Wastes in LM	Applied to new product development process	Tools and techniques
Over production or early production	– Too many products or projects – Over design or providing unnecessary features – Over specifications/tolerance or too much details provided for a part – Redundant development—i.e. reuse not practiced	– Strategic management – Project selection tools like Gantt chart, NPV, IRR etc. – QFD, Value Engineering – Robust design, Design for dimensional control – Product data management (PDM)
Transportation or movement	– Information system and information formats: has a lack of common/ compatible standards resulting in repeated communication, manual transfer and conversion – A long hierarchical structure in an organization for decision making – Too many data interfaces or multiple sources – Poor interface control (with respect to departments) or management of design data	– Standardization of documentation, hardware, software etc. – Heavy weight team structure – Cross-functional teams – Integrating mechanisms – Early supplier involvement – Co-location – Information technology systems like email, intranet and software like 3D tools, PDM etc.
Waiting or Delays	– Information created too early – Unavailable information – Unclear decision criteria or lack of needed information – Late delivery of information due to approval procedures, meetings etc. – Waiting for test results or process information	– Cross-functional teams – Communication tools like Internet, Emails, product data management etc. – Flat organization structure – Early supplier involvement – 'Pull' system as in LM
Inappropriate processing or poor process design	– Stuck up with too many meetings yielding no results or consensus – Failure to identify and manage design risk – Technology acquired, but not used – Unnecessary serial effort – Too many iterations or re-work due to changing priorities or requirements – Unnecessary data conversions – Excessive verification and testing – Underutilization of design knowledge, for example in costly parts – Working with wrong level of detail or defective information	– Concurrent Engineering – Modularity – FMEA – Use of stage-gate systems (Cooper, 2001) – Training in latest technology – Design for X tools – Designs based only on approved documents and not based on verbal communication – Modeling and simulation

Contd...

Unnecessary inventory	– Too much information – Poor configuration management – Working on designs never used, completed or delivered – Unnecessary documents, physical prototypes – Lack of use of standard parts or lack of commonality;	– Proper information storage and retrieval – TRIZ (Theory of inventive problem solving) – Reuse, Recycling – 5S – Design for minimal unique parts – Value engineering
Unnecessary motion or inefficient performance of design	– Lack of standardization of processes – Out of sequence working (poor integration) – Information forwarded to wrong people – Poor re-use resulting in unnecessary development activities – Complicated retrieval – Over authorization – Over or inappropriate tolerance – Working with wrong/incomplete information – Inappropriate changes (changes not customer driven or not of benefit to business) – Data acquired then not used – Use of immature and inappropriate technology	– Process mapping, process flow charts – Modular designs – DSM – Use of closed group in emailing – Reuse, Recycling – Product data; management tools like key characteristic database – Design for dimension control – Technology forecasting
Defects	– Warranty issues – Under specifying – Poor make or buy decisions – Poor supplier identification – Poor design for X - manufacture, assembly, cost, reliability, and supply – Inadequate use of tools generating poor design – Using an outdated/immature technology – Failure to understand and capture requirements or poor understanding of customer needs – Conversion errors	– FMEA, FTA – Quality engineering tools and techniques – Early supplier involvement – Supplier certification – Design for X tools – MADM Approaches – Focus group study, Market survey and QFD – Technology forecasting

3.7. LIMITATIONS

The DRIVER framework for LNPD is a conceptual framework, and it needs to be validated for which a clinical approach can be adopted by deploying the framework in an organization and studying its performance difference through case studies. Though an example of value stream mapping for a traditional NPD process has been demonstrated, an in-depth study is required for analyzing every activity and sub-activity of NPD process from which value added and non-value added activities could be found out. Similarly, the appropriate tools and techniques identified could be utilized with respect to real life case situations to enable the practitioners appreciate the benefits obtained from LNPD.

REVIEW QUESTIONS

1. Why product design process is considered to be complex? Give reasons.
2. What are the traditional methods that were used to resolve the complexity of the product design process? What are the shortcomings of these traditional methods?
3. Why should the organizations adopt the philosophy of lean manufacturing in their product development process? Give adequate reasons to justify your answer.
4. What are the 'tenets of lean'? Explain in detail.
5. Draw a typical new product development (NPD) process. Identify the different activities that are performed in each stage of the NPD process using necessary internet information.
6. What are the different types of activities that are generally carried out in an organization? Give an example for each type from the perspective of product development process.
7. What are the objectives of lean new product development (LNPD)? How the 'tenets of lean' can be applied for the NPD process?
8. What are the different types of wastes? Give an example for each type of wastes from the perspective of NPD.
9. Identification of wastes in the NPD process is considered to be difficult when compared to the manufacturing process: Why? Give reasons.
10. What are the benefits of applying the LM principles for the NPD process?
11. List out the differences between the traditional product development and lean product development process.
12. Explain briefly about the different tools that can be used for making the NPD process leaner.
13. What are the different techniques that can be used for making the NPD process leaner?
14. List out the differences between the concurrent engineering and lean product development process.
15. Discuss about the role of 'Design for X' concepts in LNPD.
16. Discuss about the role of 'value engineering' in LNPD.
17. What do you mean by modularity? How does it help in achieving LNPD?
18. Explain briefly about the importance of 'information technology' in the LNPD process.
19. Define framework. What is the purpose of a framework?
20. Discuss in detail about the DRIVER framework for LNPD.
21. What is value stream mapping (VSM)? What are the different types of VSM?
22. Using internet information, identify the function of symbols that are used in the VSM.
23. Explain briefly about the different tools and techniques of LNPD which can be used to eliminate the different types of wastes.

REFERENCES

1. Maylor H. 1997, "*Concurrent New Product Development: An Empirical Assessment*", International Journal of Operation and Production Management, Vol.17, No. 12, pp. 1196–1214.
2. Womack and Jones (1997) (The machine that changed the world).

3. Cooper R.G. 2001, "*Winning at New Products: Accelerating the Process from Idea to Launch*", 3rd edition, Perseus Books, Mass~chusetts.
4. Smith P.G. and Reinertsen D.G. 1991, "*Developing Products in Half the Time*", Von Nostrand Reinhold, Inc, New York.
5. McManus H.L. and Millard R.L. 2002, "*Value Stream Analysis and Mapping for Product Development*", Proceedings of the International Council of the Aeronautical Sciences, 23rd ICAS Congress, 8–13 September, 2002, Toronto, Canada, pp. 6103.1–6103.10.
6. Haque B. and James-Moore M. 2004, "*Applying Lean Thinking to New Product Introduction*", Journal of Engineering Design, Vol. 15, No. 1, pp. 1–31.
7. Kearney A.T., 2003, "*The line on design—How to Reduce Material Cost by Eliminating Design Waste*", White paper, A.T. Kearney, Inc., Available at: www.atkearney.com/shared_res/pdf/The_Line_On_Design_S.pdf
8. Karlsson C. and Åhlström P. 1996, "*The Difficult Path to Lean Product Development*", Journal of Product Innovation Management, Vol. 13, pp: 283–295.
9. Cusamano M.A. and Nobeoka K. 1998, "*Thinking Beyond Lean: How Multi Project Management is Transforming Product Development at Toyota and Other Companies*", The Free Press, New York.
10. Morelli M. 1993, "*Evaluating Communication in Product Development Organization*", Technology & Policy Program, Massachusetts Institute of Technology, Cambridge.
11. McCoy T. 1996, "*Creating an Open Book Organization: Where Employees Think and Act Like Business Partners*", AMACON Publishers, New York.
12. Bicheno J. 2000, "*The Lean Tool Box*", 2nd edition, PICSIE Books, Buckingham.
13. Fleischer M. and Liker J. 1997, "*Concurrent Engineering Effectiveness—Integrated Product Development Across Orgnization*", Hanser Gardner Publications, Cincinnati.
14. Cohen L. 1995, "*Quality Function Deployment: How to Make QFD Work for You*", Addison Wesley Publishing Inc., Massachusetts.
15. Boothroyd G., Dewhurst P. and Knight W.A. 1994, "*Product Design for Manufacture and Assembly*", Marcel Dekker, Inc., New York.
16. Leaney P.G. 1996, "*Design for Dimensional Control*", In Huang, G.Q. (ed), Design for X – concurrent engineering imperatives, Chapman & Hall, pp. 173–195.
17. Taguchi G. 1993, "*Taguchi on Robust Technology Development: Bringing Quality Engineering Upstream*", ASME Press, New York.
18. SAE. 1997, "*Value Based Decisions for Automotive Engineering*", Ref. SP-1266, SAE Publications.
19. Bernstein J., 1998, "*Design Methods in the Aerospace Industry: Looking for Evidence of Set-based Practices*", Technology and Policy Program, Massachusetts Institute of Technology, Cambridge.
20. Pahl G. and Beitz W. 1996, "*Engineering Design: A Systematic Approach*", 2nd edition, Springer Verlag Ltd, London.
21. Bakkilla M., 1996, "*A System Dynamics Analysis of the Interaction between US Government and the Defense Aerospace Industry*", Aeronautics and Astronautics, Massachusetts Institute of Technology, Cambridge.
22. Aalbregtse R.J., Hejka J.A. and McNeley P.K., 1991, "*TQM: How do you do it?', Automation*", August, pp. 30–32.
23. Freire and Alarcon (2002). "Achieving lean design process: improvement methodology", Journal of construction Engineering and Management, Vol. 128, No. 3, pp. 248–56.

MATHEMATICAL MODELING OF DESIGN DECISION MAKING

4.1. INTRODUCTION

In most product development projects, there are many important life cycle issues to be considered. Mathematical models in current literature present a framework for making important decisions on design and development. Those models assumed that the design has been chosen. The mathematical framework explained in this chapter allows the possibility of having more than one design alternatives. In this model, the user would choose the optimal design among several alternatives. Different alternatives would be considered based on performance metrics such as manufacturability, reliability, cost and maintainability. Towards the end of the chapter, a simple case study will be used to demonstrate the proposed framework.

Early design decisions (EDD) during product development process have become an increasingly important prerequisite competency to ensure corporate success in today's global market environment. Companies are under severe pressure to deliver high quality and reliable products under shorter development lead-times and at lower costs. Many in design community accept the notion that more than 70% of the final product quality and cost are determined in the conceptual design phase, as early design decisions are made. Therefore, crucial tasks in product development process are to generate different concepts (or alternatives), and then select a single best-fit design concept to fulfill particular needs. It further requires capturing product life cycle issues in decision making process in order to make concept selection process more effective. Decisions made early in the product development process significantly influence an organization's capability to compete in the global market, but the required information is not used effectively. With this in mind it seems particularly imperative to make early design decisions in the most optimal manner possible.

Engineering design decision problem is a problem of decision with multiple criteria, and can be stated as follows [1]:

Given several performance criteria which are to be simultaneously optimized, determine a method for comparing any two or more design alternatives that depends only on the values of the individual criteria for each alternative.

The multiple criteria engineering design decision problem has been addressed by various decision-making frameworks, such as analytical hierarchy process (AHP) [2], quality function deployment (QFD) [3], and Pugh concept selection methodology [4]. The mathematical structure of decision-matrix-based methods is a summation of weighted concept ratings. Optimization methods based on sums of weighted criteria have been outwardly criticized in the multi-objective optimization community [5]. Pugh's concept selection method is another popular method being widely used in early trade-off decisions in product development processes. In Pugh method, criteria are not weighted and each concept is rated as "better than," "equal to," or "worse than" a datum concept. The Pugh method avoids some of the problem associated with other decision-matrix-based methods.

Multi-objective optimization techniques have proven to be extremely powerful in resolving conflicting objective, and therefore can be viewed as a viable tool in the hands of the designer. A recent trend in the multi-criteria decision making has been to increase the use of optimization in conceptual design [6, 7]. Crossley *et al.* [7] uses a genetic algorithm and combinatorial optimization for concept selection in conceptual air-craft design. Mattson and Messac [6] introduce the notion of s-Pareto optimality and show how it can be used to improve concept selection in engineering design. Other researchers have also used optimization techniques in concept selection [8].

Interestingly, most of these frameworks select best concept based on multiple design criteria, which are related to design attributes (or parameters) and functionality. They rely on the availability of analytical models that describe the concept's performance during the conceptual design phase. Unfortunately, none of these methods address the life cycle issues at conceptual design phase. There is no system level perspective through which to understand and/or improve the EDD process. Although these methods have their own merits, they are not flexible enough in dealing with the vague information of life cycle parameters such as reliability, quality, maintainability, manufacturability, and life cycle cost. Nepal et al. [9] present a framework for optimizing the cost while modularizing the product architecture early in the concept development phase. It is important to evaluate the design concepts based on life cycle issues, which otherwise are identical or of equal value.

Our motivation to develop effective methodology for selecting the best design concept is based on the realization that economically changing a design concept once it has been developed into detail design is often costly and affects the whole product development process. Therefore, we seek to reduce the risk of costly design changes, by capturing life cycle issues up-front in decision making and by implementing an effective early design decision framework for conceptual design.

4.2. PROBLEM STATEMENT

In the new mathematical framework explained in this chapter, the focus is on Early Design Decision (EDD) in Product Development. EDD framework deals with selection of product design and product architecture based on multiple criteria/objectives. The proposed framework begins with product performance measurement systems that feed information to the optimization model as a tool for making the selection decisions. The framework utilizes Analytical Hierarchical

Process (AHP) and Multi Objective Optimization (MOO). To be precise, the objective is to achieve these goals:

1. Develops a mathematical framework that is suitable for aiding modular product architecture decisions.
2. Develops a framework that is practical, systematic and user-friendly.
3. Systematically uses expert evaluation and includes multiple (often conflicting) desired criteria to objectively make the early design decision.

4.3. PROPOSED FRAMEWORK

From the flowchart given in Fig. 4.1, the steps for Early Design Decision (EDD) in Product Development Framework are as follows:

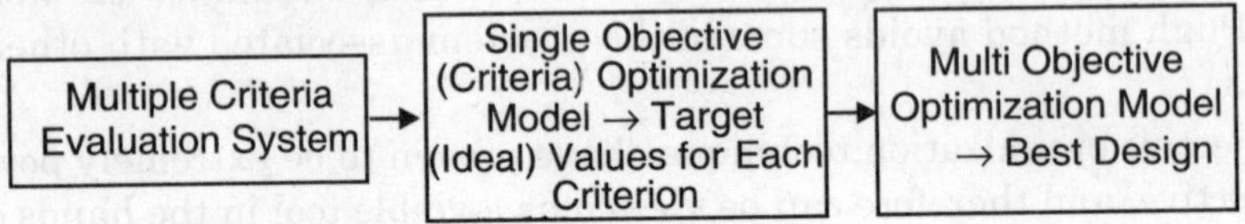

Fig. 4.1: Flow chart of proposed framework

Step 1: Develop expert evaluation system

We create expert evaluation system for evaluating different designs considered involving multiple criteria or goals using AHP. We then use results of AHP (step 1) as the performance metrics (input) for the next step.

Step 2: Develop target values for each criterion

With the information provided from step 1, develop a single objective/criteria optimization model to obtain target value of each criterion or goal. Target value is the best theoretical value that can be obtained for each criterion.

Step 3: Develop multi objective optimization

The target values obtained in step 2 would be incorporated to the Multi Objective Optimization (MOO) model. The MOO model would make the selection decision based on multiple criteria considered.

In order to explain our EDD framework, we use an example as follows. Assume that we have three possible designs that can be chosen. We will designate these three designs as design X, Y and Z respectively. And our goal is to make a decision of whether or not design X, Y or Z. This decision will be made based on several criteria: manufacturability, reliability, cost and maintainability.

Step 1: Develop Expert Evaluation System

Analytic Hierarchy Process (AHP) has been used as a tool for decision making process that entails subjective evaluation. It provides a systematic way in evaluating multiple alternatives that have different criteria. AHP systematically utilizes expert evaluation to produce the rating system. We now explain how AHP works by applying it to our example.

In our EDD framework example, the goal is to achieve the best design alternative based on several attributes: manufacturability, reliability, cost and maintainability. For each of these attributes, we create a pair wise comparison matrix. We develop scales for pair wise comparisons between two alternatives. The scales are shown in Table 4.1 where we have five different values.

Table 4.1: Scales for AHP pairwise comparison

Value	Preference
1	Equally preferred
2	Moderately preferred
3	Strongly preferred
4	Very strongly preferred
5	Extremely preferred

We first create a pairwise comparison matrix for our manufacturability attribute. Let us denote V_{ij} as the extent to which design alternative i is preferred to design alternative j. As an example, if the expert extremely prefers design X to design Y then $V_{xy} = 5$. Similarly, $V_{xz} = 3$ means that design X is strongly preferred to design Z. One observation to note is that preference between design X and Y is the reciprocal of the preference between design Y and X that is $V_{xy} = 1/V_{yx}$. Table 4.2 shows the complete pairwise comparison matrix.

Table 4.2: Pairwise comparison matrix

	X	Y	Z
X	1	3	5
Y	0.33	1	4
Z	0.2	0.25	1
sum	1.53	4.25	10

Having completed the pairwise comparison for all alternative designs we then normalize the matrix. Let denote S_i to be the sum of each column i in our comparison matrix. Therefore, S_x, the sum for column X is 1.533, S_y (column Y) is 4.25 and S_z (column Z) is 10. We normalize our V_{ij} by dividing V_{ij} with S_i. Table 4.3 shows the complete normalized pairwise comparison. Using these normalized values, we then calculate the manufacturability score for each design alternative. Let denote M_i be the manufacturability of design i. M_x is calculated by averaging normalized V_{xx}, V_{xy} and V_{xz}. The manufacturability scores for each design are also shown in Table 4.3.

Table 4.3: Normalized pairwise comparison

	X	Y	Z	Manufacturability rating
X	0.65	0.71	0.50	0.62
Y	0.22	0.24	0.40	0.28
Z	0.13	0.06	0.10	0.10

One important observation of using AHP is to make sure consistency in preference ratings given by the decision makers. It would be inconsistent to have one prefers X to Y, and Y to Z but prefers Z to X. In order to make sure that the preference values given are not inconsistent the consistency measure is also calculated. This is indicated by Consistency Ratio (CR). The following is the formula for CR:

$$CR = (\lambda - n)/\{(n-1)(RI)\}$$

RI = Random Index from Table 4.4.

n = Number of alternatives

λ = Average consistency measures for all alternatives: X, Y, Z

Consistency measure for X = $(M_x V_{xx} + M_y V_{xy} + M_z V_{xz})/(M_x)$

Consistency measure for Y = $(M_x V_{yx} + M_y V_{yy} + M_z V_{yz})/(M_y)$

Consistency measure for Z = $(M_x V_{zx} + M_y V_{zy} + M_z V_{zz})/(M_z)$

Table 4.4: Value of RI for AHP [10]

n	RI
2	0
3	0.58
4	0.9
5	1.12
6	1.24
7	1.32
8	1.41

If you have a perfect comparison matrix that is perfectly consistent matrix, your CR score would be zero. Ragsdale [10] has shown that if your CR score is less than 0.10 then the matrix is satisfactorily consistent. Table 4.5 shows the consistency measures for all design and the consistency ratio for manufacturability attribute is calculated to be 0.075. Therefore, the consistency is considered satisfactory.

Table 4.5: Consistency measure for manufacturability criteria

	X	Y	Z	Manufacturability rating	Consistency measure
X	0.65	0.71	0.50	0.62	3.16
Y	0.22	0.24	0.40	0.28	3.08
Z	0.13	0.06	0.10	0.10	3.02
				CR =	0.07

Following the same procedure as outlined above, we obtain the evaluation of all designs in regard with their other criteria: reliability, cost and maintainability as shown in Tables 4.6, 4.7 and 4.8. We can see that for all of those criteria, the matrices are considered consistent and satisfactory

ble 4.6: Normalized matrix and consistency measure: Reliability

Y	Z	Reliability	Consistency
0.11	0.16	0.14	3.01
0.22	0.21	0.24	3.01
0.67	0.63	0.62	3.03
		CR	0.02

Table 4.7: Normalized matrix and consistency measure: Cost

	X	Y	Z	Cost	Consistency
X	0.63	0.69	0.50	0.61	3.13
Y	0.21	0.23	0.38	0.27	3.07
Z	0.16	0.08	0.13	0.12	3.02
				CR	0.06

Table 4.8: Normalized matrix and consistency measure: Maintainability

	X	Y	Z	Maintainability	Consistency
X	0.13	0.14	0.11	0.13	3.00
Y	0.38	0.43	0.44	0.42	3.01
Z	0.50	0.43	0.44	0.46	3.01
				CR	0.01

Table 4.9: Summary of AHP results

Criterion	X	Y	Z
Mfg	0.62	0.28	0.10
Rel	0.14	0.24	0.62
Cost	0.61	0.27	0.12
Maint	0.13	0.42	0.46

Table 4.9 summarizes our evaluation on design X, Y and Z based on four criteria: manufacturability, reliability, cost and maintainability. Now in order to answer which design has the best overall rating, we need to address the issue of criteria's importance. How important is manufacturability criteria to reliability, to cost or to maintainability? To answer that question we will use the same method as outlined previously. Tables 4.10 and 4.11 show the pairwise and normalized pairwise comparison matrices for criterion weights. Those tables show that reliability is the most important criterion followed by cost and maintainability and the least importance is the manufacturability criterion. As usual, consistency is important and Table 4.11 shows that the consistency ratio is within the satisfactory range.

Table 4.10: Pairwise comparison matrix for criterion weights

	Manufactur-ability	Reliability	Cost	Maintainability
Manufactureability	1	0.2	0.5	0.5
Reliability	5	1	1	1
Cost	2	1	1	1
Maintainability	2	1	1	1
Sum	10	3.2	3.5	3.5

Table 4.12 shows the final rating for all three designs X, Y and Z incorporating the importance or weight of different criterion. Table 4.12 is gotten by multiplying Table 4.9 with the weight criterion obtained in Table 4.11. For example, the final performance matrix of design X on manufacturability is calculated by multiplying 0.62 with the 10% that is the weight importance of manufacturability criterion. The table shows that design X is best in manufacturability criterion and design Z has the worst manufacturability performance although it has the best performance in other three criteria. The question is then how can we resolve this problem? Our next steps would help us in making this tough selection decision.

Table 4.11: Normalized pairwise comparison matrix for criterion weights

	Normalized Comparison					
	Mfg	Rel	Cost	Maint	Criterion	Consistency
Mfg.	0.10	0.06	0.14	0.14	0.10	4.33
Rel.	0.50	0.31	0.29	0.29	0.37	3.84
Cost.	0.20	0.31	0.29	0.29	0.265	4.14
Maint.	0.20	0.31	0.29	0.29	0.265	4.14
					CR	0.04

Table 4.12: Final performance matrix for three designs after incorporating the criterion weights

Criterion	X	Y	Z
Mfg.	0.06	0.03	0.01
Rel.	0.05	0.09	0.23
Cost.	0.16	0.07	0.03
Maint.	0.03	0.11	0.12

Before continuing to steps 2 and 3, we now define the notation we use in steps 2 and 3.

Notation :

$$X_j = \begin{cases} 0 & \text{if design } j \text{ is not chosen} \\ 1 & \text{if design } j \text{ is chosen} \end{cases}$$

M_j = Manufacturability index of design j

R_j = Reliability index of design j

C_j = Cost index of design j

N_j = Maintainability index of design j

Q = Deviational variable

τ = Target value

α = Actual value

Step 2: Develop Target Values for Each Criterion

We will use the result of step 1 as summarized in Table 4.12, to develop our target value for each criterion. The target values are developed by solving independent single objective optimization models. For obtaining the target value for manufacturability criterion, we create the following optimization model:

$$Max : \sum_j M_j X_j$$

$s.t$:

(1) $\sum_{j=1} X_j = 1$

Note that the example has three designs (j = 1,2,3) to consider. The single objective optimization model tries to find the best theoretical value for the manufacturability criterion using the results of previous step as shown in Table 4.12 as input for the manufacturability index, M_j. For example, manufacturability index for design X, M_1 would be 0.06, M_2 would be 0.03 and M_3 would be 0.01. The following three independent models are solved similarly, as the previous model. Note that the model for cost criterion is to find the lowest theoretical value instead of the highest value. Also note that all four models have one single hard constraint that is to make sure that one design is chosen while obtaining the target value for all criteria. The model to find the target value for the reliability criterion is as follows.

$$Max : \sum_j R_j X_j$$

$s.t$:

(1) $\sum_{j=1} X_j = 1$

The model to find the target value for the cost criterion is as follows.

$$Min : \sum_j C_j X_j$$

$s.t$:

(1) $\sum_{j=1} X_j = 1$

The model to find the target value for the maintainability criterion is as follows.

$$Max : \sum_j N_j X_j$$

$s.t$:

(1) $\sum_{j=1} X_j = 1$

Table 4.13: Target values for all criteria

	Manufacturability	Reliability	Cost	Maintainability
Target Value	0.06	0.23	0.03	0.12

Step 3: Develop Multi Objective Optimization

In developing our Multi Objective Optimization, we create a Minimax variable Q, which is also a deviational variable. The objective function (1) indicates that our goal is to minimize Q the deviation variable. To implement this objective we create additional constraint for each goal or criterion. Constraint (2) indicates that the deviation from the target value (obtained in step 2 as shown in Table 4.13) of manufacturability index has to be less than or equal to Q. Constraint (3) shows that the deviation from the target value (obtained in step 2 as shown in Table 4.13) of reliability index has to be less than or equal to Q. Constraint (4) indicates that the deviation from the target value (obtained in step 2 as shown in Table 4.13) of cost index has to be less than or equal to Q. Constraint (5) indicates that the deviation from the target value (obtained in step 2 as shown in Table 4.13) of maintainability index has to be less than or equal to Q. With all these four additional constraints, as the model minimizes Q, it simultaneously also minimizes the deviation from target values for each goal: manufacturability, reliability, cost and maintainability. In other words, the maximum deviation from the target of each criterion or goal is minimized, thus the term Minimax.

$$Min : Q$$

$$s.t :$$

$$(1) \quad \sum_{j=1} X_j = 1$$

$$(2) \quad \{(\tau_m - \alpha_m)/(\tau_m)\} \leq Q$$

$$(3) \quad \{(\tau_r - \alpha_r)/(\tau_r)\} \leq Q$$

$$(4) \quad \{(\alpha_c - \tau_c)/(\tau_c)\} \leq Q$$

$$(5) \quad \{(\tau_n - \alpha_n)/(\tau_n)\} \leq Q$$

where $\alpha_m, \alpha_r, \alpha_c, \alpha_n$ are defined as follows:

$$\alpha_m = \sum_j M_j X_j$$

$$\alpha_r = \sum_j R_j X_j$$

$$\alpha_c = \sum_j C_j X_j$$

$$\alpha_n = \sum_j N_j X_j$$

Using our example, we found out that the results of our Multi Objective Optimization model is shown in table below:

Table 4.14: Results of the current example multi objective optimization model

	X_1	X_2	X_3	Q
Value	0	0	1	0.084821

The model shows that given four different criteria to consider design Z is the best design. Note that, this design as shown from the results of step 1 has the lowest performance in manufacturability but since it performs very well in other criteria and possibly that the weight or importance of other criteria is quite high, the multi objective optimization model decides that design Z is the best among all three designs considered.

It is possible further to refine the mathematical model presented in this chapter. This framework utilized Analytical Hierarchical Process (AHP) and Multi Objective Optimization model with a case study example. Since there are multiple criteria to consider, a design that performs best in one criterion is not necessarily going to be selected. The framework shows that a design that has the best overall criteria performance is the one to be selected. Future research should include additional constraints like time constraints or other design related constraints; in the multi objective optimization model.

REVIEW QUESTIONS

1. Why the decisions made at the conceptual stage of product design are so important? What bearing those decisions may have on the various stages of the product life cycle?
2. Why the optimization problem at the conceptual design stage is of multi-objective type?
3. What is the importance of consistency matrix?
4. How the relative importance of the different criteria is dealt with?
5. How the target values of difference objectives are developed?
6. Apply the algorithm explained in this chapter to a practical design problem and prepare a document with comments on the results.

REFERENCES

1. Scott M.J. and Antonsson E.K., Arrow's Theorem and Engineering Design Decision Making, "*Research in Engineering Design*", Vol. 11. No. 4, 2000, pp. 218–228.
2. Saaty T.L. 1980, "*The Analytic Hierarchy Process: Planning, Priority Setting, Resource Allocation*", McGraw Hill, New York, N.Y.
3. Hauser J. and Clausing D. 1988, "*The House of Quality*", Harvard Business Review, Vol. 66, No. 3, 63–73.
4. Pugh S. 1990, "*Total Design*", Addison-Wesley, New York, N.Y.
5. Messac A. and Ismail-Yahaya A., "*Required Relationship between Objective Function and Pareto Frontier Orders: Practical Implications*", AIAA Journal, Vol. 39, No. 11, 2001, pp. 2168–2174.
6. Mattson C.A. and Messac A., "*Concept Selection in n-Dimension using s-Pareto Frontiers and Visualization*", 9th AIAA/ISSMO Symposium on Multidisciplinary Analysis and Optimization Paper, No. AIAA–2002–5418, Atlanta, GA, Sep. 2002.
7. Crossley W.A. and Laananen D.H., "*Conceptual Design of Helicopters vis Genetic Algorithm*", Journal of Aircraft Vol. 33, No. 6. 1996, pp. 1060–1070.
8. Wang J., Ranking Engineering Design Concepts Using a Fuzzy Outranking Preference Model, "*Fuzzy Sets and Systems*", Vol. 119, No. 1, 2001, pp. 161–170.
9. Nepal B., Monplaisir L. and Singh N., "*Integrated Fuzzy Logic-based Model for Product Modularization during Concept Development Phase*", International Journal of Production Economics, Vol. 96, 2005, pp. 157–174.
10. Ragsdale C., "*Spreadsheet Modeling Decision Analysis*", South-Western College Publishing.

ASSESSING INNOVATION THROUGH PRODUCT ANALYSIS

5.1. INTRODUCTION

Rapid product development forms the backbone of today's industry. Substantial increase in competition compels design firms to develop new products at an increasingly rapid pace. This situation pressurizes engineering teams to develop better products and at the same time develop products faster [1]. Continuous innovation is a key factor to enable a company to generate profit on a continued basis, through the introduction of new products in the market – a prime intention for Product Life Cycle Management. A central goal of product development is to create products that are sufficiently novel and useful.

This research focuses on the determination of novelty of engineering products. Determination of novelty is important for ascertaining the newness of a product, to decide on the patentability of the design, to compare designers' capability of solving problems and to ascertain the potential market of a product. Few attempts at measuring novelty is available in literature [2, 3, 4], but more in-depth research is required for assessing degree of novelty of products.

This research aims to determine the novelty of a product by enabling a person to determine the degree of novelty in a product. A measure of novelty has been developed by which the degree of 'novelty' of products can be ascertained. An empirical study has been conducted to determine the validity of this method for determining the 'novelty' of the products.

Eder [5] claims that creative problem solutions, cause innovation in all design areas. Amabile [6] states that individual creativity is vital for organizational creativity and innovation to occur; later Nystrom [7] linked these to performance and survival.

During product development, creative solutions can give a product advantage over other competing products. Creative products can therefore be used to increase the price of the products and to get a larger market share [8]. Without creative problem solving, products will be traditional, which can cause loss [9]. A substantial part of the future income for a company will come from new products; thus, companies should introduce new products continuously in order to sustain in this competitive market and make profit [10]. This situation pressurizes engineering teams to develop better products and at the same time develop products faster [11]. Creativity is the core ingredient of innovation and it enhances the possibility of generating superior products. Continuous innovation is a key factor to enable a company to generate profit on a continued basis, through the introduction of new products in the market – a prime intention for Product Life Cycle Management.

'Product design', in which the idea of a product is conceived and embodied, costs are committed and other qualities like manufacturability etc. determined, is the crucial stage for any product in its entire life cycle. Miles and Moore [12] mentions that compared to the overall cost of a scheme, the design cost are a relatively small part; and yet they have a fundamental bearing on the overall costs, durability, serviceability and utility of the product. Good products are the result of good design. A central goal of product development is to create products that are sufficiently novel and useful. Thus, in the modern age we come across many so called 'new' products arising out of product development by various companies; this creates the requirement for identification of good products from the chafe, enabling us to recognize the good products and also identify better inventors/designers.

However, in spite of many research articles being published over the last few decades on creativity, although many authors have written about creativity, there is no recipe, they claim, for creative thinking. Developing and finding effective methods for enhancing creativity is essential.

Though creativity has been used in education for over 50 years, we still struggle to define and identify it [13]. Frankenberger and Birkhofer [14] state that, 'Creativity is commonly accepted as an important characteristic of innovative problem solving. What does 'creative' mean and how creativity shows up? A definition of creativity relevant for engineering design is still difficult despite enormous amount of research on creativity in the last few decades.' This in turn makes identification of a suitable method to measure creativity difficult [15].

In a recent work, Sarkar [16] proposed a 'common' definition and related 'common' measures of creativity. With an initial compilation of 164 definitions of creativity, 50 'expert' definitions formulated by creativity researchers were subsequently selected and analyzed for commonalities across them, leading to the construction of the 'common' definition, using two different methods viz. relationship analysis and majority analysis. The common definition is **'Creativity in engineering design occurs through a process by which an agent uses its ability to generate ideas, solutions or products that are novel and useful'.** The measures of creativity are, therefore, using measures of **'Novelty'** as well as **'Usefulness'**.

In this chapter, a method for assessing product novelty has been discussed in Sarkar [16]. Using these two methods together, one could determine the creativity of products.

5.2. NEED FOR NOVELTY DETERMINATION

The essential element that separates a product from any of its predecessors is its 'novelty'. Identification of novelty is important for the following reasons:

1. To identify products those are more creative.
2. To determine the creativity of designers in order to recognize or recruit them.

3. To select a better product from similar products.
4. To assess the patentability of a recently generated product.

Again, in the day-to-day life we come across many products manufactured by several companies that are quite similar to each other (like pens, speakers etc.); and it might be difficult for us to identify which product is novel, since the difference among these products may be very small. Thus, apart from identifying novel products, this work also aims to ascertain the qualitative degree of novelty viz. very high, high, medium and low degree of novelty. Next section deals with the tem 'novelty' in greater detail.

The inspiration of this work came from the following observations:

1. While going through Altshuller's work, the inventor of TRIZ methodology [17] it is noticed that there are many levels of innovative products *i.e.* conventional inventions (32%), small invention (45 %), substantial invention inside technology (4%), invention outside technology (4%) and discovery (1%) as found by Altshuller. Hence, we feel that if products are different from each other in terms of the level of innovation and thus novelty, it follows that the products should also differ from each other in terms of the relative degree of novelty. Even though in all these above cases the patented idea is novel, the level of novelty must be different. The products or ideas that fall under 'invention outside the technology' should be more novel than that of those falling in the first category that is 'conventional invention'. The current approaches to assess novelty seem inadequate to honour this distinction.
2. One more observation is that, for similar products like pens, scientific calculators etc., any latest product might come with only 2–3 new functions and so this latest product should be taken as a novel product. Again there are products like new medicines for curing life-threatening diseases such as cancer or AIDS for the first time in history – definitely a novel solution. In both the cases, the most recent products are novel, but the degree of novelty of the products mentioned in the second case is much higher than in the first case. So, novelty detection alone will not be able to differentiate between them—the degree of novelty should also be established.

5.3. UNDERSTANDING NOVELTY

'New' is something that has been recently created. 'Novel' products are those that are new to the entire human race.

'Novelty' encompasses both new and original. Novelty is 'not resembling something formerly known' [18]. Novelty may also be defined with reference, either to the previous ideas of the individual concerned, or to the whole of human history. The former definition concerns P-creativity (P for Psychological), the latter H-creativity (H for Historical). H-creativity pre-supposes P-creativity, for if someone has a historically novel idea, then it must be new to the person as well as to others [19]. Thus a 'new' product can be termed as 'new' (this should not be confused with our general notion of 'new product' as something that is recently introduced or manufactured or used for the first time), when it comes through P-creativity. Later this product must be checked with all other available products in that category to assess its absolute novelty. Generation of novel products requires H-creativity.

For novelty detection the common characteristics of products could be compared and differences among these characteristics should indicate whether the new product is novel or not as compared to the old products that perform the same or similar task or function. On the other hand, if a new product fulfils a need for the first time in the history, the product must be taken as novel.

Identification of a 'novel' product is difficult since we are not aware of all the products available in all the countries. A database containing the names and characteristics of all the products in all domains from each country would be an ideal database that would have enabled one to assess with greater ease the novelty of a newly generated product. Absence of such a search system compels us to depend upon the knowledge base of the experts in that domain to which the newly generated product belongs. Amabile [6] suggests using a few experts to assess the novelty of a new product. If the product is completely new to them and also satisfies some form of need for the society, then the product can be taken as novel. It is imperative that to assess the novelty of a product, one should know both the time line of similar inventions (the sequence of invention of products belonging to the same category – in terms of the domain and functionality) and the characteristics of the previous similar products belonging to the same product line.

As incremental innovations take place more often than radical innovations, the number of products which have slight improvements over their predecessors are many in numbers, making assessment of novelty of these products hard. The methodology as discussed here deals with a process by which both an expert and a novice designer can assess the novelty with its degree, of a new product, by comparing the new product with the old ones.

The objectives of this chapter are:

1. To develop a method to assess the novelty of a product.
2. To improve the same method to assess the degree of novelty of a product.
3. To evaluate this methodology (initial evaluation).

5.4. DEVELOPMENT OF A METHODOLOGY TO ASSESS THE NOVELTY OF A NEWLY GENERATED PRODUCT

To judge and compare various products it will be useful to compare the characteristics of the products. So, any methodology which breaks a product into its characteristic components should be suitable for comparing products. The widely used model in this regard is the Function, Behaviour and Structure (FBS) model. Many researchers [20, 21, 22, 23, 24, 25] have developed definitions and methodology for determining the FBS of products. The definitions of function, behaviour and structure, derived from the above-mentioned references, as taken for this work are as follows:

- *Function* : Description of what a system does—what is intentional and at a higher level of abstraction than Behaviour.
- *Behaviour* : Description of what a system does—which is generally non intentional and at a lower level of abstraction than Function. It is the way in which the function is achieved.
- *Structure*: Description of what the system constitutes—described by the elements and interfaces of which the system and its immediate interacting environment are made.

Since novel products are those that are new (recently generated) and original (appearing for the first time in human history); it is implicit that if the function(s) of a new product are different from all other available products, then the new product should be very highly novel product—the need that it fulfills was not previously fulfilled by any other available products at that time, else the product may or may not have been novel. Some examples of highly novel products are (when introduced for the first time): television — to broadcast video and audio data over long distances without any physical connection between the sender and the receiver; camera (pin hole)—to capture image for future use, similarly multi-utility systems such as X-ray machines or drugs such as penicillin are also highly novel products.

Next, if the new product structure matches with that of any other product then the new product cannot be novel, else it should be taken as novel (see the initial steps of the novelty detection method in Fig. 5.2).

The above method should help us to identify novel products, but it will not enable us to assess the degree of novelty of products, which is required to distinguish among similar products as opposed to individual radical innovations. Thus, a more comprehensive FBS model should be used. In a comparatively recent study [26] the product characteristics (of an FBS model) have been subdivided into seven elementary constructs. We found that this model can be employed to assess relative degree of novelty of products. The seven elementary constructs are:

- *Parts* : A set of physical components and interfaces constituting the system and its environment of interaction.
- *State* : The attributes and values of attributes that define the properties of a given system at a given instant of time during its operation.
- *Organ* : The structural context necessary for a physical effect to be activated.
- *Physical effect* : The laws of nature governing change.
- *Input* : The energy, information or material requirements for a physical effect to be activated; interpretation of energy/material parameters of a change of state in the context of an organ.
- *Physical phenomenon* : A set of potential changes associated with a given physical effect for a given organ and inputs.
- *Action* : An abstract description or high level interpretation of a change of state, a changed state, or creation of an input.

The relationships between these constructs are as follows: *parts* are necessary for *creating organs*. *Organs* and *inputs* are necessary for *activation* of *physical effects*. Activation of *physical effects* is necessary for *creating physical phenomena* and changes of *state*, changes of state are *interpreted* as *actions* or *inputs*, and *create* or *activate* parts. Essentially, there are three relationships: *activation, creation* and *interpretation*.

Using the above constructs and links, a model of causality has been proposed and discussed [26]. The causal description language is acronymed as SAPPhIRE model, SAPPhIRE standing for **S**tate-**A**ction-**P**art-**Ph**enomenon-**I**nput-o**R**gan-**E**ffect, Fig. 5.1.

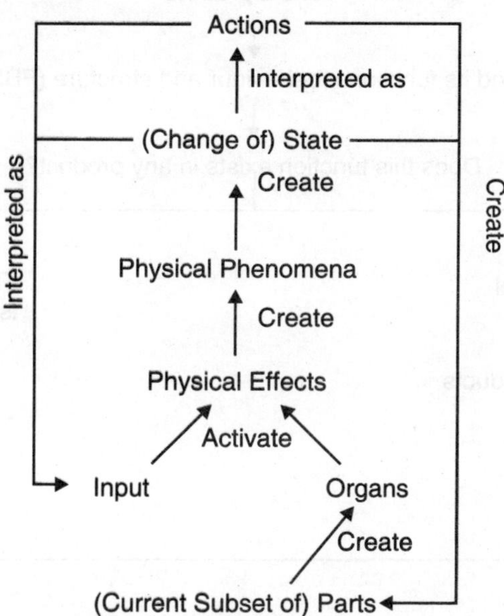

Fig. 5.1: The SAPPhIRE model of causality

Now, coming back to the novelty detection issue, we see above that *physical effect* is the main construct that governs the output of a product. So, products that are different in the physical effect level will be more novel than those that are different only at the *part* level. Hence, if two similar products works are based on the same physical effect and does the same action but one is only a structural improvement of a previous product, then the later product should be taken as a low novelty product.

Considering the above discussion, a method for novelty detection has been developed as shown in Fig. 5.2. In the initial part of the method, the FBS model has been used to identify novel products. Next the SAPPhIRE model has been employed to assess the relative degree of novelty among products.

If a product has many functions, then each of the functions should be compared with the existing products as many times as the number of functions. Apart from products, we believe that solutions and ideas could also be compared given that the ideas are matured enough for the FBS and SAPPhIRE models to be used.

Let us take a few examples to explain this model in greater details:

Very high novelty products: The first safety lamp to be used in mines, sewing machines, staplers, dynamites etc. would fall into this category, when introduced for the first time. These products have very high novelty because no such products existed doing the same function at that time.

High novelty products: At the time when microwave cooker was invented there were electric cookers, gas cookers and kerosene fuel stoves. The electric cooker is more similar to the microwave since both runs on electricity, and yet it was found to have high novelty compared to the electric cookers. The FBS and the SAPPhIRE models of these cookers were found out and represented in brief.

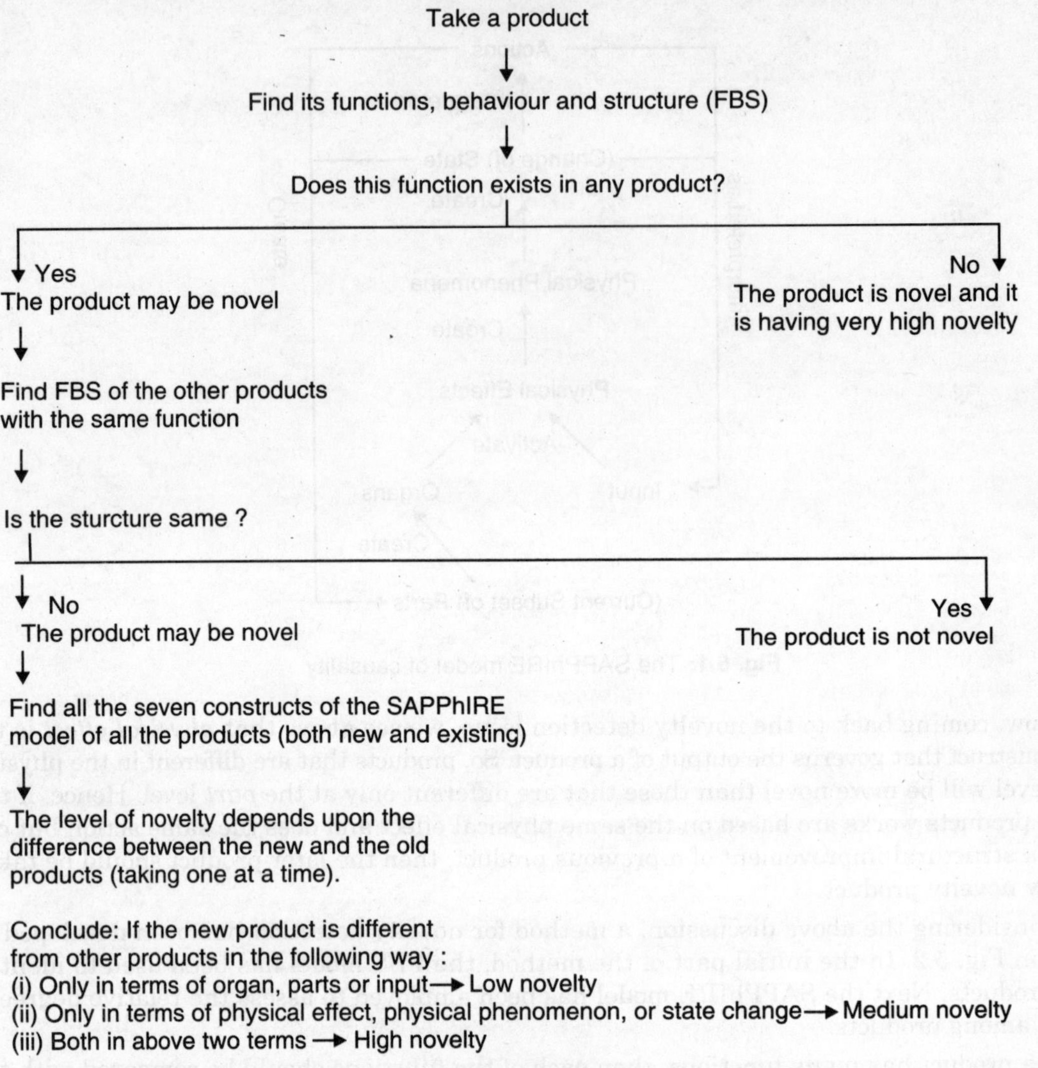

Fig. 5.2: The proposed approach for novelty detection

For electric cooker:

- Function: To produce heat to cook food.
- Behaviour: A coil heats up when electricity passes through it, thus producing heat to cook food in a container placed over it.
- Structure: A container housing a coil placed inside a non conducting material. The two ends of the nichrome wire are connected to the electric plug.
- Action: Generate heat
- State change: The wire turns from cold to red hot
- PP: Due to resistance in the wire the coil generates heat
- PE: Ohms law, heat transfer laws

- Organ: Ohmic resistance, specific heat capacity.
- Parts: Coil, holder
- Input: Electric power

For microwave oven:

- Function: To produce heat to cook food.
- Behaviour: Microwave generated in one part of the oven goes inside the food particles and these particles vibrate internally producing heat.
- Structure: Magnetron – the microwave generator, a closed container, controls and safety systems.
- Action: Generates heat
- State change: Rise in temperature
- PP: Vibration of the molecules.
- PE: Heat generation principles when micro wave is used
- Organ: Oscillation of polarized food molecules, eddy current
- Parts: Microwave generator, enclosure
- Input: Electric power.

Following the above method of novelty detection, we see that compared to electric heater microwave oven is highly novel since it differs from its contemporary products both in terms of *organ*, *parts* or *input* and *PP*, *PE* or change of *state*. The other comparisons, of micro wave oven with gas cookers and kerosene fuel stoves have not been shown here, but since they are even more different from the micro wave oven than electric cookers, such comparison should have shown that micro wave oven is highly novel also compared to them.

5.5. EVALUATION

To use this model one should be aware of the followings:

1. The sequential innovation/invention of the products to be compared
2. The constructs of the FBS and SAPPhIRE models.

One important requirement for the successful evaluation of this model is that the output of the model (that is the degree of novelty as determined), should match with the output of the experts (the degree of novelty as perceived/determined by the experts of the same field in which the product belongs to).

Initial evaluation: Before evaluating the model with experts it is necessary to perform pilot and initial evaluation of this method. Initial evaluation has been done with three in-house researchers. Each person is asked to select any two sets of products and note them on a piece of paper. Next, they are requested to analyse these products themselves and identify those products that they think is novel and by what degree. Later, they are required to assess the relative novelty of the same set of products using this method. In all the cases, the novelty degree as determined by them without (using their own feeling and analysis) and with the software matched. This initially validates the model and calls for a comprehensive evaluations by experts especially from companies.

5.6. OTHER AVAILABLE MODELS FOR ASSESSING PRODUCT NOVELTY

Few attempts at measuring novelty are available in literature [2, 3, 4], but more in-depth research is required in this area.

Saunders [3] work deals with finding the novelty of patterns restricting to mainly aesthetic novelty of patterns (1) How often similar patterns have been experienced. (2) How similar these patterns have been and (3) How recently these patterns have been experienced. Computationally, novelty is detected using processes that estimate one or more of these properties for a given stimulus pattern and a representation of previous stimuli.

Shah and Hernandez [4] discussed about ideation effectiveness and proposed measures for novelty (i) using function of products (ii) using physical principles of products. Equations for assessing novelty has been proposed and explained. Two methods for novelty determination have been proposed. The first technique is based upon the grading of description of the functions that it satisfies. The second method is based upon posterior classification and counting of distinct solution ideas with respect to prior knowledge.

Chakrabarti and Khadilkar's [2] works deals with finding the novelty using the following criteria- Vertical Criteria Weightages: Need, Task, Sub-system structure (principle), Technology, Sub technology, Implementation and Horizontal Criteria: main, supplementary, additional. The overall method for assessing product novelty is as follows: (i) First compare, with the reference product, the product whose novelty has to be assessed, and identify the differences, (ii) Calculate the novelty value of each difference and add -for calculating novelty value for a difference, first multiply the weightage for novelty at the vertical level and horizontal level. Then multiply this value with the horizontal-level weightage at one level above in the vertical direction and then repeat this process to get an overall value.

Except this last work no other work above aims to assess the degree of novelty of a product. The proposed work is distinct from all available work since

(i) It is based upon the well known product categorization systems (viz. the FBS and the SAPPhIRE model) giving a strong foundation for the development of this method and

(ii) With this method one will be able to identify novel products as well as their degree of novelty.

However, a detailed benchmarking of the proposed approach against the available approaches is underway.

5.7. DISCUSSION ON NOVELTY

Comparison of chronologically close inventions belonging to the same category (i.e., having similar functions) shows that the difference between two consecutive inventions is usually incremental. This means that if the invention sequence is taken into account, consecutive inventions are unlikely to be found highly novel. They, in general, would have either low or medium novelty, and only in rare cases would have high or very high novelty (mainly in major discoveries). Thus, the natural sequence of invention proceeds with incremental change in novelty. Let us consider a few examples.

Example 1: History of Clock.

The sequence of inventions is:-

- Sun dial and water clock (2 million BC).
- Tower clock (1094 AD).
- Domestic clocks (15th century AD).
- Self winding clock (1923).

Example 2: History of pens

- Scribing by Romans on sheets of wax (1300 BC).
- Finger as a pen with plant juice or blood as ink (4000 BC).
- Bamboo reels, Egyptian turning Papyrus into thin sheets (500–300 BC).
- Quill pen (7th century).
- Bronze pens (1465).
- Gold and silver nib pens (16th century).
- Steel pens (1803).

In each of the above examples, it appears that in most cases, if any two consecutive products from the same invention sequence (same product line), are compared then the later would not be possibly a highly novel product. For example, tower clock and the domestic clocks are different only in terms of size or bronze pens were similar to the gold and silver nip pens. These are low novel products. But, if two products are chosen such that one is from the beginning and the other from the end of the list, then that the later product will be highly novel. For example, self winding clocks are highly novel compared to sun dial or steel pens are medium novel compared to quill pens.

By analysing the product changes incurred in an invention sequence, it might be possible to say, in a statistical sense, how different products are — in terms of action, state change, physical principle etc. It might be possible to use this to predict the invention sequence of future products. This could be used as a guideline for altering products with an aim to improving products easily by changing their action, physical principle etc.

5.8. SUMMARY

A method for novelty detection of products and determining their degree of novelty has been developed. It is noted that product characteristics can be employed to ascertain the relative degree of novelty of products. FBS model has been used to determine the novelty of products and later SAPPhIRE model has been made to find out the relative degree of novelty of products. A methodology that enables assessment of both the novelty of a product and the degree by which it is novel has been proposed. Initial evaluation supports effectiveness of this method. Further research is aimed at eliminating the limitations of this method to make it more general and fool proof, and benchmark it against other available approaches.

REVIEW QUESTIONS

1. Why it is needed to determine the novelty of a product?
2. Creative solutions in product development give way for increasing the price. Do you agree? Justify your position.
3. What should be the relative standing of *usefulness* vis-a-vis *novelty* of a new product? Explain.
4. With reference to a new product, do the concepts of 'new' and 'novel' mean the same? Justify your answer.
5. Describe the FBS model of classifying the characteristics of a new product. Apply it to a new type of vacuum cleaner.
6. What is SAPPhIRE?
7. How the *physical effect* is related to novelty of a product?

REFERENCES

1. Eppinger D. Steven, Whitney E. Daniel, Smith P. Robert and Gebula A. David, 1994, "*A Model Based Method for Organizing Tasks in Product Development*", Research in Engineering Design, Springer-Verlag, 6:1–13.
2. Chakrabarti Amaresh and Khadilkar Promod, 2003, "*A Measure for Assessing Product Novelty*", International conference on Engineering Design, 2003, Stockholm.
3. Saunders Rob, 2002. "*Curious Design Agents and Artificial Creativity—A Synthetic Approach to the Study of Creative Behaviour*", 2002 Doctor of Philosophy, Department of Architectural and Design Science, Faculty of Architecture, University of Sydney.
4. Shah J.J. and Varghas-Hernandez N., 2003, "*Metrics for Measuring Ideation Effectiveness*", Design Studies, Vol. 24, pp. 111–143.
5. Eder W. Ernst, 1995, "*Introduction. In International Workshop: Engineering Design and Creativity*", State Scientific Library, Pilsen, Czech Republic, 16–18 November, 1995.
6. Amabile, T.M., 1996, "*Creativity in Context*". Westview Press, Boulder, Colorado.
7. Nystrom H., 1990, "*Organizational Innovation*". In JLFEMS West (Ed.), Innovation and creativity at work: Psychological and organizational strategies, pp. 143–162. New York: Wiley.
8. Ottosson Stig, 1995, "*Boosting Creativity in Technical Development*", in International Workshop: Engineering Design and Creativity, State Scientific Library, Pilsen, Czech Republic, 16–18 November, 1995.
9. Larry W., Zimmerman P.E and Glen D Hart, 1988, "*Value Engineering*" — A Practical Approach for Owners, Designers and Contractors, CBS Publishers and Distributors, ISBN no. 0-442-29587-1.
10. Roozenburg N.F.M., and Eekels J., 1991, "*Product Design: Fundamentals and Methods*", John Wiley and Sons, ISBN no. 471-94351-7.
11. Urban Glen L and Hauser John R, 1993, Design and marketing of new products, Prentice hall, ISBN no. 0-13-201567.
12. Miles J.C. and Moore C.J., 1994, "*Practical Knowledge based System in Conceptual Design*", Springer-verlag, London.
13. Forum Expert's, 2002, "*Defining and Encouraging Creativity*", Vol. 6, Issue 4/Summer 2002, Web: http://www.dukegiftedletter.com/articles/vol6no4_ef.html.

14. Frankenberger and Birkhofer, 1995, "*In International Workshop: Engineering Design and Creativity*". In: State Scientific Library, Pilsen, Czech Republic, 16–18 November, 1995.
15. Christiaans H., 1993, "*In International Workshop: Engineering Design and Creativity*". In State Scientific Library, Pilsen, Czech Republic, 16–18 November, 1995.
16. Sarkar Prabir, 2007, "*Development of a Support for Effective Concept Exploration to Enhance Creativity of Engineering Designers*". Ph.D. Thesis, Indian Institute of Science, Bangalore.
17. Terninko John, Zusman Alla and Zlotin Boris, 2000, "*Systematic Innovation, An introduction to TRIZ*" (Theory of Inventive Problem Solving), St. Lucie press.
18. Sternberg, Robert J. and Lubart, Todd I., 1999, "*The Concept of Creativity: Prospects and Paradigms*". In Robert J. Sternberg (Ed.) Handbook of Creativity. Cambridge: Cambridge University Press.
19. Boden A. Margaret, 1999, "*Computer Models of Creativity*", Robert J. Sternberg (Ed.) Handbook of Creativity. Cambridge: Cambridge University Press.
20. Chakrabarti A., 2001, "*Towards Hybrid Methods for Synthesis*". International Conference on Engineering Design, ICED 01 GLASGOW, pp. 379–386.
21. Chandrasekaran B., 1994, "*Functional Representation and Causal Processes*", in Advances in Computers, M.C.Yovits, Editor., Academic Press pp. 73–143.
22. Deng Y.M., 2002, "*Function and Behavior Representation in Conceptual Mechanical Design*", AI EDAM, Vol. 16, pp. 343–262.
23. Goel A.K.,1997, "*Design, Analogy and Creativity*". AI in Design.
24. Qian L. and Gero J.S., 1996, "*Function-Behavior–Structure Paths and Their Role in Analogy Based Design*". Artificial Intelligence for Engineering Design, Analysis and Manufacturing, Vol. 10, No. 4, Sept., pp. 289–312.
25. Umeda et al., Y., 1996, "*Supporting Conceptual Design Based on the Function-Behavior-State Modeler*", AIEDAM, Vol. 10, No. 4, Sept., pp. 275–288.
26. Chakrabarti Amaresh, Sarkar Prabir, Leelavathamma, and Nataraju B.S., 2005, "*A Functional Representation for Aiding Biomimetic and Artificial Inspiration of New Ideas*", Artificial Intelligence in Engineering design, Analysis and Manufacturing, 19, pp. 113–132, CUP.

6

PRODUCT CANNIBALIZATION

6.1. INTRODUCTION

Cannibalization is to make use of a part taken from one thing in repairing or creating something else. Product development and market development are two important areas of product life cycle engineering and management. To survive and be successful in a worldwide highly competitive market, industries are continuously trying to design and launch innovative products with new features and new technologies. We have a demanding customer expectation and rapid innovations that promote a life cycle of a product becoming shorter and shorter. In such a situation product cannibalization is the key of success for any manufacturing industry. The focus of this chapter is to propose a solution for optimum product development for effective product cannibalization so that industry can accrue maximum benefit by launching innovative products at right times. To launch an innovative product in the market at the right moment requires proper market survey along with proper design, R and D, prototype product development and testing as well as product aesthetics and ergonomic aspects concurrently. These stages and their interrelationship have also been highlighted here.

All products and services have certain life cycles. In manufacturing firm, it is very essential to launch the product as early as possible in the market and get more advantages in terms of sale. For rapid launching of product, optimum utilization of all recourses during product development phase is a very important parameter [1]. Kristjjansson & Hildre [1] and Norman [2] stated very clearly that for manufacturing industries, it is necessary to find out proper launching time and withdrawal time of product due to high competitiveness so, product design, research and development departments have many responsibilities. For up- gradation of any new product it is very necessary to find out the methods by which effective product development is possible [3, 4]. Concurrent engineering is one of the effective techniques in product development phase [5] by which product can be launched early in the market. This chapter will give the very important direction in the field of product life cycle for effective cannibalization.

6.2. PRODUCT CANNIBALIZATION IN PRODUCT LIFE CYCLE

Product cannibalization occurs when a company decides to replace an existing product with a newly introduced product in its place. In the normal case of cannibalization, an improved version of a product replaces an existing product, as the existing product reaches its sale maturity peak in the market. The new product is sold at a competitive price to sustain the sales, as the old product approaches the end of its life cycle. Nevertheless, at times companies have introduced new version of a product, when the existing product is only start to grow. In this way company sustains peak sales all the time and does not wait for existing product to start growing. Cannibalization is a very important tool to success in the competitive world. Cannibalization at right time is the key to success for any industry. Late implementation or early cannibalization shows very bad results during product life cycle.

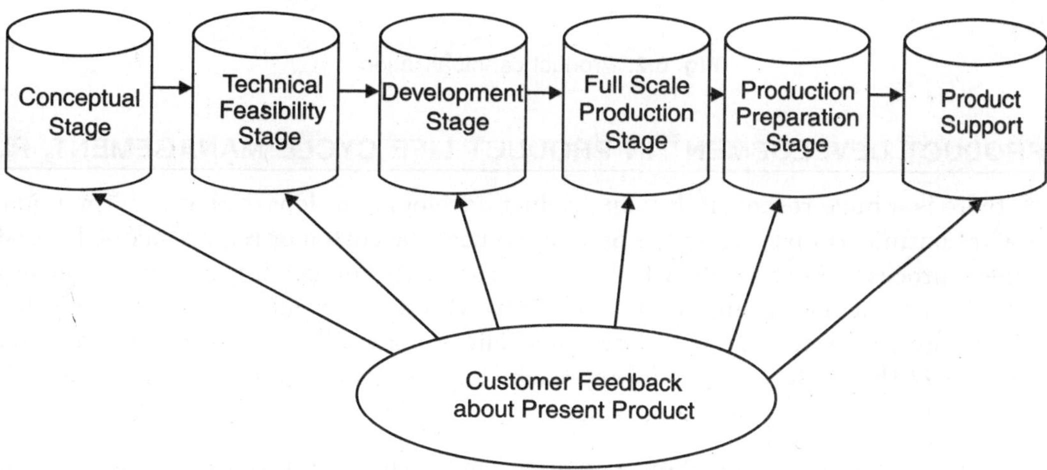

Fig. 6.1: Product development and customer support

For product cannibalization, continuous upgradation of product is very important. Upgradation is possible by continuous research work in product development phase. As shown in Fig. 6.1, continuous feedback from customer at every stage of product development phase during product life cycle of existing product gives vital ideas. So, experts from every stage can take corrective action for upgrading the product. Every department will get the feedback from customers and the real problems or lacunae can be found out for updating the product [6].

Figure 6.2 shows the product cannibalization. When the old product is at maturity stage, it is necessary to launch the new updated product. Time between O to O' plays vital role in product development. Before the declination of older product, it is necessary to start developing a new product. For upgradation of old product, product development department has to work continuously. So in product development phase, *concurrent engineering* plays a vital role for reduction in time O to O'.

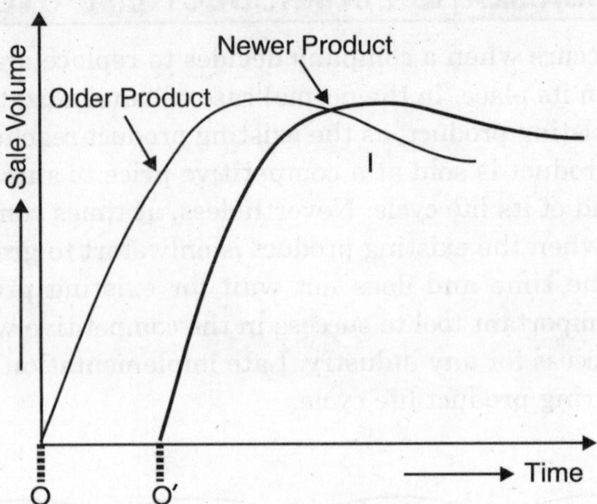

Fig. 6.2: Product cannibalization

6.3. PRODUCT DEVELOPMENT IN PRODUCT LIFE CYCLE MANAGEMENT, PLM

In PLM, there is a huge responsibility on product development department to form a quality product with continuous updating of the product to meet the customer requirements. In product development process, there is need to work concurrently for getting maximum output in minimum time. Concurrent engineering is a new approach applied to engineering design philosophy where technical and non-technical disciplines work simultaneously in the design and manufacturing of the product. This new approach is in response to and an improvement of the intrinsic problems of traditional engineering.

Traditional product development consumes more time for research while a iterative designing and redesigning process require a lot time spent in designing the product. This can put a company in disadvantage in comparison to a more efficient competitor with a faster time-to-market strategy. Specialist team is required to handle the different activities in the product life cycle. Product development stages like conceptual, technical feasibility, development, commercial validation, and production preparation, full-scale production and product support, each is involved with a team of engineers, each team located in various regions within a plant or across plants in different geographical locations, within a country or across the world. Proper coordination between all those teams is very important for getting maximum output in minimum time.

As we discussed previously, product development contains six important stages and teams of these different stages need to work with each other concurrently to optimize the product with respect to various life cycle issues and launch the product early in the market. Concurrent engineering is, however, difficult to practice when specialist teams are located in different parts of world, an increasing trend with economic globalization. For such problem, digital communication network is the best solution. As shown in Fig. 6.3, the stages of product development working concurrently to each other by using a digital communication network and videoconferences so every specialist member is easily accessible to others in the team and across the teams. Together they can make important decisions about product. The model and result are stored in digital form in a central or distributed server and accessible to all team members over a local area network or Internet. Collaborative design discussed in an earlier chapter is one effective solution to achieving this coordination. Most PLM tools available in the market nowadays have an embedded collaborative design workbench.

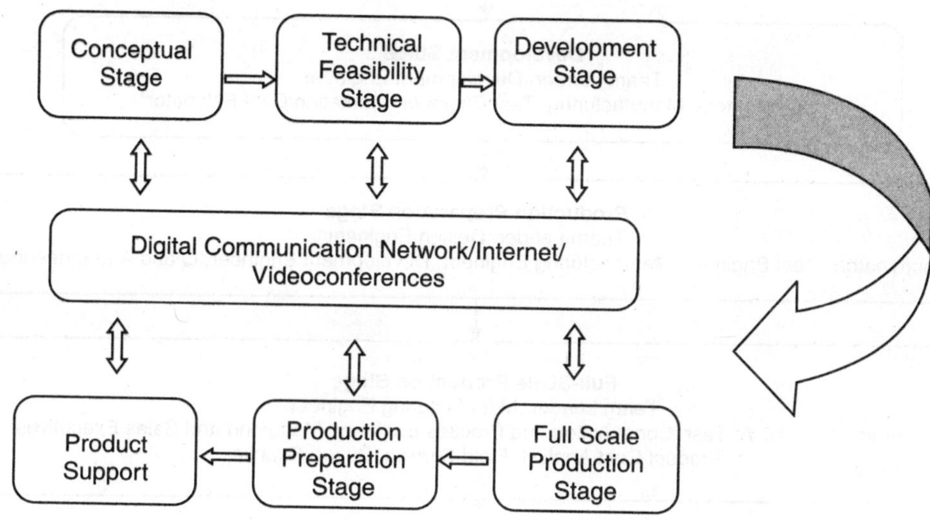

Fig. 6.3: Concurrent approach to new product development

6.4. FORMULATION OF A MULTI SPECIALITY PROJECT TEAM

Each stage of the product development requires a team leader and coordination of other staff. Figure (6.4) shows the team leaders and coordinators in each stage of product development.

It is very common for manufacturing department to make changes on the design of a product to ease its assembly, without consulting the research and development department, which can result in a malfunctioning of the product [7].

Many times purchasing department finds parts that seem to comply with the engineering specification at a much lower price than the ones preferred by engineering. However, those parts may not perform as well, or last as long as expected causing machine down-time, costly repairs and a bad reputation or those parts may not follow the maintenance standards resulting in poor customer service and costly retraining.

As shown in above Fig. 6.4, every stage of product development contains the team leader and coordinators. In every stage experts in related field form a team. So, before making any decision related to product, team leader has to discuss problems with coordinator for effective decision-making. Each and every expert in the process of decision-making plays vital role.

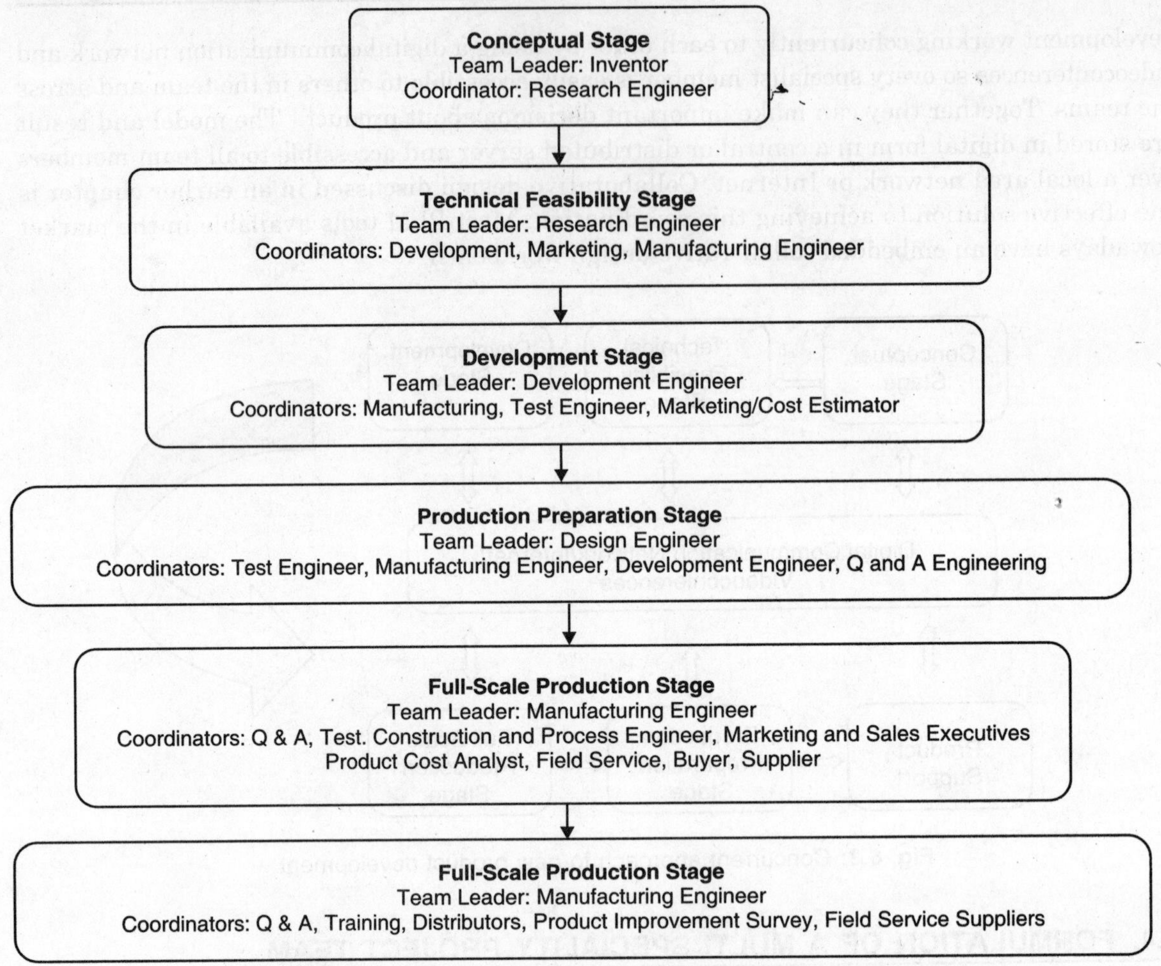

Fig. 6.4: Concurrent engineering in new product development

REVIEW QUESTIONS

1. Define product cannibalization. Is it advisable for a company to resort to product cannibalization? If yes, under what conditions?
2. What are the six important stages of product life cycle, as defined in the discussion of this chapter?
3. What is the functionality of teams in each stage of the product life cycle?
4. What is concurrent engineering?
5. How concurrent engineering enables continuous improvement?
6. What benefits accrue to the product development team by following the concurrent engineering?
7. How collaborative design can help to achieve the goals of concurrent engineering? How this is relevant to product cannibalization?

8. Search in the corporate examples of consumer products for suitable case studies of product cannibalization. Trace out the reasons and conditions leading to product cannibalization by those companies. Did those companies invariable benefited from it?
9. Do research on how currently companies take decision on product cannibalization. Can you propose a more rationale methodology?

REFERENCES

1. Kristjjansson A. H. and Hildre H. P., *"Framework for Evolution Platforms in Product Developing Organization"*, 7th Workshop on Product Structuring - Product Platform Development, Chalmers University of Technology, Goteberg, March 24–25, 2004.
2. Norman (1988).
3. Day G., "*The Product Life Cycle Analysis and Application Issue*", Journal of Marketing, Vol. 41, 1981, pp. 60–67.
4. Scacchi (1997).
5. Krishnan V. and Karl Ulrich, Product Development Decisions: A Review of Literature, "*Management Science*", Vol. 47, No. 1, 2001, pp. 1–21.
6. Drucker P., "*The Discipline of Innovation*", Harvad Business Review, Vol. 63, 1985, pp. 67–72.
7. Paulson Gjerde, Slotnick and Sobel, New Product Invention with Multiple Features and Technology Constrain", "*Management Science*", Vol. 48, No. 10, 2002, pp. 1268–1284.

SOFTWARE MODEL INTEGRATIONS IN PRODUCT DESIGN

7.1. INTRODUCTION

For better and efficient functioning of a product during its life cycle, it is important that the product is designed by taking into account various aspects like mechanical failure, vibration, controllability, etc. In order to consider all these aspects, the designer has to integrate several software models, for example, a model developed in ADAMS for dynamic analysis needs to be exported to another one, e.g., MATLAB, for its control simulation. It is, however, important that the designer is careful about the data transfer from one software to another, as many characteristics, e.g., a revolute joint between two objects in ADAMS, may get lost when transferred to the other software. Besides, the options used in a software, e.g., the solver in ADAMS or MATLAB, should be chosen appropriately. Otherwise, the interpretations of the result will not be appropriate. For example, a diverging response of a system obtained from MATLAB may be interpreted as the instability of the system, which may not really at all due to the system characteristics but due to the inappropriate choice of the options. Hence, a proper understanding of the issues related to the software data exchange and options, etc., must be looked into before one just does it mechanically. These aspects are emphasized in this paper. Two case studies, namely, ADAMS/MATLAB and ANSYS/ADAMS integrations, are presented here as illustrations.

A mechanical system modeled in one software needs to be exported to another software(s) for different analysis purposes. In order to make the mechanical model accessible to various software environments, it is desirable to have one common software-independent format that all software understands using their pre-processors. After analysis the results are stored again in another neutral format from where they can be forwarded to different environments in which a post-processor interprets the results through plots and animations. Due to professional competitions, true neutral formats do not exist, even though efforts through STEP (Standard for the Exchange of Product Data) [1] under the ISO (International Organization for Standards) have started. Besides, there are inter-company collaborations to exchange model data, e.g., ADAMS/MATLAB or ANSYS/ADAMS, etc. Two case studies are reported in this paper to illustrate the concept of model integrations in order to perform analyses based on a common model generations. In the first case, an automobile suspension model is developed in ADAMS for its dynamic analysis, which is exported to MATLAB for its control simulation. ADAMS [2] is a powerful modeling and simulating software environment where we can build and simulate the models of any mechanical system that has moving parts. We can test our model in the same conditions as our actual product will be subjected to. We can also validate our model against test data, and refine our model with flexible parts, realistic geometry, and controls. We can also find the important parameters that affect our product design, and find the right combination of design values to arrive at the best product design. Best of all, we can do all this over and over again, very quickly, say, in minutes or hours, where as a real test will require at least days, weeks or months depending on the complexity of the system at hand. Hence, a lot of time and money is saved. MATLAB [3], on the other hand, is the software for high-performance numerical computation and visualization. It provides an interactive environment with hundreds of built-in functions for technical computation, graphics and animation. Moreover, it provides easy extensibility with its own high-level programming language. SIMULINK is a module of the MATLAB software for modeling, simulation, and analyzing dynamic systems. It supports linear and nonlinear systems, modeled in continuous time, sampled time, or a hybrid of the two. Systems can also be multi-rate, i.e., have different parts that are sampled or updated at different rates. We can easily build models from scratch, or take an existing model from other software environments, say, ADAMS, and add to it. Simulations are interactive so that we can change the parameters on the fly and immediately see what happens. We have instant access to all the analysis tools in MATLAB so that we can take the results and visualize them. A goal of SIMULINK is to give us a sense of fun of modeling and simulation through an environment that encourages us to pose a question, model it, and see what happens. It is also practical. ANSYS [4], the third software used in the integration examples presented in this paper, is the general purpose finite element modeling package for numerical solution of a wide range of mechanical problems. These problems include static/dynamic and structural analyses (both linear and nonlinear), heat transfer and fluid problems, as well as acoustic and electromagnetic problems. In the second case study, a thin cantilever beam is modeled in ANSYS and exported to ADAMS. In ADAMS, the fixed end is replaced with a revolute joint to simulate its as a flexible pendulum. The paper is organized as follows: Section 2 presents the related issues in model integrations, followed by the case studies in Sections 3 and 4. Finally, the conclusions are presented in Sec. 5.

7.2. ISSUES IN MODEL INTEGRATIONS

The surge of different software for various purposes, e.g., motion, stress and strain, control, and vibration analyses, etc., has made the task of today's designer simpler. However, if there is a necessity to perform more than one analyses the designer or analyst generally needs to repeat the system's model, as one software format may not be understood by the format used by another software. Hence, there is a need for common data format, which can be interpreted by different software. Such formats are referred as Neutral Data Format in [5]. Such formats are interpreted by individual software using their pre-processors. Similar neutral data formats can be used to generate outputs, which then can be exchanged amongst different software and interpreted using their post-processors. The standardization of such data for model integrations requires international cooperation within STEP (Standard for the Exchange of Product Data) (1) under the ISO. There are also efforts by individual software companies to, e.g., ADAMS has collaborations with ANSYS and MATLAB developers to exchange data with them [6]. Such facilities make the modeling work really hassle-free.

Irrespective of the data exchanges, whether it is STEP or company specific, it is still important to be able to interpret the options appearing in the pre/post-processing toolboxes. For example, ADAMS or MATLAB software, have, besides many, options of choosing a solver or an integrator. This is basically a numerical integration methodology to be used. For certain systems, e.g., flexible body mechanical systems, there are stiff systems [7], which are very sensitive to the initial configurations of the system. For such systems, Gear method works better than the most popular method of Runge-Kutta (RK). However, using the RK method if one gets the system's behaviour as diverging, he or she may conclude that the system is unstable. Such conclusions are not correct. Alternatively, for the same flexible system if the finite element models of the components are developed in the software like ANSYS, then the proper choice of the type of elements, e.g., beam, shell, etc., are very important. An improper choice may lead to completely false results, which the user may not recognize unless he or she is aware of them. Numerical stiffness also arises due to wide variations of the numerical data used as the system parameters. For example, in a controlling system the controller gains may vary in 1s or 10s, whereas the system's inertias may be in 0.001s or 10-5s depending on the units used. Such wide variations of numbers in the system matrix cause ill-conditioning leading to instability in the simulation results, similar to those mentioned above. Such behaviour is also not due to the nature of the physical system at all but purely due to the numerical procedure adopted in analyzing them. Remedial actions like normalizations, use of different numerical algorithms, etc., are necessary to overcome such problems. Without which the use of any software by anyone is like a driver without any training. However, this is an arduous task for the user to understand the meaning of all the options available in software, particularly, when the company-specific terminologies are used. In order to circumvent such problems, it is suggested that a relatively simpler problem is taken up for the analyses whose analyses and effect due to variations of the input parameters are well established.

7.3. ADAMS/MATLAB INTERFACE

In this section an automobile suspension system is modeled as a spring-mass-damper system in ADAMS. The steps to generate the model are explained. Its control analysis results in MATLAB environment are also presented.

ADAMS Modeling

A simple automobile suspension modeled is developed in ADAMS, using the ADAMS/View [2], as shown in Fig. 7.1(a). The model is consisting of a mass, spring, and damper. It is fixed at the top, and hanging freely. If there is no damping, the simulated results from ADAMS are shown in Fig. 7.1(b), which is as expected.

If we include the damping, the obvious changes occur in the output, which is shown in Fig. 7.1(c). This figure shows how the amplitude will decrease with respect to increase in damping. ADAMS generated model is then exported to MATLAB-SIMULINK environment, as shown in Fig. 7.2. Simulated results from the ADAMS exported MATLAB-SIMULINK results are exactly same as those in Fig. 7.1(c).

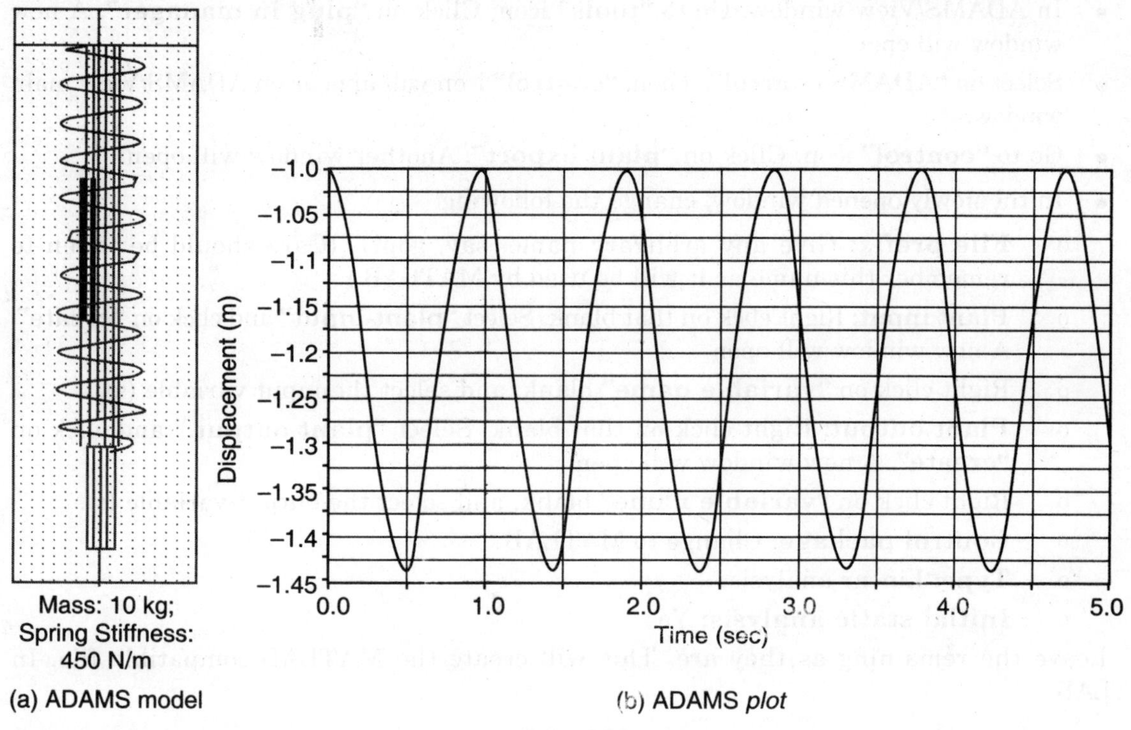

(a) ADAMS model (b) ADAMS *plot*

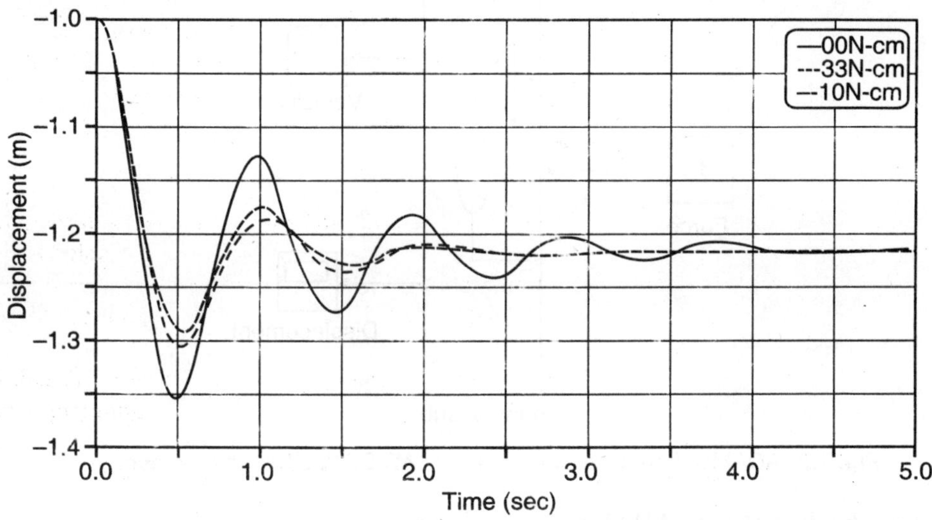

(c) Behaviour of the damped system (Damping coefficients: 20, 33, 40 N-s/m)

Fig. 7.1: A simple automobile suspension model in ADAMS

The steps to transfer the ADAMS data to MATLAB requires some pre-processing in ADAMS and same post-processing in MATLAB, as explained in the following steps:

- In ADAMS/View window, Go to **"tools"** icon; Click on **"plug in manager".** A new window will open.
- Select on **"ADAMS control".** Then, **"control"** icon will appear on ADAMS/View main window.
- Go to **"control"** icon. Click on **"plant export".** Another window will open.
- In the newly opened window, change the following:
  - **File prefix:** Give any arbitrary name, say, gopi1, (Care should be taken to remember this name as it will be used by MATLAB.)
  - **Plant input:** Right click on that blank. Select **"plant-input"** and click on **"create".** A new window will open.
  - Right click on **"variable name"** blank, and select the input variable (s).
  - **Plant output:** Right click on that blank. Select **"plant-output,"** and click on **"create".** A new window will open.
  - Right click on **"variable name"** blank, and select the output variable (s).
  - **Control package:** Change to MATLAB.
  - **Type:** Linear analysis.
  - **Initial static analysis:** Yes.

Leave the remaining as they are. This will create the MATLAB compatible file. In MATLAB,

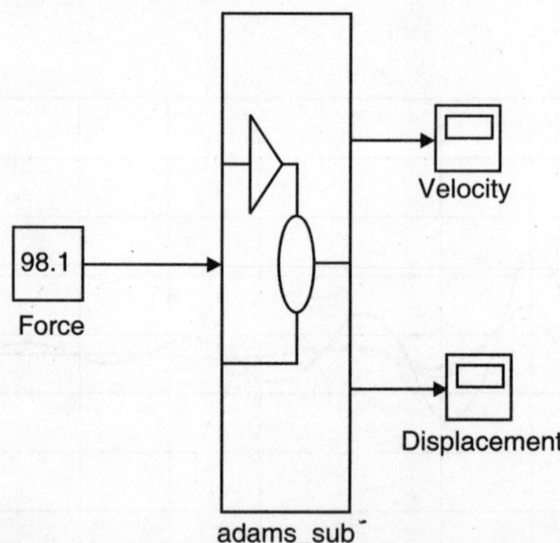

Fig. 7.2: ADAMS model exported to MATLAB-SIMULINK environment

- Change the directory to ADAMS file directory.
- Type the file prefix name on the command window. Then the **"inputs"** and **"outputs"** of the model will appear in the command window.
- Type **"adams_sys."** A new window will open, as shown in Fig. 7.2, which will contain the exported ADAMS model.

- Then, open SIMULINK, and prepare the figures as shown in Figs. 7.3 (a), 7.4 (a), and 7.5 (a) for control simulations.

Control Simulations in MATLAB

In order to study the proportional (P), derivative (D), and proportional-derivative (P-D) control [8] simulations in MATLAB on the automobile suspension system, arbitrarily gain values are taken as input to check their effects. As shown in Fig. 7.3 (a) proportional gain of K = 450 N/m is given. Corresponding results are shown in Fig. 7.3(b).

Later, only derivative gain of 33 N-s/m is given, Fig. 7.4 (a), and the results are shown in Fig. 7.4 (b). Finally, both proportional and derivative gains of 450, 33, respectively, are given, Fig. 7.5 (a), whose response is shown in Fig. 7.5 (b). The stability of the mass displacements for the gain values has been verified from the fundament pole placement of a control system [8].

7.4. ANSYS/ADAMS INTERFACE

ANSYS/ADAMS interface is essential in real-life applications when flexibility is to be considered in the moving members of a mechanical system at hand. It is an important aspect in a multibody system, e.g., to recognize the resonance point, finds the stresses in a member while in motion, etc. As an example, consider a robot arm, which behaves as a flexible beam if its length is long compared to its cross-section. In such situations, the arm will not be able to reach its targeted position due to its vibration during motion. Such vibration aspect of the tip of the flexible robot arm is presented here.

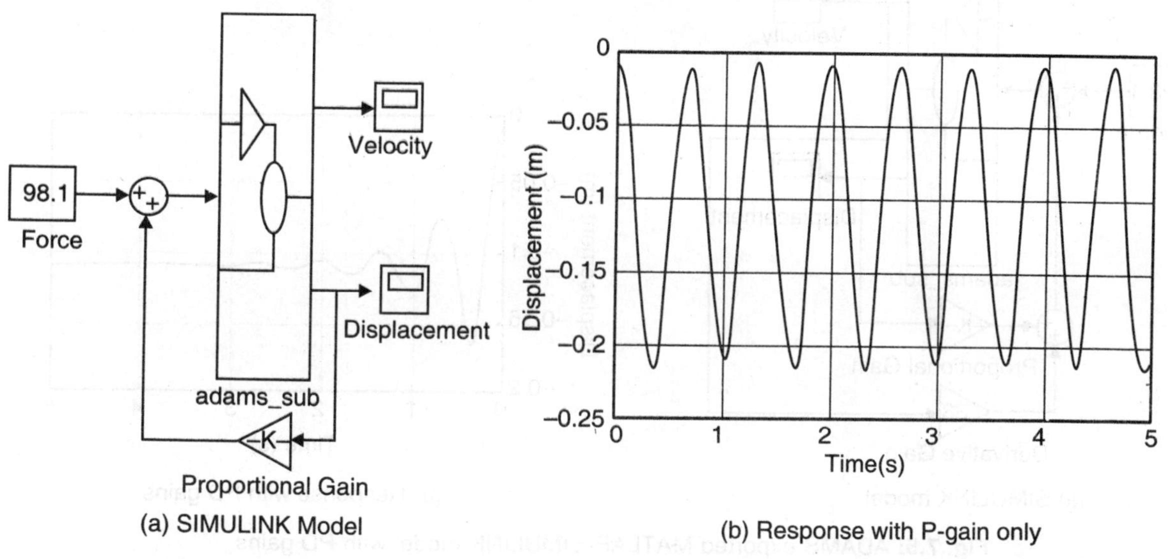

Fig. 7.3: ADAMS exported MATLAB-SIMULINK model with proportional (P) gain

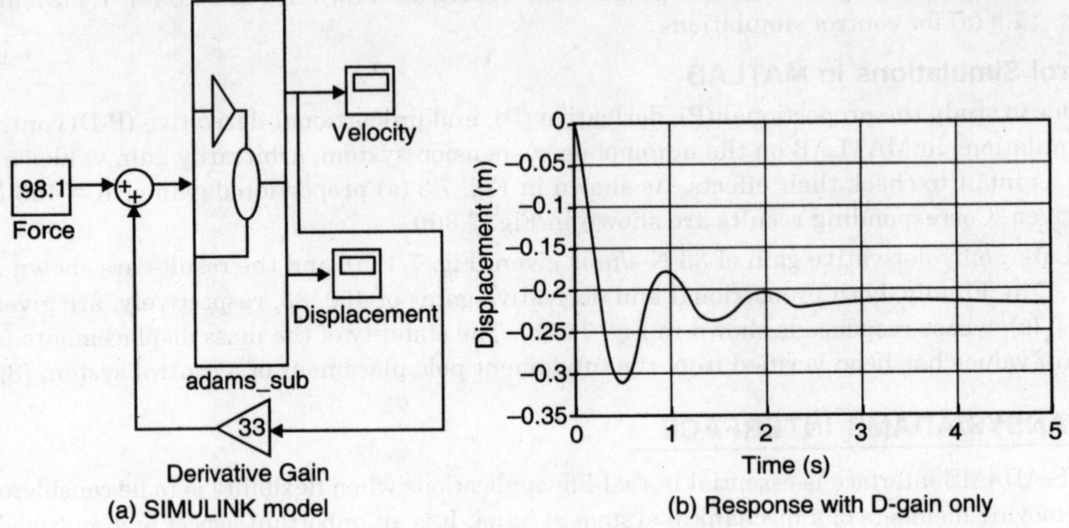

Fig. 7.4: ADAMS exported MATLAB-SIMULINK model with derivative (D) gain

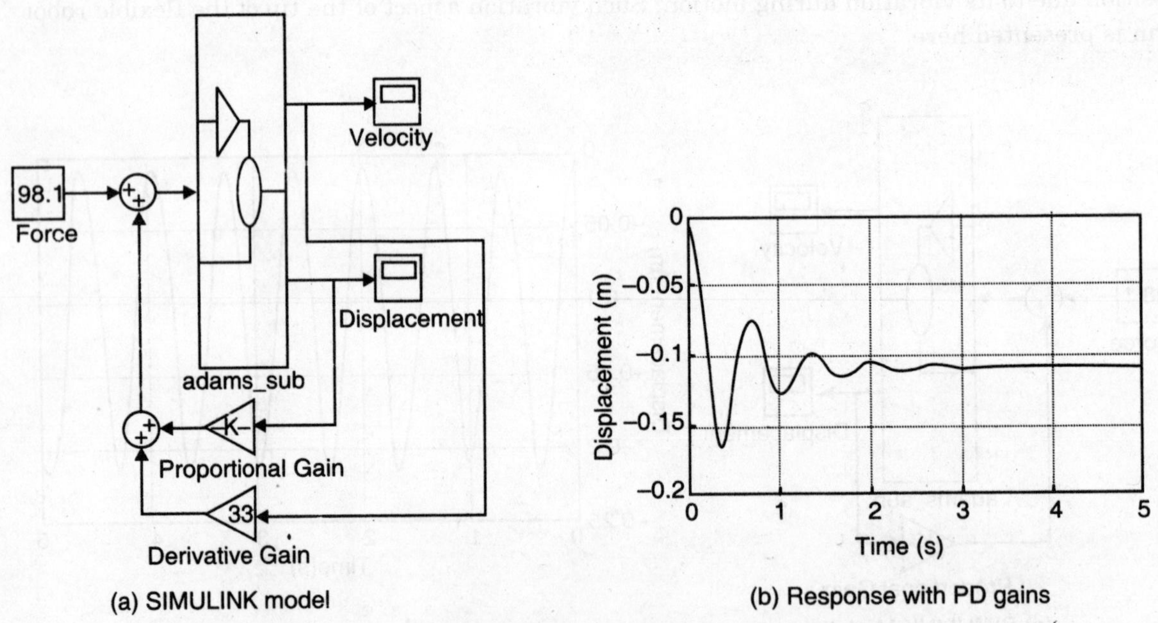

Fig. 7.5: ADAMS exported MATLAB-SIMULINK model with PD gains

ANSYS Modeling

A steel beam of 0.33 m length, 0.024 m width and 0.001 m of thickness hinged at one end is taken here as the model of flexible robot arm, which vibrates while in motion, say, falling freely under gravity. Hence, a flexible pendulum is analyzed in this section. Initially, the beam is modeled in ANSYS, as shown in Fig. 7.6. The static simulation is performed in ANSYS. The results are matching with those obtained theoretically. For example, the tip deflection due to self weight is obtained in ANSYS as 6.624 mm, whereas theoretically it is 6.798 mm. The first natural frequency is obtained from ANSYS as 32.028 Hz, while from the formula available in any textbook on vibration [9] is 32.08 Hz. Next, the model is exported to ADAMS for the analysis of its dynamic behaviour. In ADAMS, the beam falls under gravity only.

The steps followed to create an ADAMS compatible file in ANSYS software are as follows:

- Using ANSYS user-interface, the beam is modeled taking the shell element (as the thickness is very small compared to other dimension)
- Then the boundary conditions are applied (i.e., one end is free and other end is hinged).
- After incorporating boundary conditions to the model the analysis is performed in modal module and static module to crosscheck its natural frequency and the static deflection respectively.
- Then the model is exported to ADAMS through the **"ADAMS connection"** module. In this process, the solver generates the **".mnf file"** (modal natural file) and saves all the required information in that file.

In ADAMS,

- In the first step, a rigid body with the above dimensions is modeled in the ADAMS/View.
- Then, the modeled rigid body is converted to the flexible body by superimposing the imported **.mnf** file generated by ANSYS.
- After making the rigid body flexible, the boundary conditions are applied (i.e., one end is free and other end is hinged through a revolute joint) keeping the gravity on.

The superimposition of the **.mnf** file on the rigid body is checked by ADAMS' pre-processor using the connection of "**marker connectivity.**" Now the flexible structure is ready for the simulation study, as shown in Fig. 7.7 (a), whose results are shown in Fig. 7.7 (b). The results of Fig. 7.7(b) have been verified from those obtained using the analytical methodology proposed in [9]. As seen from Fig. 7.7 (b), the path of the tip of the flexible beam is sinusoidal, along with the beam vibration superimposed on it, as expected.

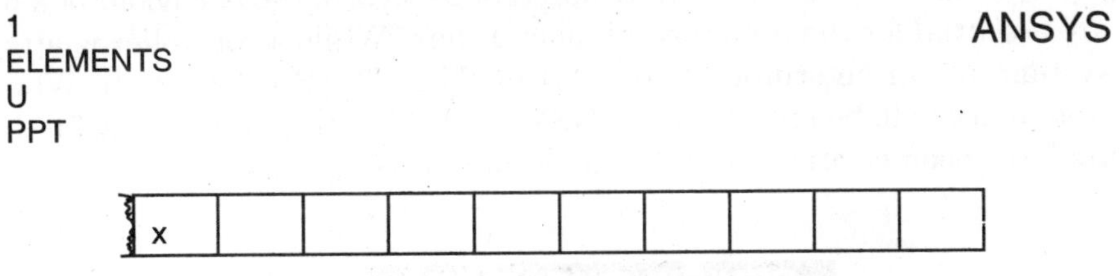

Fig. 7.6: Modeling of a thin beam in ANSYS

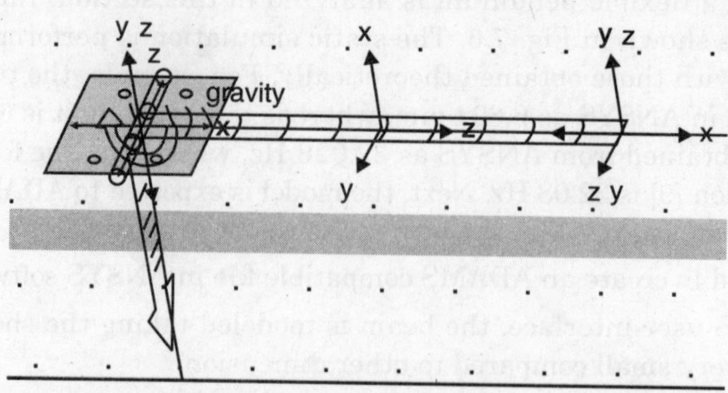

Length: 0.33 m; Width: 0.024 m; Thickness: 0.001 m

(a) ADAMS window for the flexible beam model

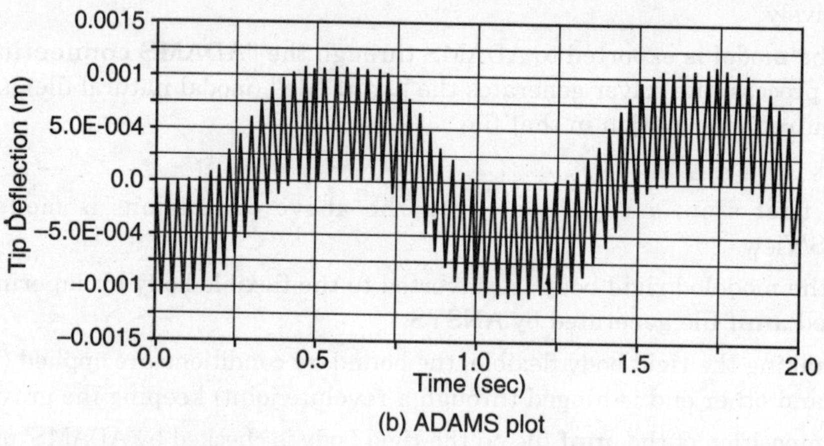

(b) ADAMS plot

Fig. 7.7: ANSYS exported ADAMS model and its result [10]

7.5. CLOSURE

In this chapter, the ADAMS/MATLAB and ANSYS/ADAMS interfaced models are presented. The above integrations made if possible to study different aspects of the behaviour of a mechanical system, which are essential for the improved product design. While a controller is attached with a mechanical system, it can be studied using ADAMS/MATLAB integration. Whenever the flexibility of a system needs to be studied, the ANSYS/ADAMS integration is useful. In both the cases, the results have been verified from the fundamental equations.

REVIEW QUESTIONS

1. Why a company may have to deal with more than one software package for design or analysis?
2. What is a *neutral format*? Why it is needed?

3. What are the challenges in designing a neutral format?
4. What are the most popular neutral formats available? Is STEP the most complete as well as most convenient? Justify your answer.
5. Why occasionally there is a need to build a model in a CAD package and export it to a CAE package? Can not the geometric model building also be done in CAE package itself?
6. For which applications the control simulations are useful?
7. Is there any need to transfer the mesh of a particular problem created in one CAE package to another CAE package? Explain.

REFERENCES

1. http://www.steptools.com/library/standard/step_1.html, Retrieved on Nov. 06, 2005.
2. ADAMS Manual, Getting Started Using ADAMS/View, Version 10, USA, 1999.
3. Rudra Pratap, "*Getting Started with MATLAB: A Quick Introduction for Scientists an Engineers*", Oxford University Press, New York, 2002.
4. ANSYS Manual, ANSYS Tutorials, University of Alberta, Canada, 2001.
5. Schiehlen W., "*Multibody System Dynamics: Roots and Perspective*", Int. J. Multibody System Dynamics, Vol. 1, pp. 149–188, 1997.
6. Ryan R.R., "*ADAMS—Multibody System Analysis Software*", *Multibody Systems Handbook* (Ed. W. Schiehlen), Springer-Verlag, New York, pp. 361–402.
7. DiPrima B., "*Elementary Differential Equations and Boundary Value Problems*", John Wiley and Sons, New York, 2001.
8. Nakra B.C., "*Theory and Applications of Automatic Controls*", New Age International (P) Limited, New Delhi, 1998.
9. William T. Thomson and Marie Dillion Dahleh, "*Theory of Vibration with Applications*", Pearson Education, 5th edition, New Delhi, 2003.
10. Mohan A., Saha S.K., and Singh S.P., "*Hybrid Formulation for Robot Dynamic Models*", Proc. of the Nat. Conf. on Product Development with Mechatronic Systems for Global Quality, Madurai, May 2–3, pp. 59–65.

8

KNOWLEDGE GENERATION IN DESIGN PROCESS

8.1. INTRODUCTION

In the changing global scenario, knowledge reuse is considered as one of the most effective ways to meet the current demands of product development. Even though literature contains many knowledge reuse approaches, representations, and capture and retrieval methods, adoption of these methods and tools in the industrial set-up is minimal. Possible reasons for this status are that the knowledge needs of designers and industries are not appropriately understood and addressed. This research attempts to bridge this gap by undertaking a descriptive study in an industry in order to understand the processes of knowledge generation during the product development process. We propose a KRIT model in which interactions of the designers with people and tools play the central role in this knowledge development process. The chapter will describe the need and understanding obtained by this model.

In this globalizing era, customer's requirements are stringent in terms of innovation, quality and cost. Industries have to face those challenges with decreasing time to market and product development personnel frequently changing jobs. As experienced by many industries today, both knowledge and expertise are getting geographically distributed along with resources and equipment. To overcome these challenges, companies should utilize all competence available, work in teams and scout for talent across the globe.

The competence developed within a company can be synergistically used and leveraged only if the necessary knowledge produced during the design process is efficiently captured, structured and made available for reuse across its projects and units. The perceived benefits of retaining the required knowledge are enhanced innovation through cross pollination of ideas, avoiding duplication of tasks, reduction of learning time, better training for new designers, and improved retention of core competence even with low retention of product development personnel.

In literature, many knowledge reuse approaches, representations, and capture and retrieval methods are proposed. But adoption of these methods and tools in an industrial set-up is minimal. Possible reasons for this status are that the knowledge needs of designers and industries are not appropriately understood and addressed. This research attempts to bridge this gap by undertaking a descriptive study in an industry in order to understand the processes of knowledge generation during the product development process.

Chakrabarti et al. [1] argue that glossary is important for engineering design research because it will foster unambiguous communication among the research community. To emphasize their argument, terms used in this chapter are defined with examples from the observed case study in Appendix. In the rest of this section, we will elaborate the need and present understanding of knowledge processing activities and interactions.

Need to Understand Knowledge Processing Activities

In design research, information and knowledge are often used interchangeably and defined rarely. To avoid confusion in this section, information and knowledge are considered as used in the following research papers. Ehrlenspiel [2] notices that designers need 20–30% of their working time for information gathering. Marsh [3] observes that designers spend on an average 24% of their time in information acquisition and dissemination, and the majority of the information is obtained from personal contacts rather than formal sources. Crabtree et al. [4] point out that project delays are mainly due to time spent in information acquisition and information access. The associated delays range from a single day to a year. MacGregor et al. [5] observe that engineers use company systems and colleagues in the same office to get information and engineers perceive that 34% of their time is taken in sourcing and locating relevant information. Frankenberger & Badke-Schaub [6] argue that availability of information is a central factor for the success of design. They observe that designers spend more time individually than in teams but critical situations occur in collaboration.

Understanding and Models of Knowledge Processing Activities

Court and Culley [7] use Information Access Diagrams to provide an understanding of how engineering designers access design information. They identify memory as a significant area source of information; and irrespective of the type and stage of design, similar access paths are followed. Clarkson and Hamilton [8] observe that knowledge is often localized and resident in key personnel and not documented or retrieved easily by other designers.

Wu and Duffy [9] develop a model to present information flow in design based on Situation Theory. The model includes input information of sender(s) and receiver(s), interaction between agents, output knowledge of agents, the goal of interaction, and the goal of sender and receiver. They argue that this model helps to analyze design information systems and provides a basis for investigating the situatedness of design information flow. McMahon et al. [10] speculate an information-connection model for design activities and argue that the quality of resultant designs is dependent on the quality of information connection. The model consists of information connection activities and information generation activities with their respective costs and times.

Interactions

Larrson et al. [11] observe that one to one conversations are common in co-located teamwork and they serve as a natural part of creative teamwork. MacGregor et al. [12] observe that interactions between engineers for information exchange and collaborative design are in a ratio of approximately 5 : 1 and these two interact cyclically. Eppinger & Salminen [13] found that even where the development process shows uni-directional information transfer, the actual communication between individuals are predominantly bi-directional exchanges. Minneman [14] addresses ways that design work emerges from interactions among individuals and groups as they establish, develop, and maintain a shared understanding. Negotiating understandings, conserving ambiguity, tailoring engineering communication for recipients and manipulating mundane representations are identified as crucial group activities.

Brereton et al. [15] investigate how social interactions shape a product: the content of an evolving design depends upon negotiation strategies and other subtle and ubiquitous social interactions. Team members' orientation to a solution or process is demonstrated by the levels of commitment in utterances; team members continuously engage in monitoring multiple issues at multiple levels of attention. Harvey and Koubek [16] argue that to communicate successfully, one person must 'mutually accept' the other's references before the conversation proceeds.

Eckert and Stacey [17] categorize the variety of interaction in design by the "dimensions of communication situations": form of communication, form of task, subject expertise, tool expertise, organization, and representation of information. They argue that no single approach to support communication is sufficient to handle the richness and variety of possible communication acts. Beyer et al. [18] argue that basic patterns of communication include information object communicated, its means of communications, sender and recipient. The objectives are to inform oneself, to ask, to provide information, to forward information, to inform somebody, and to exchange information. Lockledge and Salustri [19] provide a formal process to construct a mechanism for structuring design communications using a variation of the Design Structure Matrix that may keep participants updated on the design status. Frankenberger et al. [20] develop a model to describe the interaction between individual prerequisites, group prerequisites, external conditions, task, design process and result.

Summary and Relevance

On average designers spend 30% of their working time in knowledge acquisition and dissipation during the design process. The efficacy of the designers will be improved significantly only if the knowledge generated during the design process is properly organized for later use. The current understanding of the knowledge processing activities is not detailed enough to support this process. Also the models do not cover the whole spectrum of knowledge processing activities. The exhaustive list of interactions occurring in the design process and its influences on knowledge processing activities is not discussed in the current literature survey. The subsequent sections will address the above gaps found in the literature survey.

8.2. A KRIT MODEL AND RESEARCH OBJECTIVES

A KRIT model has been proposed to illustrate the designing and knowledge operations through interactions. Figures 8.1 and 8.2 describe the KRIT model and influence diagram.

In this model, requirements satisfaction has been considered as a primary objective of an engineering design process, because satisfying design requirements achieve the customers' needs, apart from enabling the development of the design into a product [21]. Each requirement consists of a set of tasks with purposes and outcomes, and is executed through a complex variety of interactions. For example, a designer may have to interact with another designer, clients, tools, groups of designers or group of clients. Each interaction may lead to new tasks, and will involve various knowledge operations: knowledge production, sharing, storage, structuring or reuse. Requirements, tasks, interactions and knowledge will influence each other in the consequent order.

The hypotheses are that the purpose of the tasks will be satisfied only if a designer is able to get, produce, share, reuse knowledge properly (knowledge satisfaction) through effective interactions (interactions satisfaction); the satisfaction of every purpose of the tasks will influence the requirement satisfaction of the artifact. In this competitive scenario the requirement of the artifact should get satisfied in a resource effective way. The resources can be clubbed into two categories: time and cost. The set of tasks, interactions and knowledge will and get influenced by available resources. We can say that the quality of interaction is good, only if it satisfies the purpose of the tasks in a resource effective way.

In this chapter, we discuss the links numbered 1 and 2 shown in the Fig. 8.1. The focused questions are,

- What are the various types of tasks that occur and their proportions?
- What are the various types of interactions that occur and their proportions?
- What are the types of knowledge produced and their proportions?
- How are the tasks, interactions and knowledge related to each other during product development processes?

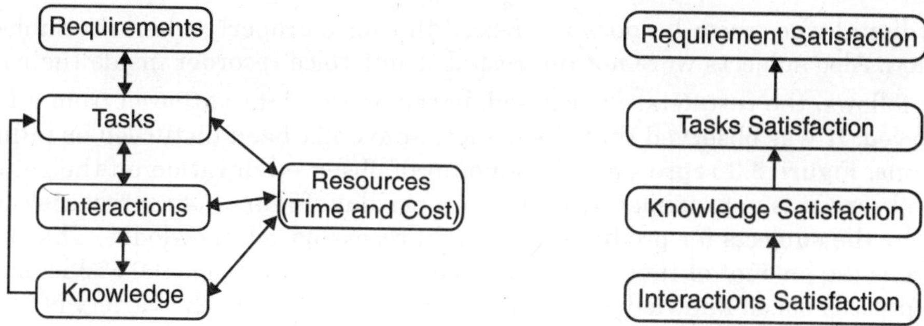

Fig. 8.1: KRIT model

Fig. 8.2: Influence diagram

8.3. DATA CAPTURE METHODS AND ITS LIMITATIONS

To answer the research questions a case study was undertaken in a product development organization. Three designers involved in different projects were observed for five, three and seven days respectively. The data capture methods employed to collect the required data are questionnaires, unstructured interviews, voice recordings, digital snap-shots, video recordings, desktop sharing and data sheets. The questionnaires were used to collect information about organization, projects and subjects involved in the observations.

Unstructured interviews were conducted with the observed subjects whenever necessary to understand the subjects' activities or problems that occurred during observation. Voice recordings were employed whenever there was an interaction between the observed subject and other people. Digital snap-shots were used to capture the final outcomes of the subject when they were interacting with their notebooks or paper.

Video recordings were used to capture the data generation during the complex interactions that involved two or more people with documents or other information sources. Desktop sharing was used to capture the subjects' interactions with the computer. The CAD package

UNIGRAPHICS was used mainly by the observed subjects. For desktop sharing VNC software developed by RealVNC was used. To capture the shared desktop Hycam2 software developed by Hyperionics was used. Data sheets gave details about the purpose of the tasks, interactions, place of interactions, duration of the interactions and whether interactions were satisfying or not. The limitations and hindrances that occurred during the observations were,

- The data not produced but eventually understandable in the interactions are context and incomplete sentences which is later fulfilled in analysis.
- For desktop sharing first we employed NetMeeting software developed by Microsoft Corporation. But during observations we found that there were interoperability issues between NetMeeting and Unigraphics CAD software.
- Linguistics constraint. The subjects have interacted in languages which the observer was not able to understand.
- Desktop capture, snap-shot and audio do not synchronize properly due to the complex nature of the interactions as well as the constraint of the observer to operate all the devices concurrently. Hence, we used video recording to capture the complex interactions.
- The digital camera with digital zoom of 4 × was not able to take a snap-shot of written documents clearly. Hence, we used a scanner to digitize the written documents.
- The digital voice recorder does not record the voice properly when the subjects are in motion. Also subjects were not interested to put voice recorder inside their pockets.

In what follows, the results are discussed, based on the data gathered from all the three subjects observed. It was observed that the subjects have not been disturbed or influenced by the observations. Figure 8.3 represents the amount of observed duration on the subjects. The average time observed is approximately four hours per day. Figure 8.4 represents the amount of time spent by the subjects for product-based and process-based knowledge. The observation from the figure is the amount of time spent for product-based knowledge (90.8%) is much higher than that for process-based knowledge (9.2%). Figure 8.5 represents the variety of interactions performed by the subjects in order to satisfy their tasks. In the nineteen interactions found, the most frequently observed interactions were One + Computer (33.8%), One to One + Comp (25.9%) and One to One (12.2%).

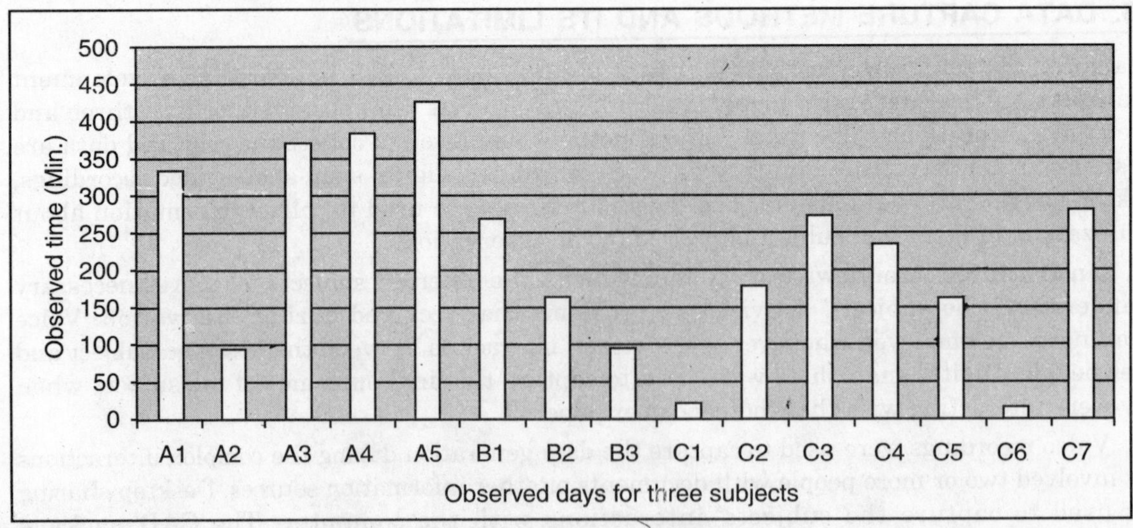

Fig. 8.3: Observed durations

□ Product related knowledge ■ Process related knowledge
500
450
400
350
300
250
200
150
100
50
0
Duration (Min.,)
A1 A2 A3 A4 A5 B1 B2 B3 C1 C2 C3 C4 C5 C6 C7
Observed days for three subjects

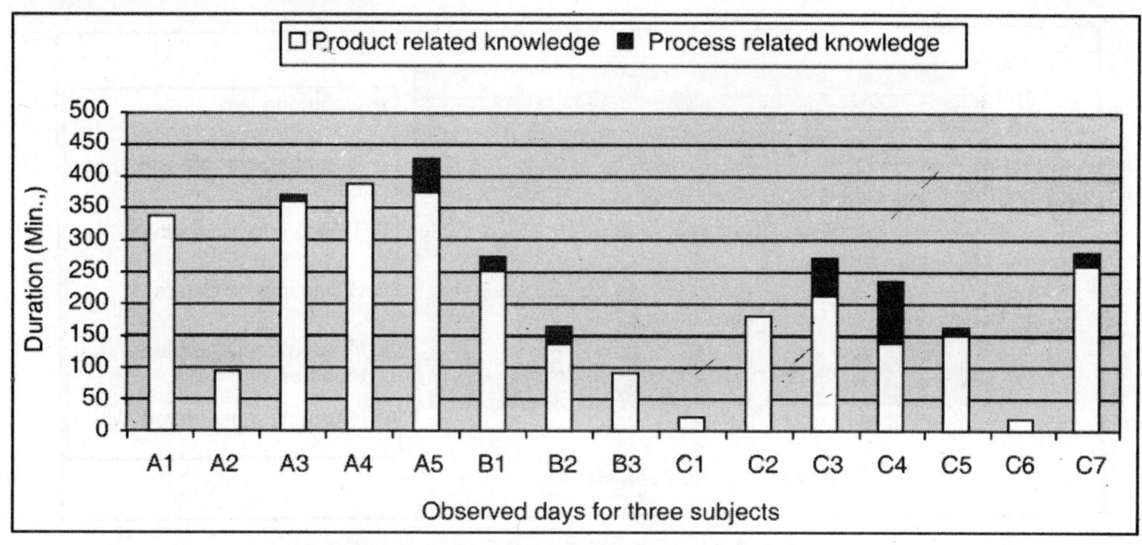

Fig. 8.4: Amount of time spent on types of knowledge

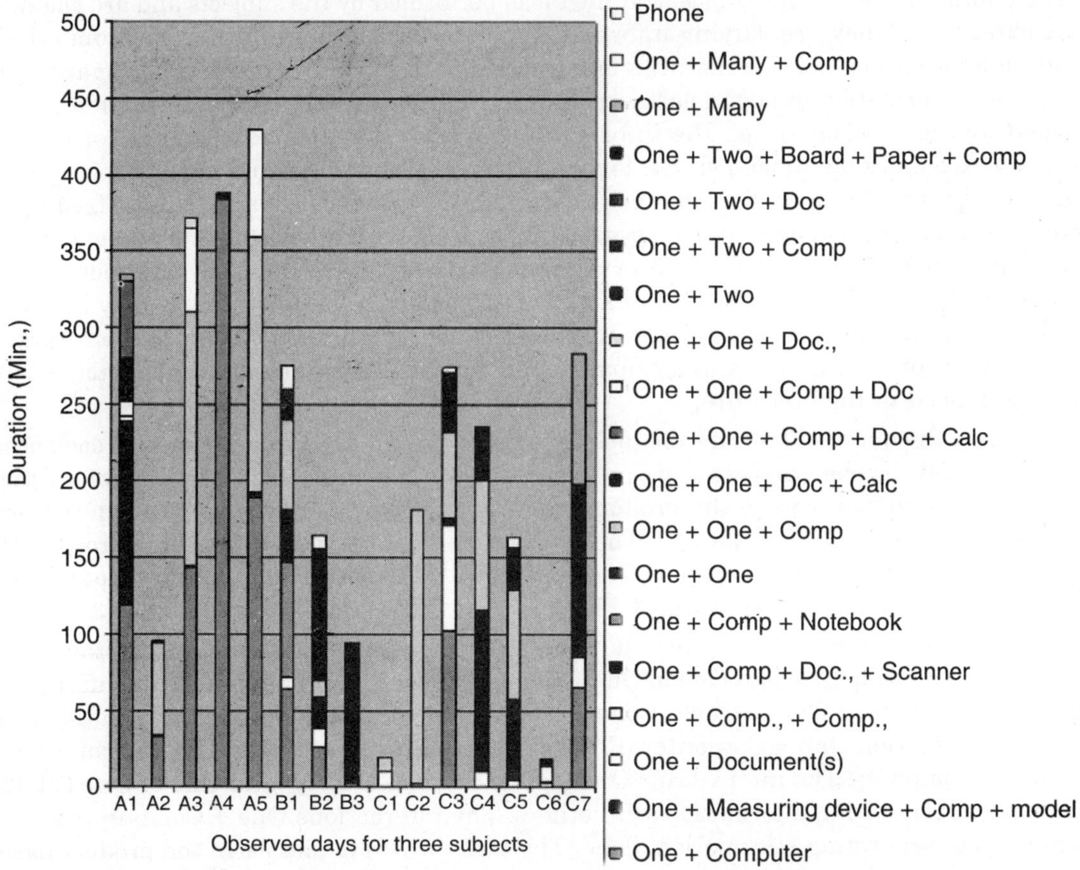

500
450
400
350
300
250
200
150
100
50
0
Duration (Min.,)
A1 A2 A3 A4 A5 B1 B2 B3 C1 C2 C3 C4 C5 C6 C7
Observed days for three subjects
Phone
One + Many + Comp
One + Many
One + Two + Board + Paper + Comp
One + Two + Doc
One + Two + Comp
One + Two
One + One + Doc.,
One + One + Comp + Doc
One + One + Comp + Doc + Calc
One + One + Doc + Calc
One + One + Comp
One + One
One + Comp + Notebook
One + Comp + Doc., + Scanner
One + Comp., + Comp.,
One + Document(s)
One + Measuring device + Comp + model
One + Computer

Fig. 8.5: Amount of time spent on each interactions

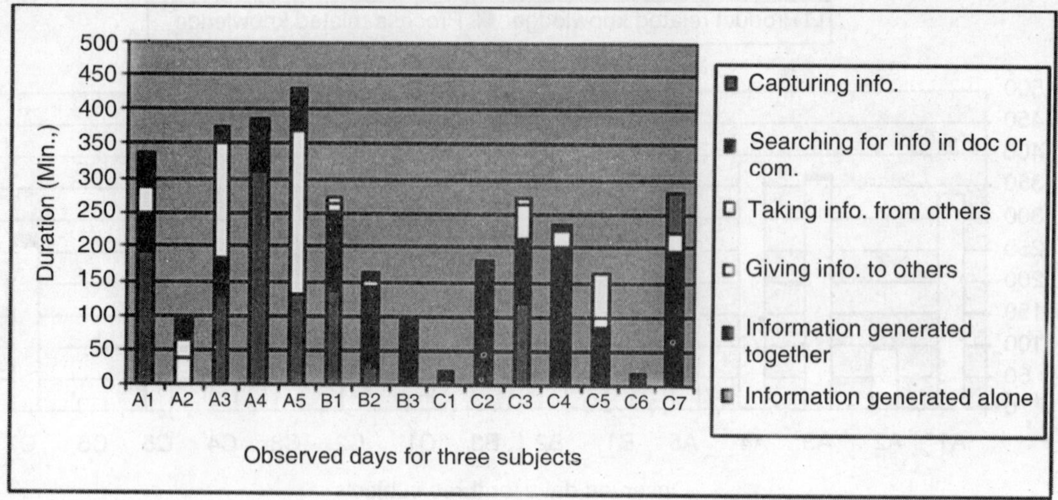

Fig. 8.6: Amount of time spent on types of tasks

The information exchange is based on the tasks performed by the subjects and are classified into six categories. They are: Giving information to others, Taking information from others, Generating information alone, Generating information with others, Capturing information and Searching for information in documents or computer. The tasks capturing information is least performed by the designers. i.e., the information capturing happened as a part of the five information exchanges mentioned above. Figure 8.6 represents the amount of time spent by the subjects on various information exchanges on each day of observations. The most frequently observed information exchanges are generating information with others (36.3%) and generating information alone (32%). It has been observed that nearly 22.3% of the time was spent by the designers in taking information from others and giving information to others. It may be due to the novice nature of the observed designer. In this study, designers spend 22.3% of their working time in information acquisition and dissipation during the design process that matches with 20–30% mentioned in the literature.

To answer the question how are the tasks, interactions and knowledge related to each other during product development processes, we analyzed the link between each other. The interactions that played an important role in the production of product based knowledge are One + Comp (30%) and One to One + Comp (23.1%). The interaction that played an important role in the production of process based knowledge is One + Comp (3.8%). Interaction that does not produce product based knowledge is One to One + Phone.

The following paragraph describes the link between types of tasks and interactions. The interactions One + Comp (2%) and One to One + Comp (2.07%) predominate in giving information to others. The interaction One to One + Comp (12.6%) predominates in taking information from others. The interaction One + Computer (6.7%) predominates in searching for information in documents or computer. The interactions One to One + Comp (11.2%) and One to One (11.1%) predominate in generating information with others. The interactions One + Computer (23.2%) predominates in generating information alone. The tasks which dominate in the product based knowledge are generating information together (34.2%) and generating information alone (30.5%). The task that dominates the process-based knowledge is giving information to others (3.6%).

From these studies we can conclude that focusing on designer interacting with computer, designer interacting with another person and designer interacting with another person in front of a computer may lead to achieve the primary objective of capturing much of the necessary knowledge generated during the design process without intruding the designer's activities.

8.4. SUMMARY

In this chapter, we discussed our observations of knowledge generation during design process in the observed industry. A KRIT model is proposed to understand the knowledge generation process better. We have shown the various types of tasks that occur and their proportions, the various types of interactions that occur and their proportions, and the types of knowledge produced and their proportions. We have shown how tasks, interactions and knowledge generated are related. In order to validate the KRIT model and the results we are doing another case study in larger product development organization. In the knowledge processing activities, knowledge generation is a basic activity. With the help of this we plan to study knowledge capture and reuse aspects. The most important question to be answered in this study will be what knowledge is developed that is not captured but should be otherwise.

REVIEW QUESTIONS

1. What are the primary requirements of an engineering design?
2. Explain about the different kinds of information exchange between two designers collaborating on a design.
3. Explain about the different data capture mechanisms.

REFERENCES

1. Chakrabarti A., Murdoch T. and Wallace K., (1995) "*Towards a Glossary of Engineering Design Terms*", International conference on engineering design (ICED), pp. 185–186.
2. Ehrlenspiel K., 1997, "*Knowledge-explosion and Its Consequences*", ICED, Vol. 2, pp. 477–484.
3. Marsh J.R., 1997, "*The Capture and Utilisation of Experience in Engineering Design*", Ph.D. Thesis, St. John's College, Department of Engineering, University of Cambridge.
4. Crabtree R.A., Fox M.S. and Baid N.K., 1997, "*Case Studies of Coordination Activities and Problems in Collaborative Design*", Research in Engineering Design, Vol. 9, pp. 70–84.
5. MacGregor S.P., Thomson A.I. and Juster N.P., 2001, "*Information Sharing within a Distributed Collaborative Design Process: A Case Study*", Proceedings of DETC.
6. Frankenberger E. and Badke-Schaub P., 1999, "*Information Management in Engineering Design – Empirical Results from Investigations in Industry*", ICED, pp. 911–916.
7. Court A.W. and Culley S.J., 1995, "*A Methodology for Analyzing the Information Accessing Methods of Engineering Designers*", ICED, Vol. 2, pp. 523–528.
8. Clarkson P.J. and Hamilton J.R., 1999, "*Signposting–The Design Process*", ICED, pp. 107–112.
9. Wu Z. and Duffy A.H.B., 2001, "*Using Situation Theory to Model Information Flow in Design*", ICED, 155–162.
10. McMahon C.A. Lowe A. and Culley S.J., 1999, "*An Information-connection Model for Design*", ICED, pp. 1651–1656.

11. Larsson A., Torlind P., Karlsson L., Mabogunje A., Leifer L., Larsson T. and Elfstrom B-O., 2003, "*Distributed Team Innovation–A Framework for Distributed Product Development*", ICED.
12. MacGregor S.P., Thomson A.I. and Juster N.P., 2002, "*A Multi-level Process based Investigation of Distributed Design*", Proceedings of the Engineering Design Conference, Kings College London, England, July 9–11 2002.
13. Eppinger S.D. and Salminen V., 2001, "*Patterns of Product Development Interactions*", ICED, pp. 283–290.
14. Scott L. Minneman, "*The Social Construction of A Technical Reality: Empirical Studies of Group Engineering Design Practice*", Ph.D. Thesis, Department of Mechanical Engineering, Stanford University.
15. Brereton M.F., Cannon D.M., Mabogunje A. and Liefer L.J., "*Collaboration in Design Teams: How Social Interaction Shapes the Product*", Analysing Design Activity, Eds Nigel Cross, Henri Christiaans and Kees Dorst, pp. 319–341.
16. Harvey C.M. and Koubek R.J., 1998, "*Toward a Model of Distributed Engineering Collaboration*", Computers in industrial engineering, Vol. 35, No 1–2, pp. 173–176.
17. Eckert C. and Stacey M., 2001, "*Dimensions of Communications in Design*", ICED.
18. Beyer N. and Weber F., 1999, "*Concepts and Prototype for a Practical Communication Environment for Supporting and Managing Concurrent Product Development*", ICED, pp. 1431–1436.
19. Lockledge J.C. and Salustri F.A., 2001 "*Design Communication using a Variation of the Design Structure Matrix*", ICED, pp. 27–34.
20. Frankenberger E., Badke-Schaub P. Birkhofer H., 1997 and "*Factors Influencing Design Work Empirical Investigations of Teamwork in Engineering Design Practice*", ICED, Vol. 2, pp. 387–392.
21. Nidamarthi S., 1999, "*Understanding and Supporting Requirement Satisfaction in the Design Process*", Ph.D. Thesis, Gonville and Caius College, Department of Engineering, University of Cambridge.
22. Ahmed S., Blessing L. and Wallace K., 1999, "*The Relationships between Data, Information and Knowledge based on a Preliminary Study of Engineering Designers*", ASME Design Theory and Methodology, Las Vegas, Nevada.

APPENDIX

Design: A complete and detailed description of the product. CAD drawing describing the design of single cavity injection mould with bill of materials.

Requirements: The technical and non-technical issues of the intended product considered by the designers during the engineering design process are termed as requirements. Technical issue observed is to design a single cavity mould for a base component. Non-technical issues observed are vendor follow up and scheduling of the project.

Requirement satisfaction: The degree to which the design satisfies each requirement. For non-technical issues, how well the project was completed with respect to the scheduled target dates. For technical issues, it will be adjudged with the help of repeatability or correction occurs after the requirements were satisfied. No repeatability occurred during the observed period. It will also be adjudged with the help of outcomes and duration and it involves subjectivity. This analysis is not carried out in the present observation.

Tasks: A piece of work to be done to satisfy requirements. Some of the observed tasks are "modifying existing mould design", "measuring dimensions from physical model" and "asking for particular file or information".

Tasks satisfaction: The degree to which the purpose of the piece of work to be done gets satisfied. It will be adjudged with the help of repeatability or correction occurred after the tasks were satisfied. No repeatability of tasks occurred during the observed period. It will also be adjudged with the help of outcomes and duration and these involve subjectivity. This analysis was not carried out in the present observation.

Interactions: Mutual or reciprocal action or influence of objects to produce/exchange or intended to produce/exchange knowledge. Some of the interactions observed are One + Computer, One to One + Comp and One to One + Comp + Doc + Calc. The representation of interactions such as One + One + Comp + Doc + Calc and One + Computer should be read as observed subject interacting with another through computer, document and calculator, and observed subject interacting with computer alone respectively.

Interactions satisfaction: Interactions satisfaction is expressed in terms of communication satisfaction (ability to express or share and understand the discourse), Generate-Evaluate-Select-Verification cycle performed and intensity of the interactions (number of issues or proposals per duration of the interaction). Designers are able to express their thoughts clearly to others but misunderstandings were found during discourse.

Data/Information/Knowledge: Data, information and knowledge are relative concepts that cannot be defined in absolute terms. An awareness stage and an interpretation stage differentiate between data, information and knowledge (22). The definition and example are explained in Figs. 8.7 and 8.8 respectively.

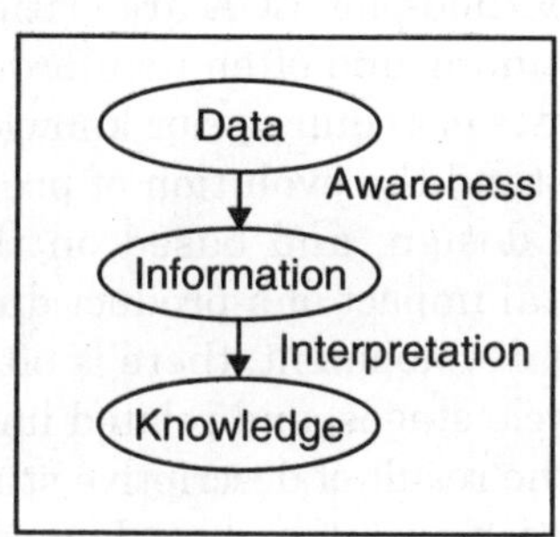

Fig. 8.7: Definition

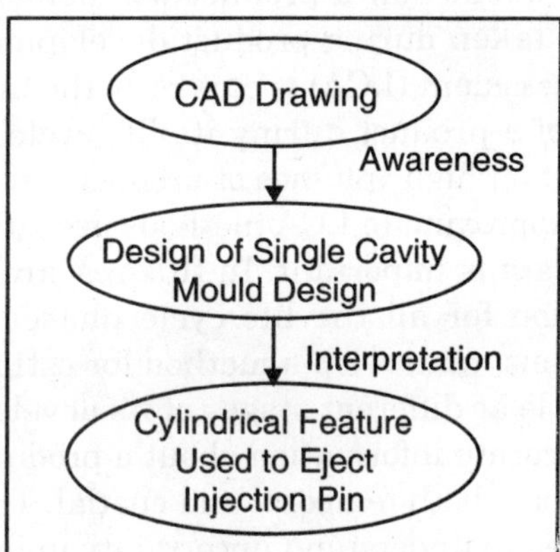

Fig. 8.8: Example

Knowledge satisfaction: The degree to which the knowledge required for the designers for the particular task is available within himself or in the organization. Detailed analysis is elaborated in Section 5.

Product-based knowledge: Design knowledge concerned about objects being designed. The statement such as "That draft?" and "Which line?" are classified into product-based knowledge.

Process-based knowledge: Design knowledge concerned about how to design. The statement such as "without that line you do once", "cut till the inside surface" and "now I will make this way" are classified into process-based knowledge.

9

LIFE CYCLE ASSESSMENT IN DESIGN

9.1. INTRODUCTION

Products make substantial impact on the environment. Literature suggests that earlier phases of product development i.e., design can play a major role in reducing these impacts because decisions about how a product will behave during its life cycle stages are constrained by the decisions taken during product development, causing unintended environmental impact. Life Cycle Assessment (LCA) is currently the most promising technique for estimating environmental impacts of a product during its life cycle. Currently, detailed methods for LCA are critically dependent on high volumes of product specific data, are time consuming and often unaffordable. Current approximate LCA methods are either incomplete, inaccurate or require prior knowledge of what data is important. In this research, our goal is to understand the evolution of product information for all the life cycle phases in different stages of design, and based on these observations, to develop a method for estimating the environmental impact of a product during its life cycle at different stages of its development. During product development, there is often a lack of accurate information about a product's structure and life cycle stages, and related impact information – both temporal and spatial. The chapter will discuss the result of descriptive studies undertaken to understand uncertainty in product life cycle information evolution, based on which requirements for the assessment method were identified.

Products make a substantial impact on environment. The ratio of product mass to waste mass directly or indirectly produced as a result of the product during its life cycle is about one to twenty [1]. These wastes are thrown into the environment in each stage of the product life cycle from raw material extraction to product retirement. The lifecycle principle, where the whole impact of a product across its life is to be examined (from 'cradle to grave') is gaining importance in product development [2]. We need to consider environment also as one of the criteria along performance, quality and cost in product development. Environment is gaining importance as an evaluation criterion because of government regulations, competition and customers' requirement.

Traditionally research has been carried out on how to treat the wastes produced, while subsequently the focus shifted to reduction of waste by utilizing the waste produced as by-products. More recently strategy has shifted to eliminate waste in the first place itself and this led to the idea of sustainable development. Sustainable development is defined in the Brundtland report [3] as 'development which meets today's needs without placing the ability of future generations to meet their needs at risk'. For sustainable development, design can play a major role as the decisions taken in this stage will affect the later phases of the life cycle.

In this chapter we try to understand the uncertainty in product life cycle information evolution in various stages of design so as to identify the requirements for an uncertainty assessment method for use throughout design.

The objectives are to.

- Identify shortcoming of current practices.
- Understand the uncertainty in product life cycle information evolution.
- Identify requirements for the uncertainty assessment method.

Based on the literature, the requirements of a method to generate, evaluate and select a product proposal in the context of Design for Environment throughout the design process were identified. A series of three design experiments were conducted to understand the uncertainty in product life cycle information evolution throughout the design process. This understanding was used to identify the requirements for the uncertainty assessment method.

During product development there is a need to consider the whole life cycle of a product rather than only single isolated phases. In other words, it is necessary to design the whole life cycle of the product [4, 5]. Early stages of product development are the key in doing this because if we know the environmental impacts of potential designs while designing, we can make changes to these designs then and there so as to reduce their environmental impacts [6]. Since over 80% of the product costs are committed during the early planning and product development stages, design can play a central role in reducing this environmental overloading by products [7].

Life Cycle Assessment (LCA) [8] is the most promising method for estimating environmental impacts of a product during its life cycle [9, 10]. Currently, detailed methods for LCA are critically dependent on high volumes of product-specific data, and are time consuming and unaffordable. Current approximate LCA methods are either incomplete, inaccurate or require prior knowledge of what data is important [11]. LCA tools are not well integrated with other design tools [5]. They are used to validate the design rather than to truly feed into the design process, to shape or steer design decisions [2]. At all stages of design it is necessary to make choices based on various criteria [12]. Consequently, there is a need for an LCA method integrated into the design process that can be applied to early as well as detailed design stages, and that can be driven with generic data available.

Internationally there are many methods developed like *Design for Energy Efficiency* [9], which are useful for specific phases of the lifecycle of a product. However there is no comprehensive method that can be useful for the whole life cycle of a product in various stages of its design. Even though guidelines [13] for various phases exist, the guidelines are sometimes too abstract for designers to know what to do in actual design of their products [14]. A critical need for a product developer is a method that can be used to estimate environmental impacts of a product during its whole life cycle, even if approximately, while the designer works through the various stages of its development, so that design changes can be made concurrently to the product proposal to make it more eco-friendly. This is relatively inexpensive since one would be able to do so without having to make, use or dispose off the product to see its environmental impacts and understand the trade-offs available between different product life-cycle phases.

With the progress in capability of computational tools, it appears beneficial to develop an overall support system for the entire design process. This would involve development of both synthesis and analysis support tools. The utilization of the product life cycle for this task offers good opportunities to achieve an efficient support for both analysis and synthesis [15].

Few tools were found which mapped to iterative changes required in product development and this is an area which needs further research [16]. To discourage mass production, mass consumption, and mass waste, we need to establish design methodologies for closed life cycles [17]. We need to think about the whole life cycle rather than single phases in all stages of the design so that sustainable development products can occur.

We need to automate the extraction of LCI data elements from product documentation as completely as possible so as to make LCA a viable analysis tool during the design process [18]. As it takes a huge amount of time and work for LCA, it will be good if we can automate the process of extraction without any extra effort. That is different from the usual working process.

9.2. LIFE CYCLE ASSESSMENT

Life Cycle Assessment is a process for evaluating the environmental impacts associated with a product, process, or activity by identifying and quantifying energy and materials used and wastes released to the environment, and for identifying and evaluating opportunities to effect environmental improvements [8]. It consists of four main activities.

Goal Definition and Scope

This activity consists of specifying the focus of the study. We have to specify the functional unit; which consists of description of product, life cycle and performance. It is important to compare products with same performance. For example to compare vacuum cleaners the functional unit will be to clean the house (dry, wet) for a period of 5 years.

Inventory Analysis

This activity consists of developing all the inventories necessary for the life cycle phases specified in previous activity. It is the most difficult and time taking activity in this method. For example for vacuum cleaners we need to collect the data of amount of materials used like ABS plastic and the processes used for extraction and transportation, manufacturing, assembly processes used for the component manufacturing, energy required, consumables in the usage phase and the disposal type and path.

Environmental Impact Assessment (EIA)

This activity consists of assessing the environmental impact of the inventories developed in previous activity. There exist different methodologies for calculating EIA of products. For example in Eco-indicator 99 [19], the overall environmental impact of a product is shown in a single number that combines the impacts on (a) human health, (b) eco-system quality, and (c) resources.

Improvement Assessment

This activity consists of structuring the EIA results and identifying the areas that have substantial potential for improvement in terms of sustainable development.

Table 9.1: Comparison of requirements of DFE tools among current tools

Tool Requirement ↓		**Current LCA Tools**	**DFE Workbench**	**Ecodesign PILOT**
Proactive	Yes	No	Yes	Yes
Support Collaboration	Yes	Yes	Yes	No
Easy to learn, Understand and Use	Yes	No	Yes	Yes
Allow to understand Rationale	Yes	No	No	No
Act as Checklist	Yes	No	No	No
Self Documenting	Yes	No	Yes	Yes
Help in fulfilling specific requirements	Yes	No	No	Yes
Reduce the risk of forgetting important elements in PD	Yes	No	No	No
Reduce total time	Yes	No	Yes	Yes
Computer based	Yes	Yes	Yes	Yes
Facilitate various kinds of communication	Yes	Yes	Yes	No
Store knowledge and experience as know-how backup	Yes	No	No	No
Useful in all stages of design	Yes	No	No	No
No Extra effort for analysis	Yes	No	Yes	No
Integrated to CAD	Yes	No	Yes	No
Trade off between choices	Yes	No	No	No
Uncertainty Analysis	Yes	No	No	No
Standards and Regulations	Yes	No	No	No
Quantitative and Qualitative Evaluation	Yes	No	No	Yes
Analysis and Improvement	Yes	No	Yes	Yes
Consider total life cycle	Yes	Yes	Yes	Yes

Substitution of object information: This activity is a combination of two activities: subtraction of already available object information and addition of new object information. For example, in a single activity, the designer removed the rope and modified the handle part. Figure 9.5 shows the substitution of an object (rope).

Fig. 9.5: Substitution of objects

Focus to object or information: In this activity, a designer concentrates on a particular object or information. For example, while designing workout equipment for executives, the designer drew a sketch representing a skipping rope with handles. In the next sketch, he drew only the handles without drawing the rope because he wanted to focus on the handle. Figure 9.6 shows this focus activity.

Fig. 9.6: Focus to object

Fig. 9.7: Defocus from object

Defocus from object or information: Here a designer defocuses from a focused object or information, by representing the outline. For example, in the defocus activity shown in Fig. 9.7, the designer sketches the details of the handle and then the outline of the handle.

Substitution of object/information: This activity is a combination of two activities; subtraction of already available object/information and addition of new object/information. For example, in a single activity, the designer removed the rope and modified the handle part. Figure 9.5 shows the substitution of an object (rope).

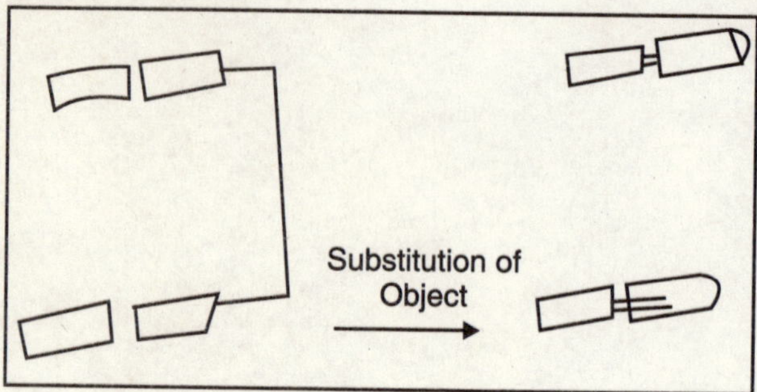

Fig. 9.5: Substitution of objects

Focus to object or information: In this activity, a designer concentrates on a particular object or information. For example, while designing workout equipment for executives, the designer drew a sketch representing a skipping rope with handles. In the next sketch, he drew only the handles without drawing the rope because he wanted to focus on the handle. Figure 9.6 shows this focus activity.

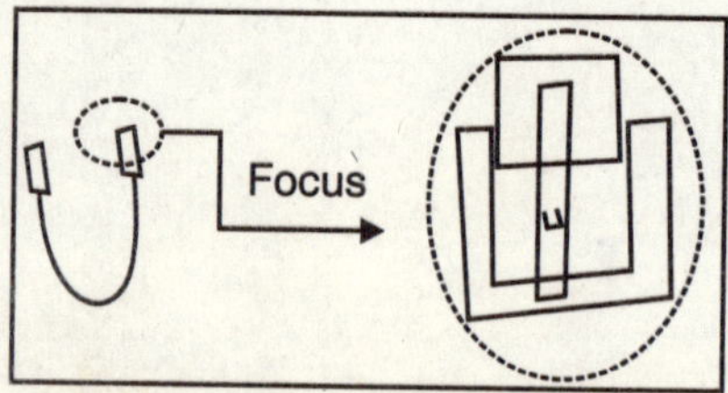

Fig. 9.6: Focus to object

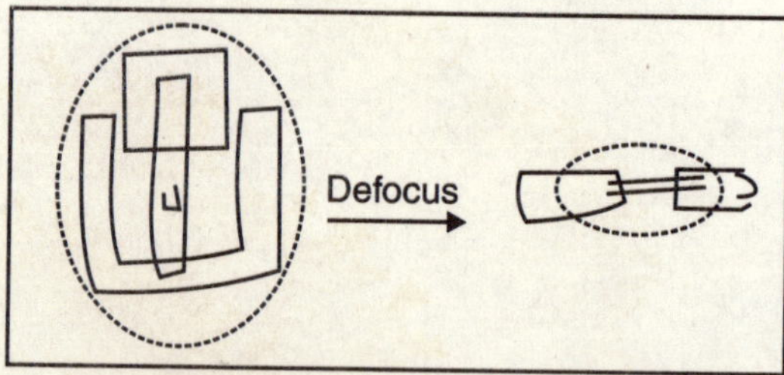

Fig. 9.7: Defocus from object

Defocus from object or information: Here a designer defocuses, from a focused object or information, by representing the outline. For example, in the defocus activity shown in Fig. 9.7, the designer sketches the details of the handle and then the outline of the handle.

Change of the view or focus: This activity is a combination of two activities; defocusing from the already focused object/information and focusing on others. For example in Fig. 9.8, the designer was initially interested on the internal object (spring) within a rope assembly. Afterwards he changed his point of interest to the outside object (casing).

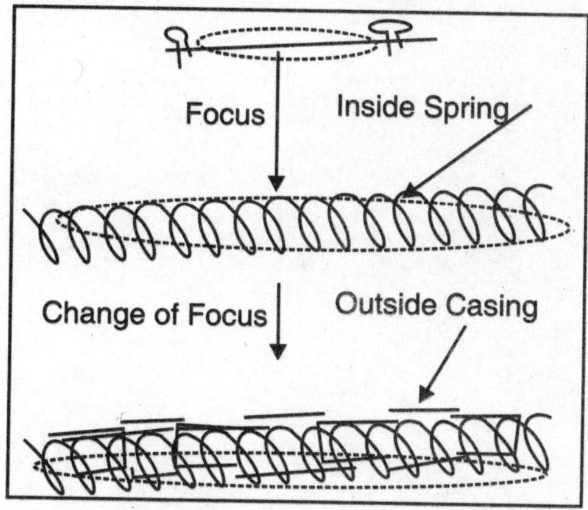

Fig. 9.8: Change of focus

Change of orientation of the objects: Here a given object is orientated in a different way as a result of the activity. For example, the designer in Fig. 9.9 initially sketched the object vertically and then changed this to be horizontal.

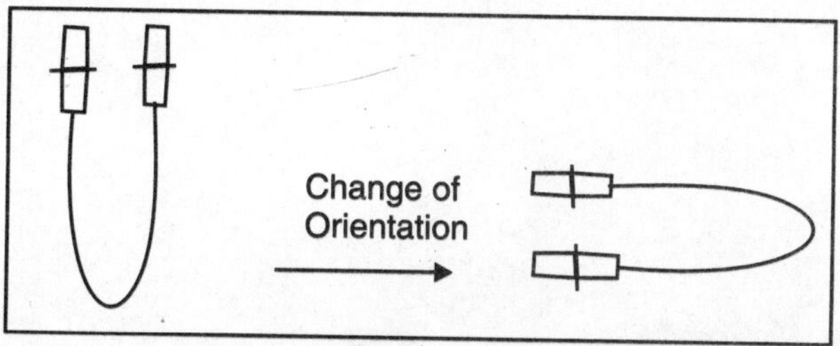

Fig. 9.9: Object rotation

There are some activities that are spoken only, and cannot be represented using drawings or as associations between objects with information. There should be some mechanism for capturing these activities, while allowing a designer to do the activities fast and with ease.

9.6. EVOLUTION OF INFORMATION ABOUT PRODUCT CONFIGURATION

The information in the first phase was about identification, analysis and selection of design problem and tasks. It is mainly in this stage, designer worked around the specification given and tried to identify the actual problem definition and its requirements.

The information in second phase is on finding the principles, global configuration (main assemblies, function etc.) of the concept, generating, associating the ideas with the existing ones and primary evaluation. For example while solving a problem for exercising equipment, designer

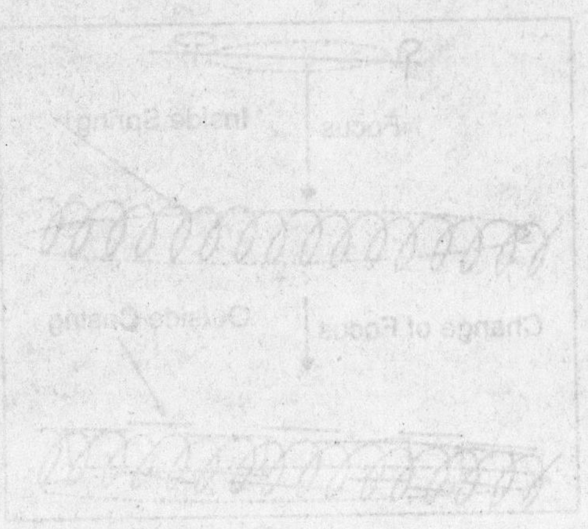

the uncertainty in product life cycle information evolution with respect to LCA and requirements for the uncertainty assessment method were identified. There are mainly four stages of design. In these stages the information about the product life cycle changes from abstract for example material of a component as a class to the concrete as a specific material. We also found that designers often evaluate their design with respect to requirements, it's an iterative process. There is a need of a method to incorporate the product information transition. We propose that there are mainly four types of uncertainties with respect to Life Cycle Assessment in Design; Product Structure, Life Cycle phases, Information Quality, and methodological choices. The requirements of a DFE tool are also identified. Subsequent tasks are design and implementation of the method and its evaluation using in-house design experiments to assess its effectiveness and efficiency.

REVIEW QUESTIONS

1. Why Environmental concerns become important in product development?
2. Relate the concept of sustainable product development to the environmental protection.
3. Describe how LCA can be a method of estimating the environmental impacts of a product.

REFERENCES

1. Züst R., "Sustainable Products and Processes", ECO-Performance — 3rd International Seminar on Life Cycle Engineering CIRP '96, pp. 5–10.
2. Chris Sherwin, Tracy Bhamra, "Beyond Engineering: Ecodesign as a Proactive Approach to Product Innovation", Ecodesign 99, Feb 1–3, 1999, Tokyo, Japan, pp. 41–46.
3. Brundtland G.H., "Our Common Future", World Business Council for Sustainable Development, 1992.
4. Schott, H. et. al., "Sustainable Product Development — A Challenge for Design Science", Proc. ICED 97, Tampere, August 19–21, 1997, pp. 665–668.
5. Reiner Anderl et al., "Life Cycle Modelling — A Cooperative Method Supports Experts in the Entire Product Life Cycle", Proc. ICED 99, Munich, August 24–26, 1999, pp. 1801–1804.
6. Mathias K. Lindqvist M., "Framework and Information for Appropriate Life Cycle Assessment", Proc. ICED 01, Glasgow, August 21–23, 2001, pp. 669–676.
7. Stephen J. Turoskaya, "Integrating Environmental Design in process to Implementation", Proc. ICED 99, Munich, August 24–26, 1999, pp. 571–574.
8. Frank Consoli, et. al., "Guidelines for Life Cycle Assessment: A 'Code of Practice'", SETAC, Brussels, 1993.
9. Brezet J. Crul, "Design for Environment in Practice", Proc. ICED 97, Tampere, August 19–21, pp. 611–616.
10. Sanna Caluguis, Guido La Rosa, Antonino Risitano, "Indicators for Environmentally Conscious Product Design", Ecodesign99, Feb 1–3, 1999, Tokyo, Japan, pp. 71–76.
11. Lagerstedt, C Luttropp, "Functional Priorities in Eco-Design — Quality Function Deployment, Value Analysis and Functional Profile", Proc. ICED 01, Glasgow, August 21–23, 2001, pp. 725–732.
12. T.A. Bhamra, S Evans, T.C. McAloone, M Simon, S Poole, A Sweatman, "Integrating Environmental Decisions into the Product Development Process: Part 1 The Early Stages, Ecodesign99, Feb 1–3, 1999, Tokyo, Japan, pp. 329–333.

the uncertainty in product life cycle information evolution with respect to LCA and requirements for the uncertainty assessment method were identified. There are mainly four stages of design. In these stages the information about the product life cycle changes from abstract for example material of a component as a class to the concrete as a specific material. We also found that designers often evaluate their design with respect to requirements, it's an iterative process. There is a need of a method to incorporate the product information transition. We propose that there are mainly four types of uncertainties with respect to Life Cycle Assessment in Design; Product Structure, Life Cycle phases, Information Quality, and methodological choices. The requirements of a DFE tool are also identified. Subsequent tasks are design and implementation of the method, and its evaluation using in-house design experiments to assess its effectiveness and efficiency.

REVIEW QUESTIONS

1. Why environmental concerns become important in product development?
2. Relate the concept of sustainable product development to the environmental protection.
3. Describe how LCA can be a method of estimating the environmental impacts of a product.

REFERENCES

1. Züst R., "*Sustainable Products and Processes*", ECO-Performance — 3rd International Seminar on Life Cycle Engineering CIRP 92, pp. 5–10.
2. Chris Sherwin, Tracy Bhamra, "*Beyond Engineering: EcoDesign as a Proactive Approach to Product Innovation*", Ecodesign 99, Feb 1–3, 1999, Tokyo, Japan, pp. 41–46.
3. Brundtland G.H., "*Our Common Future*", World Business Council for Sustainable Development, 1998.
4. Schott, H., et. al., "*Sustainable Product Development — A Challenge for Design Science*", Proc. ICED 97, Tampere, August 19–21, 1997, pp. 665–668.
5. Reiner Anderl, et. al., "*Life Cycle Modelling—A Cooperative Method Supports Experts in the Entire Product Life Cycle*", Proc. ICED 99, Munich, August 24–26, 1999, pp. 1801–1804.
6. Mueller K, Lampérth M, "*Parameterised Inventories for Approximate Life Cycle Assessment*", Proc. ICED 01, Glasgow, August 21–23, 2001, pp. 669–676.
7. Stephan J. Clambaneva, "*Engineering Environmental Design awareness to Implementation*", Proc. ICED 99, Munich, August 24–26, 1999, pp. 571–574.
8. Frank Consoli, et. al., "*Guidelines for Life Cycle Assessment: A 'Code of Practice*", SETAC, Brussels, 1993.
9. Joost F Prins, "*Design for Environment in Practice*", Proc. ICED 97, Tampere, August 19–21, pp. 611–616.
10. Fabio Guiudice, Guido La Rosa, Antonino Risitano, "*Indicators for Environmentally Conscious Product Design*", Ecodesign99, Feb 1–3, 1999, Tokyo, Japan, pp. 71–76.
11. J Lagerstedt, C Luttropp, "*Functional Priorities in Eco-Design — Quality Function Deployment, Value Analysis, and Functional Profile*", Proc. ICED 01, Glasgow, August 21–23, 2001, pp. 725–732.
12. T A Bhamra, S Evans, T C McAloone, M Simon, S Poole, A Sweatman, "*Integrating Environmental Decisions into the Product Development Process*", Part 1 The Early Stages, Ecodesign99, Feb 1–3, 1999, Tokyo, Japan, pp. 329–330.

13. Brezet H., Van Hemel, C., "*EcoDesign—A Promising Approach to Sustainable Production and Consumption*", UNEP 1997.
14. Yasushi Umeda, "*Key Design Elements for the Inverse Manufacturing*", Ecodesign99, Feb 1–3, 1999, Tokyo, Japan, pp. 338–343.
15. Grüner C., Birkhofer H., "*Decision Support for Selecting Design Strategies in DFE*", Proc. ICED 99, Munich, August 24–26, 1999, pp. 1089–1092.
16. T.C. McAloone, S Evans, "*Using Empirical Data to Build An Advisory Tool for Eco Design*", Ecodesign99, Feb 1–3, 1999, Tokyo, Japan, pp. 52–55.
17. Tetsuo Tomiyama, "*Reversible Reconfiguration: A Key for Reuse*", Ecodesign99, Feb 1–3, 1999, Tokyo, Japan, pp. 310–315.
18. Harald E Otto, Karl G. Mueller, Fumihiko Kimura, "*A Framework for Structured Data Retrieval in LCA Using Feature Technology*", Ecodesign 2001, Dec 11–15, 2001, Tokyo, Japan, pp. 250–255.
19. Mark Goedkoop, Renilde Spriensma, "*The Eco-indicator99—A Damage Oriented Method for Life Cycle Impact Assessment: Methodology Report*", PRé Consultants B.V., 2000.
20. Amaresh Chakrabarti, Srinivas Kota, Nageshwar Rao and Sekhar Chowdary, "*Product Development Platform for Real-time Capture and Reuse of Evolving Product Information*", Int. J. Product Life Cycle Management, Vol. 2, No. 3, 2007, pp. 207–227.
21. Srinivas Kota and Amaresh Chakrabarti, "*Understanding the Evolution of Product Life Cycle Information with Respect to Life Cycle Assessment in Design*", in A. Bouras, B. Gurumoorthy, S. Han (Eds.), *Product Life Cycle Management: Going beyond Product Development and Delivery,* Inderscience Enterprises limited, 2006, pp. 300–313.

- Least manufacturing cost.
- To keep introducing innovative designs more quickly than your competitors
- Shorter lead times and reliable delivery performance
- Customer Service' performance that makes your customers bend over backwards to buy from you rather than from your competitors.

As per Kybrich et al. [6], Manufacturing Excellence is commonly understood as "advanced manufacturing techniques that can be adapted and used to elevate a facility's manufacturing performance to world-class levels.

Gilgeous et al. [7], defines world class manufacturers are those companies which:

- Continuously outperform the industry's global best practices,
- Know intimately their customers and suppliers,
- Know their competitor's performance capabilities, and
- Know their own strengths and weakness.

Importance of Manufacturing Excellence: The need for manufacturing excellence has arisen as Global market scenario; Competitive equations and multi-faceted constraints that are propelling manufacturing in several directions; Revolution in IT and telecommunications; High customer expectation of responsiveness, service and satisfaction; Availability of wide choice of branded and unbranded products; Short product cycle; Reduced lead time from concept to materialization of the product; and Quality and costs have necessitated reinventing manufacturing. Hence the importance and need for "Manufacturing Excellence" in the present era.

Various researchers have given different characteristics of the concept of ME, here are some of them.

As per Shigeo Shingo [8], manufacturing excellence is characterized by:

- TQM and productivity,
- Leadership and empowerment,
- Manufacturing vision and strategy,
- Process integration,
- Quality and productivity integration,
- Business integration,
- Productivity improvement,
- Quality enhancement.

Richard J. Schonberger [4], gives the characteristics as:

- CAD, TQM, CAE, CIM, FMS, JIT,
- Supplier development,
- Employee involvement,
- Manufacturing strategy.

- Least manufacturing cost
- To keep introducing innovative designs more quickly than your competitors
- Shorter lead times and reliable delivery performance
- 'Customer Service' performance that makes your customers bend over backwards to buy from you rather than from your competitors.

As per Rubrich et al. [6], Manufacturing Excellence is commonly understood as "advanced manufacturing techniques that can be adapted and used to elevate a facility's manufacturing performance to world-class levels".

Gilgeous et al. [7], defines world class manufacturers are those companies which:

- Continuously outperform the industry's global best practices,
- Know intimately their customers and suppliers,
- Know their competitor's performance capabilities, and
- Know their own strengths and weakness.

Importance of Manufacturing Excellence: The need for manufacturing excellence has arisen as Global market scenario; Competitive equations and multi-faceted constraints that are propelling manufacturing in several directions; Revolution in IT and telecommunications; High customer expectation of responsiveness, service and satisfaction; Availability of wide choice of branded and unbranded products; Short product cycle; Reduced lead-time from concept to materialization of the product; and Quality and costs have necessitated reinventing manufacturing. Hence the importance and need for "Manufacturing Excellence" in the present era.

Various researchers have given different characteristics of the concept of ME, here are some of them.

As per Shigeo Shingo [8], manufacturing excellence is characterized by:

- TQM and productivity,
- Leadership and empowerment,
- Manufacturing vision and strategy,
- Process integration,
- Quality and productivity integration,
- Business integration,
- Productivity improvement,
- Quality enhancement.

Richard J. Schonberger [4], gives the characteristics as:

- CAD, TQM, CAE, CIM, FMS, JIT,
- Supplier development,
- Employee involvement,
- Manufacturing strategy.

K.C. Ng and Ip Hung [9] gives the characteristics as:

- Manufacturing strategy,
- Open management culture,
- Leadership,
- Systems perspective,
- Supplier and CR management,
- Better communication channel development,
- FMS, JIT,
- Environmental constraints,
- Consistency and dependability,
- Empowerment ,
- ERP, Concurrent engineering, IT, New technology.

As per Vic and Maria Gilgeous [7], the following factors characterize manufacturing excellence:

- Innovations and change,
- Empowerment,
- Customer focus and commitment,
- Learning organization,
- Quality,
- Management focus on team management,
- Technology,
- Supplier relationship

The world that we know of is an evolving and a fascinating organism moving forward on human inventions, innovations, creativity and consequent change and learning. While studying global competitiveness of firms, the one element that stands as the key to competitive excellence is the ability to continuously bring out innovation as an *organized activity*. These innovations ambit the entire value and cost chain including organization, manufacturing, accounting and cost allocation, logistics, *new product development,* employee training and skill building. Innovation may not be only in tangible product form. Toyota's JIT manufacturing system based on kanban and pull (rather than push) concepts are major organizational innovations that provided Japan a strong foothold in tough US markets. If the entrepreneurial process for transformational 'catching up' and establishing global 'leadership' is characterized by the process of moving through three states. Where the three states are defined as **State 1:** as the present state of competitiveness in recently liberalized economy like India; **State 2:** A state to qualify up to a certain level of competitiveness which can be reached through usual methods of cost cutting, new models introduction, rationalization, etc. and **State 3:** To be a successful global competitor. To catch up and supersede the global leaders i.e., to reach the state 3 from state 2, it is imperative that firms be innovative, entrepreneurial, and flexible, and develops a capability to learn and unlearn faster than present leaders [10].

not twice but continuously, effectively and efficiently. There is a need to take some steps for continuous and systematic product excellence.

The model of continuous product innovation was proposed by Chapman et. al. (1) as shown in Fig. 10.1.

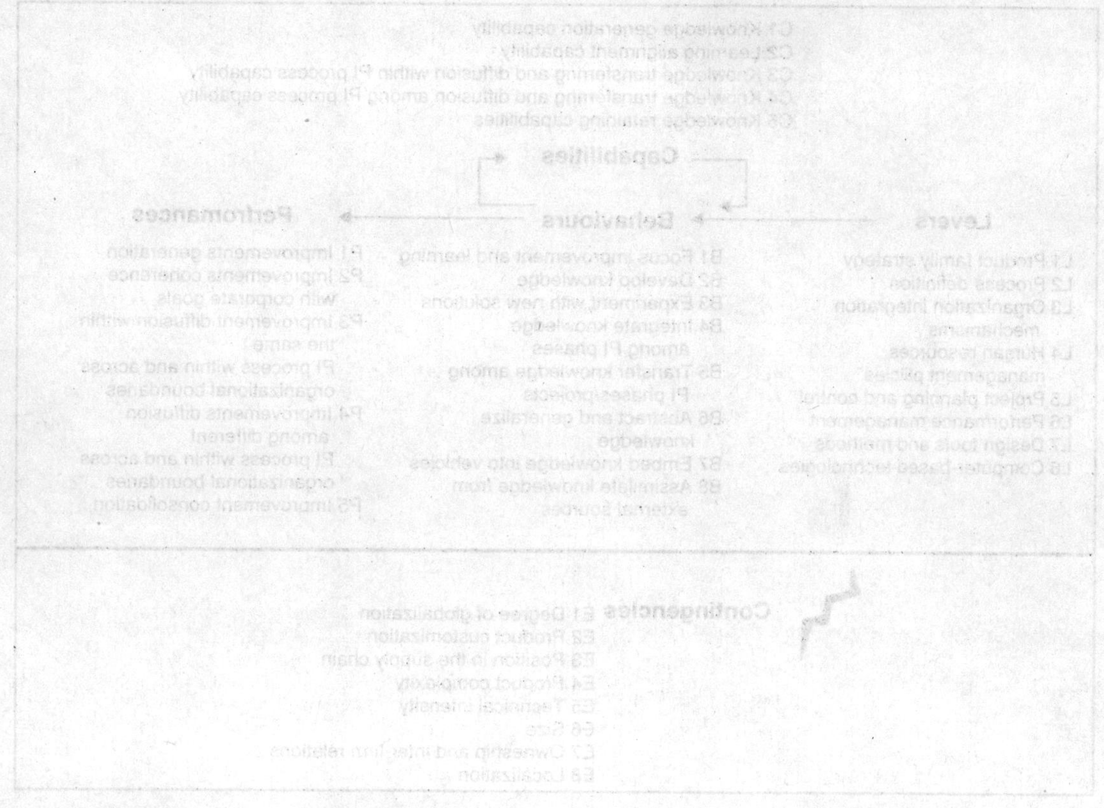

Fig. 10.1: A model of continuous product innovation [13]

The model identifies certain key levers or mechanisms that manager uses when managing the product innovation process, these lever categories and examples of specific levers are as below:

1. Product family strategies:
   a. Product family plans
   b. Carry over policies
   c. Standardization policies.
2. Innovation process definition:
   a. Stage gate processes
   b. Company innovation procedures.
3. Organizational integration mechanisms:
   a. Teamwork
   b. Organizational structure
   c. Committees.

not twice but continuously, effectively and efficiently. There is a need to take some steps for continuous and systematic product excellence.

The model of continuous product innovation was proposed by Chapman et. al. (1) as shown in Fig. 10.1.

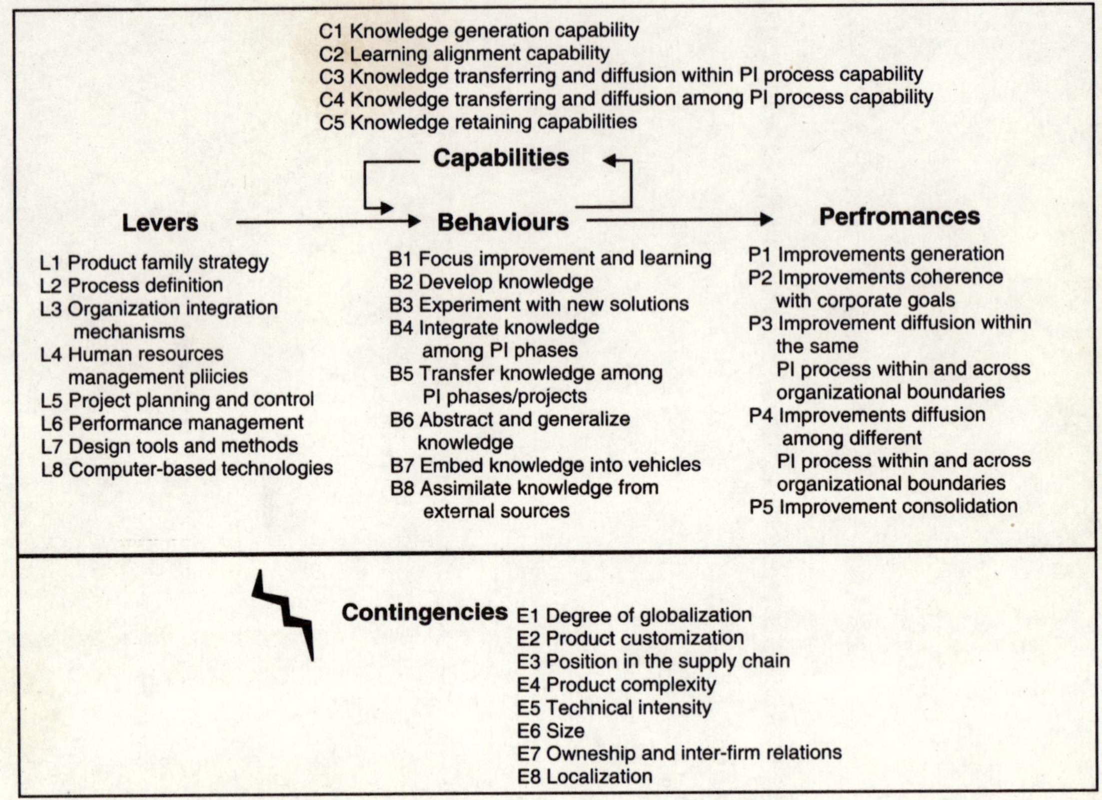

Fig. 10.1: A model of continuous product innovation [13]

The model identifies certain key levers or mechanisms that manager uses when managing the product innovation process, these lever categories and examples of specific levers are as below:

1. Product family strategies:
   a. Product family plans;
   b. Carry over policies;
   c. Standardization policies.
2. Innovation process definition:
   a. Stage gate processes;
   b. Company innovation procedures.
3. Organizational integration mechanisms:
   a. Teamwork
   b. Organizational structure
   c. Committees.

4. Human resource management policies:
   a. Personnel relations;
   b. Departmental assessment and staff development plans;
   c. Remuneration and reward systems;
   d. Empowerment programs.
5. Project planning and control:
   a. Project termination reports;
   b. Design reviews.
6. Performance measurement:
   a. Comparison of measurement against previous results or with other subsidiaries or leading organizations.
7. Design tools and methods:
   a. Standardized design methodologies and procedures;
   b. Libraries of standard design solutions;
   c. Integration procedures (e.g., QFD, DFM).
8. Computer based technologies:
   a. IT systems;
   b. Computer aided technologies (CAD, CAM, CAE)
   c. Prototyping technologies.

Along with the levers the manager uses certain performance indicators used to measure the performance of the product innovation process, they are:

1. Time to market the product
2. Product performance
3. Design performance
4. Impact on firms competitiveness
5. Impact on product portfolio
6. Others like: ROI, R&D expenses etc.

Research indicates that all the relevant variables, levers and performance measures are influenced by contingencies. This contingency approach is based on the premise that there can be no single best way to manage and improve organizational performance. This is especially true for the product innovation process. One very important clichés is that the product innovation process should be market led, i.e., all staff should be involved in analyzing the innovation opportunities.

Systematic Product Innovation

Top management must lead the way by clearly communicating the criteria for the selection of innovation opportunities. This company wide alertness to market opportunities and stimulates a flow of ideas from the whole organization. Companies get the better at spotting opportunities as their corporate culture changes to one of market orientation.

Understanding the value of a product to the end user and to the intermediate customer is vital. For many products value has an emotive as well as a functional component. Scientific reasoning can be applied to product innovation reducing risk and lifting profit potential.

A comprehensive approach for companies that want to bring a continuous stream of successful new products to the market. New innovation opportunities must be systematically analyzed and developed through a process so called "scientific" reasoning.

A company needs to think in a broader perspective and focus of key internal and external performance drivers such as customers' needs, competition and the capabilities of the company with following components:

1. Functional physical needs
2. Functional non physical needs
3. Source and amount of economic value that customers will get from having the above needs met.
4. Source and amount of emotive value to customer from having the above needs met.
5. Technical resources needs for R&D and design
6. Resources needed to produce the product or service
7. Promotion and distribution
8. Basis of competitive advantage
9. Basis of sustainable competitive advantage.

10.4. ROADMAP TO PRODUCT EXCELLENCE

Product Excellence is a contingent on a well-conceived roadmap to build a foundation of visibility and control of all systems and processes within an institution. The people, systems and processes put into place must be agile and flexible enough to adapt to the changing needs of the business. The roadmap to product excellence focuses on four major objectives:

1. Accelerated time to market— Introducing new products and services to the marketplace is the central to a financial institutions business. Introducing products and services ahead of competitors gives a company a considerable leadership position. People, systems and processes must be aligned to accelerate time to market new products and services.
2. Provide "in sourcing" to customers: as margins on products continue to decline additional services offered on the core product needs to be raised.
3. Outsource non core processes: Cost reduction can be achieved by outsourcing non customer facing non differentiating services by out sourcing.
4. Rationalize systems: Duplicated systems need to be eliminated and system rationalized.

10.5. CLOSURE

Innovation provides the cutting difference between mediocrity and excellence. The chapter gives a brief introduction about the topic in the first section. In the second section of the chapter a brief overview about the concept of manufacturing excellence and its importance in today's competitive age is discussed. The concept of product excellence and product innovation is dealt in the third and consequent sections to support the topic of discussion i.e., innovative product planning as an important pillar to reach summit of manufacturing excellence. The roadmap gives the path to achieve product excellence. Developing and introducing new products and services is vital to most manufacturing companies. Well managed processes for bringing in new products to markets can lead to a significant competitive advantage. In addition to new product development the same processes and resources are applied to product improvements, corrections and variations throughout the product life cycle. The companied need to focus on making sure the new products meet market needs and satisfies the production and service needs. Hence success on today's markets can only be achieved through innovation, agility and continuous product excellence.

REVIEW QUESTIONS

1. What is meant by 'manufacturing excellence'? How it is different from the term 'productivity', which has been already widely recognized as a measure of performance of an industrial output?
2. Should the path of achieving manufacturing excellence be 'process-centric' or 'product-centric' or 'layout-centric' or all of them? Justify your answer.
3. Cannot 'incremental improvement' in the existing product lead to manufacturing excellence?
4. Is innovative product development for achieving manufacturing excellence an option, a conditional imperative or all-time mandatory activity? Explain.
5. What role can the top management play in achieving the manufacturing excellence for a company?
6. Does continuous as well as need-based training of the staff help to achieve manufacturing excellence in a company?
7. Examine the need for focus on benchmarking for establishing a framework of achieving manufacturing excellence in a company.
8. Are parameters for achieving manufacturing excellence same for manufacturing industry and services industry? Discuss.
9. Does the just-in-time (JIT) manufacturing philosophy help to achieve manufacturing excellence?
10. How manufacturing excellence can be achieved in a company by adopting a well-framed supply chain management system?

REFERENCES

1. Knuckey, S. , Leung-Wai, J., Meskill, M., [1999], Gearing Up : A Study of Best Manufacturing Practice in New Zealand, "*Ministry of Commerce*", 64–68.
2. Kepner-Tregoe [1999] *"The Path to Manufacturing Excellence"* , Princeton, NJ.
3. Roth A.V., Giffi C.A., Seal G.M . [1992], Operating Strategies for the 1990s: Elements Comprising World-class Manufacturing in "*Manufacturing Strategy – Process and Content*", Voss C.A., Eds., Chapman and Hall, London, 1st Edition.
4. Schonberger R.J., [1986] "*World Class Manufacturing—The Lessons of Simplicity Applied*", The Free Press, New York,.
5. Todd, J., [1994] "*World Class Manufacturing*", (McGraw Hill Book Company, London) 1 Ed. pp. 2–23.
6. Rubrich L., Watson M., [1998] "*Implementing World Class Manufacturing*" WCM Associates, Port Wayne, U.S.

7. Gilgeous, V., Gilgeous, M. [1999], A Framework for Manufacturing Excellence, *"Integrated Manufacturing Systems"*, 10/1 pp. 33–44.
8. Guidelines for *"Shigeo Prize for Excellence in Manufacturing"* Available at: http://www.gbmp.org/lean/2004%20%20NE%20Shingo%20Guidelines.pdf.
9. Ng, K.C., Hung, I.W. [2001] *"A Model for Global Manufacturing Excellence"* Work Study, Volume 50, number 2, pp. 63–68.
10. Jain, A.K., [2000] "*Competitive Excellence: Critical Success Factors*" Vikas Publishing House (P) Ltd., New Delhi, pp. 144–171.
11. Hardaker, G., [1998] *"An Integerated Approach towards Product Innovation in International Manufacturing Organizations"* European Journal of Innovation Management 1,2, pp. 67–73.
12. Corso, M., Pavesi, S., [2000] *"How Management can Foster Continuous Product Innovation"* Integrated Manufacturing Systems , 11/3, pp. 199–211.
13. Chapman, R.L., O'Mara, C.E., Ronchi, S., Corsa, M., [2001] *"Continuous Product Innovation: A Comparison of Key Elements Across Different Contingency Sets"* Measuring Business Excellence Volume 5,3 pp.16–23.

CONCEPTUAL DESIGN

11.1. INTRODUCTION

In this chapter, the design and implementation of a Conceptual Design Environment System (CoDE) to enable modeling of 3D freeform objects from 2D sketch strokes drawn online is presented. Robust algorithms have been written to 'understand' the intented shape expressed in terms of the Sketch Strokes drawn by the designer. CoDE enables modeling of both axi-symmetric and non axi-symmetric object shapes and provides tweaking utilities to create Cutouts (freeform holes), Depression and Protrusion features on them.

Product modeling is the heart of any Computer Aided Design/Modeling CAD/CAM activity as it creates a central repository of product information required by various down-the-line application tasks in its lifecycle. Present day CAD/CAM systems are exact geometry modelers, which need precise dimensions and geometric information for part design. This information may not, however, be available at the conceptual design stage, where flexibility of shape creation and manipulation is more important to the designer. Conceptual design needs interactivity and ease to create a flexible, manipulatable model capturing designer's intent. Freehand sketching is a familiar, efficient and natural way of expressing ideas, which the designer finds very useful. In sketch based design, wide range of design alternatives can be quickly produced, evaluated, modified and compared [1]. Most fields like engineering, architecture, software design etc. share a tradition of sketching as an essential part of main design process [2]. Despite this, sketching behavior is largely unsupported in the current CAD systems for the Conceptual Design Stage.

Literature reports very scant research on product (geometric) modeling systems using online sketch and gesture strokes. Reported systems target on specific object shapes such as toys [3], fashionable cloth [4] etc. and are mostly proof of the concept type.

Igarashi et. al., [3] reported a Sketch Stroke based system for modeling of 'stuffed animal' toys. User draws sketch-strokes online, which are either for gross shape creation and/or for editing the existing model. Constrained Delaunay Triangulation and Vertex Elevation algorithms are used to construct the model. Teddy can model rotund objects with spherical topology only.

A similar work by Charlie C.L. et. al., [4] tried to overcome the limitation of spherical topology to model objects with open surface for cloth fashion designing. "Quick Sketch" developed by Eggli L. et. al., [5] is a sketch based modeling system in which a sequence of user drawn points are sampled to interpret geometric primitives like 2D lines, circular arcs, full circles, B-Spline etc. and their relationships like Perpendicularity, Parallelism etc. It can handle freeform surfaces like ruled, surface of revolutions, sweep. Another system called FSD (Fast Shape Designer), developed by Casper G.C. et. al., [6] attempts to interpret designer's intent from two sketch-strokes called as "old line" and "added line". The FSD system then displays different strategies of merging those two strokes and user can select any one to output a single resultant stroke. The system is just a proof of concept type. SKETCH [7] is a gesture based conceptual modeling tool reported, which can draw 3D primitives like Cylinder, Cone, Cube, Frustum etc. by some definite gesture strokes. It is able to perform very limited Boolean Operations, and cannot handle freeform surfaces efficiently.

All the above systems focus on specific geometric shapes and are thus, very restrictive in their capabilities. A need thus exists to develop a user-friendly sketch based modeling system for conceptual design.

11.2. MODULAR ARCHITECTURE OF CoDE

Conceptual Design Environment (CoDE) is an intelligent 'Online' Sketch-Stroke based Geometric Modeling system designed and developed to aid product designer/artists during the conceptual design phase.

Figure 11.1 shows the architecture of CoDE indicating various functional modules. CoDE has been designed to handle axi-symmetric and non axi-symmetric (asymmetric) object shapes, which need freeform surfaces. Such shapes are commonly encountered in the artistic design (styling) of pottery items like Vases, Bowls, Containers, fancy art items etc. CoDE provides a White Canvas to the designer to sketch his/her ideas, which are immediately transformed into a 3D model by the system. User can tweak this CAD model interactively to create Cutouts, Depressions, and Protrusions to further refine the shape.

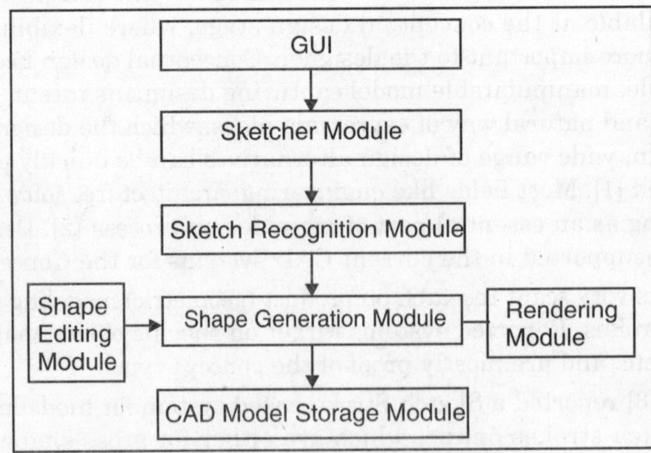

Fig. 11.1: Modular architecture of CoDE

CoDE enables object modeling in a user-friendly fashion and finally stores the model in standard CAD format (.STL file). It thus, provides portability to other CAD softwares. CoDE has been developed in Java 1.4 [8] using Java3D.

Various functional modules of CoDE are discussed one by one.

Graphical User Interface, GUI

CoDE presents a very user-friendly environment to the designer in the form of a "White Board Canvas" and a 3D canvas.

Several menu options provided on the White Board Canvas are enumerated below:

File ------ I----Save model
I----Exit

Design---I----Symmetric
I----Asymmetric

Sketch---I----Fit Bspline curve
I----Fix Bspline curve

View------I----Clear screen
I----model

Help------I----Contents

Figure 11.2 shows a White Board Canvas along with user drawn strokes/sketches. Figure 11.3 shows the 3D model rendered on 3D Canvas. User can use either of them for drawing sketch strokes. Various viewing options such as Zooming, Panning, Rotation etc. are provided on the 3D canvas for viewing the model.

Sketcher Module

Designer uses the sketcher module for drawing freehand sketch strokes on the White Board Canvas using a mouse (Fig.11.2). Two types of strokes are recognized by the system viz. the Main Stroke that signifies the shape curve and the Auxiliary stroke which signifies displacement operator for 3D shape creation. Each mouse drawn stroke is extracted in terms of pixel data points (coordinates) and stored as a Linked List object [9].

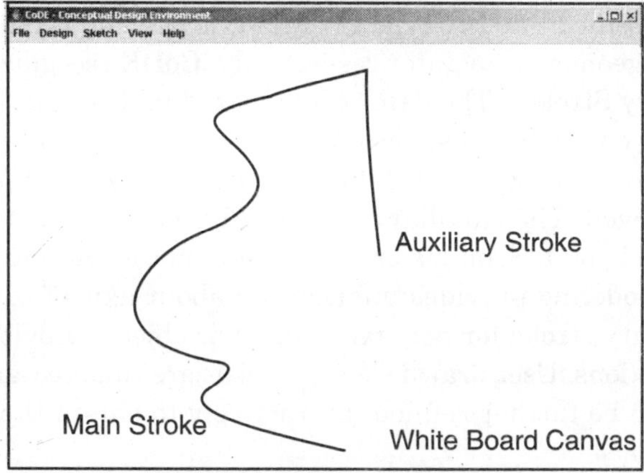

Fig. 11.2: GUI of CoDE

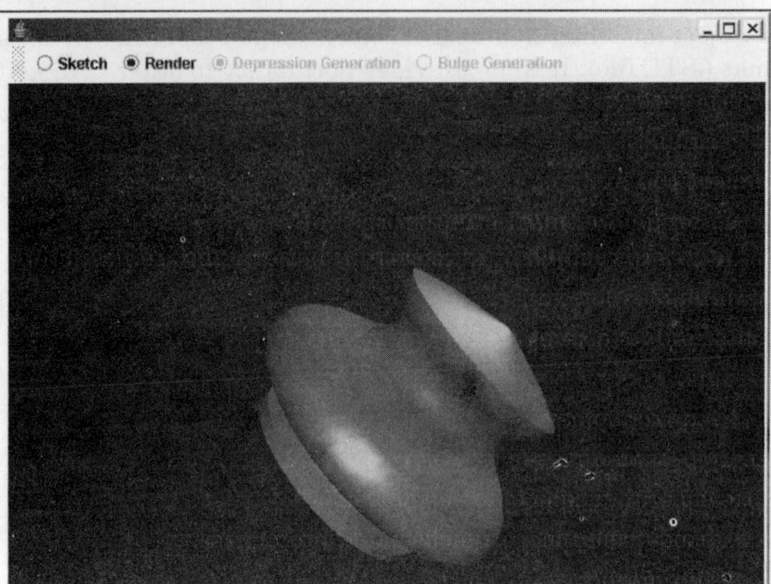

Fig. 11.3: 3D canvas with a rendered model

Slower the speed of stroke drawing, larger number of data points will be picked up by CoDE and hence the accuracy of stroke recognition will be better. Speeds of time step discretization can be setup in CoDE to suite various sketching skills of the designers.

Sketch Recognition Module

This module is the heart of the conceptual design system, CoDE. It automatically interprets the "designer's intent" by processing the user drawn sketch strokes and creates the 3D part shape.

In order to understand how a human artistic designer expresses his/her ideas through sketches, a critical survey was done at the Industrial Design Center, IIT Bombay. A large variety of sketches of axi-symmetric and non axi-symmetric object shapes (vases, bowls, containers etc.) were critically studied to identify various sketch strokes. Figure 11.4 shows sketch strokes drawn by artists for designing some typical parts.

The strokes can be classified into two type viz. Main Stroke and Auxiliary Stroke signifying the shape element and the geometric operator respectively. CoDE recognizes "Designer's intent" from the Main and Auxiliary Strokes. The data points picked up from the Main Stroke are used to represent it as a BSpline curve. It is displayed to the user along with the control points. User can modify the shape of the Main Stroke (BSpline) by dragging control points till satisfactory shape modification is achieved. The Auxiliary Stroke (Fig. 11.4) is identified from its various features like Start point, End point, Semi-major and Semi-minor radius etc. The Auxiliary Stroke for axi-symmetric object modeling provides information about axis of revolution and angle of revolution while the auxiliary stroke for non axi-symmetric objects provides information about depth between two cross sections. User drawn sketch strokes are thus recognized and represented as BSpline curve which can be further refined interactively to obtain the final shape.

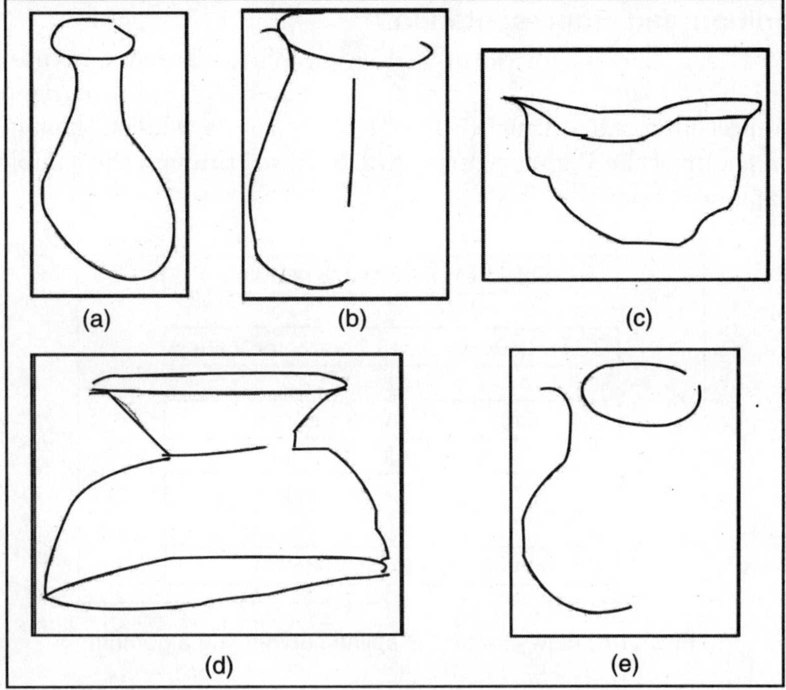

Fig. 11.4: Typical sketch strokes drawn by artists

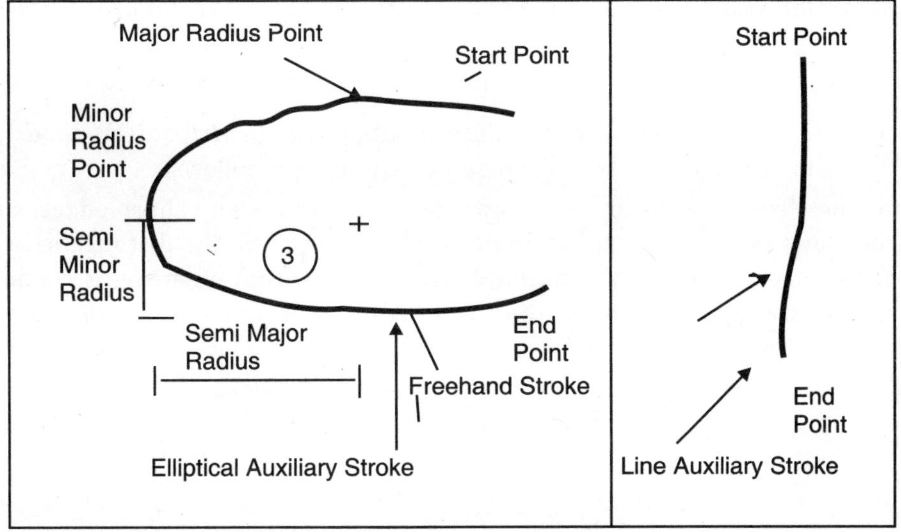

Fig. 11.5: Types of auxiliary stroke

Stroke Recognition and Representation

As mentioned earlier, user-drawn Main Stroke on White Board Canvas is sampled and represented as a Linked List storing 2D Co-ordinate points. A B-spline curve-fitting algorithm is developed to represent a smooth curve through these points. The mathematical formulation to obtain Control points of the B-spline curve, which passes through the sampled stroke/sketch data points, is explained below.

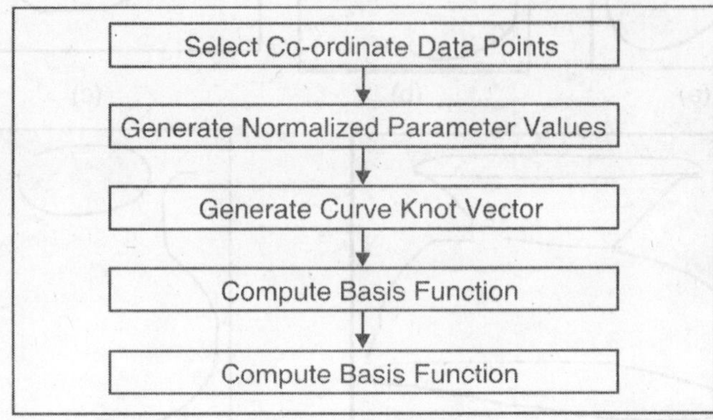

Fig. 11.6: Flow chart for B-spline curve fitting algorithm

Figure 11.6 shows flowchart for B-spline curve fitting algorithm. Various steps are enumerated in the following discussion.

Step-1:

CoDE automatically selects coordinate data points (31 data-points in our implementation) from the Linked List of user drawn sketch stroke data points. Let the curve data-points be denoted by d_0, d_1 ... d_{30}.

Step-2:

Generate the parameter values for selected data points using *Centripetal Method* [10]. The Centripetal Method for Knot Vector selection can be explained as follows:

While drawing strokes, the user often slows down on corners and sharp edges, similar to a car driver who slows down on sharp turns to avoid skidding. So the distance between two points on corners/sharp edges will be smaller as compared to that on straight portion of the curve. The distance between two points can thus, provide information about the curvature of the curve. Centripetal method approximates curve corners and smooth regions better.

Let u_0, u_1, u_2 ···, u_{30} be the parameter values at respective data points.

Then,

$$\begin{aligned} u_0 &= 0.0 \\ u_k &= \sum_{i=1}^{k} \sqrt{[d_i - d_{i-1}]} \\ u_{30} &= 1.0 \end{aligned} \qquad \text{...(11.1)}$$

d_i is the ith data point. The parameter values are stored as normalized values.

Step-3:

Generate Knot vector. In the present work, a B-spline curve with 5th degree 11 control points is found sufficient to fit to most of the hand drawn sketch strokes including some complex contours. This would need 17 knot vectors to be computed. Let t_0, t_1, t_2,···, t_{16} be the knot values. Then an open knot Vector is generated as follows:

$$t_0 = t_1 = t_2 = t_3 = t_4 = 0$$

$$t_{j+4} = \frac{j}{n-p+1} = \frac{j}{16-4+1} = \frac{j}{13}, \ [j = 1,2,3,..,7] \qquad ...(11.2)$$

$$t_{12} = t_{13} = ...t_{16} = 1$$

The knot vector values are also normalized.

Step-4:

Now compute basis function matrix,

$$N = \begin{bmatrix} N_{1,5}(u_1) & N_{2,5}(u_1) & \cdots & N_{10,5}(u_1) \\ N_{1,5}(u_2) & N_{2,5}(u_2) & \cdots & N_{10,5}(u_2) \\ \vdots & \vdots & \cdots & \vdots \\ N_{1,5}(u_{30}) & N_{2,5}(u_{30}) & \cdots & N_{10,5}(u_{30}) \end{bmatrix} \qquad ...(11.3)$$

where,

$N_{i,p}(u_k)$–Cox-de Boor basis functions enumerated below:

$$N_{i,1}(u) = 1 \qquad \text{if } x_i \le u \le x_{i+1}$$

$$= 0 \qquad \text{otherwise}$$

$$N_{i,k}(u) = \frac{(u - x_1)N_{i,k}(u)}{x_{i+k-1} - x_i} + \frac{(x_{i+k} - u)N_{i+1,K-1}(u)}{x_{i+k} - x_{i+1}}$$

where, x_i is knot vector

u is parameter value

Step-5:

Compute Control Points for the BSpline curve through P and Q matrices, where, P is a matrix of control points from P_0 to P_{10} and Q is a matrix shown below:

$$\begin{bmatrix} \sum_{k=1}^{29} N_{1,5}(u_k) * q_k \\ \sum_{k=1}^{29} N_{2,5}(u_k) * q_k \\ \vdots \\ \sum_{k=1}^{29} N_{10,5}(u_k) * q_k \end{bmatrix}$$

N stands for Basis function as given in equation (11.3) value while expression for 'q_k' is given in equation (11.5)

$$P = \begin{pmatrix} P_0 \\ P_1 \\ \vdots \\ P_{11} \end{pmatrix} Q \qquad \text{...(11.4)}$$

q_k is given as,

$$q_k = d_k - N_{0,5}(t_k)d_0 - N_{0,5}(t_k)d_{30} \qquad \text{...(11.5)}$$

Step-6:

Points on BSpline curve can be found out by,

$$P(u) = \sum_{i=0}^{30} d_i * N_{i,k}(u) \qquad \text{...(11.6)}$$

The above algorithm computes control points of B-spline curve (*P*-matrix) corresponding to the user drawn sketch stroke. Detailed mathematical formulation of this algorithm is explained in the literature [9].

Figure 11.7 shows typical End Points of the free hand strokes. Utilizing these control points the BSpline curve is generated and is shown on "White Board Canvas".

11.3. STEPS IN MODELING

Various steps in modeling of objects using CoDE are enumerated below:

Axi-Symmetric Objects

With reference to the Fig. 11.7, the following steps can be identified.

- User draws sketch of silhouette edge (Main Stroke) of the object on the white canvas.
- The CoDE system then fits a B-spline curve to this Main Stroke and then displays curve and its control points to the user (Fig. 11.7a).
- User can edit the shape of curve by dragging those control points on screen using mouse. After editing the curve, the CoDE system fixes it (finalizes the shape of curve) (Fig. 11.7b).
- Then user draws second stroke, called Auxiliary Stroke, which will provide necessary information regarding axis of revolution (Fig. 11.7c).
- CoDE then designs and renders the interpreted model on 3D Canvas.

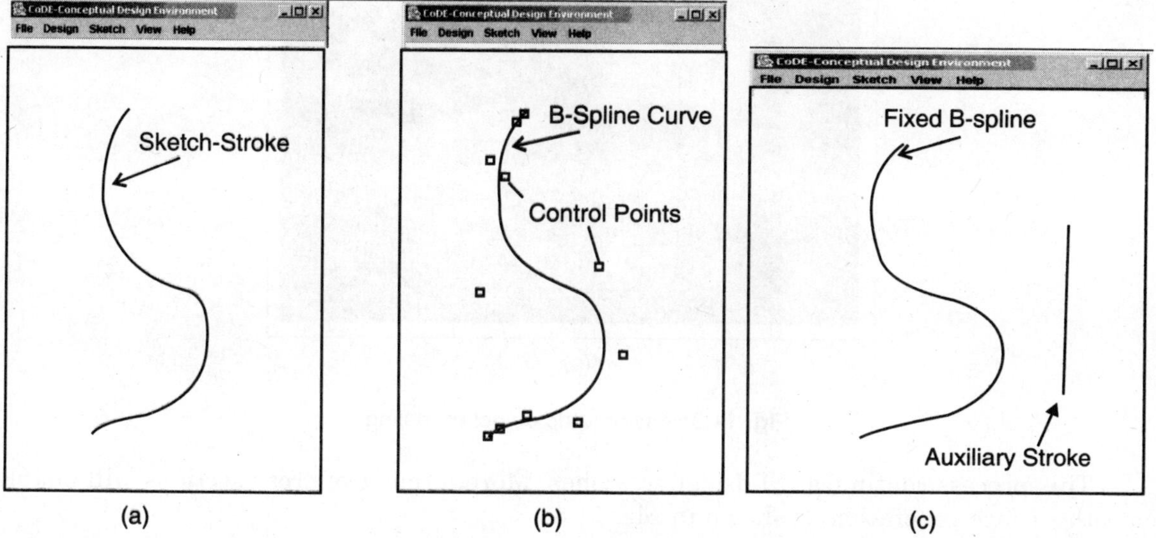

Fig. 11.7: Axi-symmetric object modeling

Modeling Non-Axisymmetric Objects

With reference to the Fig. 11.8, the following steps can be identified.

- In this mode Main Stroke is considered to be cross sections of the object and Auxiliary Stroke is considered to be the distance between two cross sections.
- User-drawn sketch stroke (Main Stroke) is fitted with BSpline curve (Fig. 11.8a), similar to axi-symmetric case.
- Then user draws an Auxiliary Stroke, a vertical line depicting depth between two consecutive cross sections.
- After this stroke, user draws another Main Stroke, which is a next cross section. Depth information from previous auxiliary cross section is retained. (Fig. 11.8b).

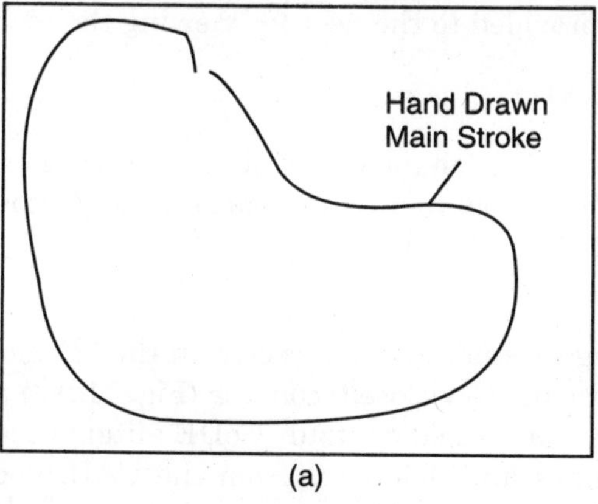

(a)

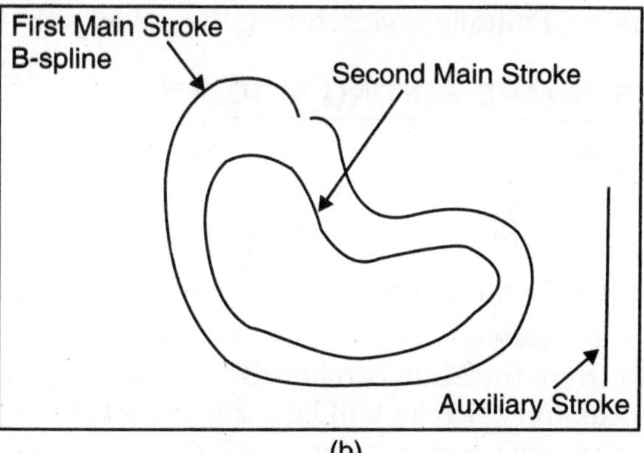

(b)

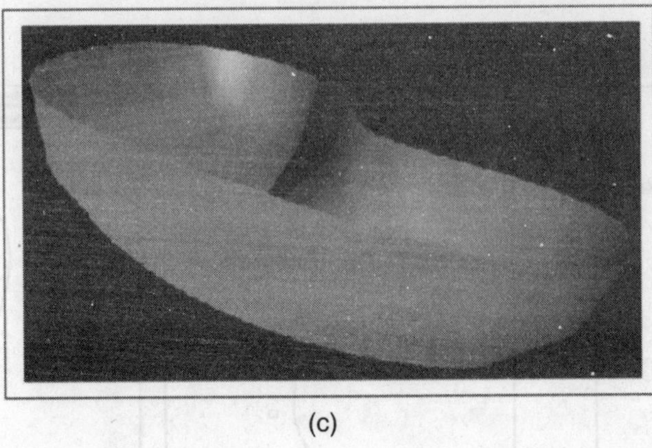

(c)

Fig. 11.8: Asymmetric object modeling

This process continues till designer wishes. More number of cross sections will enable smooth surface generation as shown in Fig. 11.8c.

11.4. SHAPE GENERATION AND RENDERING MODULE

Based on the Main and the Auxiliary Stroke inputs, by the user, Geometry Processing Module designs the object shape considering it as Rotational Sweep Surface (Surface of Revolution) or a Lofted/Ruled Surfaces.

1. Surface of Revolution is given by (assuming rotation about x axis)[10],

$$Q(u, v) = [x(u), y(u)\cos\alpha, z(t)\sin\alpha] \quad ...(11.7)$$

where,

$Q(u, v)$ is a point on Surface of Revolution and u and v are parameters in two directions.

2. Ruled Surface/Lofted Surface is represented by

$$Q(u, v) = (1 - v)^*B_1(u) + v^*B_2(u) \quad ...(11.8)$$

The geometry is mathematically computed and faceted with triangles [9]. The facets generated are used to display the object created during rendering by Java3D on 3D Canvas. Utilities like Zooming, Panning and Rotating the model are also provided to the user for viewing the model.

11.5. SHAPE EDITING MODULE

From the user drawn Main and Auxiliary Strokes, "Gross" shape of the object is created and rendered (Figs. 11.7, 11.8). Facilities for tweaking this model to produce features like Cutouts, Protrusion, Depression are provided in the Shape Editing Module.

Cutout Generation

User generates a freeform Cutout (hole) by drawing sketch strokes directly on the 3D model itself. Here the Main Stroke will be a closed (or approximately closed) contour (Fig. 11.9 a) and the Auxiliary Stroke will be a zig-zag stroke inside that closed contour. CoDE automatically identifies the region to be 'cutout' from these strokes and deletes it from the CAD model. Figure 11.9 b, c shows cutout generation [9].

Protrusion/Depression Generation

For generating Protrusion (bulge) or Depression feature on the Gross shape, the user first selects the region to be deformed on the 3D model. Maintaining the topology of the object facets, the picked facet and its neighboring region are deformed by a preset distance. Figure 11.10 shows objects with depression and protrusion type of deformation.

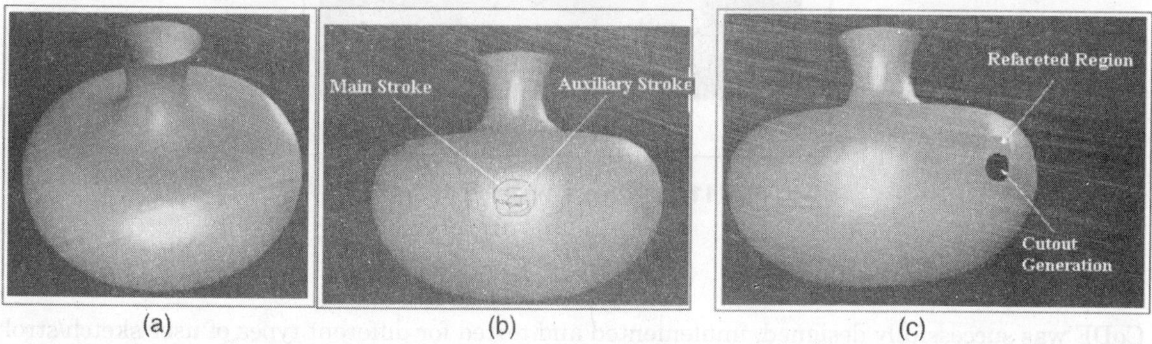

Fig. 11.9: Symmetric object modeling with cutout operation

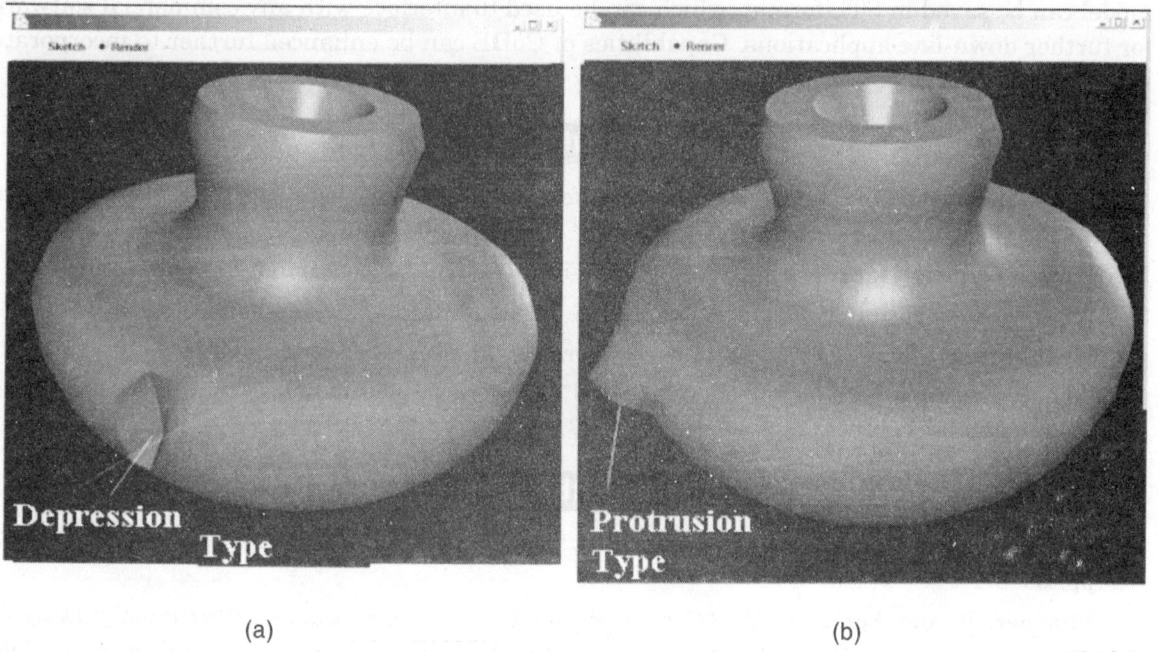

Fig. 11.10: Deformations on 3D model

Finally the generated CAD part model is stored in the STL file format. The model can thus be transferred to any commercial CAD modelers, giving interoperability and portability. Figure 11.11 shows some typical objects modeled in CoDE.

Fig. 11.11: Object modeled in CoDE

11.6. CLOSURE

CoDE was successfully designed, implemented and tested for different types of user sketch/stroke inputs. It provided a very user-friendly sketch based modeling environment capable of understanding the intent of the conceptual designer/artists. CoDE was found to be robust, consistent and accurate enough to model reasonably complex object models. The output CAD model can be saved in STL format, which can be used to interface with any commercial software for further down-line applications. Capabilities of CoDE can be enhanced further to incorporate more complex multi object models.

REVIEW QUESTIONS

1. Sketch the product life cycle and relate the different stages in it. Where does the conceptual design stage appear in the life cycle?
2. What is the importance of sketch recognition in an automated conceptual design module?
3. What is the importance of stroke recognition in the automated conceptual design workbench?
4. What is meant by normalization of parameters? What gain is made through it?
5. What is the framework for storing conceptual design stage information so that it can be used in the remaining stages of the product life cycle?

REFERENCES

1. Andries van Dam, *Post* WIMP User Interfaces, "*Communications of the ACM*", Vol. 40, No. 2, 1997. pp. 63–67.
2. Plimmer, B. and Apperley, M., "*Computer-Aided Sketching to Capture Preliminary Design*", Conferences in Research and Practice in Information Technology, Third Australasian user interface conference (AUIC), Melbourne, Australia, Vol. 7, 2001. pp. 9–12.
3. Igarashi, T., Matsuoka, S. and Tanaka, H., "*Teddy: A Sketching Interface for 3D Freeform Design*", SIGGRAPH' 99 Conference Proceedings, 1999, pp. 109-116.
4. Charlie, C.L. and Matthew M.F.Y., Freeform Extrusion by Sketched Input, "*Computer and Graphics*", Vol. 27, 2003, pp. 255–263.

5. Eggli, L., Bruderlin, B.D. and Elber, G., Sketching as a Solid Modeling Tool, *"Solid Modeling"*, 1995, pp. 313–321.
6. Casper, G.C., Van Dijk, and Mayer, A.A.C., Sketch Input for Conceptual Surface Design, "*Computers in Industry*", Vol. 34, 1997, pp. 125–137.
7. http://sun.java.com as seen on 30/09/2005.
8. Munde. S., "*Intelligent Product Modeling from Sketches*", M.Tech. Thesis, Mechanical Engg. Dept., IIT Bombay, 2004.
9. http://www.cs.mtu.edu/~shene/COURSES/cs3621/NOTES/ as seen on 30/09/2005.
10. Rogers D.F. and Adams J.A., "*Mathmatical Elements for Computer Graphics*", Tata-McGraw Hill, 2002.

CONCURRENT ENGINEERING

12.1. INTRODUCTION

This chapter discusses the definition of concurrent engineering as a systematic integrated approach for the development of products using life cycle concept. Concurrent engineering incorporates the various elements of product life cycle like manufacturability, assemblability, testability, serviceability, reliability, quality, cost, disposability, etc. during the product development phase. Need, principles and benefits of concurrent engineering have also been emphasized in this chapter.

The term Concurrent Engineering (CE), also called simultaneous engineering was coined in the U.S. in 1989. It means a way of work where the various engineering activities in the product development, production process development and field support development are integrated and performed as much as possible in parallel rather than in sequence [1].

Concurrent Engineering is a systematic/management technique used for product development, which provides an integrated approach to the design of products and their related processes from concept to disposal. The various elements of product life cycle like manufacturability, assemblability, testability, serviceability, reliability, quality, cost, disposability, user requirements etc. are incorporated during the product design and development phase. Integration is usually accomplished through computerized technical tools and cross-functional teams consisting internal (e.g., manufacturing, assembly, R&D, Service, process planning) and external (e.g., customers, suppliers) members. The basic CE has two elements: improved process and closer cooperation. The improved process provides greater clarity to the activities and focus on: concurrent process, quality, cost and delivery (QCD), customer satisfaction and competitive benchmarking. The closer cooperation, which facilitates better communication and provides unity to the cross functional team focus on: integrating organization (multifunctional team) and strategic relationship with suppliers. Concurrent engineering helps to produce quickly better and cheaper products [2].

12.2. SEQUENTIAL VERSUS CONCURRENT ENGINEERING

Traditionally, the marketing department of an organization identifies the need of a product, expected performance and the viable price range from the potential consumers/customers. This department passes these loose specifications to the design department that works on these specifications to develop technical requirements (e.g., material/size, shape etc.) and detailed design. During the design no consideration was given to the available manufacturing facilities or techniques, the way product will be assembled, the serviceability methods once the product is in the field or the effect on environment etc. In other words the design department makes a design, which is usually best from the viewpoint of design department only. This design is passed to the manufacturing department to develop the manufacturing processes necessary to produce the design. If any changes or corrections (which are usual) are required then the design is passed back to the designers for necessary modifications or corrections. Only when the manufacturing department is fully satisfied with the design, it passes the design to next department in the progression as shown in the Fig. 12.1. This traditional sequential approach to the product design which is also known by the name of over the wall engineering encourages a large number of modifications and corrections in the later stages, when it is more expensive and difficult to do so and increases the product development lead time that organizations cannot afford these days because of the prevailing market conditions as explained later in this section.

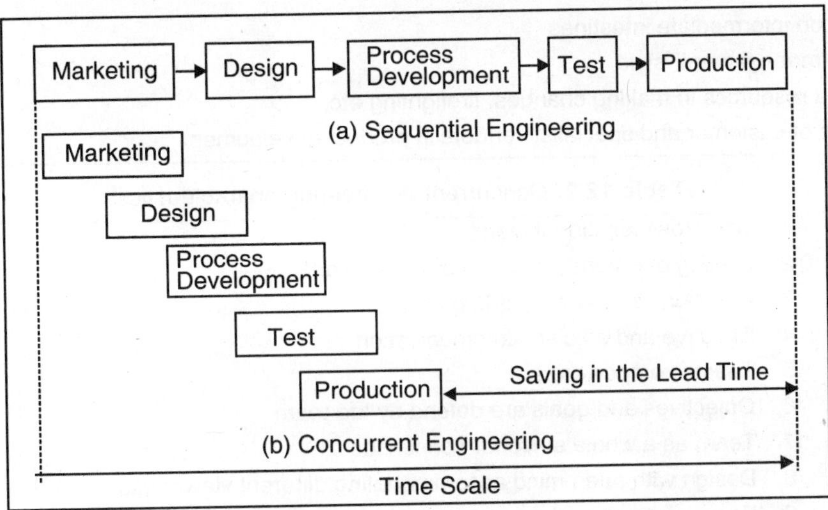

Fig. 12.1: Schematic figure showing sequential and concurrent process techniques

In CE, once the needs of the customers are identified, a multi-functional team is formed that will consider all the aspects of the product life cycle like manufacturability, assemblability, testability, reliability, serviceability, cost, user requirements, disposability etc. at the time of the design itself. There is a close integration between the various departments during the design leading to a design that requires hardly any correction or modification. This way the product can be brought to the market quickly at a low cost and with an improved quality.

The traditional approach emphasizes specialized excellence, often at the expense of overall success. Inward looking technical specialists can be overwhelming competent at a specialized conference but the results in a product development program usually lack integration and often focuses on sub optimal objectives. Individual brilliant results are of little value if they do not lead to the overall optimal solution. Take an analogy. The Indian cricket team has many star players but India keep on losing most of the games. Why? The answer to most of us is simple: lack of game plan and teamwork. The same metaphor is true in product development also. Even if an organization has brilliant individual specialists, the result will be poor unless the organization has an improved process (better game plan) and closer cooperation (team work) i.e., concurrent engineering. Main characteristics of sequential engineering and concurrent engineering are shown in Tables 12.1 and 12.2, respectively.

Table 12.1: Sequential engineering characteristics

1. Sequential activities increase development cycle time thereby increasing time to market and consequential losses
2. Weak commitment to previous decision
3. Design specifications (PDS) considered in isolation
4. Divergent interpretation of the specifications
5. Redundant and obsolescent information
6. Authoritative management
7. Ineffective communication
8. Focus on intermediate intestines
9. "Crisis management" style
10. Wasted resources in making changes, firefighting etc.
11. No role of customer and specialist vendors in product development

Table 12.2: Concurrent engineering characteristics

1. Use cross functional team
2. Lasting decisions, utilize all previous work
3. Decisions in a single trade off space
4. Effective and efficient communication
5. Team is empowered
6. Objectives and goals are defend by the team
7. Team as a whole strive for consensus
8. Design with open mind accommodating different viewpoints
9. Focus on important aspects
10. Utilize the relevant information as early as possible
11. Continuous follow up

12.3. NEED OF CONCURRENT ENGINEERING

There are many reasons that have influenced the need of CE, like: saving in lead-time, technology push, demanding customers and increased competition [3]. We will discuss them in detail now.

Lead-Time

One of the prime motivations for concurrent engineering approach is the desire to decrease or shorten the product development time or lead-time or in other words developing the products faster [4]. It is fully recognized that addressing all the problems of product life cycle in the design phase shortens the product lead-time. These days, for some products, average product lifetime is less than the average product development time. Therefore, for a company to survive and remain competitive, it has to decrease the product development time maintaining the high quality and low cost.

Technology Push

Newer technologies are being developed continuously. These newer technologies may bring down the costs, production time and even may improve the quality. But, such knowledge is often with the production engineer and not the design engineer. Therefore, to make the optimum use of newer manufacturing methods, close cooperation between design, production, and R&D departments is essential.

Demanding Customers

These days, customers are becoming increasingly more demanding. Low costs and good quality they take for granted and then they demand products, which are more closely targeted to their needs. Companies, hence, have to be not only effective but also innovative too to fulfill the demand of customized products.

Increased Competition

With the opening of the global markets, the competition has increased manifolds. To survive in the market an organization has to develop and introduce innovative products well ahead of the competitors. Any delay in the introduction of the product in time causes losses to the profits of the organization.

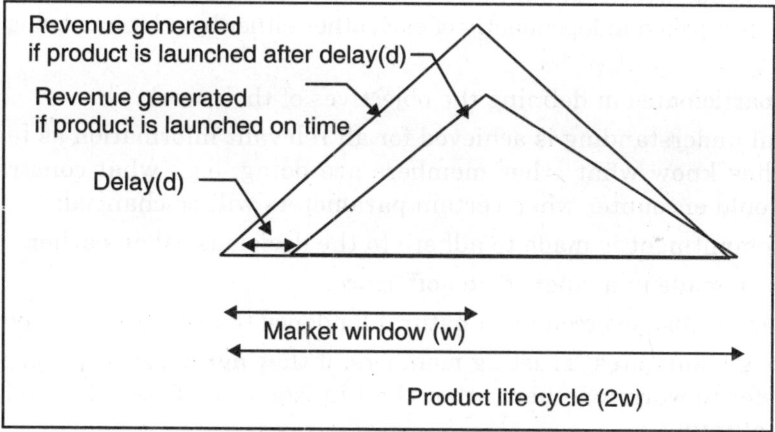

Fig. 12.2: Effect of delay on profits [5]

Carter and Baker [5], introduced a simple method to measure the impact of delay in launching a product:

Losses in profit = $d\,(3w - d\,)/2w^2$, where

d : delay in product launch

$2w$: product life cycle as shown in the Fig. 12.2.

Because of the short lifetime of the products, no time is available to the companies for re-engineering or modifications and the only alternative left is to look forward to the philosophy of 'right first time'. This can be possible by involving the people from manufacturing, finance, sales, services and specialist vendors etc. at the time of the product design itself. The impact is not limited to the design's initial release. There is significant impact of the time at which changes made during the development process. The cost of changes rapidly increases if the changes are made late as shown below for a major electronic product (Dataquest, Inc):

Design changes made during	cost ($)
Design	1000
Design testing	10000
Process planning	100000
Test production	1000000
Final production	10000000

12.4. BASIC PRINCIPLES IN CE

It is possible to propound certain basic principles for concurrent engineering to guide the design engineers and make their life a little easier [3, 6]. The same can be listed as follows:

- Start all tasks as early as possible.
- Utilize all relevant information as early as possible.
- Work structuring: Systematically structure the work or work environment so that each task can be performed independently of each other either by a human being or a machine or a computer.
- Everyone participates in defining the objectives of their work.
- Operational understanding is achieved for all relevant information as team will work better if they know what other members are doing, e.g., what constraints a team member would encounter when certain parameters will be changed.
- A strong commitment is made to adhere to the decisions taken earlier.
- Decisions are made in a single trade-off space.
- Decisions are robust, overcoming a natural tendency to resort to quick, novel decisions.
- Trust among teammates. Trusting members, if they agree to accept responsibility for a task, prefer to work together rather than in isolation. This will also lead to better teamwork affinity.
- The team strives for consensus
- Team should be empowered to make decisions in product development and should be given "ownership" of what they produce.
- The team uses a visible concurrent process.

- Constancy of purpose: This requires a change in thinking beyond the goals of one's individual department or team to the company's goals. Aiming toward constancy of purpose results in everyone contributing his/her best—working towards a common set of consistent goals.

These principles are simple and at first look seem to be unexceptionable, but in traditional process these principles have been frequently violated.

12.5. BENEFITS OF CONCURRENT ENGINEERING

Concurrent Engineering has been the focus of many industrial organizations for new product development, due to the ability of the cross-functional team to reduce the total time to design and manufacture or time to market. This reduction in the time to market is a major source of competitive advantage in the manufacturing environment we have today. Use of concurrent engineering should improve the performance of the organization in general but certain human and organizational characteristics may affect the degree to which improvement is felt in the organization. In literature, there are many companies showing benefits of concurrent engineering: Boeing reduced its product development cycle time by 40–60% AT&T reduced the process time for a new micro programmed digital switch by 46% Deere and Company reduced the product development time for construction equipments by 60% and cut manufacturing cost by 30–40%. McDonnel Douglas reduced rework costs 29%; reduced scrap cost 57%; and reduced non-conformance by 38%.

Typical benefits from concurrent engineering are [3, 7] are as follows:

- Reduction in time to market
- Improved product quality
- Increased customer satisfaction
- Reduced cost of production
- Reduction in engineering change requests
- Increased return on investment
- Increased level of group productivity
- Interdepartmental cooperation
- Increased employee satisfaction
- Reduction in development cycle time
- Increase in sales
- Reduction in life cycle costs
- Design rationalization
- Better communication
- Increased flexibility to accommodate changes
- Decreased occurrence of obsolescence
- Better use of scarce technical resources
- Other benefits: reduced lead time for creating bid proposals, reduced product development costs, parts reduction, lower inventories, fewer rework orders, less scrap.

Using CE, General Electric has cut its delivery time of customer made circuit breaker boxes from three weeks to three days. AT&T has reduced the product development time for telephones from two years to one. AT&T has also reported part counts down to one ninth of their previous levels and quality improvements up over one hundred times in surface defects. Motorola used to make electronic pagers in three weeks from order; it now produces them in two hours. The reported benefits from companies are large. In general products get to the market faster and at less cost. The quality is inherent in the product design rather than being an after thought. Employee morale and satisfaction is high as their voice is heard during the product development cycle. Projects run on time so suppliers and subcontractors too benefit and associate themselves with product success. Customers too are happy. So, who does not benefit from CE? None? Ponder.

12.6. CLOSURE

This chapter complements the previous studies by providing a clear definition of concurrent engineering and differentiating between the sequential engineering and concurrent engineering. This chapter points out the need of concurrent engineering in today's competitive environment and also gives a comprehensive list of basic principles and benefits of concurrent engineering.

REVIEW QUESTIONS

1. Define *sequential engineering* and *concurrent engineering.* What are the basic tenets of concurrent engineering that make it different from sequential engineering?
2. What is *lead time*?
3. How lead time can be saved by replacing sequential engineering with concurrent engineering? Explain with an example.
4. How the lead time can affect the profits of a company?
5. How the Internet technology can be used to achieve concurrent engineering?

REFERENCES

1. Soblenius G., (1992), "*Concurrent Engineering*", Annals of the CIRP, Vol. 41/2/645–655.
2. Loureiro G. and Leaney P. G. (2003),*"A Systems and Concurrent Engineering Framework for the Integrated Development of Space Products"*, Acta Astronautica, Vol. 53, pp. 945–961.
3. Sangwan K. S., (2001), *"Fundamentals of Concurrent Engineering"*, EDD Notes, BITS, Pilani.
4. Duffy V. G. and Salvendy G., (1999), *"Relating Company Performance to Staff Perceptions"*, the impact of concurrent engineering on time to market, Internal Journal of Production Research, Vol. 37(4), pp. 821–834.
5. Carter D. and Baker B., (1992), *"Concurrent Engineering: The Product Development Environment for the 1990's"*, Addison-Wesley Publishing Company, Massachussets.
6. Koufteros X., Vonderembse M., and Doll W., (2001), *"Concurrent Engineering and Its Consequences"*, Journal of Operations Management, Vol. 19, pp. 97–115.
7. Oliveto F. E. (2000), "*Concurrent Engineering: Evolution and Application*", IEEE Proc. of the National Aerospace and Electronic Conference, pp. 737–744.

GREEN PRODUCT DESIGN

13.1. INTRODUCTION

As environmental awareness increases, industries and businesses have responded by providing "greener" products and using "greener" processes. Many companies have found it advantageous to explore ways of moving beyond compliance, using pollution prevention strategies and environmental management systems, to improve their environmental performance. One such way is Life Cycle Assessment (LCA). This concept considers the entire life cycle of a product. The LCA is an objective process to evaluate the environmental burdens associated with a product, process, or activity. This can be done by identifying and quantifying energy and material uses and environmental releases, to assess the impact on the environment. This helps to evaluate and implement opportunities to effect environmental improvements. This chapter discusses LCA methodology applicable to industrial products and related activities.

Environmental issues have gained greater public and legal scrutiny in the past 25 years. The public has become more aware and interested in the consumption of goods and services and their impact on the natural resources and the quality of the environment. In the past 12 years, an increasing number of firms have been shifting their focus from remediation of pollutants to pollution prevention. This has included "green" design activities and replacing some materials and manufacturing processes with more ecofriendly ones. These efforts on behalf of the companies have drastically reduced the levels of industrial pollution. Industry is realizing that the impact of a product on the environment starts with the design and ends at the ultimate disposal of the product after its use. What considerations are made during the design phase of a product affects the whole life cycle. It is important to determine the environmental impacts of the manufacturing process, the product and its ability to be recycled.

While reducing the impact a company and its product make on the environment is a good idea, the cost associated with minimizing these impacts must be considered. Environmental compliance has been considered a cost of doing business, and staying out of jail. The increased use of waste minimization programmes has not only reduced the cost of compliance (which goes up over time) but has lowered the cost of producing products. In the number of cases the quality of product has improved. This in itself reduces costs and improves competitiveness of the products. In the global marketplace, the reduced impact of products on the environment has serious marketing impact. Sound environmental practices will result in designs that meet or exceed the requirement of the countries where they will be sold. It is reasonable therefore to consider the total impact on the environment of a product and plan means of reducing that impact during the planning and design phase of the product life cycle (1). This is possible by utilizing Life Cycle Assessment (LCA) methodology, which has been referred as the 'green technology' in this chapter.

This chapter discusses about the conceptual background of LCA, its framework and also the implementation of the various phase of the methodology.

13.2. LIFE CYCLE ASSESSMENT

LCA is a quantitative environmental performance tool, essentially based around mass and energy balances but applied to a complete economic system rather than a single process. In terms of system boundary definition, this represents an extension to the conventional system analysis, in which the system boundary is drawn around the process of interest only. Figure 13.1 illustrates the way in which LCA can complement conventional process analysis. While chemical or process engineering is normally concerned with the operations within the system 1, LCA considers the whole material and energy supply chains, so that the system of concern becomes the everything within system boundary 2. The material and energy flows that enter, exist in or leave the system include material and energy resources and emissions to air, water and land. These are often referred to as environmental burdens and they arise from activities encompassing extraction and refining of raw materials, transportation, production, use and waste disposal of a product or process. The potential effects of the burdens on the environment, i.e., environmental impacts, normally include Global Warming Potential (GWP), acidification, Ozone Depletion (OD), eutrophication etc.

The LCA methodology is still under development. At present, the methodology framework comprises four phases (2):

- **Goal and Scope Definition:** selecting the system boundaries (Fig. 13.1) to ensure that no relevant parts of the system are omitted.
- **Inventory Analysis:** performing mass and energy balances to quantify all material and energy inputs, wastes and emissions from the system i.e., the environmental burdens.
- **Impact Assessment:** aggregating the environmental burdens quantified in the inventory analysis into a limited set of recognized environmental impact categories, such as global warming, acidification, ozone depletion, etc.
- **Interpretation:** using the results to reduce the environmental impacts associated with the product or process.

Goal and Scope Definition

The most common LCA goal is to conduct an assessment of the environmental attributes of a specific product or process and to derive information from that assessment on how to improve the environmental performance. If this exercise is conducted early in the design phase, the goal may be to compare two or three alternative designs.

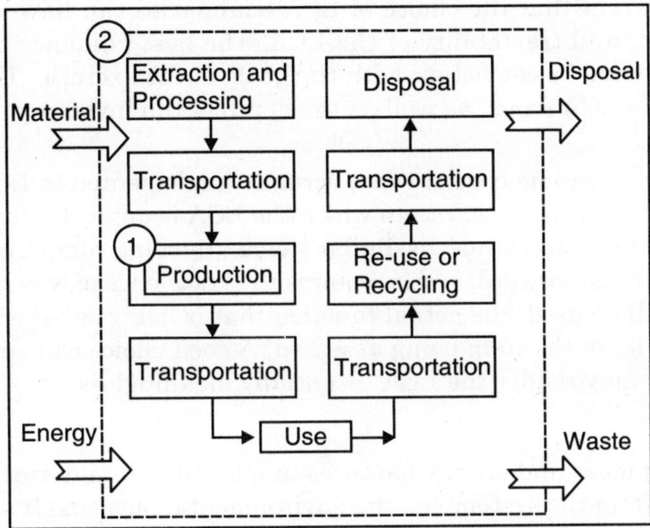

Fig. 13.1: Stages in the life cycle of a product (system boundary: 1, process analysis: 2)

If the design is formalized, the product is in the manufacture, or the process is in operation, the goal can probably be no more than to achieve modest changes in environmental attributes at minimal cost and minimal disruption of an existing operation [1]. Scope definition refers to what is included and excluded from each step in the process. The scope plays the major role in the effort, time, and expense that will be involved in the LCA. If confined to the subset of life stages or a limited selection of impacts, the LCA will be much more efficiently carried out. The results may be a report of limited usefulness, however, if potentially important factors have been omitted. Nonetheless, experienced assessors often opt for restricting the LCA scope in order to increase the tractability. Their concept is that significant benefits can often come from a limited LCA [1]. It is, in fact, not assured that more substantial benefits will emerge from more extensive LCA, and an LCA so broadly scoped that its successful completion is uncertain may turn out to yield no benefits at all. The outcome of the LCA study can have a positive impact on human health, the ecosystem, and natural resources, as shown in the Fig. 13.2.

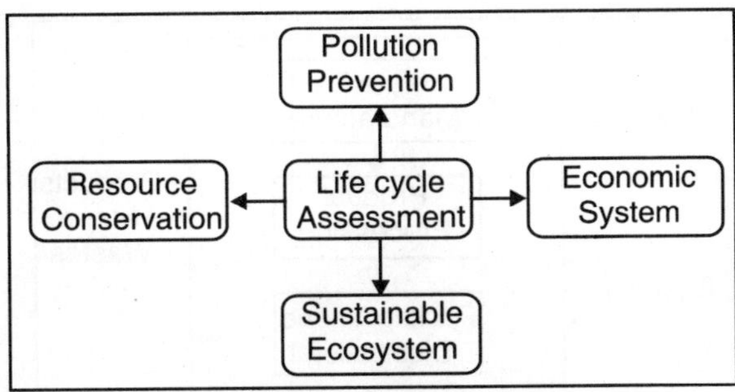

Fig. 13.2: Life cycle goals

It should be apparent that the choice of LCA boundaries can have enormous influence on the time scale, cost, and tractability of the LCA. The best guidance that can be given is that the boundaries should be consistent with the goals of the exercise. The goals of the LCA thus define much of the LCA scope, as well as the depth of the inventory and impact analysis activities.

Functional unit can become crucial when actually implemented in LCA, which is defined as amount of product, material, or service to which the LCA is applied. This definition becomes especially important when two rather dissimilar items are being compared. The choice of the functional unit is not straightforward, and it requires the knowledge of way in which the product, material, or service will be used, the actual function that is being satisfied, and perhaps even the anticipated longevity of the item being assessed. A good choice can support a useful LCA outcome; a poor choice may render the LCA essentially meaningless.

Inventory Analysis

It refers to performing mass and energy balances to quantify all material and energy inputs, wastes and emissions from the system i.e., the environmental burdens. It should be noted that inventory in LCA refers to the raw materials inputs and emission outputs rather than to a quantity of stored materials. After the scale of the LCA has been established by defining the scope, necessary data is collected. The process is begun by constructing, in cooperation with the design and manufacturing team, an inventory flow diagram. The aim is to list, at least qualitatively but preferably quantitatively, all inputs and outputs of the materials and energy throughout most or all life stages. The cataloging of inputs and outputs must have an objective. The specific objectives for conducting the inventory include, but are not limited to, the following [2]:

- Compare input and output associated with alternative products, materials, or processes.
- Determine points within the life cycle or given process where the greatest reduction in resource requirements and fugitive emissions might be achieved.
- Establish a comprehensive baseline.
- Assist in training personnel responsible for reducing the environmental impacts associated with product or processes.
- Assist in the substantiation of product claims about reducing their impact on the environment.
- Supply information for the evaluation of public policy affecting resource uses, recycling, or releases.

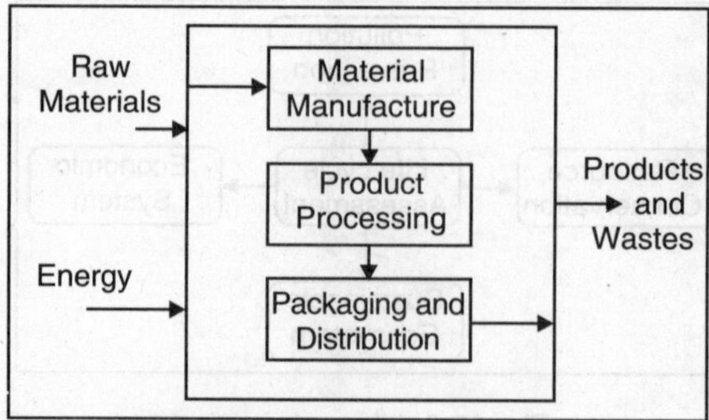

Fig. 13.3: Manufacturing stage

The obtaining of raw materials and the manufacturing stages can be and usually are complex. Because of this we can establish sub-boundary conditions for these activities. An example is shown in Fig. 13.3. In this figure the manufacturing boundary has included three fundamental steps i.e., material manufacture, product processing and packaging and distribution.

Figure 13.4 shows more specifically about preparation of inventory for sulphur dioxide gas treatment with limestone. Block diagram representations are helpful for making material and energy balances more clear and tangible. Input and output material streams are shown with the help of arrows. These block diagrams can be used for breaking the system into subsystems. The use of this type of modeling facilitates both simple and complex processes and systems [3]. The model makes use of data gathering and check-sheets for easier understanding and minimizes the missed data. A sample check-sheet is shown in Table 13.1. Because the user may want to determine where the opportunities are in the product stream to minimize environmental impacts. This method can provide a less costly means of analysis and data gathering than conventional methods.

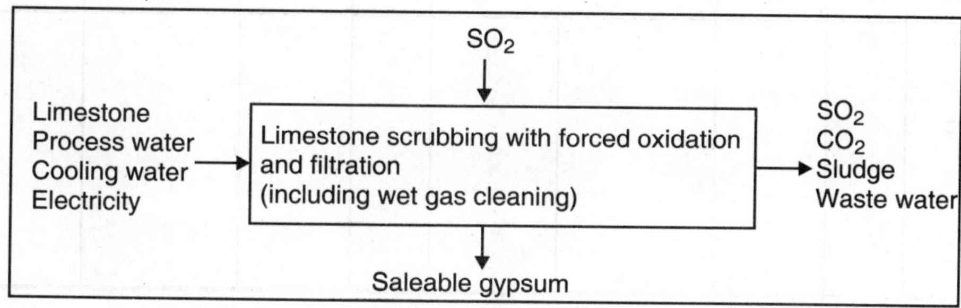

Figure 13.4: Raw materials, utilities, products and waste per tonne of SO_2 treated for limestone scrubbing options

Table 13.1: Sample check-sheet

Data Check-sheet		
Project Name: ______________________		
Investigator: ______________________		
Date: ______________________		
Reference no.: ______________________		
Project ID no.: ______________________		
Company: ______________________		
Input items	**Quantity**	**Units**
Raw materials		kgs/litres
Energy		kW/ Hp/ MJ
Other		
Output items	**Quantity**	**Units**
Materials		kgs/litres
Energy		kW/ Hp/ MJ
Other		

Scientific: Scientific based analysis is used to obtain and process the data.

Comprehensive: All significant materials and energy uses and waste releases are included. Any element missing because of data unavailability or cost and time are clearly documented.

Details: The inventory is conducted in a manner and to a level of detail commensurate with the purpose of study.

Table 13.2: Sample spreadsheet for data integration

Purpose: What is being studied? Why?									
Scope/ Limitation: What are the boundaries? Assumptions?									
	Inputs				Outputs				Other
Process no.	Air	Water	Materials	Energy	Air	Water	Materials	Energy	
Raw materials Packaging Products Wastes Energy									

Peer reviewed: If the study results are to be used in public manner, they should be peer reviewed using accepted protocol.

Useful: The users of the study can make appropriate decisions in areas covered by inventory. Any limitation regarding the utility of the study should be clearly noted. The meeting of these criteria should make the life cycle inventory report useful, technically supportable, unbiased, and applicable.

Impact Assessment

Table 13.3: Environmental Life Cycle Assessment categories for impacts [4].

Enhancement of the greenhouse effect	Depletion of ozone layer
Acidification Potential	Nutrification
Human toxicity	Ecotoxicity
Photochemical Oxidant Formation	Depletion of abiotic resources
Depletion of biotic resources	Waste heat
Noise	Damage of ecosystem and landscapes
Corrosion	Visibility
Odour	Victims

Quantitative information on materials and energy flows is acquired in the inventory stage of the LCA. Now the environmental influences on the activities revealed by the LCA inventory analysis on the specific environmental properties must be accurately assessed, and the relative changes in the affected environmental properties must be given some sort of priority ranking. Together, these steps constitute LCA's impact assessment. Assessing environmental influences is complicated procedure, but it can be, in principle at least, be performed by employing stressors, which are items identified in the inventory analysis that have the potential to produce changes in the environmental properties.

Conceptually, however, the needed stressor relationships can be imagined to established and available for use. By combining LCA inventory results with stressor relationships, a manufacturing process might be found, for example, to have a minimal impact on local water quality, a modest impact on regional smog, and a substantial impact on stratospheric ozone depletion.

For example, lead in motor vehicle exhaust is a stressor because lead is the systematic poison for animals and humans. The relationship among stressors and the environment are developed by the environmental science community, and are not always available with the degree of detail and precision required. Thus, the first stage of LCA impact assessment is the identification of stressors and environmental concerns to which they are related. The result of this stage is an estimation of burdens, that is, of potential impacts. Table 13.3 shows the environmental impact categories. The impact assessment can be structured according to the following steps:

Classification: It begins with the raw data from the inventory analysis on flows of materials and energy. Given that data, the classification steps consists of identifying environmental concerns suggested by the inventory analysis flows. Some of the pollutants emitted during product manufacturing, use or reuse are given here.

These pollutants are classified as follows:

Carbon Dioxide (CO_2), Nitrous Oxide (N_2O), Methane (CH_4), Volatile Organic Compounds (VOCs), Chloroflurocarbons (CFCs), Chlorinated Hydrocarbons (HC), Sulphur Dioxide (SO_2), Nitrogen Oxides (NO_x), Hydrochloric Acid (HCl),Ammonia (NH_3), Hyrogen Fluoride (HF), Phosphate ion (PO_4^{3-}), Phosphorous (P), Alcohols, Ketones, Ethers, Olefins, Acetylenes, Aromatic and Aldehydes.

Characterization: It is process of combining different stressor-impact relationships into a common framework. For example, on a time scale of 100 years the contribution of 1 kg methane to global warming is 42 times as high as the emission of 1 kg CO_2. This means that if the characterization factor of CO_2 is 1; the characterization factor of CH_4 is 42. Thus, the impact category indicator result for global warming can be calculated by multiplying the life cycle inventory result with the characterization factor. The above-mentioned pollutants are characterized in Table 13.5.

Localization: It is the operation of comparing environmental impacts occurring in different regions with different characteristics.

Table 13.4: Classification of the pollutants [5]

Impact Categories	Pollutants
Global warming	Carbon Dioxide (CO_2), Nitrogen Oxide (N_2O), Methane (CH_4), Volatile Organic Compounds (VOCs)
Ozone depletion	CFCs, Chlorinated HCs.
Acidification potential	Sulphur Dioxide (SO_2), Nitrogen Oxides (NO_x), Hydrochloric Acid, Ammonia (NH_3), Hyrogen Fluoride (HF)
Eutrophication	NO_x, NH_4^+, N, Phosphate ion (PO_4^{3-}), Phosphorous (P).
Photochemical oxidant potential	VOCs including alkanes, halogented HCs, alcohols, ketones, ethers, olefins, acetylenes, aromatic and aldehydes.

Table 13.5: Characterization of pollutants [5]

Classification	Pollutants	Characterization
Global warming	Carbon Dioxide (CO_2), Nitrogen Oxide (N_2O), Methane (CH_4), Volatile Organic Compounds (VOCs)	Global warming is expressed relative to CO_2
Ozone depletion	CFCs, Chlorinated HCs.	Ozone depletion is expressed relative to CFC-11.
Acidification potential	Sulphur Dioxide (SO_2), Nitrogen Oxides (NO_x), Hydrochloric Acid, Ammonia (NH_3), Hyrogen Fluoride (HF)	Acidification potential is expressed relative to SO_2
Eutrophication	NO_x, NH_4^+,N, Phosphate ion (PO_4^{3-}), Phosphorous (P).	Eutrophication potential is expressed relative to Phosphate ion.
Photochemical oxidant potential	VOCs including alkanes, halogented HCs, alcohols, ketones, ethers, olefins, acetylenes, aromatic and aldehydes.	Photochemical oxidant potential is expressed relative to ethylene.

For example, the process of localization attempts to compare the emission of 1 kg of moderately toxic material into a pristine ecosystem with the impact of the same emission into a highly polluted ecosystem.

Valuation: It is process of assigning weighting factors to the different impact categories based on their perceived relative importance as set by social consensus. For example, as assessor or some International Organization might choose to regard ozone depletion impacts as twice as importance as loss of visibility, and apply weighting factors to the normalized impacts accordingly.

The most contentious aspect of LCA methodology is the evaluation phase arising from the selection of the relative weightings for various environmental effects. One approach to a single index is the evaluation phase has been described by Annema [2]. In this method, the score for each category from the classification step is divided by the national score for that category.

The resulting fractions are then added for all classification categories. Thus,

Environmental Index =

$$\sum\left(\frac{\text{Magnitude of category } i \text{ for product under study}}{\text{Magnitude of category } i \text{ from all man—made sources within specified region}}\right)$$

The national scores for each classification category are calculated using pertinent national emission consumption data. The approach involves some inherent weighting classification categories derived from the contribution of the product under study to national effects for each category.

An important consideration that plays a role in the prioritization of impacts is that a single impact generally has many stressors, and a single stressor has many impacts. As a consequence, the industry-environment interaction generally embodies a summation of impacts of different magnitudes and different spatial and temporal scales.

The result is that a choice between two design options may involve tradeoffs such as choosing between having a modest influence on a single impact of high importance or smaller influences on several impacts of lower importance. One approach of linking stressors and impacts is a series of matrix displays, the axes of matrices being Technological Activities (specific activities or industries that generate stressors such that gases or particles) and Critical Properties (specific impacts). A matrix of this type for atmospheric concerns is shown in Table 13.6. To construct this figure, a critical atmospheric concern such as "precipitation acidification" and its direct and indirect chemical causes are linked with sources responsible for initiating those interactions.

The result is a matrix that shows the impact of each potential source of atmospheric change on each critical atmospheric component. A rough qualitative assessment of significance is included. The diagram also includes estimates of reliability, an important component of an assessment effort. The matrix is constructed with separate rows for industrial operations and for the major energy generating processes, and can obviously expanded in much greater details than is shown here. The simplest impact assessments of Table 13.6 involve only a single cell of the matrix.

A typical example is the study of the impacts of a single source, such as a new coal-fired power station, on a single critical atmospheric component, such as acidification potential. For the purpose of policy and management, it is desirable to assess the impacts of a single source on several critical environmental properties. The coal combustion study noted above would fall into this category if the impacts were assessed not only on acidification, but also on photochemical oxidant production, material corrosion, visibility degradation, heavy metal emissions, etc. If desired, the columns could be summed in some way (such as assigning numbers to potential importance ratings) to give the net impact of the ensemble of sources on each critical property. Similarly, the rows could be summed in some way to give the net effect of each source on the ensemble of properties.

Interpretation

This step essentially combines the finding from the inventory analysis and impact assessment to develop conclusions and recommendations. It uses the goals and scope of LCA, as the basis for the development of conclusions and recommendations. It results in a report containing methodologies, data sources, calculations, and findings of LCA in a format consistent with the goals for the study.

Table 13.6: An initial ensemble assessment of impacts on the global atmosphere [3]

Critical properties / Technological activities	U.V. Absorption	Greenhouse effect	Regional oxidants	Acidification potential	Visibility	Corrosion	Oxidative efficiency
Crop production							
Domestic animals							
Petroleum combustion							
Coal combustion							
Natural gas combustion							
Industry							

Assessment reliability: low, moderate, high

Potential Importance			
Some		Major	
Moderate		Controlling	

13.3. CLOSURE

LCA is a quantitative environmental performance tool, essentially based around mass and energy balances but applied to a complete economic system rather than a single process. LCA has two main objectives. The first is to quantify and evaluate the environmental performance of the process or product from cradle to grave. Another objective is to provide a basis for assessing improvements in the environmental performance of the system. LCA includes "green" design activities and replacing some materials and manufacturing processes with more environmentally friendly ones.

REFERENCES

1. Caimborne D., "*Environmental Life Cycle Analysis*", Lewis Publishers, New York.
2. ISO 14040 (1997), "*Environmental Management-Life Cycle Assessment*", Part 1: Principles and Framework. ISO.
3. Graedel T., *"Streamlined Life Cycle Assessment"*, Prentice Hall NJ.
4. Heijungs, R. (final editor), 1992, *Environmental Life Cycle Assessment of Products*", Guide-October 1992 (Centre of Environmental Science, Leiden, The Neatherlands).
5. Azapagic A., R. Clift, "*The Applications of Life Cycle Assessment to Process Optimization*", Computers and Chemical Engineering, 23 (1999), 1509–1526.

Assembly and Disassembly

14.1. INTRODUCTION

Industrial companies change the paradigms of business operations from optimization of manufacturing processes to optimization of products life cycles in order to activate the value of products, taking into account the potentials of product services in all phases of each product's life. From design to the end of their life capital intensive products, like manufacturing or assembly systems, are linked to a manufacturer network by global communication systems. This network allows special services even in the phases of usage and recycling. For this new paradigm it is necessary to develop strategies, methods and technologies to manage the business processes and the information and knowledge required in all phases of a product's life and to industrialize the processes of design, assembly, usage, service and re manufacturing by disassembly and recycling. It is the objective of this chapter to define the processes for management of life cycle with a focus on assembly, service and disassembly of capital intensive products.

In the processes of design and engineering, the functionalities of products are defined, by assembly, maintenance and disassembly, the real configuration, functionality and characteristic properties for usage of products are finished or changed. In the usage phase special know-how on the design and characteristic properties, like specific process knowledge to optimize utilization and performance, is required. The increasing technical complexity promotes product-near services and assistance of manufacturers. There are new business models for selling only the functionality of capital intensive products, rather than the products themselves.

Behind these tendencies there is a new paradigm: for adding value and maximum utilization, products are linked in the Manufacturers Network from beginning to end. For this paradigm manufacturers need life cycle management systems, tools and technologies. Assembly and disassembly play the key role in the, Life Cycle Management (LCM).

14.2. VIEW OF PARTNERS IN SERVICE

We will look at the various stakeholders' point of view in the following sections.

Manufacturer's View

In general, the life cycle of products can be defined as the phase of design and engineering, manufacturing, assembly, usage, service, disassembly and recycling. Deeper dimensions are defined in and depend on the specific structure of products and production. The main objective is to fulfill the requirements of markets and customers for efficient utilization of manufacturing resources. The new view is adding value in the usage and recycling-phases by customer near services including maintenance and disassembly for reconfiguration, reuse or recycling.

This view to the usage and recycling phases makes it necessary to take more than ever, before into account aspects of life cycle design and engineering or the capability of systems for assembly, disassembly and diagnostics in all phases and especially in the phase of usage. They offer product oriented profitable services along all operations by supporting the diagnostics of the actual properties, also partial disassembly and assembly for reconfiguration, upgrading and final disassembly for recycling.

Customer's View

In general, customers are interested in achieving a high utilization in the usage phase and lowest cost even if the processes have to be changed. Flexibility of manufacturing systems is required with minimal setup time and cost and a guaranteed process performance. High efficiency of the usage of complex technical products depends on specific skills and know-how on the details of machines, mechatronic components, software and the optimization of processes. This costly tendency can be overcome by specific skilled services and assistance or support by manufacturers. Users like to buy specialized services to reduce the fixed costs of products and the cost of inspection, maintenance and reconfiguration or upgrading.

The economic efficiency of capital-intensive products in industrial manufacturing depends on the profile of products, technical requirements and capacities. Requirements are changing continuously. Manufacturing systems therefore have to be adapted. So far industries are lacking economic evaluation methods to control the life cycle and to find the optimal point in time for the end of a product's life.

Life Time

A different point of view is needed today to evaluate a product's life, marking the length of product life is the most important factor. The product life span compared to the technical progress has to be suitably and correctly interpreted. To this end the resources have to be used more effectively and the processes need to be optimized.

Figure 14.1 presents the life cycles of typical products. The four examples show the differences in the duration of life time of single life cycle stages. A manufacturing system can be operated up to twenty years if the technical possibilities for the compensation of attrition are applied. Accordingly, one fact becomes clear: the duration of life cycle stages, in particular the usage and the end stages of life, is determined by the product structure and particularly by the management of the life cycle activities.

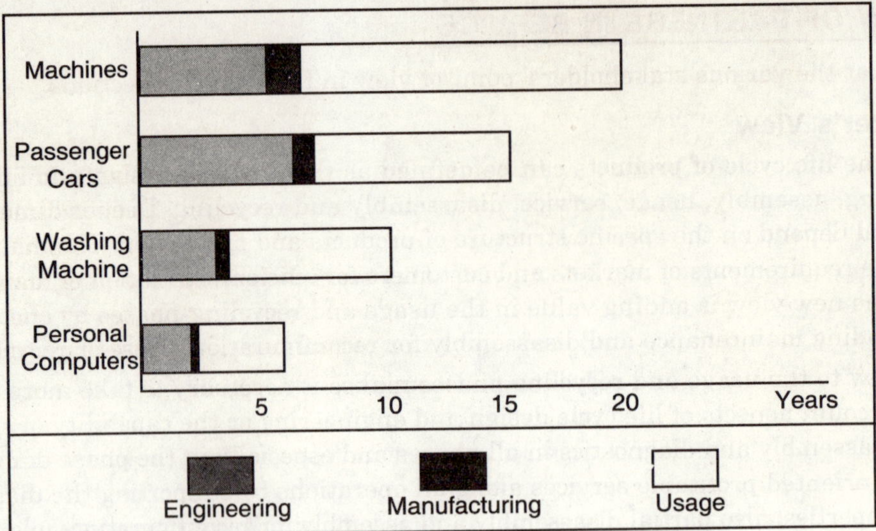

Fig. 14.1: Phases of life cycle—Typical products

The increasing quality of products allowed substantial higher utilization of technical products. Even manufacturing systems allow much higher life utilization than years before. Investigations on the life time utilization of machining centers had the result of nearly doubling hours of effective use. This may be an indicator for economic potentials of modern products.

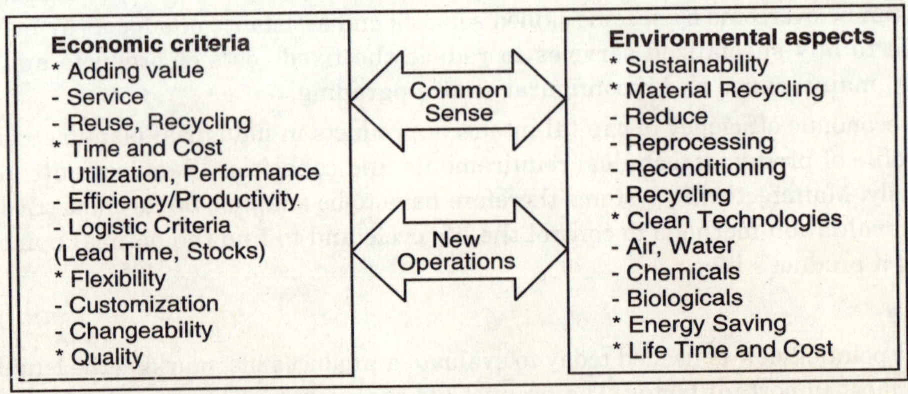

Fig. 14.2: Objectives of life cycle management

Common Objectives of All Stakeholders

The new paradigm of activating the product's cost-benefit is oriented on both economic aspects and fulfilling requirements for the environmental behavior of technical products applying ecological criteria. It assumes that the concentration on core competence and specialization offers new potentials to add value or to reduce cost of usage by industrialization of service and disassembly. The common understanding between manufacturers and users is a prerequisite to activate potentials and to get the best and maximum benefits of each technical product, during its life cycle and to fulfill the economic and environmental objectives. (See Fig. 14.2)

Common understanding and active optimization require technical solutions for linking products over their entire life cycle and at any phase to the information networks of manufacturers and users. This can be achieved by implementing technical products in global IT-Networks and electronic services. The vision following the new paradigm is the permanent link of products to the networks of manufacturers. Communication is the platform for any product oriented service with the aim of maximum benefit in the life time. Today nearly all technical products include electronic information or control systems, which give them a certain technical intelligence. Future technical products, e.g. machines and systems, include multifunctional control systems with standardized interfaces to connect them with IT-networks and links to peripherals.

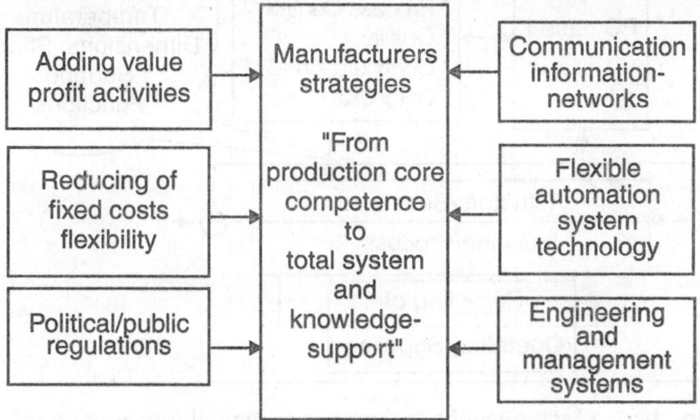

Fig. 14.3: Driving forces for life cycle management

14.3. TECHNOLOGIES FOR CONNECTIVITY

The driving technologies for the linking of products in the networks of manufacturers are:

- Global Communication Systems
- Internet/Intranet
- Multimedia
- Basic Software for Managing Data
- Virtual and Distributed Engineering Systems

Most of these technologies are globally available and offer a high innovation potential in the future. Outside of the interests of manufacturers to increase value in the after sales business and interests of users to reduce capital costs, the main technical driving forces are the standards of communication and information technologies. A strong trend in diffusion of electronics into machines and into management and engineering systems is a fact of today. Figure 14.3 shows general driving factors for the development of LCM.

Technical Intelligence of Products for LCM

High-quality technical products like manufacturing systems or cells possess internal technical intelligence which enables them to adapt to various applications and surrounding conditions.

They have been optimized in detail and are able to be flexibly configured because of their modular construction electronic elements are integrated into electrical and mechanical components in a decentralized way more or less completely as so-called "embedded systems". The synchronization of mechanical systems is replaced by digital systems with decentralized actuators and sensors. In this process, the technical functions are being realized in increasingly smaller dimensions and spaces. Cooperative mechanisms between various sub-systems are based on integrated information systems.

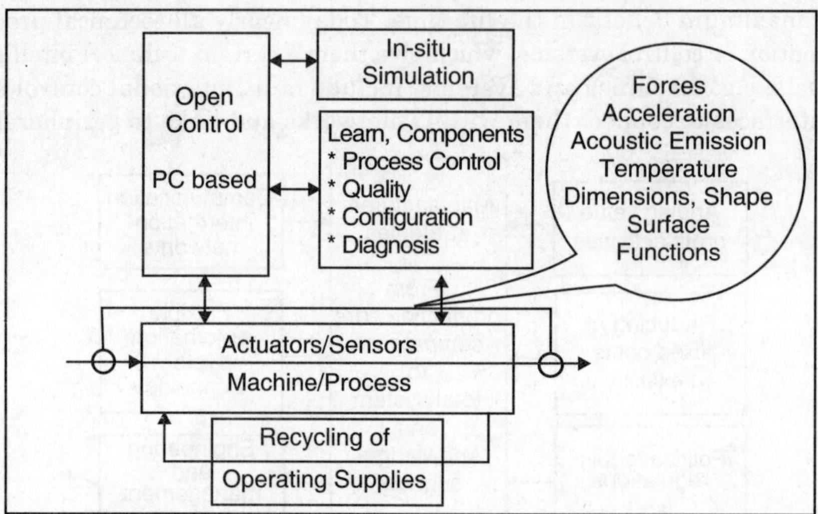

Fig. 14.4: Machines with integrated technical intelligence [1]

Figure 14.4 illustrates the functions of new generations of machines like, e.g., manufacturing systems with internal information systems and technical intelligence. The integration of technical intelligence can be used as a platform for the life cycle management and remote services [1].

Modularisation

Undoubtedly, the structuring and modularisation of products are of exceptional importance for assembly and all disassembly operations. In the automotive industry, for example, platform concepts have caught on, which have proved worthwhile for a wide range of variants, with the same technical basis and for product groups with specific standards. They equally permit customizing or the repeated utilization of single components in different performance classes. Modularization and standardization is a clear strategy to reduce the variety of parts and components in product families and to manage the customization of products, which plays important roll for disassembly and recycling.

Internet for Life Cycle Tracing

Today the Internet offers a great variety of usable tools for the life cycle management. There are world wide standards for communication and exchange of data and information. The Internet uses tools, engines and robots (behind the interfaces) to propose and search information and knowledge. There are new standards for b2b (business to business) and b2c (business to consumer). Leading companies in the automotive and machine branches use the Internet as a platform for logistics and the administration of processes between OEM and suppliers. Internet technologies offer a broad spectrum of tools for managing the link to manufacturers and users wherever they are located. Examples are the management of the logistics of part supply to assembly, the maintenance or the technical support in the usage phase.

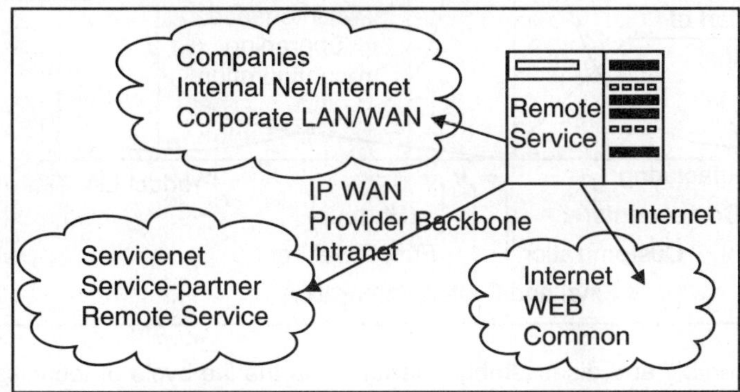

Fig. 14.5: Architecture of internet and intranet for LCM

The basic architecture of the Internet and of intranets in companies' information technology is illustrated in Fig. 14.5. Security techniques (firewall, en- or decryption) have to be adapted to the needs of manufacturers and to operate in closed areas with service partners. Both the architecture of the control systems within a product and the availability of internet at each work place are important for assembly and maintenance. The diagnosis of functions and the monitoring of the usage can be integrated by these systems and linked to the Internet.

14.4. STRATEGIES FOR ASSEMBLY AND DISASSEMBLY

Assembly and disassembly are the core processes in the life cycle of technical products. The usable functions of products are finalized by assembly and specific technical characteristics. The result is defined as the "as build" status in which products will be delivered to users. But technical products have a ramp up phase, where specific parameters have to be set and defects to be eliminated. Even in these phases assembly and disassembly operations follow the objective to reach the estimated performance.

Users try to protect capability and performance. After finalizing the products, usage causes wear and defects. By maintenance, which includes diagnostics, disassembly, assembly and tests, wear and defects are eliminated. Service is more and more supported by tele-service and a backup management of spare parts and spare components, based on the actual configuration of the product. These processes change the state of products permanently. Disassembly operations accumulate at the technical and economic end of life of a product. But remanufacturing and reuse demand cycles of operations: inspection, disassembly, compensation of wear, assembly, test and ramp-up. Flexible and integrated assembly and disassembly systems are practical solutions for the industrialisation of the remanufacturing phases.

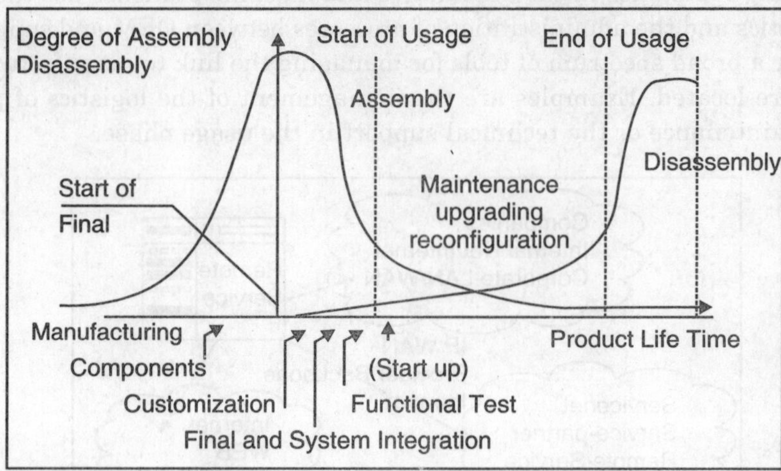

Fig. 14.6: Assembly and disassembly operations in the life cycle of technical products

Design for Life Cycle

Manufacturers are becoming responsible for the economic and environmental performance of their products throughout their life cycle. Life cycle engineering aims to integrate issues and parameters into product development throughout the life cycle of a product, taking into account the requirements of long-time usage and recycling. Objectives of the life cycle design and assessment are the manufacturability, cost and mainly the reduction of environmental losses. Many basic investigations and methods contribute to the design for life cycle. Manufacturers must reduce the use of materials and the impact on the environment, while preserving or improving the functionality, of the products. The latter must include the possibilities of life time management including the technical support or service during usage. Additionally, constructions have to be chosen which focus on operations of the life cycle partners—manufacturer, user and recycler.

The supervision of the life cycle includes a great variety of opportunities to integrate experiences with products in the fields of usage. Today, however, manufacturers are not well prepared to manage extremely high data volumes of technical (digital) data in the information supply (forward) or feedback.

Assembly and Ramp-Up

Assembly is defined by the operations to finalize products for optimal usage. Basic technologies are joining and handling operations. They have to be done in sequences, which are mainly influenced by the structure of products and the configuration of assembly systems. In the area of highly developed products like machines and systems assembly includes functional tests. After delivery these types of products often become integrated into the user's internal infrastructure, and logistic systems like transport and storage.

Additionally, they will be integrated into the user's internal information system. People have to be trained on the specific object and to learn the instructions. Many companies lose time and utilization in the ramp-up phases. In these phases the highest knowledge and skills are required by users and manufacturers. Thus, there is a shift from the usability of machines and systems as the core of actions towards the stabilization of processes measured by factors of process capability. In ramp-up phases a strong cooperation between user and manufacturer is needed. In practice, one can plan the operations and operation sequences to the point of delivery and shipping to the customer. The phase of ramp-up is rarely optimized.

The Roll of Maintenance for Life cycle Management

Increasing complexity of products increases the demand for highly skilled maintenance staff and short reaction time of service. About 5 to 6% of the products price is spent for maintenance and service yearly. The strategies for maintenance for life cycle point of view are carried out as shown in Fig. 14.7. The same figure shows recovery of resources in the view of sustainable development.

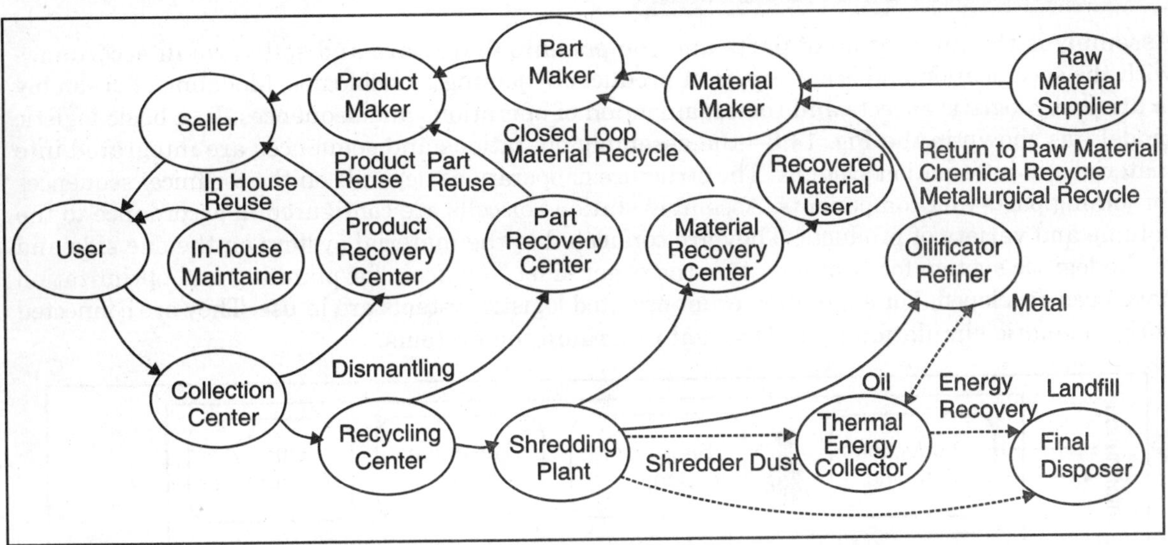

Fig. 14.7: Comet circle (Comet circle is a trademark of Richo Co. Ltd.)

Disassembly and Recycling

Disassembly operations are included in all phases of a product's the life cycle. They first appear during the product's final assembly to support assembly processes by tools or to repair defects caused by assembly. Disassembly operations are also part of maintenance and service and play a leading role in the end of a product's life. The operations and sequences depend on the complexity of products and the value of components and parts. Disassembly operations started after the diagnosis of the specific product situation and configuration. This leads to new concepts of in-situ planning of operations. It seems to be more advantageous to develop extremely flexible systems with integration of inspection, partly disassembly/assembly and tests.

Taking into account the usage phase in which products have to be disassembled and assembled partly at specific states for maintenance purposes, additional aspects like just-in-time delivery of replacement parts or minimum breakdown include even logistic objectives.

Logistics Management for Material

Products are manufactured in distributed chains with autonomous partners. In order to manage these industrial chains new solutions are being introduced to practice now. There are logistic networks for supply chain management based on methodologies for material disposition like SCM (supply chain management or Internet solutions for b2b and b2c).

The various chains can be divided into these phases: supply of material for assembly, distribution of finalised products, redistribution after usage and redistribution after recycling. All phases demand the availability of data and information as well as different methods depending on the requirements of the actual conditions and the local place of actors within the network. It is clear that basic data models are the backbone of future developments and the Internet seems to be able to integrate material and information flows.

14.5. BASIC MODEL OF ASSEMBLY

Assembly is the integration of parts and components (hardware and software) in accordance with the specification and architecture of products by joining technologies. Modeling of assembly is driven by logistic aspects and the organization of operations and sequences. In a basic logistic model—as shown in the Fig. 14.8—the assembly operations and sequences are integrated into material and information systems. The structure of operations depends on the technical sequences for joining parts and components. Assembly stations or cells are configured in accordance to the volume and variety of products. They are connected to the material system on the one side and to the logistic system for equipment on the other side. New tools for planning and optimization have been developed. For simulation event-oriented logistic systems are in use. They are connected with cinematic simulation and 3D visual programming systems.

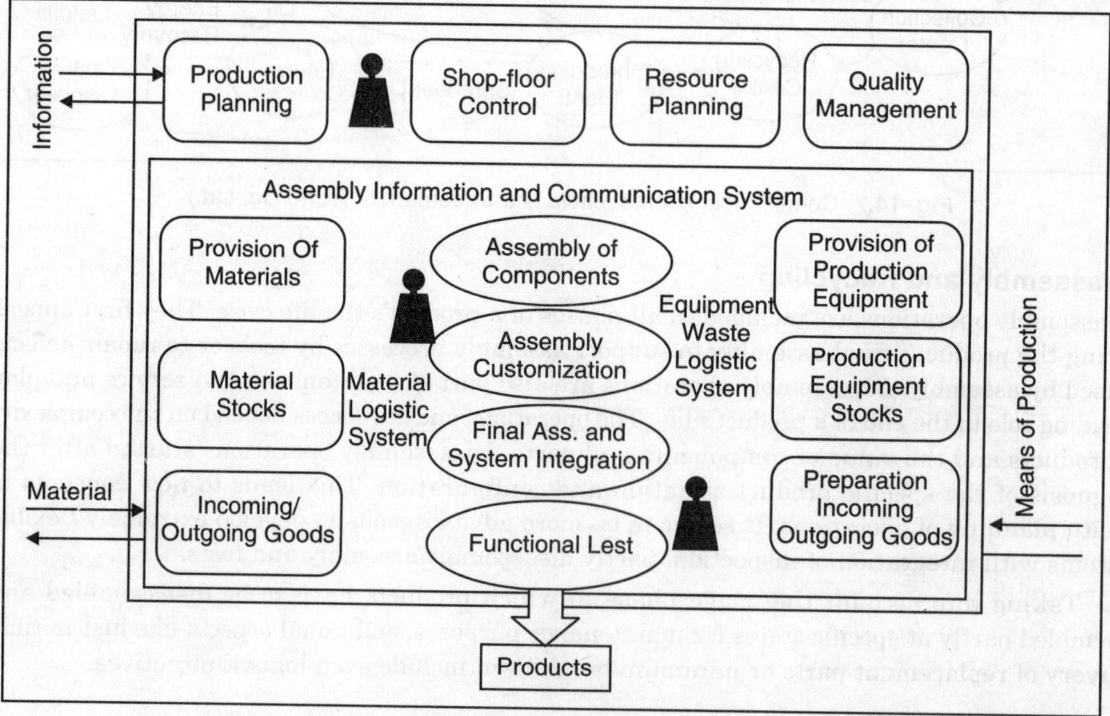

Fig. 14.8: Basic model of the assembly

14.6. E-SERVICES AND PRODUCT SUPPORT

Figure 14.9 shows the development of e-services' for technical products. In the past, the administration of operations has been supported mainly by conventional computer systems. In many applications manufacturing machines are connected by modems for upgrading control software or diagnosis of the control systems.

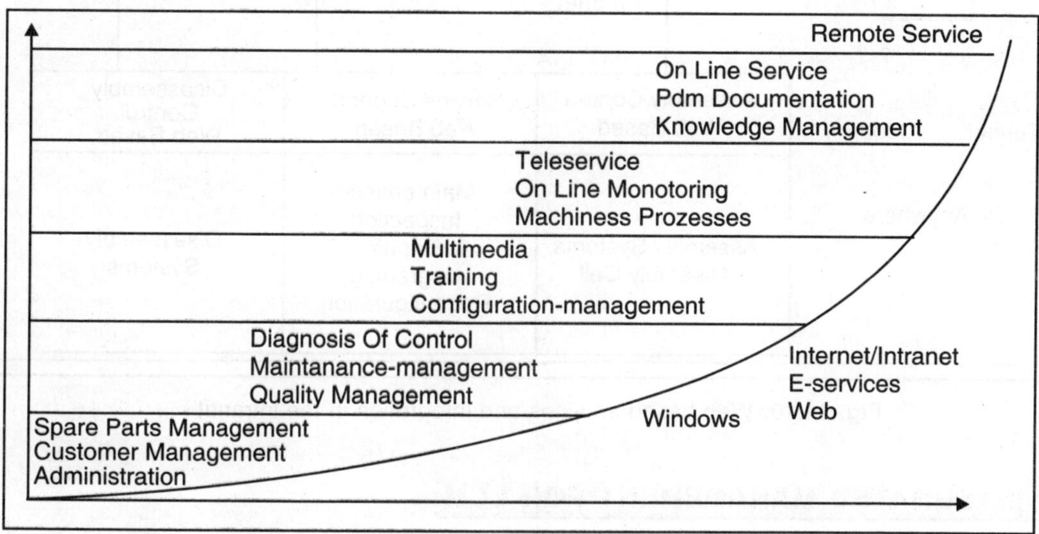

Fig. 14.9: Development of e-services for manufacturing

Today's tele-service is even operating with direct connections between machines and service stations on the side of manufacturers. This technology allows the support of maintenance staff in organizing the spare parts delivery or special services for direct help for the users. The next steps in the development take into account the connection of machines and systems to the Internet and the application of all the different tools for communication between machines, technicians and home service stations. This way it will be possible to assist the users with monitoring the machines and processes. Tele-service and e-services promote advantages like:

- Quick and efficient response
- Spare parts supply
- Reduction of breakdown time
- Monitoring of machines and processes
- Knowledge support

Direct machine services include the elimination of mainly technical deviations of the machine and the compensation of defects detected by diagnostic systems. Others, such as work-piece or part services, address the programs and operations.

Of course new models of full service around the clock are necessary. In the meantime, in areas of expensive machines some companies, like printing machine manufacturers; offer world-wide services, which are based on direct communication between the machines in usage and the service groups throughout the world.

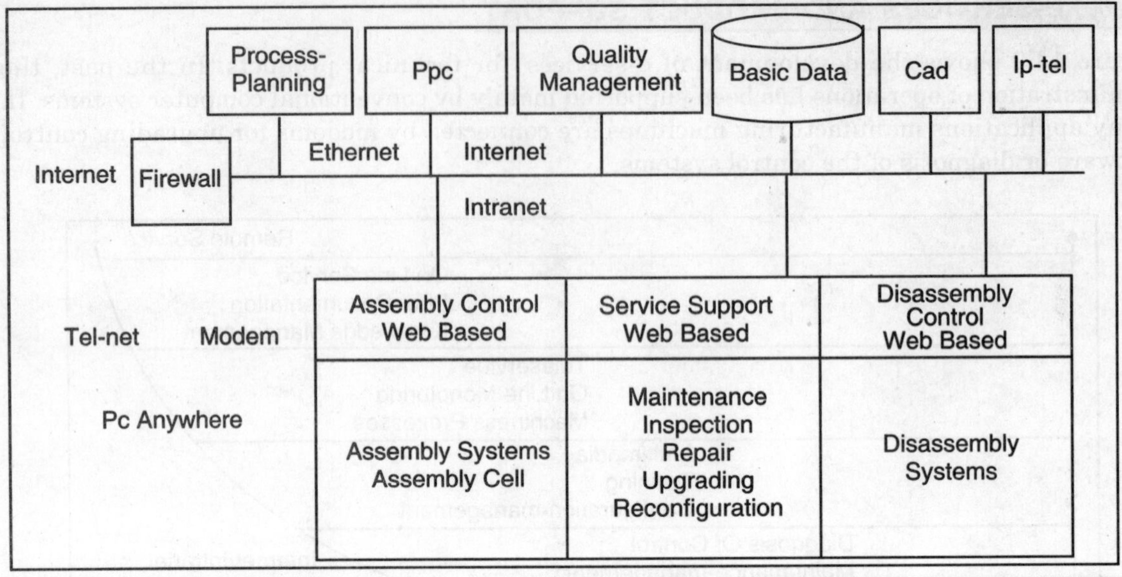

Fig. 14.10: Web based services and integration in the intranet

14.7. INTEGRATED MANAGEMENT FOR LCM

The effectiveness of a technical product is closely linked to the technical progress and also to the use of specific know-how for operation (choice of tools, adjustments, etc.). To enable longevity, the possibility for permanent upgrading has to be ensured including product operation at the limit of robust processes and system control. The crucial factors are mainly determined by the possibilities to exchange information and apply knowledge.

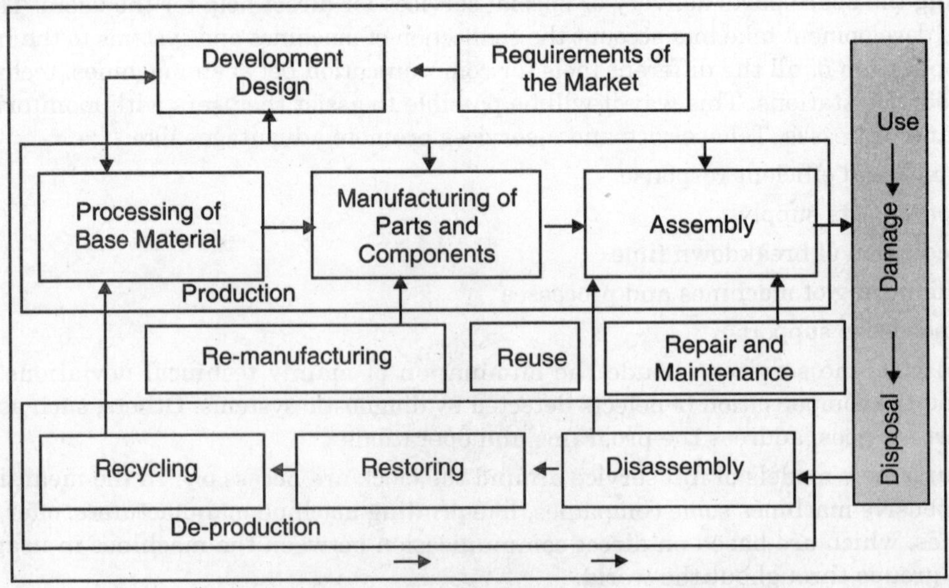

Fig. 14.11: Cycles of material and information flow

Modeling the Product Life Cycle

There have been defined different Models for the Product Life Cycle. In the industrialized view of technical products the main processes are: engineering, production with regard to the market requirements. Usage and the phases of reproduction, repair and maintenance, reuse and re-manufacturing are to develop.

In the last few years the environmental focus in industry and legislation has shifted from production processes to products and their life cycles. The demands regarding efficiency, quality and environmentally sound manufacturing, usage and disposal of modern products are further increasing. The responsibility of the producer to support the usage of robust processes and to observe environmental regulations and restrictions during the life cycle stages of manufacturing, usage and disposal are intensifying.

The product life cycle offers a number of opportunities to reduce costs or even to increase utilization and effectiveness of products during their usage phase. The producer participates in activities during both usage and support phases. Due to the latter, the accompanying follow-up costs caused by usage, service and disposal or recycling activities, etc. are increasing. At the same time, new and potentially successful economic business areas manufacturer can exploit.

The offers of companies and especially in the area of machines are driven by so called "user models" in which the manufacturer offers not the machine or the technical product but only the functional usage based on his technological know how. The physical products in this model owned by the manufacturer. Many different solutions like Tele-service or service consultancy which offer benefits and added value for manufacturer and user. Of course many questions in the adding value are open. But manufacturers has to learn to calculate cost and benefit. The aim must be an economic product life cycle with an effective use of resources regarding material, knowledge and capacity for work. This optimization inevitably follows the rules of a sustainable industrial production and is based on the life cycle of products.

Product Life Cycle Management

"Life Cycle Management" (LCM) considers the product life cycle in a holistic way with the aim of achieving the product's maximum performance. Life Cycle Engineering (LCE), Technical Support, Life Cycle Assessment (LCA), Life Cycle Costing (LCC), and last but not least Product Data Management (PDM) are means to accomplish the protection of resources and maximize the effectiveness of usage. The purpose of LCE is to design products in compliance with the key issues of sustainable development.

The most efficient use of modern technical products demands the knowledge and know-how of the manufacturer while making use of "Technical Support" i.e. modern communication networks, teleservice and teleoperations. To discover the potentials for ecological improvement, LCA uses data of the physical product life cycle for evaluation. The life cycle assessment provides quantitative basic data for a sustainable product and process management in accordance with public pressure and, last but not least, economical constraints. All these activities must follow the law of maximum economy. LCC assesses not only the life time but also the costs of operation and other costs more efficiently but also leads to higher economic effects. According to the current common agreement, cost-efficient solutions should also be environmentally sound.

Precise data of high evidence are the prerequisite for life cycle activities and for assessing ecological and economical aspects. PDM throughout the whole life cycle supports short access time and less redundancy if a well organized and structured field of data provision is realized. Industry today shows an increasing interest in this classification and especially in the development of methods and tools which can be implemented in their business activities.

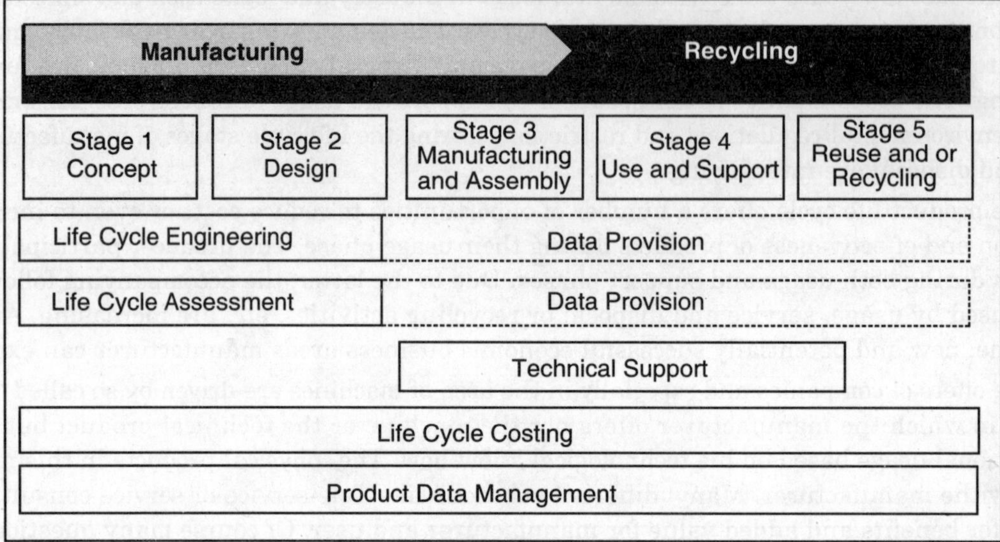

Fig. 14.12: Application of LCM methodologies

14.8. STUDY OF SOME IMPLEMENTED SYSTEMS

The main concepts on disassembly planning have been implemented in several software tools, both prototypes and already commercialized systems. In the following paragraphs, a brief review of their features is taken.

Ametide

It is a method to rationalize and optimize the design of new products from the disassembly point of view. It allows a differentiate approach to the disassembly problem based on the concept of micro disassembly (detailed study of each single disassembly operation, considering involved materials, tools, ergonomics) and macro disassembly (global study of disassembly process for disassembly sequence optimization). The user has to supply information concerning product, tools, costs and times, and the software highlights the weak points in the disassembly process.

ATROiD

It is abbreviation for Assessment Tool for Recycling Oriented Design and is mainly intended to help the designer. It is based on modules for calculation of disassembly times, costs, evaluation of the disassembly sequence. ATROiD is divided into three main levels: assessment, analysis and improvement. The assessment of the product under designing is made in terms of disassemblability and recyclability and it is measured by a score called *Eco-Potential* or *Recycling*

Potential (RP), evaluated according to several criteria concerning: material, joining techniques, subassemblies, disassembly operations, etc. The assessment does not itself provide improvements of the product design but it reveals where improvements have potential.

Dismantling Analysis Program (DemAP)

It is developed for automotive industry. It provides a component oriented description of the vehicle with a view to design for recycling and allows a uniform assessment from a combination of data fields. The software provides various analysis functions: material analysis, purity of components, fastening elements, draining masses, etc. Structural relationships among components are presented in graphical form. A photographic documentation facilitates the handling of each components as well as its design for recycling. In addition an end-of-life scenario may be simulated to help designer in finding new solutions.

Disassembly....... (DFE)

It simulates the disassembly of products at end-of-life; it reveals the associated cost benefits and environmental impacts of a product design and alternate designs can be compared. The initial disassembly sequence can be automatically generated from an assembly lists. The user can edit the initial list to form groups of items that will not be taken apart at end-of-life, but may be recycled or disposed or still assembled. Based on the disassembly sequence, two main analyses are performed: (1) financial return assessment of disassembly, disposal, reuse or recycling and (2) environmental impact assessment resulting from initial product manufacture and disposal, reuse or recycling. DFE charts these analyses together on one graph. Users can determine the cost impact of stopping disassembly at any point.

Environmental Designer (ED)

It is a simple global web-based design tool. Some of its interesting features are that during the concept stage various designs are compared for recyclability using disassembly costs. The final design is then analyzed for hazardous material content, recyclability, disassembly time and costs, energy use, use of different materials. The results are compared with the targets set at the beginning of the project.

Recyclability Evaluation Method (REM)

It is used to support the design process. It considers not only the disassembly easiness but also recycling process can be assessed quantitatively. The method is based on an evaluation scheme consisting of an automatic estimation function of the recycling process, a process model, a data base of recycling processes, and a calculation software. The recycling processes in use in the next 5–10 years from now are presumed and adopted. The output is composed of various indices useful for go/no go decisions, identifications of weakness points.

Virtual Disassembly

Manager is available on the web and allows the on line development of disassembly plan. The system is quite similar to the previous one: the user completes a series of forms where information about the product (bill of materials, mating relations, fasteners, etc.), the costs and the disassembly plant (available bins, disassembly tools, etc.) are supplied. Then the user has to specify a possible disassembly sequence and the system shows a final report where the convenience of the disassembly process can be estimated.

14.9. GLOBAL MANUFACTURING INTERFACE (GMI)

The Global Manufacturing Interface is the formal understanding between (parts of) manufacturing environments that enables the interchange of data, information and knowledge. In order to function properly, the interface must facilitate a number of aspects.

First of all, instantaneous and selective access to both current and historic information must be available, independent of the location of both the operator and the information. This should imply that immediate and transparent communication between operators. This communication has to support the interpretation of information in its context. More important however, is that the different ways of thinking and working that are typical for different (groups of) people are fully supported. Obviously, this support has to be available without any value judgement of those ways of working. Not only the mutual communication between operators has to be adequate, also the communication between operators and workstations has to be direct, efficient and effective. For all types of communication, the location of the communicators must not hamper the process. Especially decision support has to be independent of the location where the operations are carried out.

As the GMI relies on a continuous interdisciplinary sharing of data, information, knowledge and goals, the interface can become the basis for, computer supported collaborative work, concurrent engineering and dynamic control.

The modus for this lies in the fact that every piece of context information is a piece of content information for another operation. Using the GMI, therefore, the content information for a specified environment can be seen as a sliding window. This enables, the interprétation of information, the communication between humans, the communication between humans and systems.

With the increasing use of the GMI the relation between the ontologies and the manufacturing environment becomes more important. This is emphasised by the fact that the advantages of the GMI are based on the quality of the manufacturing ontology. It is obvious that the relation between an ontology and the environment has to be sound and reliable. A first step towards this relation is the use of keywords and some kind of 'in-context translation' of notions and concepts. The keywords can both be predefined as well as emerge during use. The growth of GMI requires special attention. A compromise has to be found between flexibility and ease-of-use on the one hand, and conciseness and control on the other hand. Firstly, the system must be easily extendible in such a way that proposals can be made by any operator without additional effort from anybody. However, additions have to be subject to validation. This validation and (when positive) temporary admission is instantaneously applied by the support system. Moreover, deletions or changes in the ontology must not hamper everyday use.

In order to allow support of different languages and 'dialects' in technical jargon the use of synonyms in the ontology and in the keyword search must be supported. This additionally avoids rigidity in use. Obviously, the validation of the ontologies that form the basis for the GMI require serious considerations.

14.10. CASE STUDY—RECONDITIONED MACHINERY–TEXTILE INDUSTRY

It is estimated that world over, there is total market of 22 billion per year of used machinery. Developed countries regularly sale their used machinery after approximately 10–12 years (End of Life in the view of user of machinery). Developing countries including India imports these used machinery. It is reconditioned, recycled, repaired and used for manufacturing. Predominately in process industry, machine tool industry, textile industry, used machinery is used.

Government of India even gives 20% subsidy for these types of textile machines, provides soft loans also. Most of the times these machines are imported as new machines to take advantages of 50% subsidy. It is estimated that import of about 2000–2200 Cr. of used textile machines is done from Japan, Korea, Australia, Germany, U.K., U.S.A in year 2003–2004. Approx 20% expenses, i.e. 400–500 Cr. are spend for reconditioning textile machinery. At present there is no support given by the renound manufacturers of these machinery.

These services are presently given by experienced local people. If the above mentioned life cycle management strategies are applied by manufacture and users, their will be substantial rise in operating efficiency. The typical study of industries at Ichalkarnaji—in Western Maharashtra shown following results.

Used machinery purchased worth of 100 Cr. in last 2–3 years. The buyer's of used machinery now are getting support of manufactures (in the form of technical know-how, logistic, spares, e-service) following result were observed:

- Machine down time reduced by – 30–40%
- Operating efficiency improved by – 10–12%
- Rate of production raised – 18–20%
- Energy reduction by – 5–6%
- Spares cost reduced by – 20–25%

This system of life cycle management through all Share Holder has given following indirect benefits:

- Reduced Fatigue of Direct Labour.
- Maintain Delivery Schedule
- Confidence of Buyer of Product
- Reduced Maintenance Cost
- Extended Support of Financial Institutes
- Improved Return on Investment
- Reduced Pay Back Period by 6 to 12 Months
- Improved Employee Moral

14.11. CLOSURE

This chapter summarizes the aspects of life cycle management and the role of assembly and disassembly. The central aspect is the concentration on the area of complex and high-value technical products like machines, manufacturing systems and even assembly systems. For technical consumer goods and low-value products, manufacturing has to develop processes and systems for disassembly and recycling. In the field of complex technical products the service, including maintenance, offers new aspects and the possibility to develop holistic management systems. New hybrid technical concepts, the integration into e-services and the consequences for research have been addressed. The optimisation of a product's life cycle is a strategic objective of development for industrial manufacturing. The study of textile industry supports the strategies discussed in this chapter.

REVIEW QUESTIONS

1. What is meant by the linking of products to the network of manufacturers?
2. What are advantages of modularization?
3. What is the ramp-up phase in assembly?

REFERENCE

1. Dashchenko A., "*Manufacturing Technologies for Machines of the Future : 21st Century Technologies, Issue 1*", Springer, NY, 2003.

CASE STUDY: PRODUCT LIFE CYCLE MANAGEMENT IN C.I. FOUNDRIES

15.1. INTRODUCTION

Companies work hard to increase the "customer" value of their products, accelerate time to market, increase efficiency and improve productivity of people and processes. Increasing customer value requires a keen understanding of the value chain to gain valuable insights into cost, quality and functionality, all of which help to reduce cost and enable launching of zero-defect product. Faster time to market means faster product cycle times, faster iterations of trial and error cycles and faster convergence to winning product designs. Higher efficiency and productivity means doing more with the tools available, while improving processes to minimize errors and make people and technology more effective. But in this drive to innovate faster, design and engineering teams have become "islands of automation", as they become more specialized in their respective crafts. Simulation, analysis, coordinate measuring machine and many more downstream application areas acquire specialized tools and expertise to enhance their productivity, but often can't communicate their product data or their intellect with colleagues, partners and suppliers. Products Life-cycle Management (PLM) has emerged as a solution to this challenge.

Every product has a life cycle. PLM is a tool to manage product data and related information throughout the product life cycle. PLM is a process for guiding products from idea through retirement to deliver the greatest business value to an enterprise and its trading partners. PLM employs product information and business analysis to support strategy, planning, management and execution through each phase of a product's life cycle. PLM supports an enterprise's ability to monitor activities, analyze challenges and bottlenecks, make decisions and execute decisions.

To some it is Product Data Management (PDM) while others think of it as well-defined workflow and process, wrapped around a core of data management and collaboration software. Semantics aside, the need for better communication and collaboration on a global scale is clear. Relentless competition is forcing manufacturers to react faster, work smarter, reduce cost, improve quality and compress process design and release cycles, all while developing brand new products at unprecedented rates. Products are consistently more complex as companies try to differentiate themselves. Product design cycles and time to market continue to compress, while regional advantages like quality and raw material cost drive the globalization of supply chains. At the same time, suppliers of key sub-systems are spread around the globe, as out-sourcing and off shoring have become a staple in the manufacturing diet. These are the facts of life in manufacturing today. The need could never be clearer for tools that enable a company to drive efficiency and stimulate innovation.

PDM is a set of applications and capabilities for capturing and maintaining the definition of a product and related data through all phases of a product's life.

The four most commonly used PDM applications are:

- Library functions (search and file check-in/check-out),
- Management of bills of materials (BOMs),
- Product configuration management (PCM) and
- Engineering change management (ECM).
- PDM is a core enabler of CPC and PLM.

15.2. PRODUCT LIFE CYCLE MANAGEMENT, PLM SOLUTION

The basis of an effective PLM solution starts with four fundamental technologies:

- Product Data Management
- CAD/CAE/CAM
- Collaborative Product Design
- Interoperability

In addition, many PLM solutions provide support for Product Configuration Management, Customer Needs Management and Direct Sourcing of Materials.

PDM systems provide fundamental product data management support. PDM systems have emerged in the recent decade to accommodate the need for a common repository for product data that enables the sharing of product-related information throughout the global enterprise [4]. Many companies implemented PDM systems in recent years as the core of their PLM vision, often with great success. CAD/CAE/CAM and collaborative product design products comprise the core applications that enable product design and development. Interoperability solutions are also a mission-critical, but often forgotten, enabler of effective PLM. As global business partners collaborate and react to changes in product design and process, much of their success depends on fast, efficient sharing of high-quality product data. But problems often arise during data exchange, presenting a challenge to effective collaboration in design, analysis, simulation, manufacturing, and countless other applications. Interoperability solutions help in solving these problems.

15.3. THE BENEFITS OF LIFE CYCLE MANAGEMENT

The main benefits of PLM in product development are:

- Faster time to market, due primarily to better communication of mission-critical information between functional teams, and streamlining trial and error processes between product design iterations.

- Lower product development cost, due to time savings and reduced rework and recreation of data
- Increased product quality
- Increased customer satisfaction
- Higher levels of design reuse
- Higher revenue, from more efficient processes, better design reuse and faster time to market

Successful PLM implementations require people, technology and processes. Information technology and engineering software tools can help automate functional activities in and across islands of automation, but effective PLM is more than technology. PLM is a business approach, comprised of both technology and process, which helps manufacturers derive maximum benefit from product data and IT investments throughout a product's life cycle. PLM helps drive innovation and profitability by focusing and leveraging corporate assets and product related information across the global enterprise and its value chains. Products designed with PLM cost less to produce, and have a longer, more productive lifespan. The major benefits that accrue to companies who implement PLM include: faster product development cycle; increased productivity at lower cost; higher quality; greater business flexibility in both the internal and extended enterprise, and faster time to market. PLM is making a positive difference for those leading companies who have implemented it.

According to CIM data, "PLM is a strategic business approach for collaborative creation, management, dissemination, and use of product data across the extended enterprise from concept to end of life – integrating people, processes and information." PLM helps companies derive greater benefit from their investments in IT and people. And the technology to enable PLM continues to expand as marketing people refine their product offerings.

Standardized methodologies have enabled manufacturers to cut costs while accelerating product cycle times and speeding products to market. Attention to method and process enables vastly improved communication and collaboration between designers and engineers worldwide, which yields higher productivity and higher levels of innovation. Standardized methodologies also help increase design reuse by enabling designers and engineers to better understand design intent and capture and share knowledge about their products and experience. Higher value-add work means designers can spend more time designing and engineers can do more engineering. Job satisfaction, quality and productivity increase.

PLM is a business approach, comprised of both technology and process, which helps manufacturers derive maximum benefit from product data and IT investments throughout a product's life cycle. PLM helps drive innovation and profitability by focusing and leveraging corporate assets and product related information across the global enterprise and its value chains. Products designed with PLM cost less to produce, and have a longer, more productive lifespan.

Product Life Cycle Management (PLM) is a term used for the process of managing the entire life cycle of a product from its conception, through design and manufacture to service and disposal. PLM is a set of capabilities that enable an enterprise to effectively and efficiently innovate and manage its products and related services throughout the entire business life cycle. It is one of the four cornerstones of a corporation's IT digital structure. All companies need to manage communications and information with its customers (CRM—Customer Relationship

Management) and its suppliers (SCM—Supply Chain Management) and the resources within the enterprise (ERP—Enterprise Resource Planning). In addition, manufacturing engineering companies must also develop, describe, manage and communicate information about their products (PLM). Documented benefits include:

- Reduced time to market
- Improved product quality
- Reduced prototyping costs
- Savings through the re-use of original data
- A framework for product optimization
- Reduced waste
- Savings through the complete integration of engineering workflows.

15.4. PLM PROCESSS: A GENERAL APPROACH

The core of PLM is in the central management of all product data and the technology used to access this information and knowledge. Although PLM as a discipline emerged from tools such as CAD/CAM and PDM it must be viewed as the integration of these tools with methods, people and the processes through all stages of a product's life. For simplicity the stages described are shown in a traditional sequential engineering workflow. The exact order of event and tasks will vary according to the product and industry in question but the main process are:

- Conceive
  - o Specification
  - o Concept design
- Design
  - o Detailed design
  - o Validation and analyse (simulation)
  - o Tool design
- Realize
  - o Manufacturing Planning
  - o Manufacture
  - o Build/Assemble
  - o Test (quality check)
- Service
  - o Sale and Deliver
  - o Use
  - o Maintain and Support
  - o Dispose

15.5. PRODUCT DEVELOPMENT AS APPLIED TO C.I. FOUNDRIES

In the foundries, mass production of various components is going on. The castings are of very wide variety. So, generally the gating and risering system of those castings are designed depending upon the experience and some thumb rules [1]. But this process is not always giving accurate results. In terms of gating and risering system accurate results not only mean the correct gating and risering system but also maximum yield and defect free castings. But both of these parameters cannot be achieved at the same time. Cutting down lengths of risers or runners to increase yield, we may get various defects in castings. On the contrary, to get defect free casting, we have to ensure that enough metal is poured and in the process, designer may tend to increase the dimensions related to gating and risering system. But this causes obvious decrease in the yield. So, it is very necessary to have compromise between these two most important factors [2]. To do this it is absolutely necessary to carry out the design procedure according to some predefined path.

Following points should be considered in Methoding of castings in C.I. foundries [3].

Mould Quality

The objective is to avoid enlargement of the mould from high liquid pressures exerted by the cooling and solidifying graphitic iron. Green sand and shell moulds will not withstand the solidification pressure. Chemically bonded sand moulds will resist solidification pressure if they are properly prepared. This requires mechanical compaction of sand during mould preparation and adequate curing [2]. Cement sand and dry sand moulds will normally withstand the solidification pressure.

Pouring Temperature

Pouring temperature is very important and should be specified along with casting method; the tapping temperature and mould sequences are determined accordingly. Selection of the pouring temperature depends upon composition of the alloy and the type of casting being made. Both low and excessive pouring temperatures create defects in castings. The upper limit of the pouring temperature is marked by factors such as hot tearing and lower limit by the ability of molten metal to flow in thin sections and the avoidance of skulls in ladles. Metal poured at a temperature lower than the minimum optimum would cause misrun and cold shuts whereas excessive pouring temperature tends to oxidize the metal, alloying elements get lost, gas content of melt increases and castings develop porosity; of course metal becomes more fluid at high temperatures [3]. Pouring temperature also influences metallographic structures, particularly grain size and substructure.

Solidification of Castings

The mechanism of solidification of casting and its control for obtaining sound castings is the most important problem of foundrymen. The proper understanding of solidification mechanism is essential for preventing defects due to shrinkage of the metal. As soon as the molten metal is poured in sand mould, the process of solidification starts. The mode of solidification affects properties of castings because a casting acquires a metallographic structure determined by solidification process. Besides structure, soundness of a casting also depends upon solidification mechanism. Soundness implies the degree of true metallic continuity and a casting will be sound if volume shrinkage accompanying the change of state of melt to solid is compensated by liquid metal with the help of risers, etc. Volume shrinkage or volume contraction occurs during three stages and thus contraction or shrinkage is of three types [4].

Liquid contraction (Shrinkage): When metal is poured into mould cavity, the worker should not move immediately for next mould for pouring. He should wait to see whether any cavity is observed in the mould filled with molten metal. This is because liquid contraction occurs when the metal is in liquid state.

Solidification contraction (Shrinkage) : Solidification contraction occurs during the change from liquid to solid. To compensate this, risers are provided.

Solid contraction (Shrinkage): Solid contraction occurs when the metal is solid. Solid contraction occurs after solidification. It does not influence shrinkage defects. To overcome solid contraction, pattern allowances are provided.

15.6. DESIGN FOR ROTOR OIL FILTER

The Rotor oil filter casting and the casting along with gating and risering is shown in Figs. 15.1 and 15.2 respectively. The parameters used for designing of this casting are as follows:

- MATERIAL: Grey Cast Iron
- DENSITY: 7250 kg/m^3
- TYPE OF GATING: Parting Line Gating
- GATING SYSTEM: Unpressurised Gating System
- BOX SIZE: 16' × 16' × 10'
- NUMBER OF CASTING PER MOULD: 6
- TYPE OF MOULD: Green Sand Mould
- FACING SAND: Graphite powder.

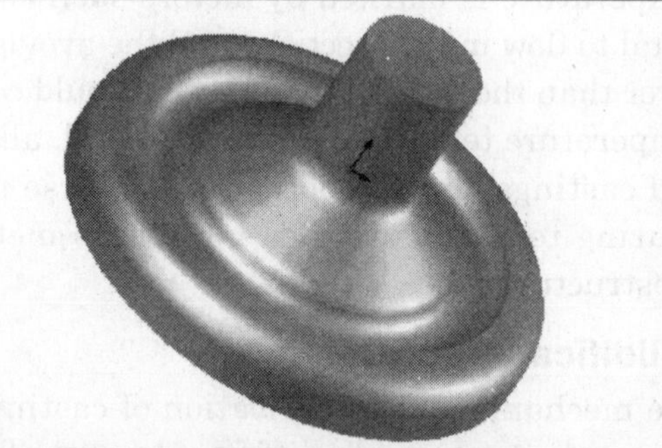

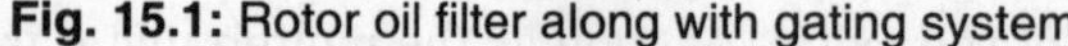
Fig. 15.1: Rotor oil filter along with gating system

Fig. 15.2: Rotor oil filter casting

15.7. DESIGN FOR INTERMEDIATE FLANGE

The intermediate flange casting and the casting along with gating and risering is shown in Figs. 15.3 and 15.4, respectively. The parameters used for designing of this casting are as follows:

- MATERIAL: Grey Cast Iron
- DENSITY: 7250 kg/m^3

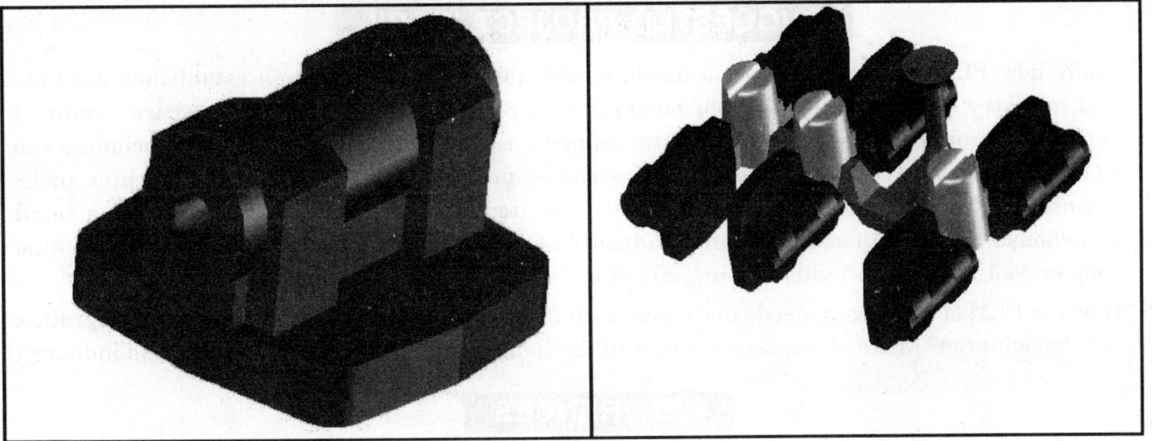

Fig. 15.3: Intermediate flange casting

Fig. 15.4: Intermediate flange along with gating system

- TYPE OF GATING: Parting Line Gating
- GATING SYSTEM: Unpressurised Gating System
- BOX SIZE: 21' × 21'× 14'
- NUMBER OF CASTING PER MOULD: 4
- TYPE OF MOULD: Green Sand Mould
- FACING SAND: Graphite Powder

The current work is done by using Solid Works 2003 for modeling and Visual Basic as the programming language for ease in calculations and better representation of the results. The modeling is done in **"*.stl"** format. Then these surface models are used for further analysis using heat transfer parameters and conditions. But due to large data per **"*.stl"** file there are limitations on weight of castings to be analyzed.

15.8. CLOSURE

The major benefits that accrue to companies who implement PLM include: faster product development cycle; increased productivity at lower cost; higher quality; greater business flexibility in both the internal and extended enterprise, and faster time to market. PLM is making a positive difference for those leading companies who have implemented it. Many good methodologies are used in manufacturing today, and countless success stories have been told of the huge business benefits that result from implementing standardized methods and processes.

Development carried out in C.I. foundries, effectively highlighted potential for implementation of PLM methodologies in foundry business. This will enable foundries to implement PLM process with focus on Engineering Change Management (ECM), Bill Of Materials (BOM), PLM and related data management for effecting modular development and design change, assessing various design options and related cost implications with productivity rise.

SUGGESTED PROJECT WORK

1. Study how PLM can be applied in a machine tool industry. A typical well established machine tool industry (for example, Batliboi machine tools, Surat, India) covers the entire gamut of operations, from casting in the foundry to complete assembly of the machine tools including the electrical/electronic controls. Often such companies produce complete variety of machine tools, beginning from a simple and mostly manually operated medium size machine meant for a small scale/house hold manufacturing unit to advanced CNC machine tools for a full-fledged machine shop or tool room of an industrial organization.
2. How the PLM of a company needs to connect with its suppliers and retailers? How such integration is to be achieved? Make an assessment of it in the light of an automobile manufacturing industry.

REFERENCES

1. Iyengar S.N., *"Indian Foundry"* Journal, Jan 2002, pp. 37–38.
2. http://www.cimdata.com.
3. www.delphi.com/dti/dcce/hm_dpd.
4. Heine R.W., Loper C.R. and Rosenthal P.C., "*Principles of Metal casing*", Tata McGraw Hill Publications, 1978.

CASE STUDY: COLLABORATION OF CAD AND VB-6 FOR SIMULATED DESIGN OF SPEED BOXES

16.1. INTRODUCTION

Speed boxes are usually designed to provide different speeds that are useful for functioning of machines for different requirements. Designing of speed box involves lengthy calculations, especially when numerical methods are employed. Computers can facilitate this effort. Repeated calculations of the same nature can be performed safely in the short time with high reliability. Further more, the design parameters can be improved to yield better design, such as lower cost, higher reliability. The chapter discusses how effectively the utilization of software skills can lead to reduce the computational efforts in designing a Speed Box. It further emphasizes on the other advantages such as lower cost and higher reliability. The chapter discusses a case study of a software specially designed for this purpose using Solid Edge 14, VB-6, MS-Access and MS-Excel. The Simulation is achieved by using Solid Edge 14, if the required data such as no. of speeds, Geometric progression ratio, lower speed and higher speed is provided the software helps in taking decision by showing ray diagram, structure diagram, speed chart and gearing diagram. This output can be readily used to create the working drawings of each part of Speed Box.

Computer Aided Design (CAD) can be defined as the use of computer system in the analysis and optimization of a design using a combination of computer graphics and programming techniques. The preliminary design, once conceived by the designer, needs to analyzed. This is to find whether the design satisfies the requirements set by the initial specifications or not.

Designing involves lengthy calculations, especially when numerical methods such as finite element analysis are employed. Computers can facilitate this effort. Once the calculations initially verified, repeated calculations of the same nature can be performed easily in short time with high reliability. To access the actual operating conditions of a machine, it is often necessary to approximate it's operation by the use of mathematical models, expressions by a set of equations. Then the solution is usually obtained in computers. Further the design parameters can be improved to yield better satisfying requirements of design such as lower cost, higher reliability etc.

Speed boxes are generally used for obtaining different speeds on machine tools as the speed reducers or stepping up the speed and can be manufactured as separate unit.

1. Design of gear sets i.e. no. of teeth on each gear etc.
2. Design of shaft i.e. dia. of shaft etc.
3. Design of bearing i.e no. of bearing etc.

It is natural for the designers to employ the emerging CAD technique to the area of gear design using various optimization strategies. This procedure essentially includes criteria like minimization of the space requirement, and thus the cost. The major emphasis is, therefore, on the finding of no. of teeth with software so that time consumption is less and ready made production drawings will be made by computer for further action.

The present chapter essentially aims at developing a general purpose software package for the design of the speed box. The user will specify the lowest speed required and highest speed required for the machine tool. Also he will specify the no. of speed steps required. The user will specify the shaft material.

Roller and ball bearings are used to support the shafts at two ends. Automatic selection of the bearings, keys on the shaft will be ensured by software. We can see the all-possible structure diagrams, ray diagram also. The linking of Solid Modeling software i.e. solid edge will show us the simulated view of speed box. So that we can get idea about the speed box what we are going to built. Also it is possible to go for finite element analysis on the same speed box using the Solid Edge software as a future development.

16.2. AIM OF SPEED AND FEED RATE REGULATION

A machining operation should be conducted at such values of cutting parameters (speed, feed, depth of cut, etc.) that ensure the minimum cost price of the machined component. The machining cost can be expressed by the equation given below [1]:

$$C = C_{mt} + C_{npt} + C_{tc} + C_r$$

Here $C_{mt} = (W + E)t_m$ represents the cost of machining time, W is the wage rate, E is cost of operating the machine tool per unit time and t_m the machining time.

$C_{npt} = (W + E)t_{npt}$ represents the cost of non-productive time, t_{npt} is the total time of non-productive operations, such as loading and unloading, idle travel of cutting tool to provide for approach and over travel, etc.

$C_{te} = (W + E)t_{tc}/Q$ represents the tool changing cost per component; t_{tc} is the time required for replacing (regrinding) a blunt tool and setting the new (reground) one and Q the number of components machined during the period of tool life.

$C_t = T/Q$ represents the cost of the tool per component, T is the cost of the tool for a period equal to the tool life and can be determined as the tool cost divided by the number of permissible regrinding.

From the generalized tool life equation $T = C/(V^a s^b d^c)$, it follows that the optimum tool life can be achieved on a particular operation only by working at the optimum values of cutting speed V, feed rate s and depth of cut d. So the provision of regulating the spindle RPM and feed rate is an essential requirement of machine tools to ensure economic machining of work pieces of different materials and sizes by cutting tools of different shapes and compositions.

16.3. DESIGN OF SPEED BOX

In this section the principles of designing speed boxes, i.e., gear boxes that are employed for stepped regulation of the RPM of the main drive are dealt with.

Laws of Stepped Regulation

Following are the criteria to decide the steps of speeds for given lower and higher speed of the speedbox [2, 3].

The RPM values constitute an arithmetic progression: The realization of arithmetic progression is based upon the idea that the difference between adjacent RPM values is constant.

$$n_1 = n,\ n_2 = n_1 + a,\ n_3 = n_2 + a = n_1 + 2a,\ \ldots,\ n_Z = n_1 + (Z-1)\,a$$

The RPM values constitute a geometric progression

$$n_1 = n$$
$$n_2 = n_1 \acute{\varnothing}$$
$$n_3 = n_2 \acute{\varnothing} = n \acute{\varnothing}^2$$
$$\ldots$$
$$n_Z = n_{(Z-1)} \acute{\varnothing} = n \acute{\varnothing}^{(z-1)}$$

Here $\acute{\varnothing}$ is the geometric progression ratio.

The RPM values constitute a harmonic progression: The harmonic progression is developed from the basic idea that the diameter range served by each RPM of the progression is equal, i.e.,

$$\Delta d_x = d_x - d_{x+1} = 1000V/\Pi\ [(1/n_x)\ (1/n_{x+1})]$$

Therefore,

$$[(1/n_x)(1/n_{x+1})] = \text{constant} = C$$
$$n_{x+1} = n_x / (1 - Cn_x)$$

The harmonic progression can be written as follows:

$$n_1 = n$$
$$n_2 = n_1/(1 - Cn_1)$$
$$n_3 = n_2/(1 - Cn_2) = n_1/(1 - 2Cn_1)$$
$$n_Z = n_{(Z-1)}/[1 - Cn_{(Z-1)}] = n_1/[1 - (Z-1)Cn_1]$$

The RPM values constitute a logarithmic progression: In this progression the diameter range is a function of the diameter.

$$\Delta d_x = 2M\,(d_x)^p$$

Here M is known as the location coefficient. Generally, $p = 0.5$.

Out of all these progression ratios, geometrical progression is commonly used in machine tool drives owing to the following advantages.

- Constant loss of economic cutting speed in the whole RPM range
- Constant loss of productivity in the whole RPM range
- Better design features.

As we know the fact that speed steps in a speed box should be obtained not through a single transmission between two shafts but through a group of transmissions among a number of shafts. This feature can be realized in an actual speed box design only if the RPM values lie in a geometric progression and may be explained by the following properties of a geometric progression.

(i) If from a geometric progression having progression ratio $\varnothing$, members are so removed that only every x^{th} member remains, then the resulting series is a geometric progression having progression ratio $\varnothing^x$.

(ii) If a geometric progression having progression ratio $\varnothing$ is multiplied by a constant A, the resulting series is a geometric progression having the same progression ratio, but whose members are A times greater.

(iii) If a geometric progression having progression ratio $\varnothing$ is multiplied by a factor $\varnothing^x$, the resulting series is a geometric progression which is shifted by x members.

Information Required for Speed Box

We will now examine the different inputs needed.

Selection of range ratio: The following information is essentially required before we can start designing speed box.

- The highest output RPM, n_{max}
- The lowest output RPM, n_{min}
- The number of steps Z into which the range between n_{max} and n_{min} is divided, and
- The number of stages in which the required number of speed steps are to be achieved

An important parameter in designing speed boxes is the range ratio R_n given by

$$R_n = n_{max}/n_{min} = (V_{max}\ d_{max})/(V_{min}\ d_{min}) = R_v\ R_d$$

After selecting the proper values of R_v and R_d, the range ratio R_n is determined. The number of speed steps in which the total range of RPM values available on the machine tool is divided, is determined as follows:

Suppose the RPM values $n_1, n_2, \ldots, n_Z$, constitute a geometric progression, then

$$R_n = n_Z / n_1 = \varnothing^{(Z-1)}$$

Here $n_1 = n_{min}$ and $n_Z = n_{max}$.

The equation can be represented as

$$\varnothing = (R_n)^{1/(Z-1)}$$

where from,

$$Z = \log (R_n\ \varnothing)/\log \varnothing$$

The value of Z can be determined provided R_n and $\varnothing$ are known.

The most commonly used values of $\varnothing$ are 1.26, 1.41 and 1.58. Smaller values of $\varnothing$ ($\varnothing$ 1.12 and 1.06) complicate the drive to an extent where it cannot normally compete with stepless regulation. Large values of $\varnothing$ ($\varnothing$ = 1.78 and 2.0) result in very rough regulation leading to large productivity losses. They are rarely used and that too only in special-purpose machine tools.

Break up of speed steps: The value of the number of speed steps Z found from formula is rounded off to the nearest whole number, preference being given to the number which can be broken into multiples of 2 and 3. If P_1 is the number of speed steps in the first transmission group and P_2 the number of speed steps in the second transmission group, then after the second stage we have P_1P_2 speed steps. Similarly, after the third stage we have $P_1P_2P_3$ speed steps, and finally, we have Z speed steps after u number of transmissions, i.e.,

$$Z = P_1 \times P_2 \times P_3 \times \ldots \times P_u$$

For obtaining a particular number of speed steps using minimum number of gears, it is necessary that $P_1 = P_2 = P_3 = \cdots = P_u$.

Hence the number of speed steps in each transmission group can be found from the expression $P = (Z)^{(1/u)}$. Then the number of speed steps should be divided in such a manner that

$$Z = 2E_1 \times 3E_2$$

Here E_1 and E_2 are whole numbers. Equation above is satisfied by the numbers $Z = 2$, 3, 4, 6, 8, 9, 12, 16, 18, 24, 27, 32, 36, etc.

Structural Diagram

Suppose a speed on one shaft yields two speed values on the next shaft, i.e., the number of speed steps of the particular transmission group is $p = 2$. If the transmission is through gears, the transmission ratios that provide the two new speed values must lie in the following range.

$$i_{max} = 2 \text{ and } i_{min} = 1/4$$

The maximum reduction of speed is limited to four times to keep the radial dimensions of the speed box within reasonable limits, while the maximum increase of speed is restricted to two times due to limitations of the pitch line velocity. The transmission range of the group is given by

$$i_g = i_{max} / i_{min} = 8$$

As earlier explained in earlier section, the number of speed steps is represented in the form:

$$Z = P_1 \times P_2 \times P_3 \times \ldots \times P_u$$

This expression can be written in a number of ways by arranging P_1, P_2, P_3, ...,

$$Z_u = P_1(X_1)\ P_2(X_2)\ P_3(X_3) \ldots P_u(X_u)$$

where $X_1 = 1$, $X_2 = P_1$, $X_3 = P_1P_2$, ..., $X_u = P_1P_2P_3 \times \ldots \times P_{(u-1)}$

Equation is known as the structural formula of the speed box. By rearranging X_1, X_2, X_3 ..., X_u, we can obtain $u!$ number of combinations.

Let us consider an example. Suppose the number of speed steps of a speed box *is* $Z = 12$ and they are to be realized in three stages, i.e., $u = 3$. The number 12 may be written as a multiplication of 2 and 3 in six different ways. Let us consider one such combination, $Z = 2 \times 3 \times 2$.

For this arrangement, we may write six structural formulae as follows:

(a) $Z = 2(1)3(2)2(6)$ (b) $Z = 2(1)3(4)2(2)$

(c) $Z = 2(3)3(1)2(6)$ (d) $Z = 2(6)3(1)2(3)$

(e) $Z = 2(2)3(4)2(1)$ (f) $Z = 2(6)3(2)2(1)$

In an identical manner 30 more structural formulae can be written for the remaining five combinations of $P_1 \cdot P_2 \cdot P_3$. The Fig. 16.1 shows the structural diagrams. The structural diagram gives information about:

- The number of shafts in the speed box,
- The number of gears on each shaft,
- The order of changing transmissions in individual groups to get the desired/spindle speed, and
- The transmission range and characteristic of each group.

From among all the possible structural diagrams, the selection of the best version is guided by the following two factors:

- Transmission ratio restriction
- Minimum total shaft size

Besides factors above, the following guidelines are important in selecting the best structural diagram.

The number of gears on the last shaft (spindle) should be the minimum possible.

The transmission ratio between the spindle and the shaft preceding it should be the maximum possible, i.e., speed reduction should be the maximum possible, and the number of gears on the shafts should not generally be more than three, though in exceptional cases it may be equal to 4. $i_{max} \times i_{min} = 1$ favors the least radial dimensions of the gear box.

16.4. SPEED CHART

The structural diagram only depicts the range ratio of transmission groups but gives no information about transmission ratios, so it is necessary to plot the speed chart. The lines joining points of adjacent shafts in a speed chart depict the transmission ratios. If:

- The line is horizontal, it corresponds to transmission ratio $i = 1$,
- The line is inclined upward, it depicts $i > 1$, i.e., speed increase, and
- The line is inclined downward, it depicts $i < 1$, i.e., speed reduction.

16.5. GEARING DIAGRAM

While developing the gearing diagram on the basis of a given structural diagram, attention must be paid to the following two aspects.

Firstly, all requirements that are essential for the proper functioning of a gear transmission must be satisfied and gearbox dimensions must be minimum. These essential requirements for proper functioning of a gear transmission may be summed up as:

- The number of teeth on the smallest gear with an uncorrected profile and 20° pressure angle, $Z_{min} = 17$.
- If gear pairs on parallel shafts have the same module, the sum of the number of teeth of mating gear pairs must be the same.
- The spacing between adjacent gears on a shaft should be such that one gear pair gets completely disengaged before the next begins to mesh.
- The number of teeth of adjacent gears must differ by at least four.

Secondly, specific features peculiar to the functioning of the machine tool for which the gearbox is being designed should be taken into account.

16.6. DETERMINING THE NUMBER OF TEETH

The number of teeth on the smallest gear should be as small as possible because an increase in the size of this gear leads to an increase in the overall dimensions of the speed box. In general, it is recommended that $Z_{min} \geq 20$ in speed boxes and ≥ 16 in feed boxes.

However, if a smaller value of the number of teeth has to be assigned due to space limitations, then Z_{min} should be calculated from the standard relationship that precludes undercutting or interference. These relations are available in the standard mechanical design textbooks [4]. If the calculated values of the number of teeth are fractions, they must also be rounded off to whole numbers.

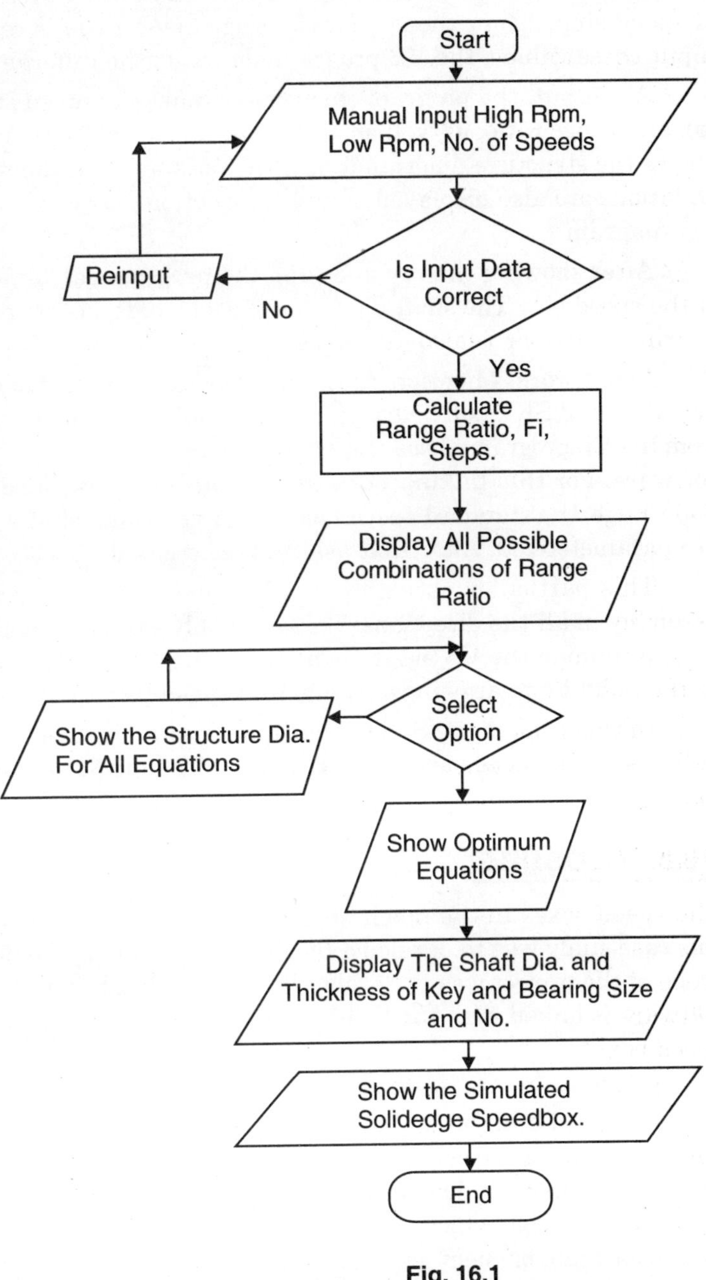

Fig. 16.1

The flowchart of Fig. 16.1 gives the sequence of computations that need to be done.

16.7. DESCRIPTION OF SOFTWARE

For this work Visual Basic (VB) version 6 was used as front end. Using VB the whole program was prepared. In this program, high RPM and low RPM are required, as input and the number of speed steps (*Z*) or the geometric progression ratio is required to be also supplied. Once we input these values, the VB program shows us the different related outputs.

As output, the program shows the number of speed steps, the geometric progression ratio (ϕ), and range ratio, as well as all the possible combinations of factors of *Z* value. The software shows the structure diagram for all the possible equations. The optimum equation(s) out of all equations are also displayed. For that equation, one can see the structure diagram and also the ray diagram.

After showing the ray diagram, the program calculates the number of teeth on each gear of the speed box. The shaft diameter of each shaft, bearing's inner and outer diameter and SKF bearing's number is also displayed. Also the size of the square key is obtained.

In our work, Microsoft Access and Microsoft Excel have been used as Back-end. The square key sizes and SKF ball bearings data is stored into the database. After getting all these outputs from the VB program, these outputs are connected with the Solid Edge, which is 3D Solid Modeling software. For this linking purpose, the output values are stored into Microsoft Excel. In the Solid Edge, the standard speed box parts (i.e. gears, shafts, plate.) are drawn and then all parts are parameterized. After that assembly is created.

This particular assembly can be rendered using rendering tool. Also we can see this assembly in all the directions. This assembly can be exploded to see the details of all parts. As we are running the VB program, as the values of VB program output changes, the same effects in the Solid Edge drawings. And actual speed box for those particular outputs can be viewed.

In this project the Solid Edge simulation is restricted to only for 6 , 8, 12, 16 , 18 speeds only, as these speeds are generally used. The output of the program at different stages is as follows:

16.8. CLOSURE

The speed boxes in the machine tools are used for obtaining various speeds. The objective of this case study was to develop a package, which helps the designing of the speed box. Hence the scope of the project was to develop user-friendly software for the developing the speed box. The software is linked with the Solid Edge version 11, so that we can see the simulated view of the speed box.

So by using this package the laborious calculations required for designing the speed box can be avoided and the whole designing is possible within the very short time. In this software, while simulating the speed box, the designing of gear is overlooked. The importance is given for the simulation only. The software can also generate the two dimensional drawings of all the parts of speed box, which can be useful for further production purpose. In the future scope, one can go for finite element analysis of the assembly also. In such a way the software can be helpful for the machine tool manufacturer. In this software the simulation of speed box is shown with the solid edge, Fig. 16.2.

(a)

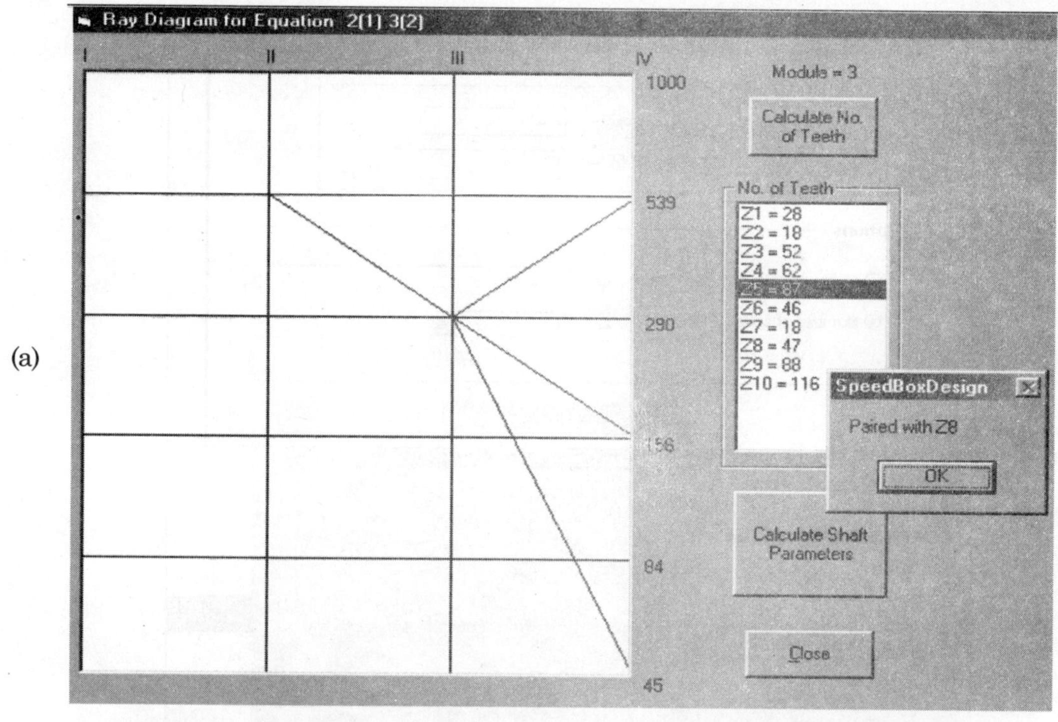
Ray Diagram for Equation 2(1) 3(2)
I
II
III
IV
1000
539
290
156
84
45
Module = 3
Calculate No. of Teeth
No. of Teeth
Z1 = 28
Z2 = 18
Z3 = 52
Z4 = 62
Z5 = 87
Z6 = 46
Z7 = 18
Z8 = 47
Z9 = 88
Z10 = 116
SpeedBoxDesign
Paired with Z8
OK
Calculate Shaft Parameters
Close

(b)

Shaft Parameters
Choose Shaft: Shaft-IV
N (hp) = 2
n1 = 45
Tau (kg/cm2)= 120
Store Values
Close
Shaft Diameter = 50.9889533853214 mm
Design Of Key
Size = 13
Length = 80.10364576834
Bearing Information
Bearing Number (SKF) = 6009
Shoulder Diameter (mm) = 45
Shaft Diameter = 50
Outer Diameter of Bearing = 75
Width of Bearing = 16

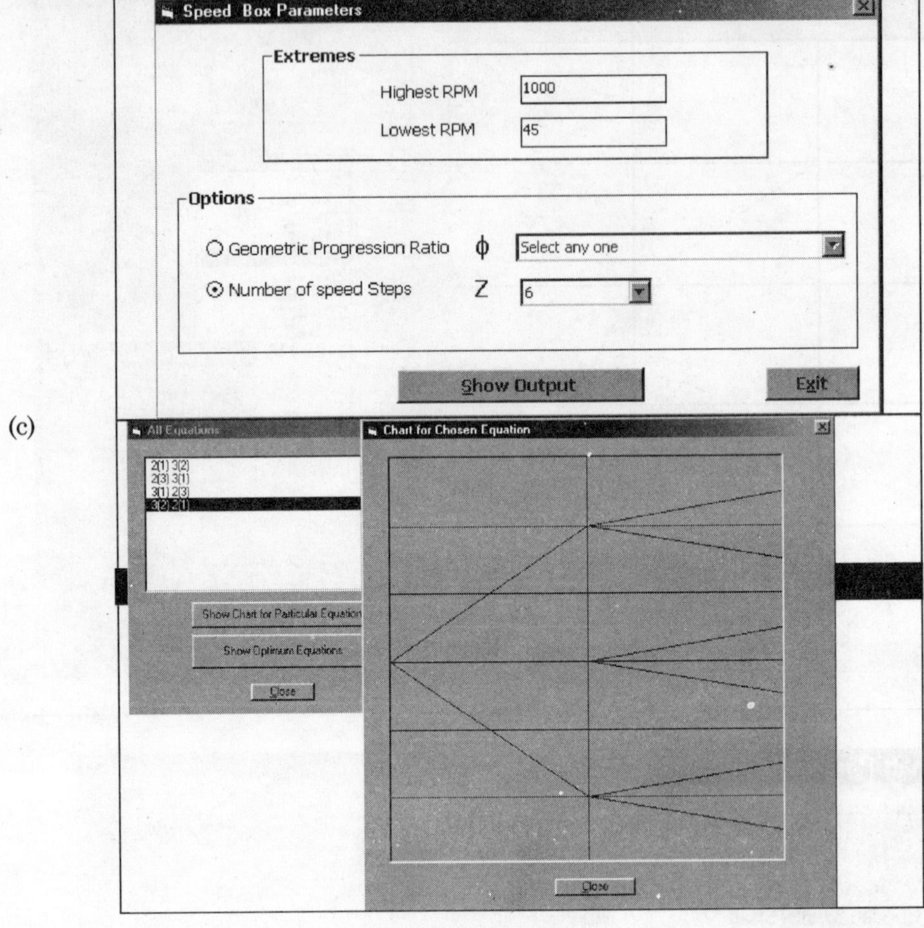

(c)

Fig. 16.2

REFERENCES

1. G.C. Sen and A. Bhattacharya, "*Principles of Machine Tools*", Second Edition , New Central book Agency (P) Ltd., page no. 99–175.
2. N.K. Mehta, "*Machine Tool Design and Numerical Control*", Second Edition, Tata McGraw-Hill , New Delhi, 1996.
3. Koingsberger F.J., "*Designing Principles of Metal Cutting Machine Tools*", Pergamon Press, Oxford 1964.
4. Shigley J.E., Mischke C.R., Budynas R.G. and Nisbett K.J., "*Mechanical Engineering Design*" (In SI units)", 8th Edition, Special Indian Edition, Tata McGraw-Hill, Delhi, 2008.

CASE STUDY: FBMod: A WEB BASED FEATURE MODELER FOR PRISMATIC COMPONENTS

17.1. INTRODUCTION

The chapter presents the design and implementation of a Web Based Feature Modeler (FBMod) for prismatic components commonly manufactured on 3-axis CNC machining centers. FBMod is developed as a Thin Client-Thick Server architecture and uses Design-By-Features philosophy for modeling. It provides a very user friendly modeling environment to enable clients to Create, Edit, Validate and View a prismatic part model on the web without the need of any proprietary CAD software at their end. The CAD model is stored in a standard format and provides very good interoperability for linking to other standalone/web-based CAD/CAM applications.

Product modeling is the heart of CAD/CAM technology as it provides a central repository to satisfy the informational needs of several down line activities of the product life cycle such as design, analysis, machining, assembly, inspection etc. In the past two decades, a wide variety of Product Modeling tools have been developed to assist a designer to model and visualize complex parts and assemblies. Today Features Technology has become the de-facto standard for Product Modeling in CAD/CAM [1]. This has helped in shortening the product life cycle to a considerable extent by achieving quick turnaround in Design and Manufacturing.

Though robust, these Product modeling systems are stand-alone types, which need powerful hardware, proprietary software and are often beyond the reach of smaller customers. The emergence of SMEs (small and medium sized enterprises) with geographically distributed centers, call for Global Product Development scenario that requires seamless integration and interoperability between geographically distributed systems. Today Internet technology has made such global connectivity a feasible task. The current research in CAD/CAM is, thus, focused on the development of Web Based CAD systems using Features Technology, which would provide "anywhere anytime" connectivity for globally distributed users. Further these web-based systems should be linked to various downstream activities to provide agility and interoperability for tele-manufacturing.

Literature documents that very few tools are currently available for Web Based CAD that support Feature based modeling of prismatic parts and provide a framework for seamless integration with the downstream application tasks.

Last two decades have seen the emergence of various schemes for object modeling such as Constructive Solid Geometry (CSG), B-Rep and Hybrid modeling. These schemes require feature extraction using different techniques like rule-based [1], syntactic pattern [2] and graph based recognition [3] for integration with downstream applications. Feature Based Modeling, on the other hand, captures part geometry and process specific feature data at the design stage itself and have been found to be very well suited for integration with down line activities such as automated process planning [4]. They have thus become a defacto standard in the industry today for CAD part modeling. Various domain-specific feature-modeling systems, such as IFPP [4], CFACA [5] and VITool [6] have been reported with a focus on prismatic parts.

Compared to the work on stand-alone modeling systems, scant research work is reported on the development of web based CAD systems for part modeling and design. WebCAD [7], CSM –Collaborative Solid Modeling system [8], CyberView [9], webSPIFF [10], cPAD [11], and Palantir [12] are few systems reported in this direction. These systems however, provide limited functionality and user interaction and support limited part shapes. They are not efficient in Internet environment as they operate slowly for client operations like modeling, visualization, file transfer etc.

With rapid growth and accessibility of Internet Technology, a need thus exists to develop a robust user friendly web based CAD system for feature based modeling and integration with downstream applications that may be provided as services over the www.

This chapter reports the attempts towards the design and development of such a Web Based Feature Modeler system named "FBMod" Various design and implementation issues of FBMod are presented in the sections to follow.

17.2. DESIGN OF FBMod

FBMod is a Web Based Feature Modeler indigenously developed at the CAM laboratory, IIT Bombay. It is aimed to provide a very user friendly interactive environment for the Modeling, Validation and Visualization of Prismatic components commonly produced on 3-axis CNC machining centers in manufacturing shops.

It is designed to provide the following functional utilities to the client:

- Client—Server Communication to provide "Anywhere-Anytime" connectivity.
- Web Based Part Modeling environment using 'Design by Features' approach.
- Part Model Validation using Design for Manufacture (DFM) approach.
- Web based Part Visualization (Faceted and Wireframe)
- CAD part model Storage and Transfer
- Interfacing with down line application tasks.

17.3. MODULAR ARCHITECTURE OF FBMod

Figure 17.1 shows the modular architecture of FBMod. It is based on Thin Client - Thick Server architecture. It provides various functional modules for Authentication, Session Manager, Feature Modeler, Database and other communication modules between the client and the server.

At the client end, FBMod demands minimal resources and most of the processing as well as the storage takes place at the server end. The data is transferred over the Internet for communication between the clients and the server providing "Anywhere-Anytime" global connectivity.

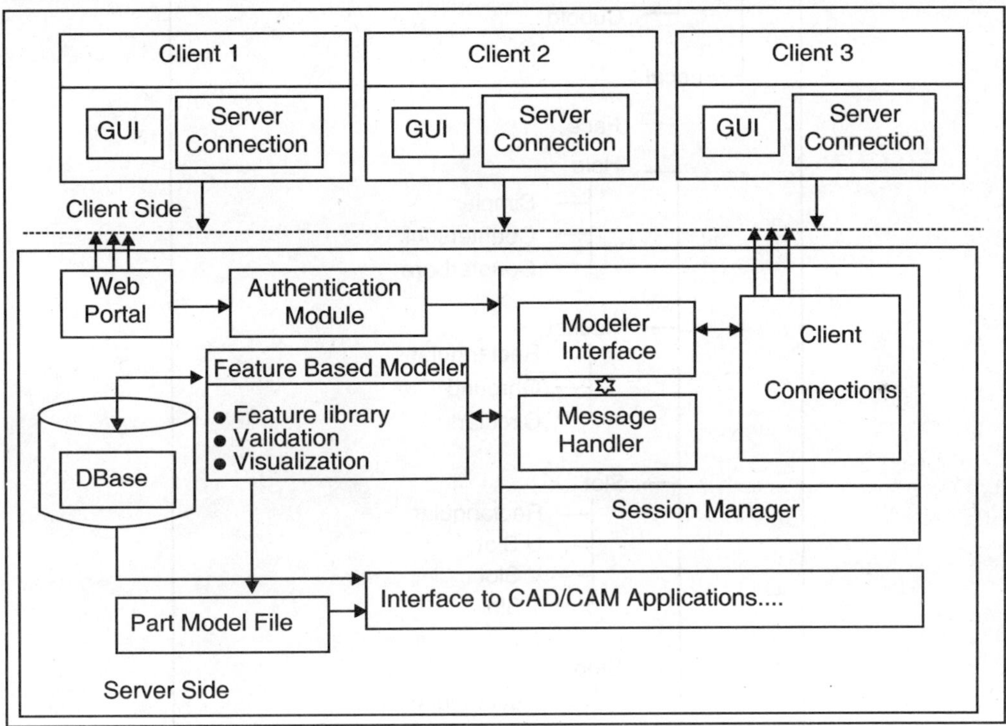

Fig. 17.1: Modular architecture of FBMod

Visualization of the ongoing modeling process (CAD model) is provided through VRML files dynamically generated and updated by FBMod. User can thus, view the synthesized model throughout the design process. On completion of the design process, the model is stored in a native file format (*.prt), which is similar to STEP.

Features are the basic entities for modeling of the parts. As a result identification of feature taxonomy is a key issue in the design of FBMod for the chosen domain.

17.4. FEATURE TAXONOMY

CAD part model is conceived to comprise of a Gross shape (Raw stock) upon which several Local Features (Holes, Pockets, Slots etc.) are positioned in various orientations and spatial locations. An extensive study of several industrial components was undertaken to identify various types of features commonly occurring on prismatic components. Based on this study and research literature, feature taxonomy for FBMod was designed. Figure 17.2 shows various feature families and subtypes provided by FBMod. In essence they cover both Face based (Hole, Pocket) and Edge based (Step, Slot) Depression type of features and would enable modeling of 90–95% of standard prismatic part shapes.

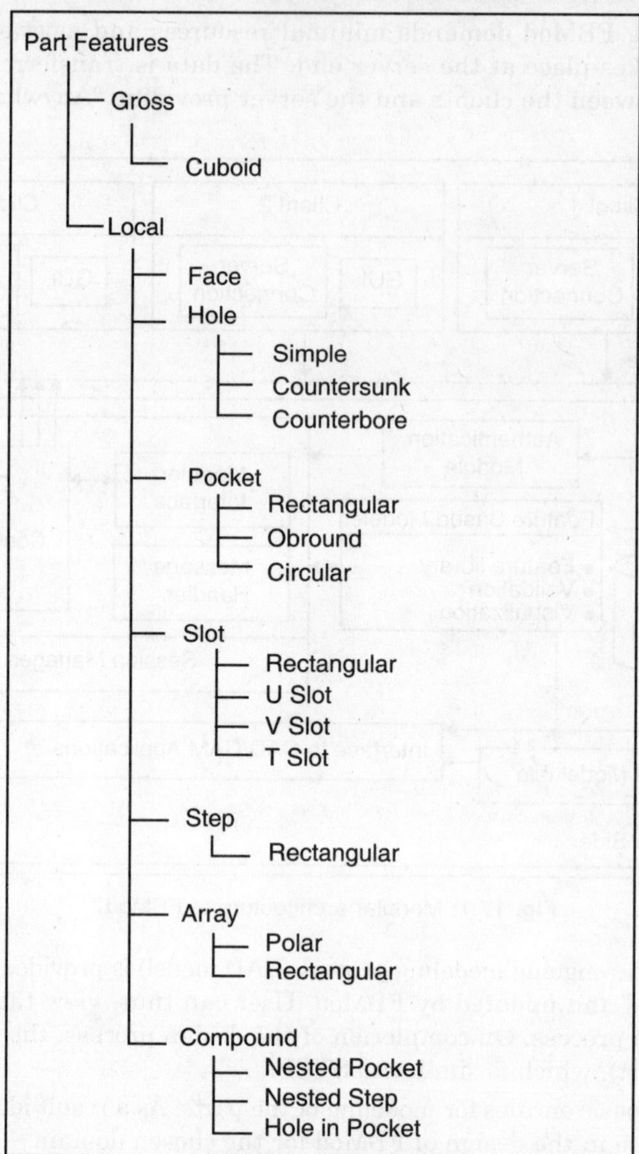

Fig. 17.2: Feature taxonomy

17.5. FEATURE REPRESENTATION USING OOPS

FBMod has been designed following the Object Oriented Programming (OOPS) design strategy. This makes the design modular and extendible for future enhancements. Each feature is essentially an Object, which represents its various design attributes. To enable user friendliness during design, specific design attributes relevant to each individual feature are identified. These are presented to the user in the form of Feature Templates. After the user enters data through the templates, it is validated for correctness and then entered into the database for further session use. Visualization module simultaneously processes and dynamically updates the CAD model after each feature input. The final CAD model is stored in the database in a native format (*.PRT) similar to STEP for future use. These steps are described in details one by one.

Feature Design Templates

The local features added on a gross part are mainly of two types: Face based features (Holes, Pockets) and Edge based features (Steps, Slots). Feature templates are designed for each feature type to acquire specific feature related information /parameters from the user.

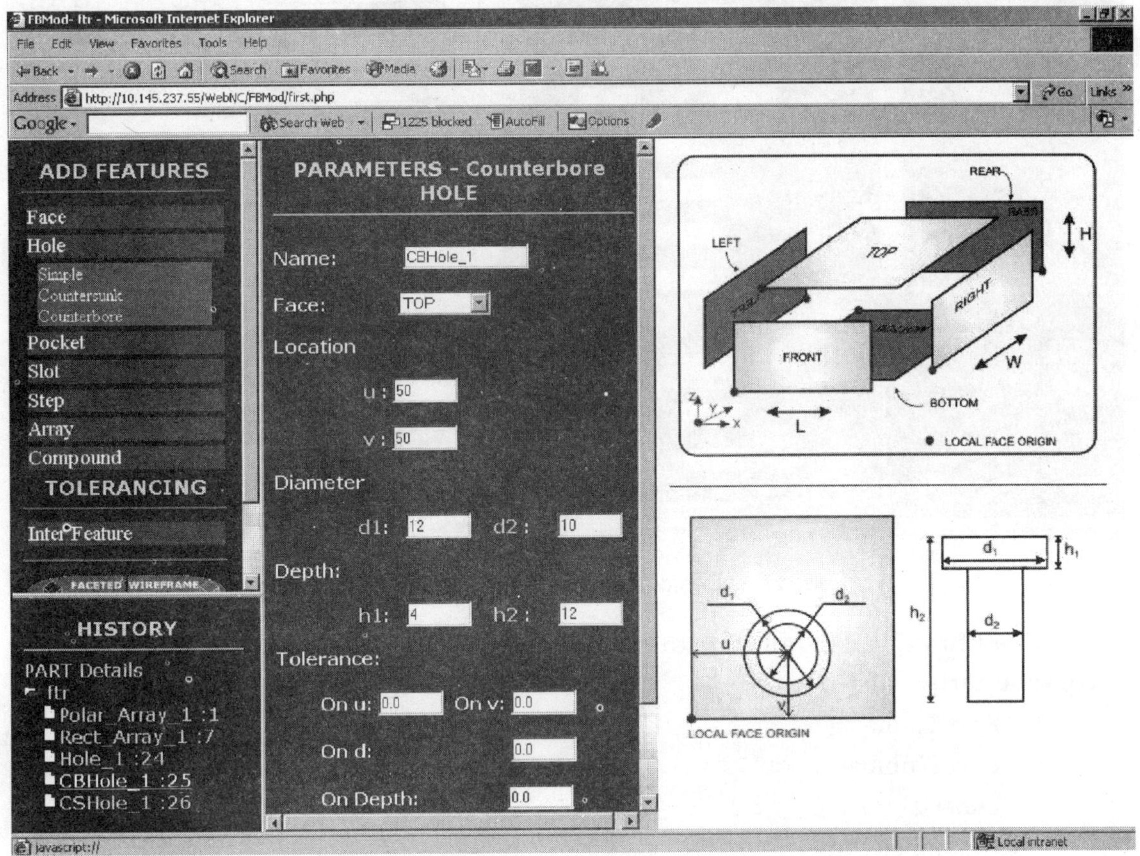

Fig. 17.3: Design template–counterbore hole

Figures 17.3 and 17.4 show typical screens showing Feature Selection Menu, Feature Parameter input Template, Feature Images and the modeling History tree. The Feature Parameter input and the associated images help user in feature based modeling in a very efficient manner.

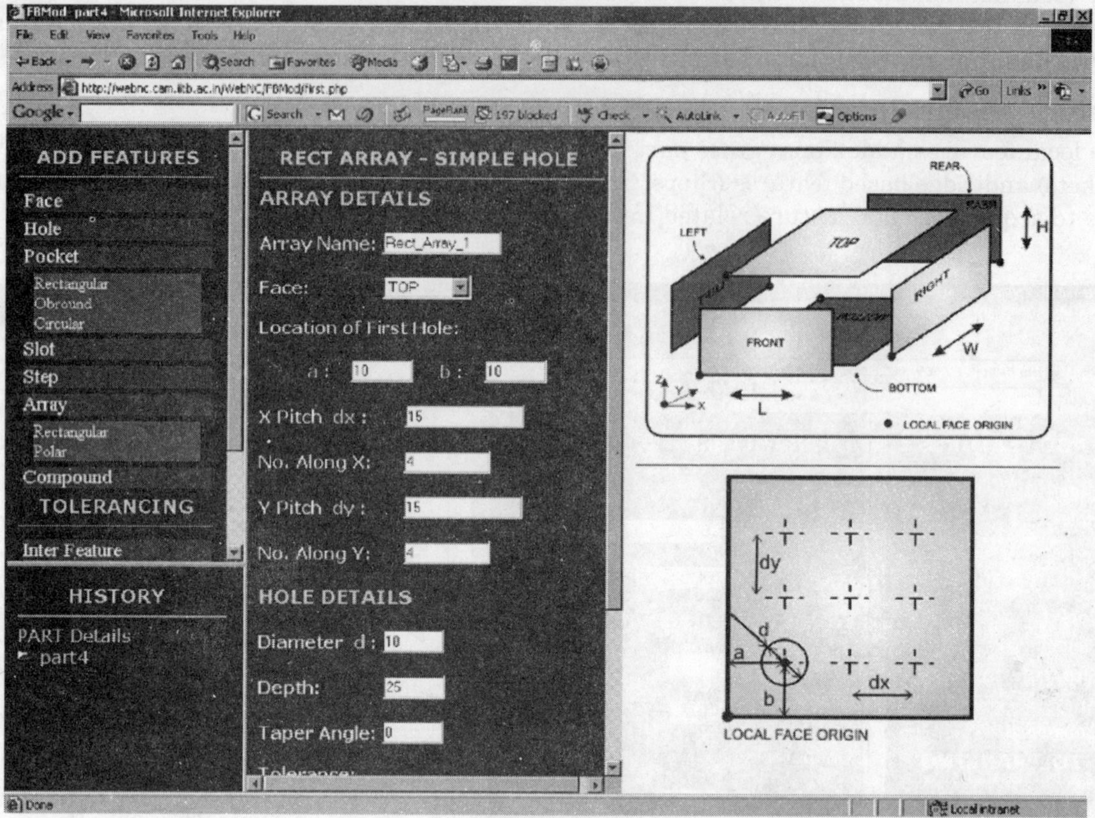

Fig. 17.4: Design template—rectangular array

Important functional parameters to be acquired for features are enumerated below:

Gross Features

- Part Name
- Part Number
- Material
- System of Units

Local Features

- Feature Name
- Face on which it occurs
- Feature Location
- Design Parameters <Generic>
- Feature Orientation <Options>
- Tolerances <Intrinsic>
- Miscellaneous attributes <Feature Specific, if any>

Inter Feature Tolerances

- Feature to be toleranced
- Datum
- Type of Tolerance
- Tolerance Value

FBMod has been designed to provide facilities for imparting feature tolerances belonging to the following two categories.

17.5.1.1. Intrinsic tolerances: These are feature specific tolerances on attributes of a feature (e.g., tolerance on hole diameter, depth, position, etc.). Designer can interactively input these in the feature design template provided.

17.5.1.2. Inter feature tolerances: These essentially comprises of tolerances, such as Parallelism, Perpendicularity, Concentricity, etc. Specific utilities have been provided in the design templates for specifying inter-feature tolerances, in terms of toleranced Feature, Datum, Tolerance Type and Magnitude.

History Tree

A Tree indicating the transaction history of the modeling session is dynamically created and displayed in the design template. This tree lists the features in a hierarchic fashion. The nodes of the tree are active links, which upon clicking enable a client to Edit feature parameters during a session to enable design synthesis. It also permits Feature Deletion.

The modeling process completes with the specification of all the feature parameters and the tolerances. To qualify as a valid part model, the features and associated parameters have to be validated for consistency and correctness. FBMod automatically carries out feature validation after every user input in the design session.

Feature Validation

Two types of validation checks need to be carried out on Features viz. Syntactic and Semantic validation checks. Feature specific syntactic validation checks are implemented at the client end. Syntactic validation carried out on Feature data ensure issues like:

- No data field is empty.
- Specific data fields do not have Illogical Values. (Zero, Negative, Alphabet, Symbol)

Semantic checks on the other hand, ensure the topological validity of features modeled. Table 17.1 shows typical semantic checks, which are carried out at the server end. Design for Manufacture (DFM) guidelines are incorporated for the range of features (Fig. 17.2) with permitted range of parameter values. In case of an inconsistency, the user is prompted to check the values of the input parameters before proceeding.

Table 17.1: Typical semantic validation checks for a simple hole

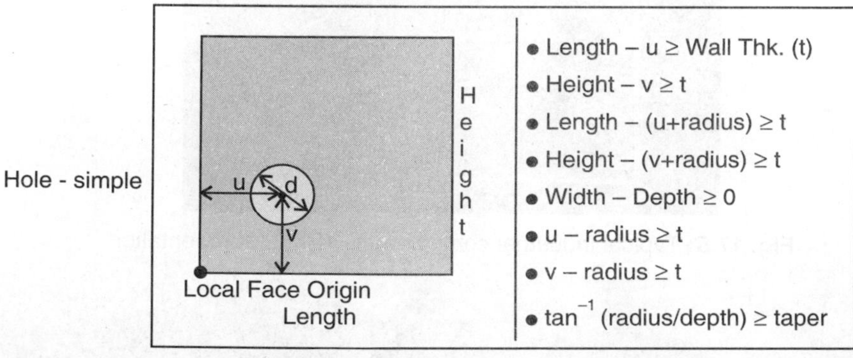

Model Representation

FBMod has been designed to provide Feature specific images, which are context sensitive. These images guide user in inputting the design parameters. FBMod gathers and processes this modeling information in a Transaction mode to capture the history of modeling. PHP scripting is employed to process this design data from the modeling session.

FBMod processes the transaction history and stores the CAD part model in the native *.prt file format which is very similar to STEP. This model file is stored on the server in the corresponding users account, which can be downloaded and further linked with down line application tasks. The *.PRT model file can be easily translated into any standard CAD format using a suitable CAD data translator.

Part Model Visualization

During the modeling session, user provides feature input using Design Templates (Fig. 17.3) to synthesize the part model. This transaction data is transferred to the server where the CAD model is dynamically created and stored in the *.PRT format.

To enable viewing of the CAD part model, the model data needs to be transmitted to the client in proper visualization format. This is a very tricky issue. Model with finer resolution (with more faces) will improve the model rendering accuracy but its verbosity will slow down rendering due to the slow network speed (bandwidth), even leading to the chocking in the data pipeline. To tackle this problem, a novel solution has been provided in FBMod. Two separate algorithms are written which run on the server to translate the CAD model into Faceted or Wireframe visualization forms. User is displayed both the options and based on the choice, the CAD part model (*.prt) is automatically translated to (*.wrl) file. (faceted or wireframe) which can be viewed by any standard VRML player. (Cartona, Cosmo, etc.) Figure 17.5 shows typical CAD model and its rendering in FBMod using VRML.

Client can carry out various operations on the rendered model such as ZOOM / PAN / 3D ROTATE / STUDY / FIT etc. FBMod provides a very high degree of immersivity for the user to plunge into the CAD model. The model is very compact in data size.

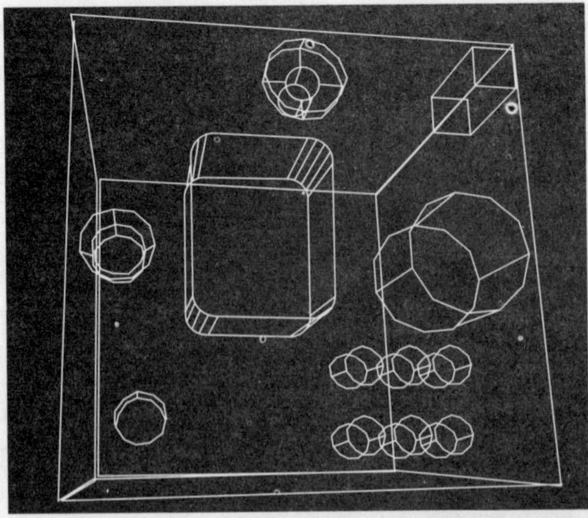

Fig. 17.5: Typical industrial components: VRML representation

17.6. IMPLEMENTATION AND TESTING

FBMod has been designed and implemented on the server setup at CAM Laboratory, IIT Bombay using a dedicated Pentium IV PC. Apache web server with PHP 5.0 has been used as a platform. At the client end, HTML and javascript is used while the server side programming is done in PHP. MySQL database is used for data handling.

FBMod has been extensively Beta tested using Internet from different client locations distributed within the country as well as overseas having different operating systems like Windows 98, Windows 2000 and browsers like Microsoft IE and FireFox. A large number of industrial components chosen from medium/large scale manufacturing industries using CNC machines were modeled successfully. The performance of the Server has been tested under various loads and has been found to be very stable.

The Beta testing clients were chosen from various industries including those using stand alone commercial CAD modelers. The client feedback during Beta testing of FBMod has been very positive in terms of its user friendliness, ease of model creation/editing, visualization, compact data storage and efficient communication over the Internet.

17.7. CLOSURE

FBMod provides a user-friendly environment for modeling complex prismatic components over the www using a feature based modeling paradigm. FBMod has been hosted on the website and extensively Beta tested from different geographical locations, both within the country and overseas, to model prismatic parts with varying complexities, network speeds and operating system requirements. It was found to provide an efficient, robust user-friendly modeling environment. FBMod has been designed to act as a front end for various manufacturing services that may be provided over the www. Further work is in progress to link it with web based CAPP system suitable for CNC machining centers.

REFERENCES

1. Qiang, J. and Michael, M.M., "*A Dempster–Shafer Approach for Recognizing Machine Features from CAD Models*", Pattern Recognition, Vol. 36, No. 6, 2003, pp. 1355–1368.
2. Kulkarni, V.S. and Pande, S.S., "*A System for Automatic Extraction of 3D Part Features Using Syntactic Pattern Recognition Techniques*", International Journal of Production Research, Vol. 33, No. 6, 1995, pp. 1569–1586.
3. Venkataraman, S., Sohoni, M. and Kulkarni, V.S., "*A Graph based Framework for Feature Recognition*", Proceedings of the sixth ACM symposium on Solid Modeling and Applications, Ann Arbor, Michigan, United States, 2001, pp. 194–205.
4. Lalit, P. and Pande, S.S., "*An Intelligent Feature-based Process Planning System for Prismatic Parts*", International Journal of Production Research, Vol. 40, No. 17, 2002, pp. 4431–4447.
5. Liu, X., CFACA: Component Framework for Feature-based Design and Process Planning, "*Computer Aided Design*", Vol. 32, No. 7, 2000, pp. 397–408.
6. Maropoulos, P.G. and Baker, R.P., "*Integration of Tool Selection with Design*", Part 1: Feature creation and selection of operations and tools, Journal of materials processing technology, Vol. 107, No. 1–3, 2000, pp. 127–134.

7. Kim, J.H., Wang, F.C., Sequin, C. and Wright, P.K., "*Design for Machining Over Internet*", Proceedings of the 1999 ASME Design Engineering Technical Conference (DETC) on Computer and Information in Engineering, Las Vegas, NY, Paper No. DETC99/CIE-9082, 1999.

8. Chan, S.C.F., A Solid Modeling Library for the World Wide Web, "*Computer Networks and ISDN Systems*", Vol. 30, No. 20–21, 1998, pp. 1853–1863.

9. Kim, C.Y., Kang, S.–H., Kim, N. and Grady, P.O., "*Internet-based Concurrent Engineering: An Interactive 3D System with Markup*", Proceedings of the 1998 ASME Design Engineering Technical Conference (DETC) on Computers and Information in Engineering, Atlanta, Georgia, Paper No. DETC98/CIE-5522, 1998.

10. Bidarra, R., Van Den Berg, E., and Bronsvoort, W.F., "*Collaborative Modeling with Features*", Proceedings of the 2001 ASME Design Engineering Technical Conference (DETC) on Computers and Information in Engineering, Pittsburgh, Pennsylvania, Paper No. DETC01/CIE-21286, 2001.

11. Shyamsundar, N. and Gadh, R., Collaborative Virtual Prototyping of Product Assemblies over the Internet, "*Computer Aided Design*", Vol. 33, No. 9, 2001, pp, 755–768.

12. Niranjan A.S., "*Collaborative Intelligent Product Modelling and Manufacturing*", Masters Thesis, Indian Institute of Technology, Bombay, 2003.

SECTION - II

COMPUTER AIDED PRODUCT DESIGN

COMPUTER AIDED DESIGN (CAD)

18.1. INTRODUCTION

Till much recently and even now in most industries of the world, the different stages of the product life cycle are understood to be discrete and disconnected 'boxes' of activity and the people in each would complete 'their' job and push the design/part to the next box in the cycle. That often used to lead to a great communication gap among the people and result in long lead times and many redesign efforts. This is because the design engineers, for example, never were concerned about the limitations and constraints of the manufacturing shop-floor and the limitations of the inspection department. They thought that their design was correct as long as it had the features that were desired and that could deliver the functionality, part was of correct size, had correct kinematics and did not have stress anywhere in it exceeding the permissible limits. However, the manufacturing facilities available may not be able to produce the product economically, the design may not have any feature that enables either inspection or maintenance or the design may not be environmentally sustainable, and so on. Due to these reasons the part would get redirected to the design department for redesign resulting in wastage of time and efforts. Better method would be to design the part by taking decisions in collaboration with people from all departments. Thus the modern paradigm of engineering design calls for a close interaction and collaboration among the personnel belonging to various stages of the product life cycle (Fig. 18.1).

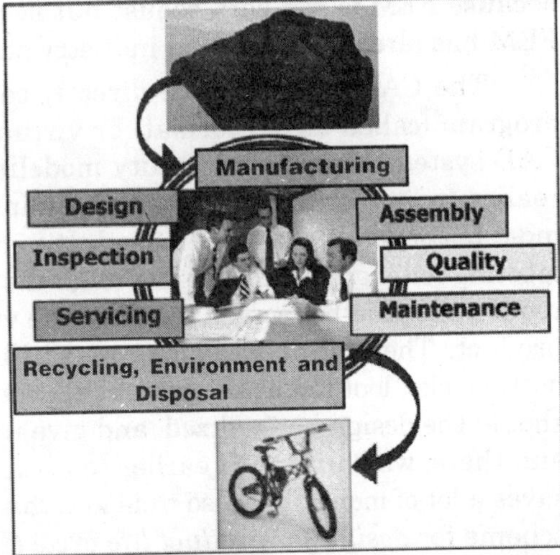

Fig. 18.1: The need for collaboration among the different stages of the product life cycle for a quality product with minimum lead time

18.2. CAD SYSTEMS

A CAD system is a complex application requiring both CAD hardware and CAD software. The CAD software is a computer program written in C, C++ Scheme or Java and has multilayered GUI providing menus and icons. It may run in client/server mode. Installing CAD software is very easy! Did you ever install AutoCAD or Pro/Engineer? Try once. A data structure is used to store both the geometry and topology. Geometry is the coordinate data whereas topology is the scheme of connectivity among the coordinate data in order to build curve, surface or solid. CAD/CAM signifies integration of them. CAD database—file(s) that stores information requires a suitable database management system (DBMS). The learning curve of CAD software is very steep signifying a high amount of orderliness in the knowledge domain of CAD.

18.3. MODULES IN CAD APPLICATIONS

The current day CAD systems have different modules, true to the need to embed all features of its product life cycle and to seamlessly port the solid model data into the needs of the other stages of the product life cycle. This maximizes not only the utility of the CAD solid model data but also increases the efficiency of operation in the downstream stages of the product life cycle and drastically reduces the lead time. The main module which is the geometric engine enables the geometric modeling activity to convert ideas into conceptual design or conceptual designs into completely synthesized design. Further, the geometric engine offers the great advantage of applying time-to-time modifications in the design on the fly with minimal rework required. It also serves as the base for new product design so that every product modeling need not be started from scratch. The application module enables analysis for critical parameters and kinematics in the identified critical zones of the part to consolidate and justify the design decisions made with regard to shape, dimensions and material. It is important to notice that the natural order of these two design steps is first synthesis and then analysis but not the other way round. The finite element method (FEM) is the most popular tool among all the existing CAD systems because FEM is not very robust but is capable of solving a wide variety of problems. In fact FEM has already become an industry standard for most types of analysis.

The CAD model can be directly converted into NC part program or rapid prototyping program (called as .stl format) or virtual reality modeling format (called as .wrl) within the CAD system. The virtual reality modeling language (VRML) is a CAD based software tool to realize manufacturing, assembly and inspection in the virtual space so that at no extra cost most bottlenecks can be removed. It is not uncommon in industries these days to convene a meeting for engineers belonging to all the stages of the product life cycle in a U-table conference room wherein a big screen displays the virtual manufacturing, assembly and inspection of the product. The engineers belonging to the various department can easily identify potential bottlenecks, looking at the virtual show of what they will have to carry out in their shop floor should the design be finalized, and give their suggestions on the necessary design changes then and there, which is much earlier to actual physical manufacturing or assembly. That not only saves a lot of money but also time and this the methodology that forms the main implementation scheme for *design for product life cycle (DPLC)*.

Once the necessary changes in the design are incorporated, the rapid prototyping program in .stl format can be generated by a click of a button and the connected rapid prototyping machine will produce the prototype overnight by a layer by layer additive process. This provides a physical part for conducting other types of check. It has been reported that prototypes made by RP after proper DPLC brainstorm were hardly ever required further changes. Programming module in the CAD system enables direct generation of the NC part program for those parts that need to be machined by a computer numerical control (CNC) machine. Most CAD systems also provide a virtual run of this NC code to show the part might be machined in a given machine. In order to generate the NC code that is specific to a particular make and model of CNC machine so that its process capability indices (for example maximum and minimum speed, feed and depth of cut settings) are correctly reflected in the code, the CAD system provides the large list of code numbers belonging to different makes and models of CNC machines. Selecting a particular code automatically customizes the NC code generated to that machine. Users can read more and try out this feature, for example, in the CATIA package of Dassault Systems. Often it is required to send the CAD solid model data to another CAD system which can be achieved by using a neutral format. Initial graphics exchange specification (IGES) and standard for exchange of product model data (STEP) are the most popular and oldest and latest in that order among the available neutral formats. For orthographic projections DXF format is suitable. Communication module in the CAD system provides these import and export facilities.

Finally, as the collaborative design is gaining wider popularity for its ability to combine multidisciplinary expertise and to enable communication among globally placed designers through their desktops in order to design products, most CAD systems are now-a-days incorporating Internet based Collaborative module to enable participative design. This module is nothing but the collaborative design workbench (CDW) we discussed in an earlier chapter. Although a lot of developments are expected in the existing collaborative modules, the emerging fast research work in this field may result in fine tuning of these tools quite dramatically. Collaborative module can be so powerful as to enable the earlier mentioned DPLC brainstorm to take place effectively without even necessitating the physical U-table conference among the engineers of various stages of the product life cycle. They can communicate right from their office chambers through their desktops. A large scale improvement in the existing form of CDW is, still needed before this can be successfully implemented, though.

18.4. ACQUIRING CAD SYSTEM

CAD systems can be low-end, mid-range or high-end types. Low-end systems are meant for beginners and less complex geometry only can be modeled in them. Examples of such systems are AutoCAD, Autodesk, Inventor, CADKEY. Mid-range systems meet the need to have complex models and support design and manufacturing needs. Examples of such systems are SolidWorks, Pro/E, and MasterCAM. High-end systems are those legacy systems that have been around since 1970s offer modeling, analysis, manufacturing, complex products such as airplanes, cars etc. For example Unigraphics, Parasolid, I-DEAS, CATIA fall in this category. Specialized CAD systems provide a very robust and universal geometric modeling and offer graphics kernel, examples for which are ACIS and Parasolid. Thus, if you are a CAD engineer incharge of procuring a suitable CAD systems for your company, you may think carefully by weighing the specific needs of your company in terms of product category, how much of the design is dependent a proper choice of the CAD system, what are the application modules required and how many people are going to use before finalizing on any one system to purchase.

The general guidelines on how to choose among various CAD systems should be based on the characteristics of a generally good CAD system that is easy to learn, easy to work with, having macro commands, having its own embedded programming language, having an open database and is compatible with other CAD/CAM and analysis program.

18.5. USING A CAD SYSTEM

Most CAD systems offer three different coordinate systems, namely world coordinate system (WCS), user coordinate system (UCS) and screen coordinate system (SCS). World coordinate system is the reference space of the model with respect to which all the model's geometrical data is stored in the system's database while saving and quitting. It uses a right handed coordinate system. Obtaining orthographic views is retrieving information from the world coordinate system planes. Data storage into and retrieval from the database is always done in world coordinate system and after converting into cartesian (x, y, z) coordinate system. The data entry can be in the work coordinate system whereas data entry by user is possible in cartesian, cylindrical or spherical coordinate system. It is useful when a desired plane of construction is not easily defined as one of the world coordinate system orthogonal planes, for example, inclined faces of a model on which a hole exists. User can conveniently define a cartesian X-Y user coordinate system that is coincident with the desired plane of construction facilitating the geometric construction. The user coordinate system can be established with its origin and coordinate directions that are conveniently coincident with a corner and edges of the face so that the feature on that face can be easily located with respect those user coordinate axes. CAD systems implement the data entered with respect to the user coordinate system after converting the same into the world coordinate system.

For example, in order to locate the centre of the circle forming the cylindrical through hole on the corner face of the block as shown in Fig. 18.2, it is much difficult to evaluate the coordinate of the centre with respect to the X-Y-Z world coordinate system. On the other hand, it is much easier to first place the block and to generate the face with respect to the world coordinate system and then to locate the centre of the cylindrical hole with respect to a user coordinate system. In order to locate the user coordinate system in a CAD system, all that you need to do is to give any three non-collinear points and tell the system which one of these you desire to be the origin of the user coordinate system. Alternately, a point and selection another existing plane will do. The CAD system automatically will generate the entire user coordinate system for the user. What happens in terms of mathematics behind such operations can be useful for the user if he or she is engaged in developing one's CAD system. If the point P_0 is chosen to act as the origin of the UCS, points P_0 and P_1 give one unit vector along the x-axis of the UCS.

$$\hat{n}_1 = \frac{\bar{P}_1 - \bar{P}_0}{|\bar{P}_1 - \bar{P}_0|}$$

In order to find the unit vector along the z-axis, $\hat{n}_3$, we define another intermediate unit vector, $\hat{n}_4$, as follows:

$$\hat{n}_4 = \frac{\bar{P}_2 - \bar{P}_0}{|\bar{P}_2 - \bar{P}_0|}$$

$$\hat{n}_3 = \frac{\hat{n}_2 \times \hat{n}_4}{|\hat{n}_2 \times \hat{n}_4|}$$

Finally, we obtain the unit vector along the y-axis of UCS from $\hat{n}_1$ and as follows:

$$\hat{n}_2 = \frac{\hat{n}_3 \times \hat{n}_1}{|\hat{n}_3 \times \hat{n}_1|}$$

The centre of the cylinder can now be easily located with respect to this UCS as follows:

$$\bar{C} = x_u \hat{n}_1 + y_u \hat{n}_2$$

Here $\bar{C}$ is the position of the centre of the cylinder with respect to the UCS, and x_u and y_u are the corresponding coordinate values. The radius to be input is a scalar quantity and if the blind hole depth of the hole can be specifed as $\bar{h} = -d\hat{n}_3$ where d is the blind hole depth.

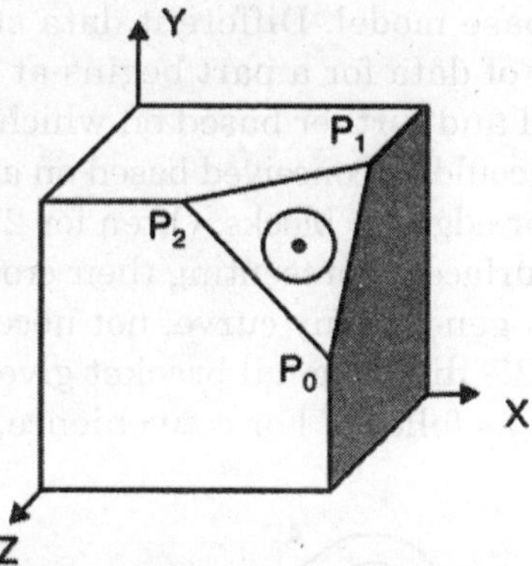

Fig. 18.2: An example of locating the user coordinate system

As it was mentioned before, the CAD system calculates UCS data back to WCS in order to store in the database. This requires a homogeneous three dimensional transformation matrix that applies on the UCS coordinate data matrix as post-multiplication.

$$[P]_w = [P_U][T]$$

$[P]_w$ = Point relative to world coordinate system (WCS) = $\begin{bmatrix} x_W & y_W & z_W & 1 \end{bmatrix}$

P_U = Point relative to UCS = $\begin{bmatrix} x_U & y_U & z_U & 1 \end{bmatrix}$

$[T]$ = The homogeneous transformation matrix, which is given by the following:

$$\begin{array}{c} \\ x_U \\ y_U \\ z_U \\ \\ \end{array} \begin{array}{c} \begin{array}{ccc} x_W & y_W & z_W \end{array} \\ \begin{bmatrix} r_{11} & r_{12} & r_{13} & 0 \\ r_{21} & r_{22} & r_{23} & 0 \\ r_{31} & r_{32} & r_{33} & 0 \\ x_d & y_d & z_d & 1 \end{bmatrix} \end{array}$$

As given above, both the points are represented in the homogeneous form with the homogeneous scaling factor being kept as 1. The r_{ij} sub-matrix in the $[T]$ is meant for rotation and its elements are determined by cosine of the angles between coordinate axes in the UCS to those in the WCS. The translational sub-matrix is substituted by x, y and z-directional translations that UCS origin has been effected upon from the WCS origin.

The major benefits of a dedicated CAD system are as follows:

- Automates repetitive tasks
- Reduces duplicate efforts
- Allows you to communicate with others via electronic means
- Makes drawing revisions easier to perform
- Enable you to customize environment
- Lets you to create accurate and high-quality drawing
- Provides impressive presentation
- Supports a wide selection of hardware

CAD systems are made up of two major constituents, software and hardware. The CAD software deals with a suitable data structure in which the solid model data is arranged and stored as per a suitable database model. Different data structures and database models are available. Since the hierarchy of data for a part begins at the lowest level of points, based on which the edges can be defined and further based on which the different portions of the part is defined, the data structure also could be conceived based on any one level. Thus the data structure could be based on vertex data or edges or blocks. Often for 2½ dimensional models solid portions can be simply represented by surfaces representing their cross-sections. These surfaces are made up of edges. An edge can be in general any curve, not necessarily always a straight line only. For example, considering the 2½ dimensional bracket given in Fig. 18.2, the vertex, edge and block data structures could be as follows. For convenience, the part's schematic is reproduced in Fig. 18.3.

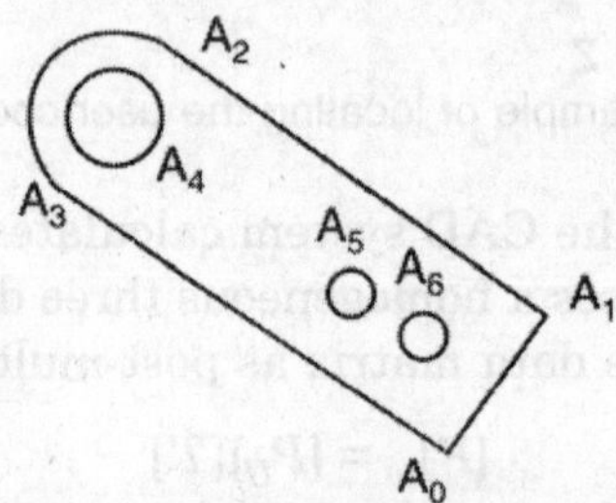

Fig. 18.3: Schematic of the bracket

The data structure based on the vertices would be formed by the vertices A_0 through A_6 describing how they form the different edges. The data structure based on the edges is formed by the different edges each of which will consist of one or two vertices. The data structure based on blocks is defined by Boolean algebra as the total block minus the three cylindrical volumes forming the holes.

A database management system (DBMS) prescribes the schema according to which the data should be stored in the computer memory for easy storage and retrieval. Thus a DBMS may be defined as an "organized schema for collection of graphics and non-graphics data to be stored on secondary storage in the computer." The objective is to collect and maintain data in a central storage that it will be available for operations and decision making. A properly developed database will eliminate redundancy, enforce standards, apply security restrictions, maintain integrity and balance conflicting requirements. The most popularly known DBMS are relational database (RDBMS), hierarchical database (HDBMS), network database (NDBMS), and object oriented database (OODBMS).

Developed in 1970s, the RDBMS is conceptually simple but versatile and was considered as a major breakthrough for both users and designers. RDBMS requires a lot of computing power but the modern computers are powerful enough to handle relational databases. RDBMS is perceived by the user as a collection of tables in which data are stored. Each table consists of series of row and column intersections. Tables are related to each other by sharing a common entity characteristic. The relationship type is often shown in a relational schema which is defined rigorously yet pragmatically by RDBMS. A table is a logical structure and yields complete data and structural independence. The distinguishable advantages of RDBMS are data independence and structural independence, easy to design the database and to manage its contents, less programming effort required and supported by powerful and flexible query capability named as Structured Query Language (SQL). SQL is a fourth generation language (4GL) and specifies "what to do" not "how to do". It was introduced by IBM in 1974. RDBMS suffers from the disadvantages that it requires substantial hardware and operating system overhead and it tends to be slower than other database systems.

In the HDBMS, data records are organized in an inverted (upside-down) tree structure, hence the name hierarchical structure. Figure 18.4 shows a typical HDBMS.

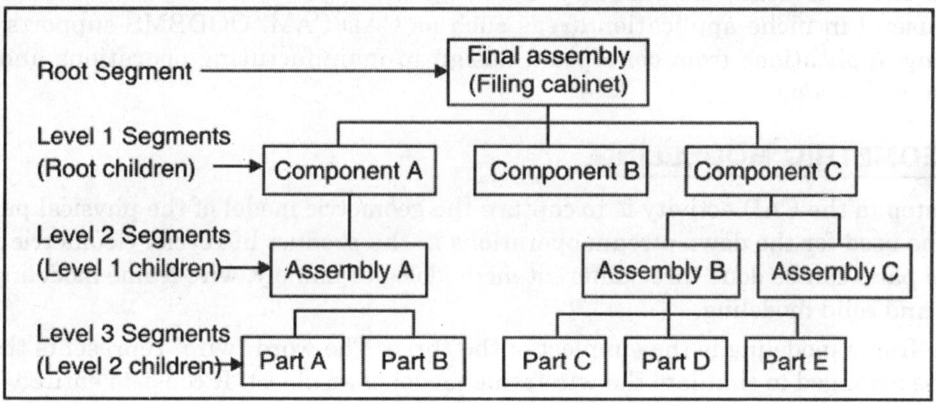

Fig. 18.4: HDBMS for a general product that is an assembly of a set of components

HDBMS can enable maintenance of one-to-many relationship in which each parent can have many children while each child has only one parent. Both top-down and bottom-up traversals are possible thus offering a choice for tracing the required data item in the shorter path. HDBMS has many advantages including data sharing and security provision, data independence resulting in reduced programming and maintenance effort, database integrity providing for the check that a child must have a parent, and efficiency in dealing with a large database containing data with predominantly 1:M relationships because many business applications fall in this category.

HDBMS suffers from certain disadvantages such as the requirement of knowledge of physical level of data storage, inability to represent relationships that do not conform to the hierarchical 1:M standard, applications programming being complex where the programmer must know the database structure and how to traverse it, lack of *ad hoc* query capability for end users, lack of standard concepts and implementation because no standard was developed, and limited portability requiring extensive programming activities to use the database. The last is perhaps the most serious of the problems.

The NDBMS defines a relationship as a set. Each set is composed of at least two record types: an owner (parent) record and a member (child) record. Thus the various relationships among the records are decomposed into a series of sets. The advantages of NDBMS include easier implementation of M:N relationships, more flexible data access compared to HDBMS because an application can access an owner and all the members within the set and one can also access owner from a member. Further, NDBMS possesses enforced data integrity according to which owner record must be defined before defining member records. The shortcomings of NDBMS are very complex structure from the application programmer's point of view, difficult to design and use properly because of the navigational nature of the data structure, and difficult to make changes in a database due to lack of structural independence.

In OODBMS, objects are abstractions of real-world entities. For example, all rotational parts can be abstracted as a cylinder with floated dimensions the different instances of which will yield the required rotational parts. Similarly student, invoice and employee are examples of objects. The objects are capable of capturing both the attribute and behavior (methods). Objects with similar characteristics are grouped together as a class and classes are organized in a class hierarchy. A subclass inherits the attributes and methods from its super class. OODBMS provides a richer modeling language by capturing attributes, methods, and class hierarchies and it is useful in niche application areas such as CAD/CAM. OODBMS supports multiple engineering applications from conceptual design to manufacturing operations and permits dynamic modification.

18.6. GEOMETRIC MODELLING

The first step in the CAD activity is to capture the geometric model of the physical part, which can then be used for the down stream operations in the product life cycle. Geometric modeling of physical parts can be done three different methodologies namely, wire-frame modeling, surface modeling and solid modeling.

Wire-frame modeling is the simplest of the three. The word "wire" represents that a bent wire can be arranged to simulate the wireframe model of an object. It consists entirely of points, lines, arcs, circles, other conics, other polynomial curves and synthetic curves. Almost all CAD packages offer wireframe methodology of building the part because it was the first to be invented and most packages began with it. Moreover, wire-frame modeling is a natural sequel to the traditional methods of drafting using a drafting instrument and French curves and the terminology is very simple. Current day wire-frame modelers support a lot of add-on functionalities also such implicit and explicit data input in both absolute and incremental modes and two dimensional as well as three dimensional part building. Geometric modifiers such as snap-on facility to points also are enabled in present day CAD systems. The disadvantages of the wire-frame modeling include the data input being laborious, ambiguity in giving the correct impression of the physical shape of the part, lack of both geometrical and topological information embedded in a single representation. Geometrical data means the coordinate or vertex or other similar numerical data. Topological information means how the various entities such as vertices, lines or edges are connected to each other.

Figure 18.5 shows the wire-frame model of a block with a slot and a hole in the slot represented wholly with edges of the object.

Fig. 18.5: Wire-frame model of a block. Is the top face of the slot visible or hidden?

Solid modelling (SM) overcomes most of these ambiguities of wire-frame modelling because SM requires only geometrical data (one corner points *P*, length *L*, width *W* and height *H*) as topological information is in-built in its representation.

Wire-frame modelling has its building blocks both analytical entities (those curves having a default mathematical equation representing the geometry) and synthetic entities (those curves whose equations have been built to either interpolate or approximate a given set of vertices). Analytic curves are line, circle, ellipse, parabola, hyperbola and other conics and synthetic curves are the Hermite cubic curve, Bezier curve, B-Spline curve and non-uniform rational B-Spline curve. Popularly, the segmentation approach is followed according to which the actual required shape of the part is obtained by suitably joining or blending a set of carefully chosen curve segments. Each curve segment could be either an analytic curve or synthetic curve having a well-defined equation.

A curve segment can be represented either non-parametrically or parametrically. Non-parametric representation in turn could be either explicit or implicit. Either way, the non-parametric representation is highly unsuitable for CAD based geometric modelling.

Explicit non-parametric representation of a general 3-D curve can be written as follows:

$$\mathbf{P} = [x, y, z]^T = [x \;\; f(x) \;\; g(x)]^T$$

Here $\mathbf{P}$ = position vector of any point *P* on the curve (Fig. 18.6). Implicit non-parametric representation of a curve is either one single equation such as like that of a circle, $x^2 + y^2 = r^2$, or the intersection of two surfaces defined by

$$F(x, y, z) = 0 \quad \text{and} \quad G(x, y, z) = 0$$

Given a value of *x*, it is a laborious process to solve for *y* and *z* which is the problem of finding the intersection of the above given two surfaces.

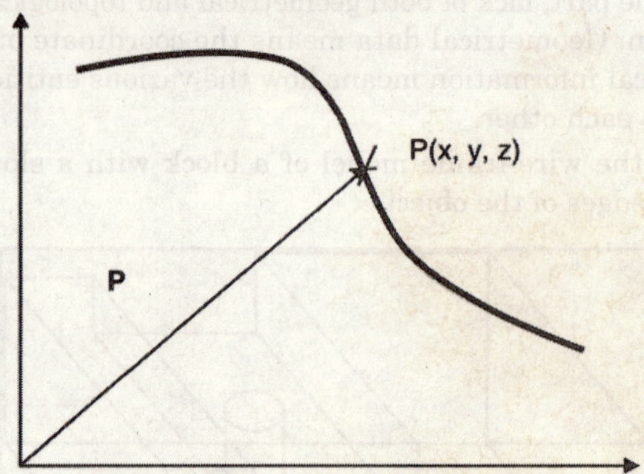

Fig. 18.6: Representation of a curve by the position vector of any point on the curve

Explicit non-parametric representation cannot be used for closed curves (like circles) and multi-valued curves (like parabolas) because it is a one-to-one relationship. The above problem is overcome by implicit representation but the latter is laborious. Infinite slope situations cannot be dealt with in computer program in non-parametric representation.

Whereas, the parametric representation overcomes all the difficulties of the non-parametric representation. It allows multi-valued and closed functions to be easily defined. It replaces the use of slopes with that of tangent vectors. The equations are polynomials and thus computationally more suitable than equations involving roots.

In parametric form, each point on a curve is expressed as a function of a parameter u. The parameter acts as a local coordinate for points on the curve. The parametric equation for a 3-D curve in space is

$$\mathbf{P}(u) = [x \;\; y \;\; z]^T = [x(u)\; y(u)\; z(u)]^T, \qquad u_{min} \le u \le u_{max}$$

Coordinates of a point on the curve are the components of its position vector. It is a one-to-one mapping from the parametric space (which is Euclidean space E^1 in u values) to the Cartesian space (E^3 in x, y, z values). To enable evaluation of slope of a parametric curve at any arbitrary point on it, the tangent curve must be evaluated Fig. 18.6. The tangent vector is a vector $\mathbf{P}'(u)$ in Cartesian space such that

$$\bar{P}'(u) = \frac{d\bar{P}(u)}{du}$$

The components of tangent vector in parametric space are

$$\mathbf{P}'(u) = [\, x' \;\; y' \;\; z']^T = [\, x'(u) \;\; y'(u) \;\; z'(u)]^T, \qquad u_{min} \le u \le u_{max}$$

The unit tangent vector is given by

$$\hat{n} = \frac{P'(u)}{|P'(u)|} = n_x\hat{i} + n_y\hat{j} + n_z\hat{k}$$

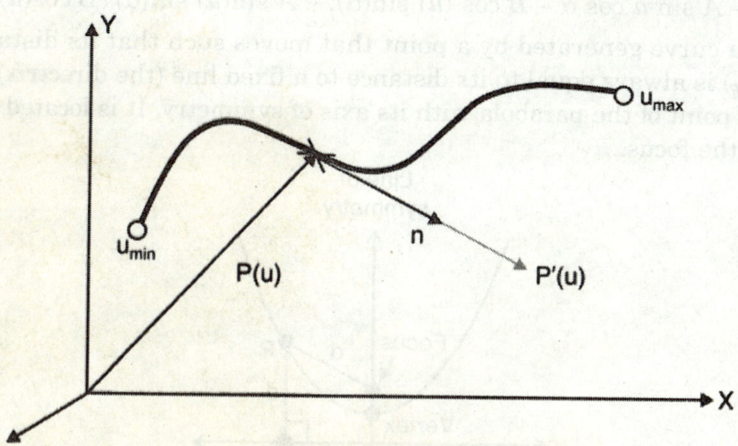

Fig. 18.7: The tangent vector, unit tangent vector and direction of the parametric curve

The simplest of all analytical curves, the straight line, can be represented either by means of its end points or by one end and direction as follows:

$$P = P_1 + (P_2 - P_1)u$$

The parametric representation of the circular arc is a special case of the full circle.

$$x = x_c + R\cos(u)$$
$$y = y_c + R\sin(u) \qquad u_s \le u \le u_e$$
$$z = z_c$$

Two concentric circles having their radii equal to the half-major axis and half-minor axis of the ellipse to be defined enable the construction of the parametric equation of the ellipse.

$$x = x_c + A\cos(u)$$
$$y = y_c + B\sin(u) \qquad 0 \le u \le 2\pi$$
$$z = z_c$$

Since an ellipse is not having circular symmetry, it has directional properties. When the major axis of the ellipse is inclined at α to the x-axis of the WCS, then it is called as a general ellipse. When α = 0, the resulting ellipse having its major axis parallel the x-axis of the WCS is called as the basic ellipse.

$$x = x_c + A\cos(u)\cos(\pm) - B\sin(u)\sin(\pm)$$
$$y = y_c + B\cos(u)\sin(\pm) + B\sin(u)\cos(\pm) \quad 0 \le u \le 2\pi$$
$$z = z_c$$

To find the tangent vector to an basic ellipse at any given point, its parametric equation needs to be differentiated with respect to u.

$$\mathbf{P}' = [-A\sin u,\ B\cos u,\ 0]$$

Similarly, differentiating with respect to u the parametric equation for general ellipse with major axis inclined at α to x-axis give its tangent vector.

$$P' = [-A \sin u \cos \alpha - B \cos(u) \sin(\alpha), -A \sin(u) \sin(\alpha) + B \cos(u) \cos(\alpha), 0]$$

Parabola is a curve generated by a point that moves such that its distance from a fixed point (the focus P_F) is always equal to its distance to a fixed line (the directrix). The vertex (P_V) is the intersection point of the parabola with its axis of symmetry. It is located midway between the directrix and the focus.

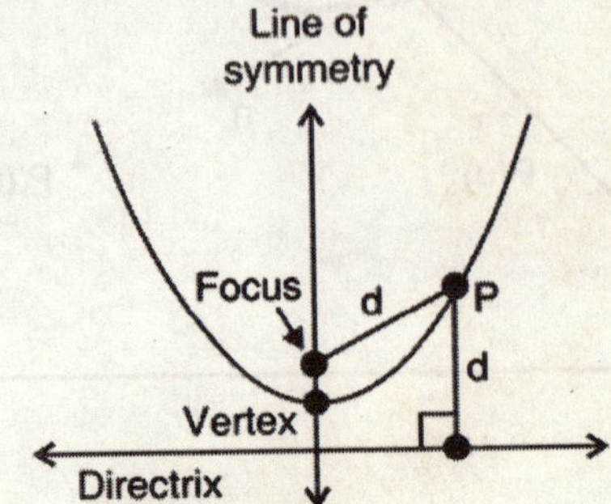

Fig. 18.8: The parametric parabola

The parametric equation of a basic parabola ($\alpha = 0$) is

$$x = x_v + Au^2$$

$$y = y_v + 2Au \qquad -\infty \le u \le \infty$$

$$z = z_v$$

Therefore the position vector of any point on the parabola is

$$\bar{P}(u) = \left[x_v + Au^2,\ y_v + 2Au,\ z_v\right]^T$$

The tangent vector is given by

$$\frac{d\bar{P}(u)}{du} = [2Au,\ \ 2A,\ \ 0]^T$$

At $u = 0$, that is at the vertex, the slope is ∞.

The parametric equation of the general (inclined) parabola with inclination $\underline{\alpha}$ is given by

$$x = x_v + Au^2 \cos\alpha + 2Au \sin\alpha$$

$$y = y_v + Au^2 \sin\alpha + 2Au \cos\alpha$$

$$z = z_v$$

$$\frac{d\bar{P}(u)}{du} = [2Au\cos\alpha - 2A\sin\alpha,\ \ 2Au\sin\alpha + 2A\cos\alpha,\ \ 0]^T$$

At $u = 0$, the slope is not unbounded unlike basic parabola, unless α is zero. The slope will be unbounded at $u = \tan \alpha$.

Finding the tangent to a parabola at any given point on it requires differentiating with respect to u the parametric equation for parabola with its axis of symmetry parallel to x-axis, we get

$$P' = [2Au, 2A, 0]$$

Differentiating with respect to "u" the parametric equation for general parabola with axis of symmetry inclined at α to x-axis, we get

$$P' = [2Au\cos\alpha - 2A\sin(\alpha), -2Au\sin(\alpha) + 2A\cos(\alpha), 0]$$

The transformation from WCS to MCS is

$$T = \begin{bmatrix} \cos\alpha & -\sin\alpha & 0 & x_v \\ \sin\alpha & \cos\alpha & 0 & y_v \\ 0 & 0 & 1 & z_v \\ 0 & 0 & 0 & 1 \end{bmatrix} \text{ for curve itself and}$$

$$T = \begin{bmatrix} \cos\alpha & -\sin\alpha & 0 & 0 \\ \sin\alpha & \cos\alpha & 0 & 0 \\ 0 & 0 & 1 & 0 \\ 0 & 0 & 0 & 1 \end{bmatrix}$$

for tangent vector.

Hyperbola is the curve generated by a point moving such that at any position the difference of its distances from the fixed points (called foci) F and F' is a constant and equal to the transverse axis of the hyperbola.

$$\frac{x^2}{a^2} - \frac{y^2}{b^2} = 1 \qquad \text{Non-parametric form}$$

$$\left.\begin{aligned} x &= \pm A\sec\theta \\ y &= \pm B\tan\theta \end{aligned}\right. \qquad \text{Parametric form: M–1}$$

$$\left.\begin{aligned} x &= x_v + A\cosh u \\ y &= y_v + B\sinh u \\ z &= z_v \end{aligned}\right\} \qquad \text{Parametric form: Method –2}$$

The General Equation for a Conic Section

$$Ax^2 + Bxy + Cy^2 + Dx + Ey + F = 0$$

Different values for combination of constants give different conic section curves. Geometrically speaking, all of the conic sections can be obtained as curves of intersection between a plane and a hollow open (means a funnel type) cone. A vertical intersecting plane containing the axis of the cone results in two straight lines whereas if it does not contain the axis of the cone, then it results in a hyperbola. An inclined plane passing through the base of the cone results in a parabola whereas if it does not pass through the base then an ellipse results. Finally, horizontal plane at different heights results in circles of different radii. The type of section can be found from the sign of: $B^2 - 4AC$.

If $B^2 - 4AC$ is...	then the curve is a...
< 0	ellipse, circle, point or no curve.
= 0	parabola, 2 parallel lines, 1 line or no curve.
> 0	hyperbola or 2 intersecting lines.

Synthetic curves represent a curve fitting or approximating (as needed) problem to construct a smooth curve that either passes through given data points or fairs or approximates the trend of the data points. Polynomials are the typical forms. Synthetic curves take up where the analytic curves leave – the latter are not that efficient at geometric design of free-form mechanical parts and hence the need for synthetic curves. Free-form parts can result from different scenarios such as like a sculpture made by a gifted sculptor, reverse engineering applications or the need to directly model a highly creative part in the CAD system by an iterative adjustment of shape. Some examples of such free form complex geometric design are car bodies, ship hulls, airplane fuselage and wings, propeller blades, shoe insoles, statue of an erstwhile national leader and aesthetically designed bottles. In free-form modeling the curve is to be represented by a collection of measured data points, called as the control points, and an existing curve must change to meet new design requirements. The best CAD system should be able to offer a real-time adjustment of the shape of the curve which requires a curve representation that is directly related to the data points. Then the curve will be flexible enough to bend, twist (tendency or change the shape by changing one or more data points in real time.

Most often, a complex is modeled by several curve segments pieced together end to end. This is the approach of "segmentation" because each time we analyzing a "curve segment" rather than the entire curve. Continuity at the joints of curve segments decides the degree smoothness of the curve. Various continuity requirements – C^0, C^1, C^2 – can be specified as indication of degree smoothness. Zero-order continuity (C^0) yields a position continuous curve whereas first (C^1) order continuity implies slope continuity. Second order (C^2) continuity implies curvature continuity.

A cubic polynomial is the minimum-order polynomial that can guarantee the generation of all three C^0, C^1 and C^2 continuities. The cubic order polynomial is the lowest degree polynomial allowing inflection within a curve segment. A cubic polynomial is the lowest degree curve that allows representation of non-planar (twisted) 3-D curves in space. Higher order polynomials (>3) are not used in CAD because of the disadvantages such as they tend to oscillate about the control points, they are computationally expensive and inconvenient, and they are uneconomical of storing curve and surface representations in the computer. The black ant is nobody's relative, neither customer's, nor vendor's nor the design engineer's! Hence there is no need to go for an order of continuity that is not warranted.

For efficient design, shape of the curve should be controllable most effectively in the easiest possible way. Type of input data plays a crucial role because the more easily the designer guess the shape of the curve for a given type of input the better the designer can control it. Control points and slope information are more easy to use than curvature information and in turn, control point information is even easier to specify that the tangent vectors. Two types of control exist, namely local control and global control. Global control is said to be present if change in one control point or tangent vector results in change of overall shape of the curve segment. Local control is said to be present if change in one control point or tangent vector results in change of shape of curve local to that point. Local control is more desirable than global control because the goal of the designer is always to adjust portions of the curve locally and achieve the desired shape.

Most commonly used synthetic curves are Hermite cubic spline. Each curve segment is defined for only 2 control points. It passes through the control points and therefore it is an interpolant and it has only upto C^1 continuity and Bezier Curve σ does not pass through the control points but only approximates the trend. It also has only upto C^1 continuity and B-Spline Curve is also most generally an approximator; an interpolating B-Spline is also sometimes possible. It has upto C^2 continuity.

Hermite Cubic Spline Curve Segment is used to interpolate the given data but not to design free-form curves. Cubic splines derive their name from "French curves or splines". Hermite **cubic** spline is one type of general parametric cubic spline–with degree equal to 3 and being determined by two data points and tangent vectors at the data points. Hermite Cubic Spline can be a 3-D planar curve or 3-D twisted curve. **Four conditions** are required to determine the coefficients of the equation–two end points and the two tangent vectors at these two points.

18.7. PARAMETRIC EQUATION OF HERMITE CUBIC SPLINE SEGMENT

$$\bar{P}(u) = \sum_{i=0}^{3} \bar{C}_i u^i, \quad 0 \le u \le 1$$

u = parameter

$\bar{C}_i$ = Polynomial(or algebraic) coefficients

In scalar form this equation is written as

$$x(u) = C_{3x}u^3 + C_{2x}u^2 + C_{1x}u + C_{0x}$$
$$y(u) = C_{3y}u^3 + C_{2y}u^2 + C_{1y}u + C_{0y}$$
$$z(u) = C_{3z}u^3 + C_{2z}u^2 + C_{1z}u + C_{0z}$$

In expanded vector form

$$\bar{P}(u) = \bar{C}_3u^3 + \bar{C}_2u^2 + \bar{C}_1u + \bar{C}_0$$

In matrix form

$$\bar{P}(u) = [U]^T[\bar{C}]$$

where

$$[U] = \begin{bmatrix} u^3 & u^2 & u & 1 \end{bmatrix}^T \text{ and } [C] = \begin{bmatrix} \bar{C}_3 & \bar{C}_2 & \bar{C}_1 & \bar{C}_0 \end{bmatrix}^T$$

$[C]$ = Coefficients vector

The tangent vector to the curve at any point is

$$\bar{P}'(u) = \sum_{i=0}^{3} \bar{C}_i i u^{i-1}, \quad 0 \le u \le 1$$

Consider the cubic spline curve segment with the two end points P_0 and P_1.

Applying the boundary conditions, $\bar{P}_0, \bar{P}_0'$ at $u = 0$ and $\bar{P}_1, \bar{P}_1'$ at $u = 1$, giving:

$$\bar{P}_0 = \bar{C}_0$$
$$\bar{P}_0' = \bar{C}_1$$
$$\bar{P}_1 = \bar{C}_3 + \bar{C}_2 + \bar{C}_1 + \bar{C}_0$$
$$\bar{P}_1' = 3\bar{C}_3 + 2\bar{C}_2 + \bar{C}_1$$

Solving the four equations simultaneously for the coefficients.

$$\bar{C}_0 = \bar{P}_0$$
$$\bar{C}_1 = \bar{P}_0'$$
$$\bar{C}_2 = 3\left(\bar{P}_0 - \bar{P}_1\right) - 2\left(\bar{P}_0' - \bar{P}_1'\right)$$
$$\bar{C}_3 = 2\left(\bar{P}_0 - \bar{P}_1\right) + \bar{P}_0' + \bar{P}_1'$$

Substituting in the parametric equation and rearranging

$$\bar{P}(u) = \left(2u^3 - 3u^2 + 1\right)\bar{P}_0 + \left(-2u^3 + 3u^2\right)\bar{P}_1 + \left(u^3 - 2u^2 + u\right)\bar{P}_0' + \left(u^3 - u^2\right)\bar{P}_1'$$

$\bar{P}_0, \bar{P}_1, \bar{P}_0'$ and $\bar{P}_1'$ are called as geometric coefficients.

$$\bar{P}(u) = \left(2u^3 - 3u^2 + 1\right)\bar{P}_0 + \left(-2u^3 + 3u^2\right)\bar{P}_1 + \left(u^3 - 2u^2 + u\right)\bar{P}_0' + \left(u^3 - u^2\right)\bar{P}_1'$$

$[V] = \left[\bar{P}_0 \quad \bar{P}_1 \quad \bar{P}_0' \quad \bar{P}_1'\right]$ = geometric coefficient vector or boundary condition vector

$$\left.\begin{aligned} F_1(u) &= 2u^3 - 3u^2 + 1 \\ F_2(u) &= -2u^3 + 3u^2 \\ F_3(u) &= u^3 - 2u^2 + u \\ F_4(u) &= u^3 - u^2 \end{aligned}\right\} \text{These are called blending functions.}$$

$$\bar{P}(u) = F_1(u)\bar{P}_0 + F_2(u)\bar{P}_1 + F_3(u)\bar{P}_0' + F_4(u)\bar{P}_1'$$
$$\bar{P}(u) = [U]^T [M_H][V], \qquad 0 \le u \le 1$$

$[M_H]$ = Hermite matrix

$$\text{Hermite matrix} = [M_H] = \begin{bmatrix} 2 & -2 & 1 & 1 \\ 3 & -3 & -2 & -1 \\ 0 & 0 & 1 & 0 \\ 1 & 0 & 0 & 0 \end{bmatrix}$$

Comparing $\bar{P}(u) = [U]^T [C]$ and $\bar{P}(u) = [U]^T [M_H][V]$,

We obtain $C = [M_H][V]$

$$[V] = \left[M_H\right]^{-1} [C]$$

$$[M_H]^{-1} = \begin{bmatrix} 0 & 0 & 0 & 1 \\ 1 & 1 & 1 & 1 \\ 0 & 0 & 1 & 0 \\ 3 & 2 & 1 & 0 \end{bmatrix}$$

Similarly the tangent vector is written as

$$\bar{P}'(u) = \left(6u^2 - 6u\right)\bar{P}_0 + \left(-6u^2 + 6u\right)\bar{P}_1 + \left(3u^2 - 4u + 1\right)\bar{P}'_0 + \left(3u^2 - 2u\right)\bar{P}'_1$$

$$= [U]^T [M_H]^u [V]$$

where

$$[M_H]^u = \begin{bmatrix} 0 & 0 & 0 & 0 \\ 6 & -6 & 3 & 3 \\ -6 & 6 & -4 & -2 \\ 0 & 0 & 1 & 0 \end{bmatrix}$$

Often we need to

Evaluation the groove for display purposes, Blend, Segment, and Truncate curves to build practical parts.

We will here consider these curve manipulations in case of all Hermite, Bezier and B-Spline curves.

Finding various points on the curve using its parametric equation is necessary for the purposes of display as well as for nodes in finite element analysis. Substituting the value of "u" in the parametric form to find the successive points is computationally inefficient. What is preferred is an incremental method using Newton's forward differences? Direct point solution is finding the coordinate values of a point for a given value of "u". The inverse problem is finding the value of "u" for a given point on the curve or near the curve.

The construction of composite curves by blending or joining either the same type of curves or different types of curves is called blending. This is required often to meet the required free form curve/contour modeling. The general problem of blending two different types of curves (say, a Bezier curve segment and a B-Spline curve segment) should be considered.

Consider two parametric curves:

$$P_1(u_1),\ 0 \le u_1 \le a, \quad \text{and} \quad P_2(u_2),\ 0 \le u_2 \le b$$

The problem is to evolve the conditions for the continuity of the two curves at the blend point, of different levels C^0, C^1 and C^2.

For C^0 continuity

$$P_1(a) = P_2(0)$$

For C^1 continuity, there should be slope continuity. Hence,

$$P_1'(a) = \alpha_1 \mathbf{T} \quad \text{and} \quad P_2'(0) = \alpha_2 \mathbf{T}$$

where α_1 and α_2 are constants.

$\mathbf{T}$ = Common unit tangent vector at the joint.

For slope continuity the last control polygonal segment of the first curve should be collinear with the first control polygonal segment of the second segment.

When curvature is to be continuous at the blend point in addition to the slope and position, the binormal vector needs to be defined [1].

$$\hat{B} = \hat{T} \times \hat{N}$$

$$\frac{1}{\rho}\hat{B} = \frac{\bar{P}' \times \bar{P}''}{|\bar{P}'|^3}$$

$\hat{B}$ = unit binormal vector

$\hat{T}$ = unit tangent

$\hat{N}$ = unit normal

$\bar{P}'$ = tagent vector of $\bar{P}$ at the given point

$\bar{P}''$ = double derivative of $\bar{P}$ with respect to "u"

ρ = radius of curvature

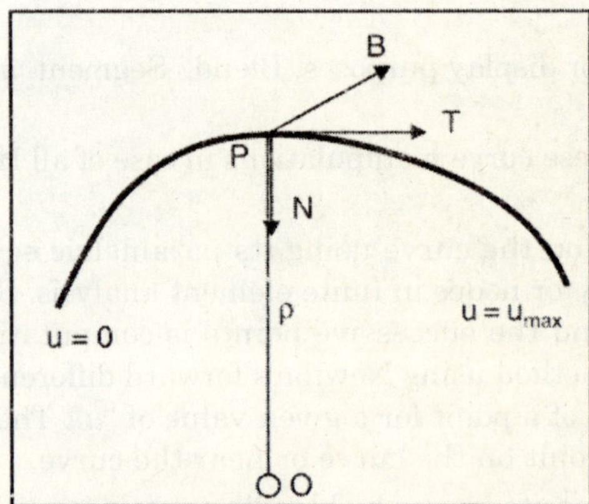

Fig. 18.9: The binormal vector

$$\frac{\bar{P}_1'(a) \times \bar{P}_1''(a)}{|\bar{P}_1'(a)|^3} = \frac{\bar{P}_2'(0) \times \bar{P}_2''(0)}{|\bar{P}_2'(0)|^3}$$

$$\hat{T} \times \bar{P}_1''(a) = \left(\frac{\alpha_1}{\alpha_2}\right)^2 \hat{T} \times \bar{P}_2''(0)$$

$$\bar{P}_2''(0) = \left(\frac{\alpha_2}{\alpha_1}\right)^2 \bar{P}_1''(a)$$

which is same as earlier conclusion of double derivatives being constant multiples.

Blending Hermite Cubic Splines

We obtain the tangent vectors at the intermediate control points as solution. Only first point and tangent vector of the first segment and second point and second tangent vector of the last segment need to be specified.

We obtain the intermediate tangent vectors by imposing continuity of curvature at the joints.

$$P''(u_1 = 1) = P''(u_2 = 0)$$

Differentiating the parametric form of Hermite cubic curve and substituting as above

$$P'_1 = -\frac{1}{4}(3P_0 + P'_0 - 3P_2 + P'_2)$$

For more than two segments a matrix equation can result from repeating this procedure, which can be solved for intermediate vectors in terms of the data points and the two end tangent vectors P'_0 and P'_{n-1}.

Generalising, $0 \le u_1 \le u_1$ and $0 \le u_2 \le u_2$, the previous equation becomes [2]

$$u_2 P'_0 + 2(u_2 + u_1)P'_1 + u_1 P'_2 = \frac{3}{u_1 u_2}\left[u_1^{\,2}(P_2 - P_1) + u_2^{\,2}(P_1 - P_0)\right]$$

This can be further generalized to a number of segments, say n, as follows:

Among n segments, let us look any two consecutive intermediate segments, k and $(k + 1)$.

$$u_{k+1}P'_k + 2(u_{k+1} + u_{k+2})P'_{k+1} + u_{k+1}P'_{k+2} =$$

$$\frac{3}{u_{k+1}u_{k+2}}\left[u_{k+1}^{\,2}(P_{k+2} - P_{k+1}) + u_{k+2}^{\,2}(P_{k+1} - P_k)\right]$$

$$0 \le k \le n - 2$$

Let for the first curve segment's data be

$$\mathbf{P}_0,\ \mathbf{P'}_0,\ \mathbf{P}_1 \text{ and } \mathbf{P'}_{11}$$

and the second curve segment's data be

$$\mathbf{P}_1,\ \mathbf{P'}_{12},\ \mathbf{P}_2,\ \mathbf{P'}_2.$$

also given is,

$$\mathbf{P'}_{11} = \mathbf{R} \text{ and } \mathbf{P'}_{12} = K\mathbf{R},$$

where K is a constant. Hence at the JOINT,

$$\mathbf{P'}_{11} = \mathbf{R} = [R_x\ R_y\ R_z]^T,$$

$$\mathbf{P'}_{12} = K\mathbf{R} = [KR_x\ KR_y\ KR_z]^T$$

Slopes at joint are

$$S_{11x} = R_y/R_x,\ S_{11y} = R_z/R_y,\ S_{11z} = R_z/R_x, \text{ and}$$

$$S_{12x} = KR_y/KR_x = R_y/R_x = S_{11x}, \text{ and so on.}$$

Thus the two curves have C^1 or slope continuity at P_1.

Looking at the curvature continuity,

$$\mathbf{P''(u)} = (12u - 6)\mathbf{P}_0 + (-12u + 6)\mathbf{P}_1 + (6u - 4)\mathbf{P'}_0 + (6u - 2)\mathbf{P'}_1$$

This equation gives the curvature vector for the two segments at $\mathbf{P}_1$ as

$$\mathbf{P}''_{11} = \mathbf{P}''(1) = 6\mathbf{P}_0 - 6\mathbf{P}_1 + 2\mathbf{P}'_0 + 4\mathbf{R}, \text{ and}$$

$$\mathbf{P}''_{12} = \mathbf{P}''(0) = -6\mathbf{P}_1 + 6\mathbf{P}_2 - 4K\mathbf{R} - 2\mathbf{P}'_2$$

Which evidently are not same.

Thus, these Hermite cubic spline curve segments do not offer C^2 continuity at the joint blending point, though they have C^2 continuity elsewhere on each segment.

Normal at a point on Hermite Cubic Spline is determined as follows:

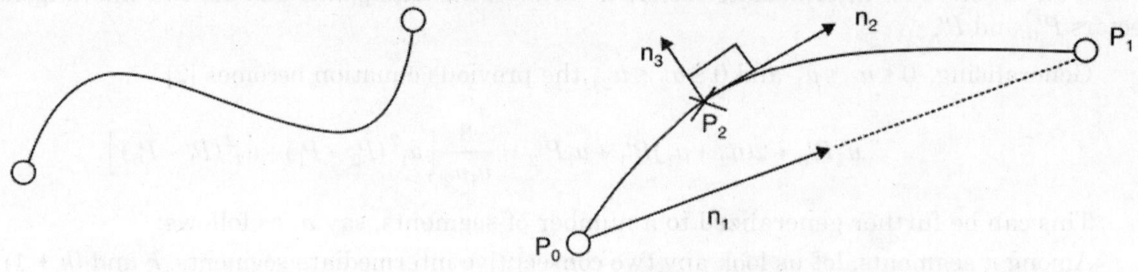

$\mathbf{n}_1 = (\mathbf{P}_1 - \mathbf{P}_0)/|\mathbf{P}_1 - \mathbf{P}_0|$ and $\mathbf{n}_2 = \mathbf{P}'_2/|\mathbf{P}'_2|$

$\mathbf{n}_4 = \mathbf{n}_1 \times \mathbf{n}_2/|\mathbf{n}_1 \times \mathbf{n}_2|$ and the normal vector therefore is $\mathbf{n} = \mathbf{n}_3 = \mathbf{n}_4 \times \mathbf{n}_2/|\mathbf{n}_4 \times \mathbf{n}_2|$

Another simpler method using our thumbrule is

$$n_{3x} = -n_{2y} \text{ and } n_{3y} = n_{2x}$$

Because $\mathbf{n}_2$ and $\mathbf{n}_3$ form a anti-clockwise pair of perpendicular unit vectors.

Bezier curve is an approximator because it passes through the end points but does not by default pass through the intermediate control points unless such a condition is imposed, adds flexibility and intuitive feel in mechanical design. It was developed by Thomas P. Bezier and P. De Casteljau independently but since the former brought it into the open domain it has been credited to him. De Casteljau's other contributions came into light later on the most important among which is the graphical sub-division algorithm for Bezier curve. Only control points (not necessarily only 2 per segment) form the input and no specification of tangent vectors is needed for Bezier curve. This feature provides for more direct relation between input (points) and output (curve shape and size). The degree of the Bezier curve can be varied (unlike HCC) and depends on the number of control points, $(n+1)$. In fact, the degree of the Bezier curve is exactly one less than the number of control points. Hence if we define the degree of the curve as $(k-1)$, then $(k-1) = n$ or in other words $(n+1)$ number of control points define an nth degree Bezier curve. Due to this Bezier curve can be smoother than HCC. The control points form the vertices of the control or Bezier characteristic polygon which uniquely defines one and only one Bezier curve. For the same set of control points, if the control polygon is changed then the curve also will be different (Fig. 18.10).

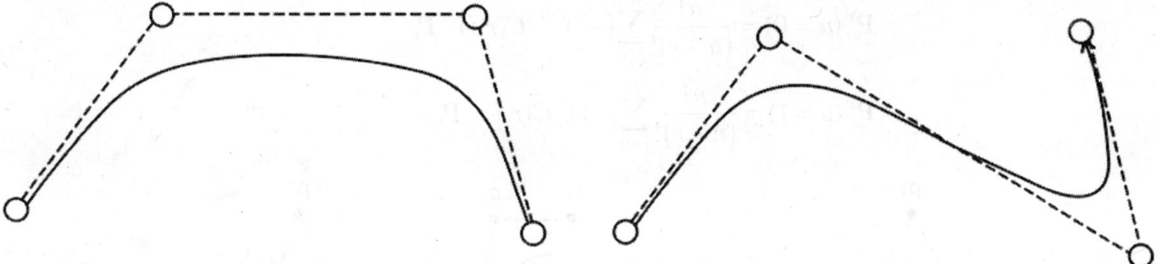

Fig. 18.10: For the same set of control points, different control polygons result in different Bezier curves

The curve shape tends to follow the polygon shape and thus easy to guess and sketch and order of the control points is very essential because changing the order can change the shape of the curve.

$$\mathbf{P}(u) = \sum_{i=0}^{n} \mathbf{P}_i B_{i,n}(u) \qquad 0 \le u \le 1$$

$$B_{i,n}(u) = C\binom{n}{i} u^i (1-u)^{n-i}$$

where the binomial coefficient

$$C\binom{n}{i} \text{ or } C(n,i) = \frac{n!}{i!(n-i)!}$$

The functions $B_{i,n}$ are the *Bernstein polynomials* of degree $(k-1) = n$. We can write the complete form of Bezier curve while observing that $C(n, 0) = C(n, n) = 1$.

$$\mathbf{P}(u) = \mathbf{P_0}(1-u)^n + \mathbf{P_1}C(n,1)u(1-u)^{n-1} + \mathbf{P_2}C(n,2)u^2(1-u)^{n-2} + \ldots + \mathbf{P_{n\text{-}1}}C(n,n-1)u^{n-1}(1-u) + \mathbf{P_n}u^n \quad 0 \le u \le 1$$

Special case, 3 control points

$$\mathbf{B(u)} = \mathbf{P}_0 * (1-u)^2 + \mathbf{P}_1 *2* u(1-\text{u}) + \mathbf{P}_2\, u^2$$

Special case, 4 control points

$$\mathbf{B(u)} = \mathbf{P}_0 * (1-u)^3 + \mathbf{P}_1 *3* u *(1-u)^2 + \mathbf{P}_2 *3* u^2 * (1-u) + \mathbf{P}_3 * u^3$$

$$\mathbf{P}(u) = \mathbf{P_0}(1-u)^n + \mathbf{P_1}C(n,1)u(1-u)^{n-1} + \mathbf{P_2}C(n,2)u^2(1-u)^{n-2} + \ldots + \mathbf{P_{n\text{-}1}}C(n,n-1)u^{n-1}(1-u) + \mathbf{P_n}u^n \quad 0 \le u \le 1$$

Substitute $u = 0$ and $u = 1$, to see that the curve is tangent to the first and last segments of control polygonal segments. Differentiating the parametric form "r" times.

$$\mathbf{P^r}(u=0) = \frac{n!}{(n-r)!}\sum_{i=0}^{r}(-1)^{r-i}\; C(r,i)\;\; \mathbf{P}_i$$

$$\mathbf{P^r}(u=1) = \frac{n!}{(n-r)!}\sum_{i=0}^{r}(-1)^{i}\; C(r,i)\;\; \mathbf{P}_{n-i}$$

Fig. 18.11: (a) Quadratic, (b), (c) and (d) cubic and (e) quartic Bezier curves

What did you observe from Fig. 18.11. The curve interpolates the first and last control points. It passes through P_0 and P_n if we substitute $u = 0$ and $u = 1$. The curve is tangent to the first and last curve segments of the control polygon [1].

$$\mathbf{P^r}(u=0) = \frac{n!}{(n-r)!}\sum_{i=0}^{r}(-1)^{r-i}\; C(r,i)\;\; \mathbf{P}_i$$

$$\mathbf{P^r}(u=1) = \frac{n!}{(n-r)!}\sum_{i=0}^{r}(-1)^{i}\; C(r,i)\;\; \mathbf{P}_{n-i}$$

Substituting $r = 1$, the first derivatives at the end points are

$$\mathbf{P}'(u=0) = n(\mathbf{P_1} - \mathbf{P_0})$$

$$\mathbf{P}'(u=1) = n(\mathbf{P_n} - \mathbf{P_{n-1}})$$

The curve is symmetric with respect to u and $(1 - u)$. That is, reversing the direction of parametrization does not change the curve shape. To check this property substitute $1 - u = v$. The symmetry of the Bezier curve is due to the symmetry (*NOT equality*) of the Bernstein basis, which in turn is due to the property of the binomial functions: $C(n, i) = C(n, n - i)$, which makes

B_i, $n(u)$ and B_{n-i},$n(u)$ to be symmetric when plotted with respect to u. The polynomial B_i,n has a maximum value of $C(n, i)(i/n)^i(1-i/n)^{n-i}$ occuring at $u = i/n$. This can be obtained by setting $dB_{i,n}/du = 0$. This means that each ith control point is most influential on the shape of the curve at $u = i/n$, where $(n+1)$ are the total control points. For example in a cubic Bezier curve defined by P_0, P_1, P_2 and P_3, these points are most influential at u = 0, 1/3, 2/3 and 1. The shape of the curve (for a given set of control vertices) can be changed by two methods. By moving a control point and by providing multiple control points at a vertex of control polygon. Multiplicity of control points makes the curve pulled towards that point (Fig. 18.12). What happens to the degree of the curve when $K > 1$? The reader can easily find the answer by writing the full parametric equation of the curve.

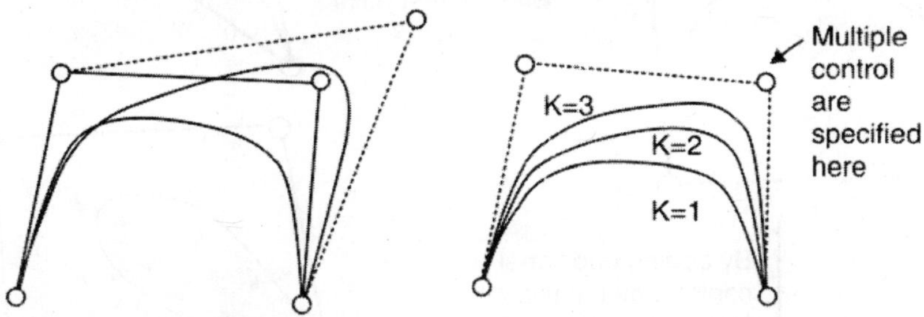

Fig. 18.12: Two ways of controlling the Bezier curve

A *closed Bezier* curve can be obtained by either making P_0 and P_n coincident or creating an additional point as last control point that is coincident with the first control point (Fig. 18.13). Will the degree of the curve change in each case? Here also, writing the full parametric equation can give the answer.

Bernstein polynomials and hence the Bezier curve possess two important properties. *Partition of unity* and *positivity*. Partition of unity means for any value of the parameter u in its range and for any $(n + 1)$ number of control points, the sum of all the Bernstein polynomials is equal to one. For example, for a cubic Bezier curve the following is true.

$$P(u) = \sum_{i=0}^{n} P_i B_{i,n}(u) = \sum_{i=0}^{3} P_i B_{i,n}(u) = P_0 B_{0,3}(u) + P_1 B_{1,3}(u) + P_2 B_{2,3}(u) + P_3 B_{3,3}(u)$$

$\sum_{i=0}^{3} B_{i,n}(u) = B_{0,3}(u) + B_{1,3}(u) + B_{2,3}(u) + B_{3,3}(u) = 1$, for any value of u in the range $0 \le u \le 1$. Indeed,

$$B_{0,3}(u) = (1-u)^3;\ B_{1,3}(u) = 3u(1-u)^2;\ B_{2,3}(u) = 3u^2(1-u);\ B_{3,3}(u) = u^3$$

$$\sum_{i=0}^{3} B_{i,n}(u) = (1-u)^3 + 3u(1-u)^2 + 3u^2(1-u) + u^3$$

$$= \left[1 - u^3 - 3u + 3u^2\right] + \left[3u + 3u^3 - 6u^2\right] + \left[3u^2 - 3u^3\right] + u^3 = 1$$

The property of partition of unity provides for two important consequences. First consequence is invariance of Bezier curve (or invariance of the relationship between the Bezier curve and its control polygon) under affine transformations. Second implication is that in the programming activity the estimation of the Bernstein polynomials for a given Bezier curve at any given u can be cross-checked for numerical accuracy by employing the partition of unity property.

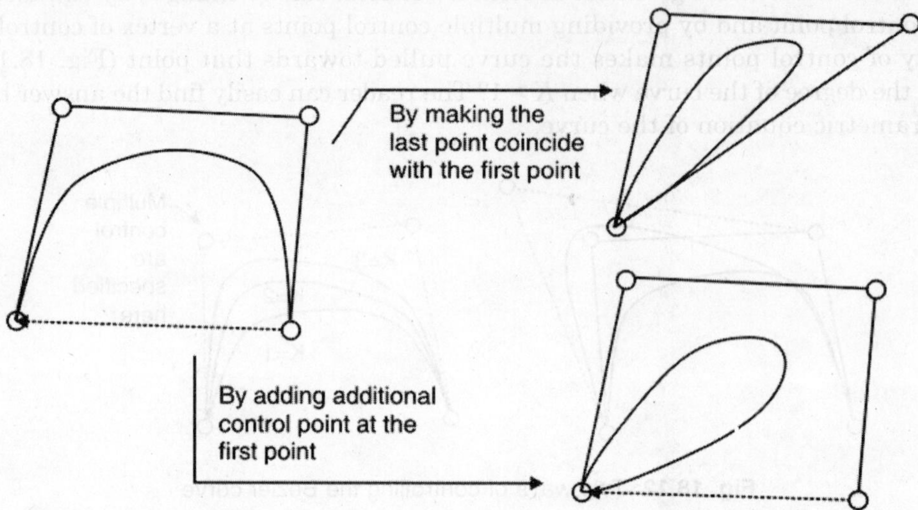

Fig. 18.13: Producing a closed Bezier curve

The property of positivity implies that each and every Bernstein polynomials is always positive.

$$B_{i,n} > 0$$

This ensures that the Bezier curve lies entirely in its convex hull. A curve is said to have *convex hull* property if it entirely lies in its convex hull defined by its control polygon. The Bezier curve defined with Bernstein polynomial basis possesses the convex hull property. The convex hull of a Bezier curve is the maximum area obtained by joining all control points to all control points (Figs. 18.14 and 18.15).

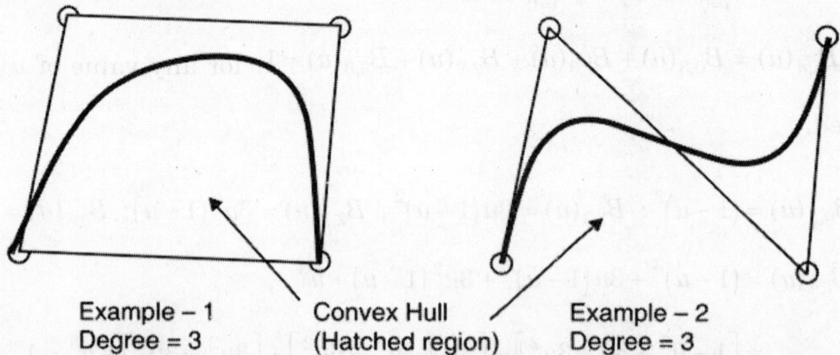

Fig. 18.14: The convex hull is a function of the control points and not the control polygon. Given a set of control points, by changing the order of the control polygon different control polygons and hence different Bezier curves can be obtained but the convex hull is the same for all of them.

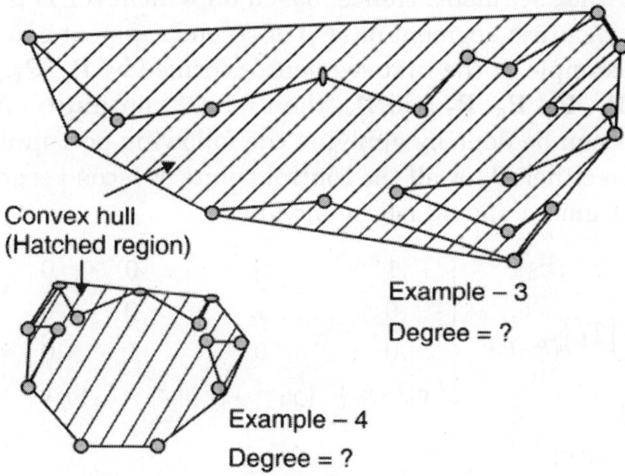

Fig. 18.15: Convex hulls for higher degree Bezier curves

Consequences of Convex Hull Property are as follows. Degeneration of the control points to become collinear will make the curve to degenerate to a straight line. Clipping type of manipulations become easy. Wild oscillations of the curve about any control point are avoided. Finally, if the convex hull is within the computer screen, then automatically it is assured that the entire curve will lie inside the screen, which is important because it is computationally cheaper to evaluate the convex hull than the curve itself in order to employ checking when 'fit the curve to the screen' is sought by the user.

In the segmentation approach of building the free-from curves using the synthetic Bezier curve segments, different orders of continuity need to be maintained at the blend points for the purpose of smoothness. Although the qualitative considerations of continuity at blend points for Bezier curve remain the same as those for the Hermite cubic curve, the actual conditions differ. For C^o continuity, the last control point of the first Bezier curve segment should be made coincided with the first control point of the second Bezier curve segment. This requires that one of the two curves needs to be translated towards the other. For C^1 continuity, the end slope of the first segment need to be equal to the starting slope of the second segment. This requires that the tangent vector at its last control point for the first curve segment be collinear with the tangent vector of the second curve segment at its first control point. Recall from an earlier discussion, that this is both necessary and sufficient condition for slope continuity at the blend point. A simpler condition that is sufficient but not necessary is that these tangent vectors be equal. We will use this condition to enforce the slope continuity. If $(n + 1)$ is equal to the number of the control points of the first segment and $(m + 1)$ is equal to the number of the control points of the second segment, equating the tangent vectors,

$$\{n(\mathbf{P_n} - \mathbf{P_{n-1}})\}_{\text{first segment}} = \{m(\mathbf{P_1} - \mathbf{P_0})\}_{\text{second segment}}$$

$$\{(\mathbf{P_n} - \mathbf{P_{n-1}})\}_{\text{first segment}} = (m/n)\{(\mathbf{P_1} - \mathbf{P_0})\}_{\text{second segment}}$$

Since the position continuity has already been enforced by carrying out the necessary translation of one or all the control points of one of the segments, $(\mathbf{P_0})$ of second segment is already equal to ($\mathbf{P}_n$) of first segment. Hence, based on whichever is permitted by the actual application, either ($\mathbf{P}_{n-1}$) of first segment or ($\mathbf{P}_1$) of second segment can be adjusted to achieve the C^1 continuity. For example, if the first segment is formed by $\mathbf{P}_0$, $\mathbf{P}_1$, $\mathbf{P}_2$ and $\mathbf{P}_3$ and second segment is formed by $\mathbf{P}_4$, $\mathbf{P}_5$, $\mathbf{P}_6$, $\mathbf{P}_7$ and $\mathbf{P}_8$, then for C^0 continuity, first $\mathbf{P}_3$ must be made coincident with $\mathbf{P}_4$. This can be done by applying the following post-multiplying homogeneous transformation matrix on either $\mathbf{P}_4$ or all the control points of second segment, if the application demands that the first segment should not be modified.

$$[T_R]\Big|_{(\bar{P}_4-\bar{P}_3)} = \begin{bmatrix} 1 & 0 & 0 & 0 \\ 0 & 1 & 0 & 0 \\ 0 & 0 & 1 & 0 \\ (x_4 - x_3) & (y_4 - y_3) & (z_4 - z_3) & 1 \end{bmatrix}$$

Next for C^1 continuity,

$$(\mathbf{P}_3 - \mathbf{P}_2) = \left(\frac{m}{n}\right)(\mathbf{P}_5 - \mathbf{P}_4^*) = \left(\frac{4}{3}\right)(\mathbf{P}_5 - \mathbf{P}_3)$$

The above condition can be used to derive for the new location of the point $\mathbf{P}_5$ for C^1 continuity for enforcing slope continuity.

$$\mathbf{P}_5 = \mathbf{P}_3 + \left(\frac{3}{4}\right)(\mathbf{P}_3 - \mathbf{P}_2) = \left(\frac{7}{4}\right)\mathbf{P}_3 - \left(\frac{3}{4}\right)\mathbf{P}_2$$

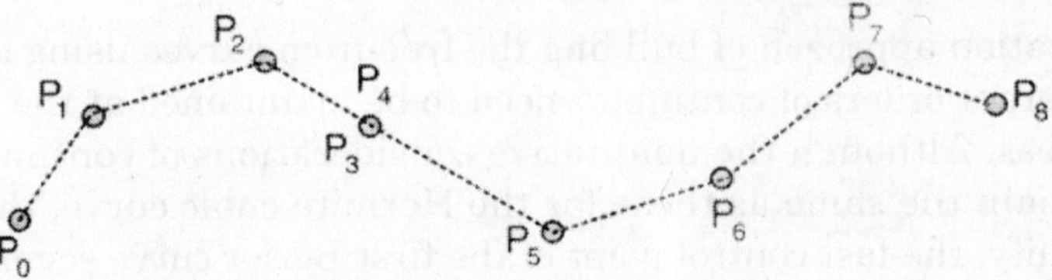

Fig. 18.16: Enforcing C^0 and C^1 continuities at the blend point of two Bezier curves

Sub-Dividing Bezier Curves Using De-Casteljau's Algorithm

This algorithm provides a convenient graphical methodology for subdividing or *segmentation* of a given Bezier curve by the required proportion as prescribed in terms of the parameter value. The control polygon plays a crucial role in the implementation of this algorithm. For example (Fig. 18.17) if a given Bezier curve is to be sub-divided into two equal length Bezier curve segments of the same degree, then the first step is to identify the middle points of the control polygonal sides and join them by straight lines. Two new straight lines will now be there. Next, the middle points of these two new straight lines are identified and are joined by a straight line. The middle point of this straight line automatically occurs on the original Bezier curve exactly at the point that divides original curve into two equal half portions.

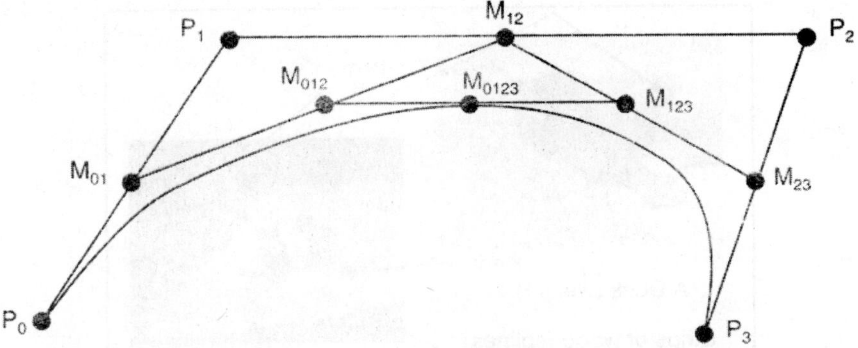

Fig. 18.17: Application of the De-Casteljou's segmentation algorithm to the cubic Bezier curve for u = 0.5

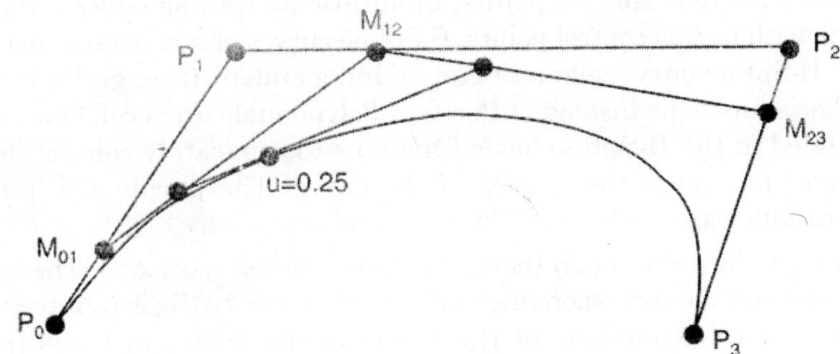

Fig. 18.18: Application of the De-Casteljou's segmentation algorithm to the cubic Bezier curve for u = 0.25

What is the importance of middle points? Because they signify that we are sub-dividing the curve at $u = 1/2$. Hence, similarly the curve can be segmented into, say, three equal length segments by choosing $u = 1/3$ which means that the control polygonal sides need to be successively processed by identifying one-third points along its length *in the correct direction*. Figure 18.18 shows the application of the De-Casteljou's algorithm for $u = 0.25$.

18.8. WHY MORE SPLINES?

B-Splines have been born out of a greater need for better and more local control on the shape of the curve to model free-form objects. Their founding ideas were based on the French splines (meaning curves) drawn using strips of wood 'knotted' by weights called as ducks (Fig. 18.19).

It is because B-Spline can offer something more than HCC and Bezier curve. Bezier and Hermite splines have global control only. *Moving one control point affects the entire curve.* B-Spline consists of a *basis* that allows an extra degree of freedom which is the degree of the curve. Degree of the curve can be independent of the number of control points. Hence local control is available. B-Spline is a powerful generalization of the Bezier curve. B-Spline can be both interpolating type and approximating type. B-Spline provides local control thus better shape

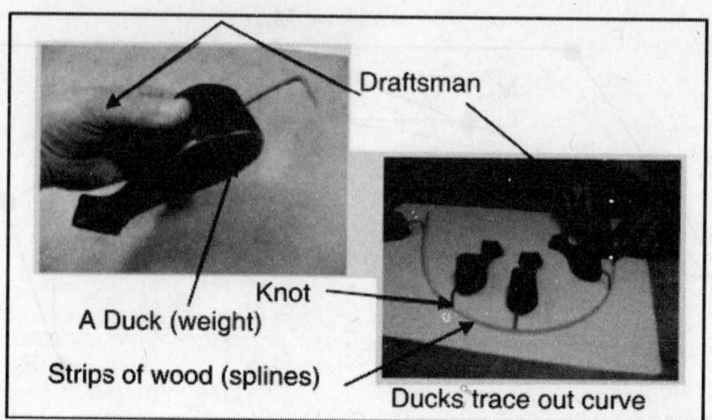

Fig. 18.19: The B-Splines have their root in the French splines drawn using ducked strips of wood

control is possible. The degree $(k - 1)$ of the curve segment is not dependent on the number control points; thus with four control points, quadratic as well as cubic B-Spline curves are possible. In general, with $(n + 1)$ control points, B-Spline curves of any degree up to n are possible. The degree of the B-Spline curve segment can be independent (though limited) of the $(n +1)$ because B-Spline basis functions instead of Berstein Polynomials are used. One additional degree of freedom is included in the B-Spline basis functions to separately control the degree of the curve, it is k, where $(k - 1)$ is the degree of the curve. The parametric representation of B-Spline curve is as follows.

B-Spline curve can be either open (non-periodic) or closed (periodic). The evaluation of the periodic curves depends on the corresponding non-periodic curve to the extent that periodic curve's basis functions are mere extensions of the fundamental forms of basis functions of the corresponding non-periodic curve. We will look at the non-periodic curve first. Its parametric curve can be expressed as follows:

$$P(u) = \sum_{i=0}^{n} P_i N_{i,k}(u) \quad 0 \le u \le u_{max}$$

$N_{i,k}(u)$ = B-Spline basis functions

$(k - 1)$ = Degree of the B-Spline curve

u = Parameter

u_{max} = Maximum value of the parameter

The control points (called de-Boor points) form the vertices of the control polygon (called de-Boor polygon). The second parameter in the basis functions, k, controls the degree $(k - 1)$ of the resulting B-Spline curve and is independent of the $(n + 1)$; Only the upper limit of the degree is decided by the $(n + 1)$. The maximum limit of the parameter is no longer unity (that is 1); it is u_{max} chosen by fixing the $(n + 1)$ and $(k - 1)$.

The parametric equation of the B-spline curve suggests that a particular level of B-spline basis functions must be evaluated before the equation is completely determined. For example, if we set the number of control points $(n + 1) = 4$ and degree of the curve $(k - 1) = 3$, following will be the parametric equation.

$$P(u) = \sum_{i=0}^{n} P_i N_{i,k}(u) = \sum_{i=0}^{3} P_i N_{i,4}(u) = P_0 N_{0,4}(u) + P_1 N_{1,4}(u) + P_2 N_{2,4}(u) + P_3 N_{3,4}(u)$$

Thus, as above, the B-spline basis functions $N_{i,4}$ need to be evaluated for i equal to 0 to 3. However, the estimation of the basis functions of the B-spline curve is not as straightforward as those of the Bezier curve and cannot be achieved in a single step. They need to be estimated out of a recursive calculation procedure, which is clear from the following the equations of basis functions.

$$N_{i,k}(u) = (u - u_i)\frac{N_{i,k-1}(u)}{(u_{i+k-1} - u_i)} + (u_{i+k} - u)\frac{N_{i+1,k-1}(u)}{(u_{i+k} - u_{i+1})}$$

where, $N_{i,1} = \begin{cases} 1, & u_i \le u \le u_{i+1} \\ 0, & \text{otherwise} \end{cases}$ This is a unit step function.

Choose $\dfrac{(\text{anything})}{0} = 0$, that is if denominator becomes zero.

The u_i are called parametric knots or knot values.

For an open or non-periodic curve,

$$u_j = \begin{cases} 0, & j < k \\ j - k + 1, & k \le j \le n \\ n - k + 2, & j > n \end{cases}$$

and the range of u is $0 \le u \le (n - k + 2)$.

The previous recursive equations show that $(n + k + 1)$ knots are needed to create a $(k - 1)$ degree curve defined by $(n + 1)$ control points. These knots are evenly (uniformly, hence the name uniform curve) spaced over the range of "u" with unit separation ($u = 1$) between knots. Multiplicity—is a standard term for indicating the multiple end knot values in the knot vector of the B-Spline.

There are four properties of B-Spline basic functions, unlike only two properties of the Bernsein polynomials. Partition of unity enables invariance under affine transformations. This was true for Bernsein polynomials also. Positivity enables the curve to lie entirely in its convex hull, which was also true for Bernstein polynomials. Local support property of B-spline basis functions enables local control. Continuity or continuous differentiability enables continuity of different levels, C^0, C^1, C^2 etc.

1. Partition of unity: $\sum_{i=0}^{n} N_{i,k}(u) = 1$

2. Positivity: $N_{i,k}(u) \ge 0$

3. Local support: $\sum_{i=0}^{n} N_{i,k}(u) = 0$ if $u \notin [u_i, \; u_{i+k+1}]$

4. Continuous differentiability: $N_{i,k}(u) \geq 0$ is $(k - 2)$ times continuously differentiable.

Partition of unity ensures that the relationship between the curve and its defining control points is invariant under affine transformations. Affine transformation is a transformation that is a combination of single transformations such as translation or rotation or reflection on an axis. Transformation is a function that changes the position or direction of the axes of a coordinate system. Positivity property guarantees that the curve segment lies completely within the convex hull of P_i. Local support property indicates that each segment of a B-Spline curve is influenced by only k control points or each control point affects only k curve segments. Continuity property tells us about the fact that B-Spline curve segment is continuous up to C^{k-2} or C^{n-1} (where $(n + 1)$ = number of control points) within the segment itself. There is a limit on the value of $(k - 1)$ that is determined by the number of the given control points. This limit is found by requiring the upper bound on the value of "u" to be greater than the lower bound for the "u" range to be valid, that is

$$(n - k + 2) > 0$$
$$[(n + 1) + (1 - k)] > 0$$
$$(k - 1) < (n + 1)$$
$$(k - 1) \leq n$$

This relation says that a minimum of two, three and four control points are required to define a linear, quadratic and cubic B-Spline curve, respectively.

Once the number of control points and the degree of the curve are decided, the next step is to estimate the knot vector before the basis functions can be estimated. For the uniform B-spline curves having increment in knot value equal to one, the estimation of the knot vector can be done directly without needing to actually use the formula suggested in the foregoing discussion. This is possible because when used along with the earlier mentioned conditions on the maximum and minimum limits on the parameter and unit increment of knot values, the condition that there exists a multiplicity of end knot values equal to exactly k, where $(k - 1)$ is the degree chosen, enables the estimation of knot vector. Following are a couple of examples.

$$(n+1) = 1; (k-1) = 0; k = 1;$$
$$n + k + 1 = 2;\ 0 \leq j \leq 1;\ n - k + 2 = 1;\ 0 \leq u \leq 1;\ u_j = [0,1]$$
$$(n+1) = 2; (k-1) = 1; k = 2;$$
$$n + k + 1 = 4;\ 0 \leq j \leq 3;\ n - k + 2 = 1;\ 0 \leq u \leq 1;\ u_j = [0,0,1,1]$$

The knot values are essential for obtaining the parametric equation of degree $(k - 1)$ for the curve. Once the parametric equation is determined, the point on the curve corresponding to any value of $0 \leq u \leq u_{max}$, can be determined by substitution. Thus it is very essential to understand the purpose of the knot values (which are discrete and finite in number, precisely $n + k + 1$) and the variation of the parameter (which is continuous between the limits of 0 and u_{max} along the curve).

The local control of B-Spline curve can be obtained by three ways, first by changing the position of the control point(s). Moving any one control point will affect only "k" number of curve portions symmetrically around it (Fig. 18.20). Secondly, by using multiple control points by placing several points at the same location, which is called as multiplicity of control points. Finally, by choosing a different degree, $(k - 1)$.

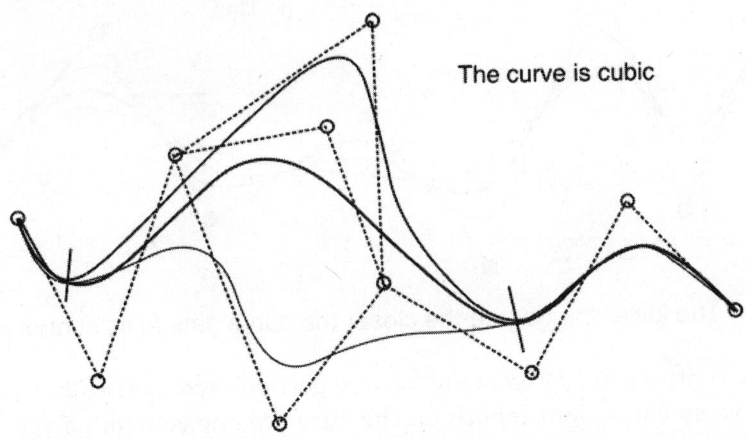

Fig. 18.20: The shape of B-Spline curve can be controlled by the movement of control point(s). When a control point is moved only *k* portions of the curve around that control point move, rest of the curve remains unchanged

Increasing the degree of the curve tightens it. The lesser the degree of the curve the closer the curve gets to the control points. For various values of $(k - 1)$, the following types of the curves result.

$(k - 1) = 0$ –> Zero degree curve (control points themselves)

$(k - 1) = 1$ –> First degree curve (control polygonal segments themselves)

$(k - 1) = 2$ –> Second degree curve (quadratic curve).

$(k - 1) = 3$ –> Third degree curve (cubic curve)

$(k - 1) = 4$ –> Fourth degree curve (quartic curve)

$(k - 1) = 5$ –> Fifth degree curve (quintic curve)

The quadratic curve is always tangent to the midpoints of all the internal polygon segments. This is not the case for other degrees (Fig. 18.21).

If k equals the number of control points $(n + 1)$, then the resulting B-Spline curve becomes a Bezier curve. As expected, u varies from 0 to 1 in this case. Multiple control points induce regions of high curvature of a B-Spline curve. When multiplicity at any point is equal to $k - 1$ the curve has a cusp or a sharp corner. When the multiplicity at any control point is "k", then the B-Spline curve will pass through it (why? But C^{k-2} continuity is maintained in both cases. Multiplicity of points maintains the degree of the curve but the knot vector is changed.

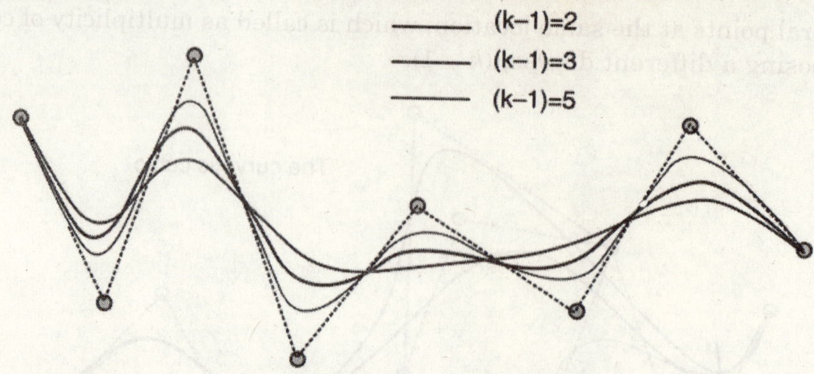

Fig. 18.21: The lower the degree the closer the curve gets to its control polygon

Changing one control point affects only k curve portions symmetrically around it (we define one portion of curve as equivalent length on the curve to one control polygonal segment. We have reserved the term 'curve segment' to mean one entire spline). In other words, each portion (or a point in it) is influenced by only k number of control points surrounding it (this is the key behind the procedure of finding the convex hull of a B-Spline curve). Since each portion (and any point on it) of the curve is influenced by "k" number of control points, the convex hull of a B-Spline for specified degree $(k - 1)$ is obtained by first joining "k" number of control points at a time repeatedly from first to the last in the control point set. This results in a set of sub-convex hulls. Then, finally the convex hull of the entire B-Spline curve is obtained as the union of these sub-convex hulls (Fig. 18.22).

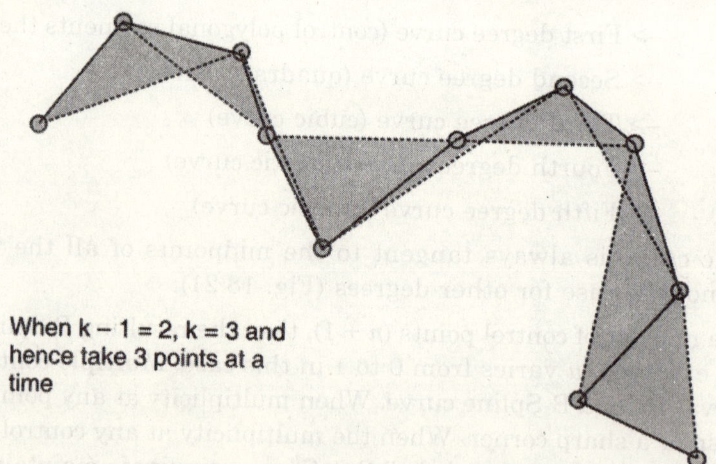

Fig. 18.22: Methodology to build the convex hull of B-Spline curve

A non-periodic B-Spline curve passes through the first and last control points P_0 and P_n and is tangent to the first $(\mathbf{P_1} - \mathbf{P_0})$ and last $(\mathbf{P_n} - \mathbf{P_n} - 1)$ control polygonal segments, similar to the Bezier curve. The different levels of B-Spline basis functions can be estimated by first using a tree structure to identify the required the basis functions at various levels as shown in Fig. 18.23.

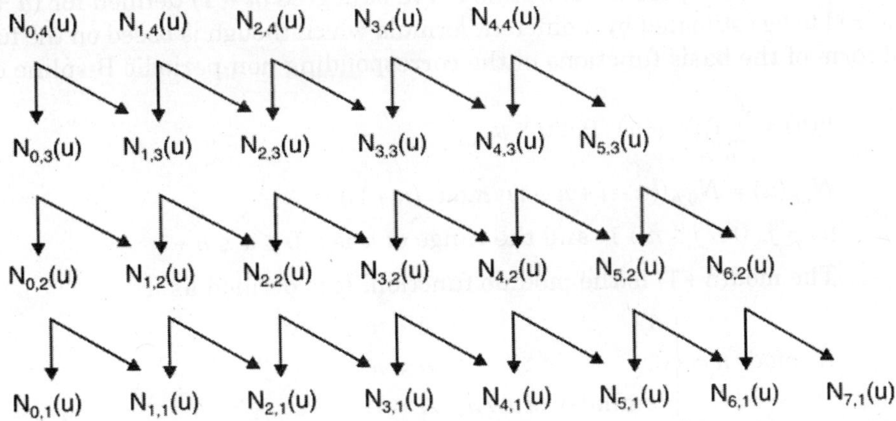

Fig. 18.23: The tree structure to find out the various B-Spline basis functions needed to evaluate the B-Spline curve

Consider a cubic B-Spline for five control points whose parametric equation will be then as follows.

$$P(u) = P_0N_{0,4}(u) + P_1N_{1,4}(u) + P_2N_{2,4}(u) + P_3N_{3,4}(u) + P_4N_{4,4}(u)$$

The Fig. 18.23 shows that one additional basis function needs to be evaluated at each lower level using the recursive relation except the last level N_i, 1, which is evaluated using the unit step function part of basis function equations.

The periodic B-Spline curve, on the other hand, is represented by the following parametric equation.

$$P(u) = \sum_{i=0}^{n} P_i N_{i,j}(u) \quad 0 \le u \le u_{\max}$$

$$N_{i,k}(u) = N_{0,k}((u - i + n + 1)\,\mathrm{mod}(n+1))$$

$$u_j = j,\ 0 \le j \le n+1$$

$$0 \le u \le n+1$$

The mod function in the above equation is defined as follows.

$$A \bmod n = \begin{cases} A, \ If & A < n \\ 0, \ If & A = n \\ \text{remainder of } A/n, \ If & A > n \end{cases}$$

The difference from non-periodic curves is the choice of knots and basis functions. Periodic B-Spline basis functions with knots as integers are utilized. These basis functions are cyclic translates of a single canonical function with a period (interval) of k for support. They also have the properties of partition of unity, positivity, local support and continuity. They do not pass through the end control points. Therefore they are not tangent to the first and last control segments.

The basis functions of periodic B-Spline curve of degree $(k - 1)$ defined for $(n + 1)$ control points is required to be estimated by a different formula which though is based on the fundamental or canonical form of the basis functions of the corresponding non-periodic B-spline curve.

$$P(u) = \sum_{i=0}^{n} P_i N_{i,k}(u) \quad 0 \le u \le u_{max}$$

$$N_{i,k}(u) = N_{0,k}((u - i + n + 1) \mod (n+1))$$

$$u_j = j,\ 0 \le j \le n+1 \text{ and the range of } u \text{ is}: \ 0 \le u \le n+1$$

The mod(n + 1) is the modulo function. It is defined as

$$A \mod n = \begin{cases} A, & A < n \\ 0, & A = n \\ \text{reminder of } A/n, & A > n \end{cases}$$

For example, 3.5 mod 6 = 3.5; 8 mod 8 = 0; 9 mod 6 = 3

The mod function enables the periodic or cyclic translation [mod(n+1)] of the canonical basis function $N_{0,k}$. $N_{0,k}$ is the same as for non-periodic curves and is to be evaluated from the Cox-DeBoor recursive equations.

A rational curve is defined by the algebraic ratio of two polynomials while a non-rational curve is defined by one polynomial only. A rational curve is invariant under projective transformation also.

The rational B-Spline curve defined for $(n + 1)$ control points P_i, has its parametric form as follows:

$$P(u) = \sum_{i=0}^{n} P_i R_{i,k}(u), \ 0 \le u \le u_{max}$$

$R_{i,k}(u)$ are the rational B-Spline basis functions and are given by

$$R_{i,k}(u) = \frac{h_i N_{i,k}(u)}{\sum_{i=0}^{n} h_i N_{i,k}(u)}$$

If we substitute $h_i = 1$, $R_{i,k}(u) = N_{i,k}(u)$

$$P(u) = \sum_{i=0}^{n} P_i R_{i,k}(u), \ 0 \le u \le u_{max}$$

$R_{i,k}(u)$ are the rational B-Spline basis functions and are given by

$$R_{i,k}(u) = \frac{h_i N_{i,k}(u)}{\sum_{i=0}^{n} h_i N_{i,k}(u)}$$

The $R_{i,k}$ (u) are a generalization of the non-rational basis functions $N_{i,k}$ (u). If we substitute $h_i = 1$ in the equation, $R_{i,k}(u) = N_{i,k}(u)$. The choice of the $H = [h_0\ h_1\ h_2\ h_3\ ...\ h_n]^T$ vector controls the behaviour of the curve.

For example, consider the control polygon given by the vertices, $P_1[0, 0]$, $P_2[1, 2]$, $P_3[5/2, 0]$, $P_4[4, 2]$, $P_5[5, 0]$. If we are interested in determining the point at $u = 3/2$ for the quadratic $(k = 3)$ open rational B-Spline curve with homogeneous weighting factors given by $[H] = [1\ 1\ h_3\ 1\ 1]$, $h_3 = 0, 1/4, 1, 5$, then we can observe that the curve gets closer and closer to the control point P_2 for increasing values of h_3.

In the segmentation approach of curve building, the process of obtaining the required free-form curve by joining segments of parametric curves is called as blending. We have already seen that blending of two curve segments requires that atleast C^1 continuity, that is slope continuity be maintained in most engineering requirements. We have also the conditions to be satisfied for slope continuity. Often, curvature continuity is also required for better smoothness in modeling of parts. It is necessary to examine how curvature continuity is related to the parametric equation of the curve or its derivatives of different orders. Such relationship can be developed by first examining what forms the magnitude and what forms the direction of curvature if the latter is to be expressed as a vector. The designer can easily specify the radius of curvature based on the needs. Hence the magnitude of the curvature is obtained as the reciprocal of the specified radius of curvature. Whereas, the direction of curvature is prescribed by the binormal unit vector (Fig. 18.24). Thus the curvature in terms of tangent vector and double derivative is defined as follows:

$$\hat{B} = \hat{T} \times \hat{N}$$

$$\frac{1}{\rho}\hat{B} = \frac{\bar{P}' \times \bar{P}''}{|\bar{P}'|^3}$$

$\hat{B}$ = unit binormal vector

$\hat{T}$ = unit tangent

$\hat{N}$ = unit normal

$\bar{P}'$ = tagent vector of $\bar{P}$ at the given point

$\bar{P}''$ = double derivative of $\bar{P}$ with respect to "u"

ρ = radius of curvature

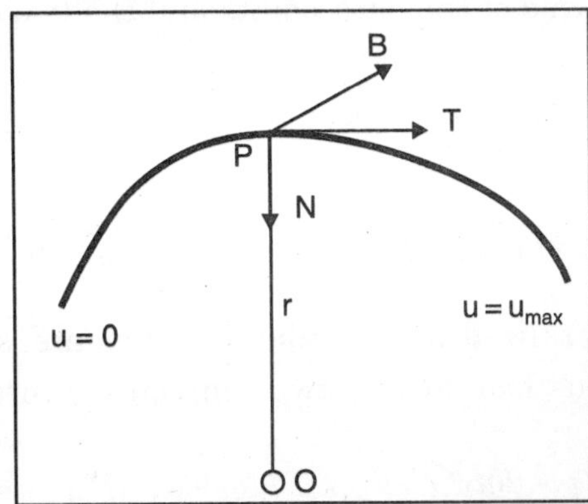

Fig. 18.24: Definition of the curvature vector

$$\frac{\bar{P}_1'(a) \times \bar{P}_1''(a)}{|\bar{P}_1'(a)|^3} = \frac{\bar{P}_2'(0) \times \bar{P}_2''(0)}{|\bar{P}_2'(0)|^3}$$

$$\hat{T} \times \bar{P}_1''(a) = \left(\frac{\alpha_1}{\alpha_2}\right)^2 \hat{T} \times \bar{P}_2''(0)$$

$$\bar{P}_2''(0) = \left(\frac{\alpha_2}{\alpha_1}\right)^2 \bar{P}_1''(a).$$

Thus curvature continuity at any point on the curve is dependent on the ratio of the magnitudes of the tangent vectors of the two curve segments at the blend point. Slope continuity would be assured even if the tangent vectors are collinear, they need not be exactly equal. When they are exactly equal the ratio (α_2/α_1) is equal to one and equality of double derivatives at the blend point is sufficient for curvature continuity.

However, it has been reported [2] that the above mentioned is only mathematically sufficient but actually the degree of smoothness even with continuity of curvature may not be found sufficient from the practical stand point. One solution proposed is that instead of the usual normalized 0 to 1 range of parameter, the maximum value of the parameter can be made equal to the chord length of the curve from its first point to the last. This has been found to improve the degree of smoothness at the blend point.

In the problem of blending Hermite cubic curve segments, often only the tangent vectors at the end control points are specified and we need to obtain the tangent vectors at the intermediate control points as solution. Only first point and tangent vector of the first segment and second point and second tangent vector of the last segment need to be specified for n number of segments to be blended. We obtain the intermediate tangent vectors by imposing continuity of curvature at the joints. In the simple situation of only two curve segments being blended,

$$P''(u_1 = 1) = P''(u_2 = 0)$$

$$P'_1 = -\frac{1}{4}(3P_0 + P'_0 - 3P_2 + P'_2)\cdot$$

That gives the intermediate tangent vector. Generalization of the parameter ranges is also possible for which we obtain the intermediate tangent vector as follows. For more than two segments a matrix equation can result from repeating this procedure, which can be solved for intermediate vectors in terms of the data points and the two end tangent vectors P'_0 and P'_{n-1}.

$$0 \le u_1 \le u_1 \text{ and } 0 \le u_2 \le u_2,$$

the previous equation becomes

$$u_2 P'_0 + 2(u_2 + u_1)P'_1 + u_1 P'_2 = \frac{3}{u_1 u_2}\left[u_1^2(P_2 - P_1) + u_2^2(P_1 - P_0)\right]$$

This can be further generalized to a number of segments, say n, as follows.

Among n segments, let us look at any two consecutive intermediate segments, k and $(k + 1)$.

$$u_{k+1}P'_k + 2(u_{k+1} + u_{k+2})P'_{k+1} + u_{k+1}P'_{k+2} =$$

$$\frac{3}{u_{k+1}u_{k+2}}\left[u_{k+1}{}^{2}(P_{k+2}-P_{k+1})+u_{k+2}{}^{2}(P_{k+1}-P_{k})\right],\quad 0\le k\le n-2$$

The problem of segmentation deals with sub-dividing a given curve segment into two or more portions and then reparameterizing each sub-divided segment in a chosen parameter range. Truncation is also similar to the segmentation problem except we would like to clip or delete all but one sub-divided segments and the retained segment needs to be reparameterized.

When a Hermite cubic curve is to be truncated and if the remaining portion of the curve is to be reparameterized from 0 to 1, the modified geometric coefficient matrix will be

$$B' = [Q_0 \;\; Q_1 \;\; Q_0' \;\; Q_1'] = [P_0 \;\; P_1 \;\; (u_j - u_i)P_0' \;\; (u_j - u_i)P_1']$$

In general, if a cubic Hermite segment obtained by truncating from u_i to u_j is to be reparameterized from v_i to v_j, the control points and tangent vectors of the remaining segment will be as follows.

$$Q_0 = P_i \;;\; Q_1 = P_j \;;\; Q_0' = \frac{u_j - u_i}{v_j - v_i}P_i' \;;\; Q_1' = \frac{u_i - u_j}{v_i - v_j}P_j'$$

This reduces to the following when $v_i = 0$ and $v_j = 1$.

$$Q_0^{\,v} = (u_j - u_i)P_i'$$

$$Q_1^{\,v} = (u_j - u_i)P_j'$$

Wireframe modeling cannot be used to model realistic physical surfaces such as those of ship hull, car bodies etc. Creating surfaces requires some quantitative data (such as points, tangent vectors) and some qualitative data (such as intuition of the desired shape and smoothness). Just like curves, surfaces can also be interpolating as well as approximating. Bezier surface is an approximator whereas B-Spline surface can be made interpolator. In addtion to modeling geometric objects, surfaces can be used to fit experimental data, tables of numbers, and discretized solution of differential equations. So the general objective here is to model a multi-dimensional surface, given points in 3-D space, and parameter value at each point. For example, constructing the surface representing the pressure distribution over a 2-D domain, $z = p(x,y)$, on an oblique airplane wing, stress distribution in a mechanical part, and contour plot is an alternative but is normally not clear in all occassions.

18.9. PARAMETRIC SURFACES

The parametric representation of surfaces using two parameters is a mapping from the two dimensional parameter space to three dimensional Cartesian space with a one-to-one correspondence of points.

$$P(u,v) = [x \;\; y \;\; z]^T = [x(u,v) \;\; y(u,v) \;\; z(u,v)]^T,\; u_{\min} \le u \le u_{\max},\; v_{\min} \le v \le v_{\max}$$

Just like the "segmentation approach" was followed for curves for building complex free-form curves by suitably blending segments of curves, in the surface modeling also the concept of "surface patches" is used. A large number of patches of surface are suitably blended to obtain the required free-form surface. Each patch can be parametrically modeled to give a topologically rectangular patch or a triangular patch. A topologically rectangular patch has four corners and four edges and when it is blended with another topologically rectangular or triangular patch, on which edges of the two patches is the blending to be done needs to be separately specified.

Two tangent vectors and one twist vector can be defined at any point on a surface patch. The twist vector at a point is the measure of twist in the surface at that point. Qualitatively, it is a measure of how drastically the curve deviates from the current plane of its existence at any point.

The tangent vector at any point $P(u, v)$ on the surface is obtained by holding one parameter constant and differentiating with respect to the other. Therefore there are two tangent vectors for each point.

Along the v = constant curve,

$$P_u(u,v) = \frac{\partial P}{\partial u} = \frac{\partial x}{\partial u}\hat{i} + \frac{\partial y}{\partial u}\hat{j} + \frac{\partial z}{\partial u}\hat{k},\ u_{\min} \le u \le u_{\max},\ v_{\min} \le v \le v_{\max}$$

and along the u = constant curve it will be,

$$P_v(u,v) = \frac{\partial P}{\partial v} = \frac{\partial x}{\partial v}\hat{i} + \frac{\partial y}{\partial v}\hat{j} + \frac{\partial z}{\partial v}\hat{k},\ u_{\min} \le u \le u_{\max},\ v_{\min} \le v \le v_{\max}$$

In matrix form,

$$\begin{bmatrix} P_u \\ P_v \end{bmatrix} = \begin{bmatrix} \frac{\partial x}{\partial u} & \frac{\partial y}{\partial u} & \frac{\partial z}{\partial u} \\ \frac{\partial x}{\partial v} & \frac{\partial y}{\partial v} & \frac{\partial z}{\partial v} \end{bmatrix}$$

The magnitudes tangent vectors:

$$\left|\frac{\partial \bar{P}}{\partial u}\right| = \left|\bar{P}_u\right| = \sqrt{\left(\frac{\partial x}{\partial u}\right)^2 + \left(\frac{\partial y}{\partial u}\right)^2 + \left(\frac{\partial z}{\partial u}\right)^2}$$

$$\left|\frac{\partial \bar{P}}{\partial v}\right| = \left|\bar{P}_v\right| = \sqrt{\left(\frac{\partial x}{\partial v}\right)^2 + \left(\frac{\partial y}{\partial v}\right)^2 + \left(\frac{\partial z}{\partial v}\right)^2}$$

and the unit vectors,

$$\hat{n}_u = \frac{\bar{P}_u}{\left|\bar{P}_u\right|}$$

$$\hat{n}_v = \frac{\bar{P}_v}{\left|\bar{P}_v\right|}$$

The twist vector at a point is the measure of twist in the surface at that point. Qualitatively, it is a measure of how drastically the curve deviates from the current plane of its existence at any point.

$$\lim_{\Delta v \to 0} \frac{\Delta \bar{P}_u}{\Delta v} = \frac{\partial \bar{P}_u}{\partial v} = \frac{\partial^2 \bar{P}}{(\partial u)(\partial v)} = \bar{P}_{uv}$$

$$\lim_{\Delta u \to 0} \frac{\Delta \bar{P}_v}{\Delta u} = \frac{\partial \bar{P}_v}{\partial u} = \frac{\partial^2 \bar{P}}{(\partial u)(\partial v)} = \bar{P}_{uv}$$

$$\bar{P}_{uv} = \left[\frac{\partial^2 x}{(\partial u)(\partial v)} \quad \frac{\partial^2 y}{(\partial u)(\partial v)} \quad \frac{\partial^2 z}{(\partial u)(\partial v)} \right]^T = \frac{\partial^2 x}{(\partial u)(\partial v)}\hat{i} + \frac{\partial^2 y}{(\partial u)(\partial v)}\hat{j} + \frac{\partial^2 z}{(\partial u)(\partial v)}\hat{k}$$

$$u_{\min} \le u \le u_{\max},\ v_{\min} \le v \le v_{\max}$$

The normal vector or surface normal at a given point on a surface is a vector that is perpendicular to both the tangent vectors at that point, $\mathbf{P}_u$ and $\mathbf{P}_v$.

$$N(u,v) = \frac{\partial \bar{P}}{\partial u} \times \frac{\partial \bar{P}}{\partial v} = \bar{P}_u \times \bar{P}_v$$

and the unit normal

$$\hat{n} = \frac{\bar{N}}{|\bar{N}|} = \frac{\bar{P}_u \times \bar{P}_v}{|\bar{P}_u \times \bar{P}_v|}$$

The unit normal can be either outward or inward which are obtained by suitably using the order of the cross product (Fig. 18.25).

Fig. 18.25: The outward and inward surface normals

$$N_{\text{outward}}(u,v) = \frac{\partial \bar{P}}{\partial u} \times \frac{\partial \bar{P}}{\partial v} = \bar{P}_u \times \bar{P}_v = -\bar{P}_v \times \bar{P}_u = -N_{\text{inward}}(u,v)$$

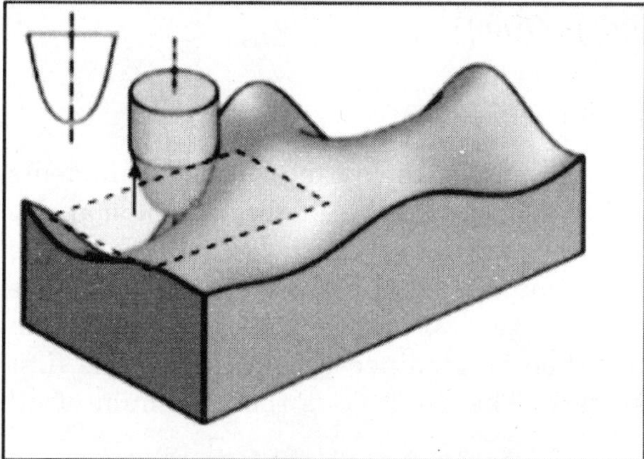

Fig. 18.26: CNC machining

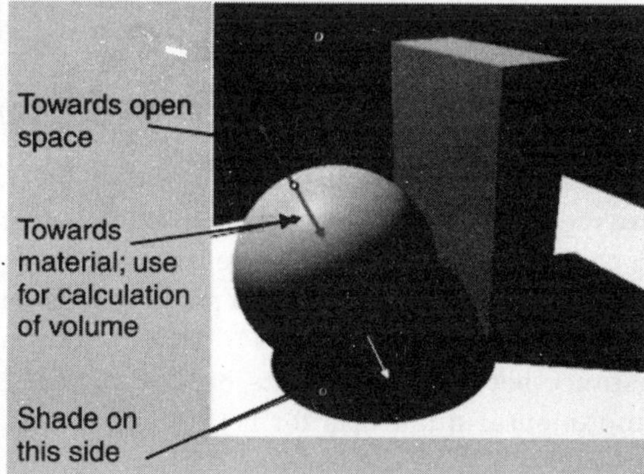

Fig. 18.27: Volume calculation and shade creation

In machining applications the positive sense of unit normal is chosen to be pointing out of the surface being machined (Fig. 18.26). In volume calculations and shading, the sense of unit normal is considered positive when pointing toward the existing material and negative when pointing to holes and open space (Fig. 18.27). The surface normal is zero when the cross product of tangent vectors is zero. It occurs at points lying on a cusp, ridge, or self-intersecting surface (Fig. 18.28).

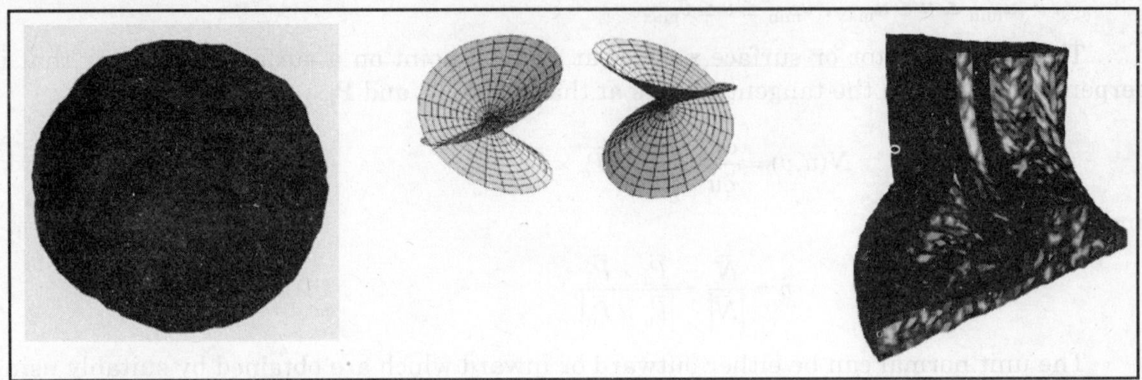

Fig. 18.28: At cusp, on self-intersecting surfaces and on ridges, the surface normal vanishes

The calculation of distance between two points ON A SURFACE is an important requirement in many practical applications. Two distinct points on the surface can be connected by many paths. The path that has minimum length is called a surface geodesic (it need not be a straight line, why?) Surface geodesics are required for optimized motion planning across a curved surface for CNC programming, robot programming and winding of coils around the rotor. The infinitesimal distance between two points **P**(u, v) and **P**$(u + \Delta u, v + \Delta v)$ is given by

$$(ds)^2 = \bar{P}_u \bullet \bar{P}_u (du)^2 + 2\bar{P}_u \bullet \bar{P}_v (du)(dv) + \bar{P}_v \bullet \bar{P}_v (dv)^2$$

When written in the following form, the above equation is called as the first fundamental quadratic form of a surface.

$$(ds)^2 = E(du)^2 + 2F(du)(dv) + G(dv)^2$$

where

$$E = \bar{P}_u \bullet \bar{P}_u \quad F = 2\left(\bar{P}_u \bullet \bar{P}_v\right) \quad G = \bar{P}_v \bullet \bar{P}_v$$

E, F and G are the first fundamental or metric coefficients of the surface. These coefficients provide the basis for length measurements on surface and specification of directions and angles on the surface. If we introduce another parameter "t", for the variation "u" and "v", that is $u = u(t)$ and $v = v(t)$, on a given path, then the movement along that path on the surface can be described by variation of this "t" in a chosen range. The geodesic, that is the path of minimum distance between two points on the surface, can then be obtained by integrating the first fundamental quadratic form (FQF-1) of the surface. The geodesic is the minimum of all S values along all the possible paths.

$$S = \int_{t_a}^{t_b} \sqrt{E\left(\frac{du}{dt}\right)^2 + 2F\left(\frac{du}{dt}\right)\left(\frac{dv}{dt}\right) + G\left(\frac{dv}{dt}\right)^2}\ dt$$

In order to investigate how the surface curves away from the tangent plane at the given point, the second fundamental quadratic form (FQF-2) of the surface is defined. For FQF-2, we need to define another infinitesimal length which is perpendicular to the tangent plane at $P(u, v)$, given by

$$\frac{1}{2}(dh)^2 = \hat{n} \bullet \bar{P}_{uu}(du)^2 + 2\hat{n} \bullet \bar{P}_{uv}(du)(dv) + \hat{n} \bullet \bar{P}_{vv}(dv)^2$$

Then the FQF-2 itself is defined as

$$\frac{1}{2}(dh)^2 = \hat{n} \bullet \bar{P}_{uu}(du)^2 + 2\hat{n} \bullet \bar{P}_{uv}(du)(dv) + \hat{n} \bullet \bar{P}_{vv}(dv)^2$$

where

$$L(u,v) = \hat{n} \bullet \bar{P}_{uu};\ M(u,v) = \hat{n} \bullet \bar{P}_{uv};\ N(u,v) = \hat{n} \bullet \bar{P}_{vv}$$

L, M and N are called the second fundamental coefficients of the surface and form the basis for defining the curvature of a surface. Surface curvature at a given point $P(u, v)$ is defined as the curvature of the normal section curve that lies on the surface and passes through the point. A normal section curve is the intersection curve of a plane passing through the surface normal **n** at the point and the surface. As you can see such normal section curves can be infinitely many. Therefore, we need to develop a general equation for curvature on any normal section curve. Let any normal section curve is defined by $u = u(t)$ and $v = v(t)$, then the curvature of that normal section curve is

$$\chi\ (chi) = \frac{L\left(\frac{du}{dt}\right)^2 + 2M\left(\frac{du}{dt}\right)\left(\frac{dv}{dt}\right) + N\left(\frac{dv}{dt}\right)^2}{E\left(\frac{du}{dt}\right)^2 + 2F\left(\frac{du}{dt}\right)\left(\frac{dv}{dt}\right) + G\left(\frac{dv}{dt}\right)^2} = \frac{\left[\frac{1}{2}(dh)^2\right]}{(ds)^2}$$

The corresponding radius of curvature then is given by $\rho = 1/\chi$. Sign of the curvature is not automatically set in the above equation but has to be chosen. Normally, the sign is taken to be positive for convex surfaces and negative for concave surfaces. One interesting question than can now arise is, if infinitely many normal section curves are possible each having its own value of χ, then there must be one normal section curve having maximum curvature, χ_{max}, and another normal section curve having χ_{min}. These are called the principal curvatures of the surface at the given point since they define the upper (χ_{max}) and lower (χ_{min}) bounds on the curvature at that point.

The principal curvatures are given by

$$\chi_{max} = H + \sqrt{H^2 - K}$$

$$\chi_{min} = H - \sqrt{H^2 - K}$$

where K is called as Gaussian curvature and H is called as Mean curvature. K and H are given by,

$$K = \chi_{max}\chi_{min}$$

$$H = \frac{1}{2}(\chi_{max} + \chi_{min})$$

The Gaussian Curvature (K) is practically more important than H. K is positive on a hill. K is negative at a saddle point. K is zero on ruled and cylindrical surfaces. The sign of K is obviously determined by the signs of principal curvatures. Surfaces having zero Gaussian curvature value everywhere on them are called as developable surfaces. A developable surface can be laid flat on a plane without stretching, tearing or distorting it. Most CAD/CAM systems have the facility of viewing the Gaussian curvature all over the surface. Principal curvatures themselves have special importance, particularly in CNC machining. In contour cutting, the radius of curvature of any contour should be greater than the radius of the cutter, otherwise gouging is said to occur. Curvature calculations are crucial to check whether gouging will occur.

Among the analytical surfaces, the planar surface is the simplest. Three non-coincident points are required to generate an infinite plane. The plane surface is useful for generating cross-sectional views by intersecting with surfaces, in mass property calculations and other applications requiring simple surfaces to be modeled.

Representation 1:

$$\bar{P}(u,v) = \bar{P}_0 + u(\bar{P}_1 - \bar{P}_0) + v(\bar{P}_2 - \bar{P}_0),\ 0 \le u \le 1,\ 0 \le v \le 1$$

It is the bilinear form of parametric equation of a line.

$$\bar{P}_u(u,v) = (\bar{P}_1 - \bar{P}_0);\ \bar{P}_v(u,v) = (\bar{P}_2 - \bar{P}_0),\ 0 \le u \le 1,\ 0 \le v \le 1$$

and the surface unit normal is

$$\hat{n}(u,v) = \frac{(\bar{P}_1 - \bar{P}_0) \times (\bar{P}_2 - \bar{P}_0)}{\left|(\bar{P}_1 - \bar{P}_0) \times (\bar{P}_2 - \bar{P}_0)\right|},\ 0 \le u \le 1,\ 0 \le v \le 1$$

Representation 2 (when two unit vectors are given):

$$\bar{P}(u,v) = \bar{P}_0 + uL_u\hat{r} + vL_v\hat{s},\ 0 \le u \le 1, 0 \le v \le 1$$

Representation 3 (when one point and the normal vector are given):

$$(\bar{P} - \bar{P}_0) \bullet \hat{n} = 0$$

The minimum distance from a point to a planar surface has a lot of important applications such as like in general reflection through a plane etc.

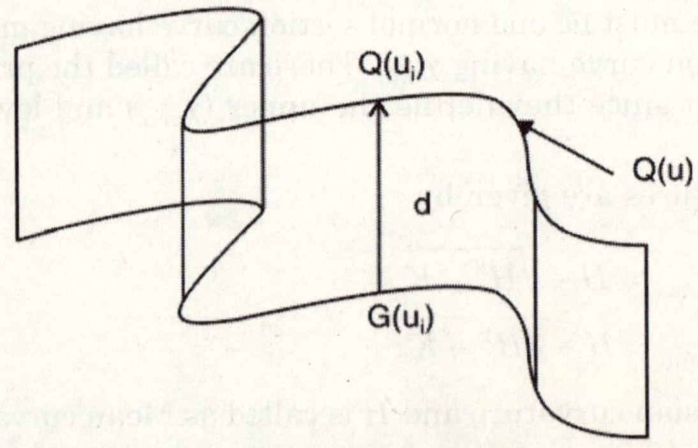

Fig. 18.29: Ruled surface

Say given is the following form of plane:

$$\bar{P} = \bar{P}_0 + u\hat{r} + v\hat{s},\ 0 \le u \le 1,\ 0 \le v \le 1 \quad ...(1)$$

$$\bar{Q} = \bar{P} + D\hat{n}$$

$$n = \text{unit normal}$$

$$\bar{P} = \bar{Q} - D\hat{n} \quad ...(2)$$

Equate (1) and (2)

$$\bar{P}_0 + u\hat{r} + v\hat{s} = \bar{Q} - D\hat{n}$$

In matrix form

$$\begin{bmatrix} r_x & s_x & n_x \\ r_y & s_y & n_y \\ r_z & s_z & n_z \end{bmatrix} \begin{Bmatrix} u \\ v \\ D \end{Bmatrix} = \begin{Bmatrix} x_Q - x_0 \\ y_Q - y_0 \\ z_Q - z_0 \end{Bmatrix}$$

The ruled surface is a linear surface. It interpolates between two given curves linearly (Fig. 18.29). The two given curves are called rails and they define the surface. Rails can be any wireframe entities. Ruled surfaces do not allow any twist or kink since it is linear.

$$\mathbf{P}(u_i,\ v) = \mathbf{G}(u_i) + v\ (\mathbf{Q}(u_i) - \mathbf{G}(u_i))$$

$$\mathbf{P}(u,\ v) = (1 - v)\ \mathbf{G}(u) + v\ \mathbf{Q}(u)$$

or

$$\mathbf{P}(u,\ v) = \mathbf{G}(u) + v\mathbf{d}$$

$$p(u,v) = (\cos(u), 0, \sin(u)) + v(0,1,0) = (\cos(u), v, \sin(u))$$

$$\Leftrightarrow \begin{cases} x(u,v) = \cos(u) \\ y(u,v) = v \\ z(u,v) = \sin(u) \end{cases}$$

Closed ruled surfaces such as cylindrical surfaces and conical surfaces can be built as ruled surfaces. When the rails of the rules surface are two parallel circles, a right circular ruled cylindrical surface results. Similarly, when a point and a circle are used as rails a right circular cone is obtained. Generalizing, for any other pair or curves a general cylinder or general frustum of cone result (Fig. 18.30). For a cylinder the parametric equation can be obtained as follows:

$$p(u,v) = (\cos(u), 0, \sin(u)) + v(0,1,0) = (\cos(u), v, \sin(u))$$

$$\Leftrightarrow \begin{cases} x(u,v) = \cos(u) \\ y(u,v) = v \\ z(u,v) = \sin(u) \end{cases}$$

Whereas, for a general cone, the parametric equation is obtained as follows:

$$\mathbf{P}(u,\ v) = (1 - v)\ \mathbf{G}(u) + v\ \mathbf{Q}(u)$$

$$\mathbf{P}(u,\ v) = (1 - v)\ \mathbf{G} + v\ \mathbf{Q}(u)$$

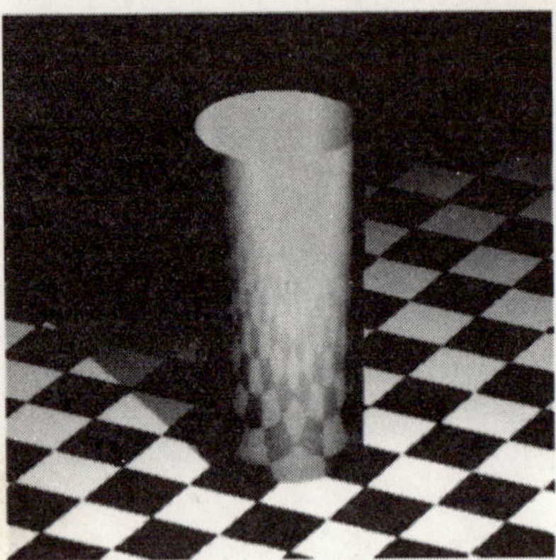
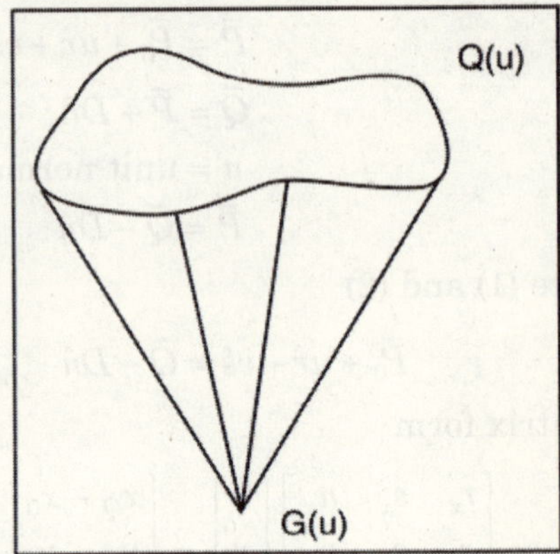

Fig. 18.30: Ruled closed surfaces

Rules bilinear surface patches are attractive for their simplicity and it can be constructed by using two non-parallel straight lines as rails (Fig.18.31).

$$p(u,v) = (1-v)\big((1-u)p_{00} + up_{10}\big) + v\big((1-u)p_{01} + up_{11}\big)$$

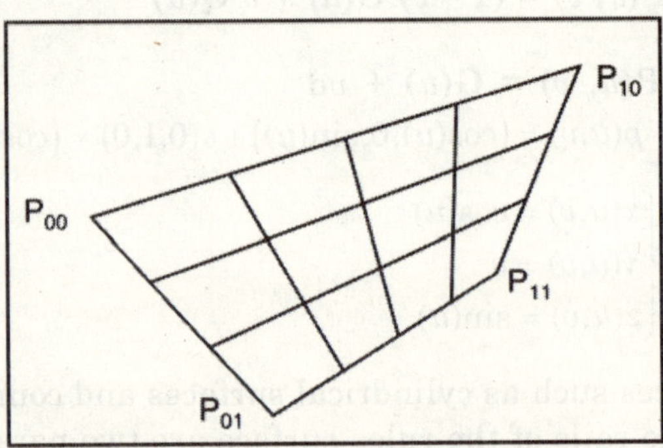

Fig. 18.31: Ruled bilinear surface patch

Surfaces of revolution (Fig. 18.32) are axisymmetric surfaces obtained by revolving a parametric curve segment around an axis by 360°. Revolving through a partial angle results in an open surface.

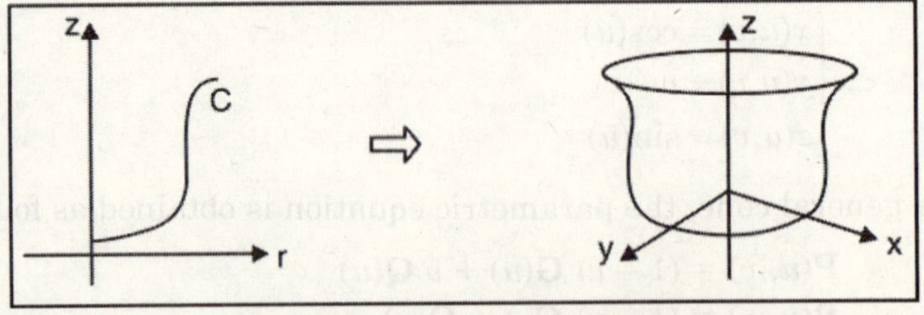

Fig. 18.32: Surface of revolution

$$(r(u),0,z(u)) \in C,\ 0 \le u \le u_{\max}, \Rightarrow (r(u)\cos(v), r(u)\sin(v), z(u)) \in Surface$$

Here, parameter v is eouivalent to angle. Or in other words,

$$P(u,v) = r(u)\cos v\hat{n}_1 + r(u)\sin v\hat{n}_2 + z_L(u)\hat{n}_3 \quad \text{where } 0 \le u \le u_{\max},\ 0 \le v \le 2\pi$$

The surfaces of revolution obtained by revolving only the analytic curves are known as quadric surfaces. Sphere, ellipsoid, paraboloid, hyperboloid and hyperbolic paraboloid are all quadric surfaces. Quadric surfaces have a lot of practical importance in real world design. For example, the solar collectors and telecommunication antenna are designed after the paraboloid. Tennis balls are spheres and many modern structures include hyperbolic paraboloids as arches.

Tabulated surfaces can be generated by using a parametric curve as the generatrix which is taken along a straight line directrix for a required length along the direction $\hat{n}_v$. The straight line always stays parallel to a fixed given vector that defines the v direction of the cylinder. The planar curve $\boldsymbol{G}(u)$ can be anyone of the wire-frame entities. The position vector of any point $\mathbf{P}(u,v)$ on the tabulated surface can be written as follows.

$$\mathbf{P}(u,v) = \mathbf{G}(u) + v\hat{n}_v,\quad 0 < u < u_{\max},\ 0 < v < v_{\max}$$

Till now, those surface patches have been discussed that have an analytical curve atleast along one direction. Tensor product surfaces are obtained by using synthetic curve description in both parametric directions. Most commonly used tensor product surfaces are the bicubic Hermite surface patch, Bezier surface patch, B-spline surface patch and the non-uniform rational B-Spline (NURBS) surface patch.

Just as the Hermite cubic curve requires four vectors, two end points and two end tangents, the bicubic Hermite surface patch requires as many as 16 vectors in order to fully define its parametric equation. These 16 vectors, including four points, eight tangent vectors (two at each corner for the topologically rectangular patch) and four twist vectors, uniquely determine the sixteen algebraic coefficients. Once these coefficients are determined, the bicubic Hermite curve can be written in a matrix as follows. Applying the boundary conditions into the parametric equation and solving for the coefficients, the following equation is obtained.

$$P(u,v) = [U]^T[M_H][B][M_H]^T[V],\quad 0 \le u \le 1,\ 0 \le v \le 1$$

Here $[M_H]$ is the Hermite matrix already defined for Hermite cubic curve. [B] is the geometry or boundary condition matrix defined as follows.

$$[B] = \begin{bmatrix} P_{00} & P_{01} & P_{v00} & P_{v01} \\ P_{10} & P_{11} & P_{v10} & P_{v11} \\ P_{u00} & P_{u01} & P_{uv00} & P_{uv01} \\ P_{u10} & P_{u11} & P_{uv10} & P_{uv11} \end{bmatrix} = \begin{bmatrix} [P] & [P_v] \\ [P_u] & [P_{uv}] \end{bmatrix} = \begin{bmatrix} \text{corner points} & \text{corner } v\text{-tangents} \\ \text{corner } u\text{-tangents} & \text{corner twist vectors} \end{bmatrix}$$

The tangent vectors and twist vectors at any point on the surface are given by,

$$P_u(u,v) = [U]^T[M_H]^u[B][M_H]^T[V]$$
$$P_v(u,v) = [U]^T[M_H][B][M_H]^{vT}[V]$$
$$P_{uv}(u,v) = [U]^T[M_H]^u[B][M_H]^{vT}[V]$$

Here

$[M_H]^u = [M_H]^v$ = The differential Hermite matrix defined already for Hermite cubic curve.

Similar to the Hermite cubic splines, the Bicubic Hermite patches give maximum C^1 continuity from one patch to the next, though they give C^2 continuity inside each of the patches. While blending two Bicubic Hermite patches, the necessary conditions are same curves (C^0 continuity) at the common edge, same direction of tangent vectors (C^1 continuity) at the common edge and the magnitudes of the tangent vectors do not have to be the same. Some scientists found the need to specify 16 vectors before the patch can be completed to be a requirement that is too demanding and hence proposed pragmatically successful forms of Hermite bicubic surface patch requiring fewer input data. Ferguson's patch is one such expediency. It is obtained by assigning the twist vectors at four corners to be zero and obtaining the corner tangent vectors from the chord length approximation using the four corner points, which are the only specified data.

Often, planar surfaces with creatively curvilinear boundaries are required. These could be topologically rectangular or triangular. For example, the visible side of the simple table drawer can have the shape of the Microsoft wave-like logo. Such surface patches can be obtained from the regular bicubic Hermite patches. One method of obtaining it is to equate the fundamental parametric equation with 16 vector algebraic coefficients of Hermite bicubic patch to the parametric plane. By equivalence the non-zero coefficients can be determined and the resulting parametric equation represents a planar patch with curvilinear boundaries.

$$P(u,v) = \sum_{i=0}^{3}\sum_{j=0}^{3} C_{ij}u^i v^j = P_0 + u(P_1 - P_0) + v(P_2 - P_0)$$

$$C_{00}u^0v^0 + C_{10}u^1v^0 + C_{20}u^2v^0 + C_{30}u^3v^0 + C_{01}u^0v^1 + C_{11}u^1v^1 + C_{21}u^2v^1 + C_{31}u^3v^1 +$$
$$C_{02}u^0v^2 + C_{12}u^1v^2 + C_{22}u^2v^2 + C_{32}u^3v^2 + C_{03}u^0v^3 + C_{13}u^1v^3 + C_{23}u^2v^3 + C_{33}u^3v^3 =$$
$$\bar{P}_0 + u(P_1 - P_0) + v(P_2 - P_0)$$

Hence

$$\bar{C}_{00} = \bar{P}_0;\ \bar{C}_{10} = (P_1 - P_0);\ \bar{C}_{01} = (P_2 - P_0)$$

rest all $\bar{C}_{ij}$ are equal to zero. The resultant $[B]$ matrix in $P(u,v) = [U]^T[M_H][B][M_H]^T[V]$

$$\text{is } [B] = \begin{bmatrix} \bar{P}_0 & \bar{P}_2 & (P_2 - P_0) & (P_2 - P_0) \\ \bar{P}_1 & \bar{P}_1 + (P_2 - P_0) & (P_2 - P_0) & (P_2 - P_0) \\ (P_1 - P_0) & (P_1 - P_0) & 0 & 0 \\ (P_1 - P_0) & (P_1 - P_0) & 0 & 0 \end{bmatrix}$$

Another way of obtaining a planar patch with curvilinear edges is to specify the twist vectors to be zero and express the position vectors and tangent vectors at three corners as linear combinations of the position vector and tangent vectors at the fourth corner, which form the only specified input.

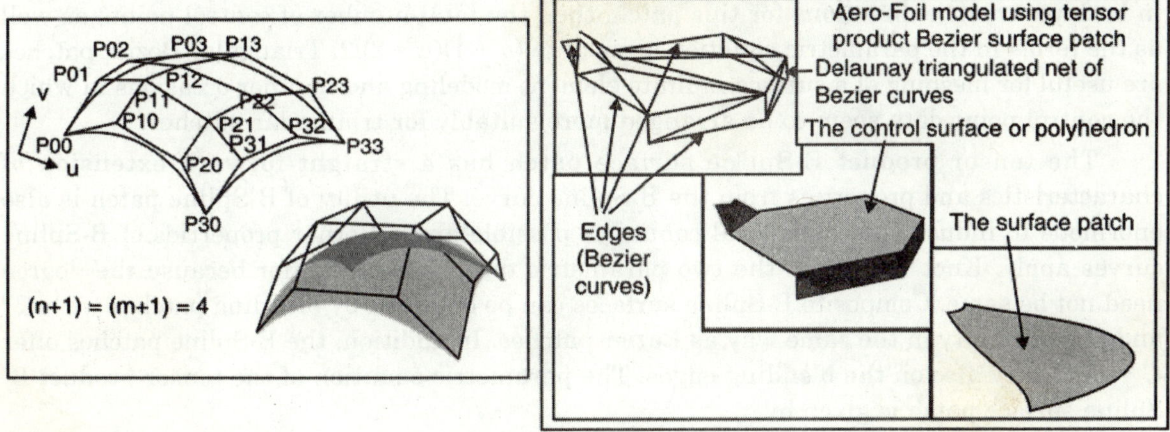

Fig. 18.33: Tensor product topologically rectangular Bezier surface patch, theory and application

Tensor product Bezier surface patches can be topologically either rectangular or triangular. Bezier surface patches do not need any tangent vector information as input and are purely based on the control point data. The designer will have a better intuitive feel of the shape of the patch follows the trend of the control surface. Each control surface results in one and only one Bezier surface patch whereas for the same set of control points many control surfaces can be built changing the order. Bezier patch is always an approximator and can have different degrees in the two parametric directions. The degree in each parametric direction is one less than the number of control points in that direction. Thus a Bezier patch with $(n + 1)$ number of control points in u parametric direction and $(m + 1)$ number of control points in the v parametric direction will have degree of n and m respectively in those parametric directions (Fig. 18.33).

The Bezier surface has convex hull property. The convex hull is the polyhedron formed by connecting all control points to all control points. The shape of the Bezier patch can be modified by either moving the control points or providing multiple control points at the same coordinate location. It is possible to express the Bezier patch equation in matrix form for generic situation of $(n + 1)$ by $(m + 1)$ control points.

$$P(u,v) = \begin{bmatrix} B_{0,n}(u) & B_{1,n}(u) & \cdots & B_{n,n}(u) \end{bmatrix} \begin{bmatrix} P_{00} & P_{01} & \cdots & P_{0m} \\ P_{10} & P_{11} & \cdots & P_{1m} \\ \vdots & \vdots & & \vdots \\ P_{n0} & P_{n1} & \cdots & P_{nm} \end{bmatrix} \begin{Bmatrix} B_{0,m}(u) \\ B_{1,m}(u) \\ \vdots \\ B_{m,m}(u) \end{Bmatrix}$$

$$\bar{P}(u,v) = [U]^T [M_B][P][M_B]^T [V]$$

Topologically triangular Bezier surface patch has three edges and three corners. It can be described by a parametric equation expressed in terms of the three barycentric coordinates sum of which at any point on the patch is always equal to one. Hence only two of them are independent true to what a parametric equation for surface patch should be. The Pascal's triangle is a convenient way of arriving at the exact forms of the polynomial blending functions constituting the parametric equation. In total, if the degree of the patch is n, which is same in both parametric directions for this patch, then the total number of control points as well as the terms in the parametric equation is equal to $(n + 1)(n + 2)/2$. Triangular Bezier patches are useful for meshing of a surface in finite element modeling and in other occasions in which the control point data seem to be arranged more suitably for triangular patches.

The tensor product B-Spline surface patch has a straight forward extension of characteristics and properties from the B-Spline curve. The utility of B-Spline patch is also enormous to industry because local control is possible and all other properties of B-Spline curves apply. Knot vectors in the two parametric directions can differ because the degree need not be same. Composite B-Spline surfaces can be obtained by blending patches with C^0 and C^1 continuity in the same way as Bezier patches. In addition, the B-Spline patches offer C^2 continuity also on the blending edges. The parametric equation of the tensor product B-Spline surface patch is given by,

$$P(u,v) = \sum_{i=0}^{n} \sum_{j=0}^{m} P_{ij} N_{i,k}(u) N_{j,l}(v), \quad 0 \le u \le u_{\max}, \ 0 \le v \le v_{\max}$$

Here $(k - 1)$ is the degree of the patch in the u parametric direction and $(l - 1)$ is the degree of the curve in the v parametric direction.

18.10. SOLID MODELING

Solids models are known to be complete, valid, and unambiguous representations of objects. A complete solid is one which enables a point in space to be classified relative to the object, if it is *inside*, *outside* or *on* the object. This classification is called as *spatial addressability* or *set membership classification*. A valid solid should not have dangling edges or faces, then only it will allow interference analysis, mass property calculations, finite element modeling and analysis, CAPP, machine vision, and NC part programming.

All commercial CAD packages offer one or both of two different solid modeling approaches, *primitives* based and *feature based*. UNIGRAPHICS (EDS Technologies), CATIA (Dassault Systems), I-DEAS (Structural Dynamics Research Corporation) offer both approaches. SolidWorks (Dassault Systems), Pro/Engineer (Parametric Technology Corporation) though earlier were mainly feature based packages, now they include CSG methodology also.

This primitives based approach allows designers to use predefined shapes (primitives) as building blocks to create complex solids (Fig. 18.34). Designers must use Boolean operations to combine the primitives. This approach is limited by the restricted shapes of the primitives.

Let A, B and C be primitive solids.

A = Block

B = Cylinder

C = Cylinder

A – B – C = D: Boolean operation

Create block A and subtract two cylinders from it using primitives approach.

D = Final solid

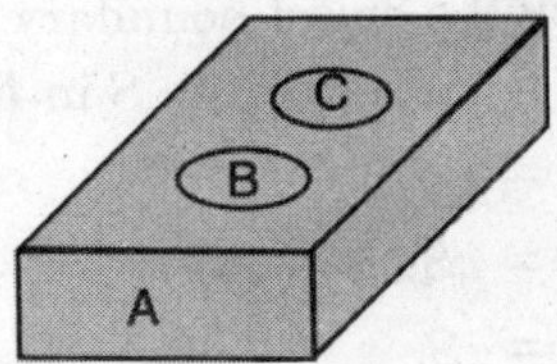

Fig. 18.34: CSG based solid modeling

Feature based modeling method is more flexible because it allows the construction of more complicated objects and more elaborate solids more readily than the primitive based modeling. Feature based modeling is in fact a generalization of primitives approach. Boolean operations are still used, but are hidden from the user. For example, creating a protrusion on the face of a cube is a Boolean union and creating a cut in the cube is a Boolean subtraction. These operations are must for creation of the final solid. Often multiple procedures may exist for creating the same solid. For example for the same bracket with two holes (Fig. 18.34) any of the following two feature based methods could be used.

Method 1: Create a rectangle ⇒ Subtract two circles ⇒ Extrude the resulting feature ⇒ The required solid is obtained.

Method 2: Create a rectangle ⇒ Extrude the rectangle to create the block ⇒ Selecting the top face of the block as sketching plane ⇒ draw two circles ⇒ Create through cuts by extrusion to obtain the final solid ⇒ The required solid is obtained.

Geometry is the actual dimensions that define the entities of the object. It is also sometimes called as metric information. Topology (sometimes called as combinatorial structure) is the connectivity and associativity of the object entities. Solid modeling procedures embed both the information in their specification. In the primitives based approach the entities are the solid entities (block, sphere, cone, torus, wedge and cylinder, Fig. 18.35) whereas in the feature based modeling the entities are the sketches themselves.

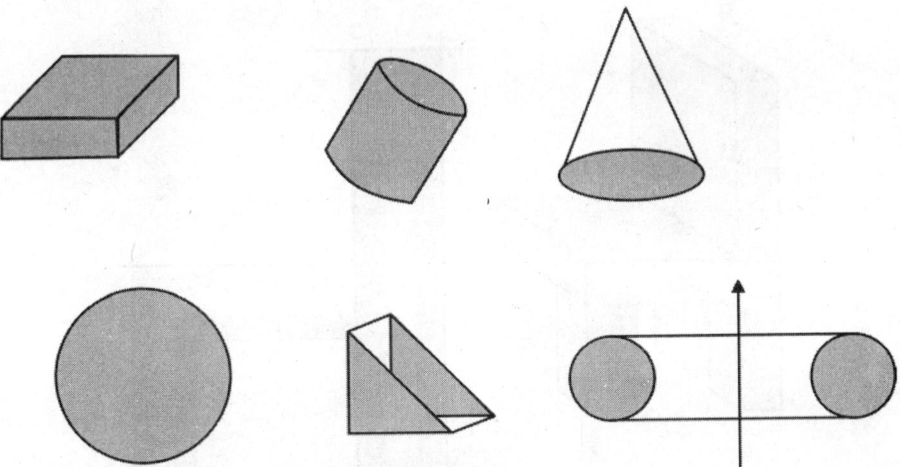

Fig. 18.35: The solid primitives

A physical object divides an E^n (n-dimensional Euclidean space) into two portions, interior and exterior, which are separated by the object boundary. Then the following can be stated.

$$\text{Object} = \text{Set of points } S \text{ in } E^3.$$

$$\text{Boundary} = bS.$$

$$\text{Exterior complement of S is} = cS.$$

$$\text{Object's interior} = iS.$$

$$S = bS \cup iS$$

The universal set of points

$$W = S \cup cS = iS \cup bS \cup cS$$

The geometric close is defined as $kS = iS \cup bS$. Ideally, therefore, kS should be equal to S. However, the geometric closure may some times be not equal to S. When $kS^1\ S$, there is said to be incompleteness and ambiguity and this can occur, due to various clipping, trimming and merging operations, if dangling edges or surfaces or created.

Regularized sets (r-sets) are those "curved polyhedra" with "well-behaved boundaries" and all representation schemes are expected to be applied only on r-sets. If they are applied on non-regularized sets, then ambiguity in the representation is very likely to occur.

Solid modeling may also be carried out using the method of half-spaces. A half-space is defined as a regular point set in E^3 as follows.

$$H = \{P : P \in E^3 \quad and \quad f(P) < 0\}$$

where

P = point set in E^3

$f(p) = 0$ defines the surface equation of the half-space boundaries.

Half-spaces could be planar, cylindrical, spherical, conical or toroidal. As an example, a chair made of half-spaces method is shown in Fig. 18.36.

$$S = \cup \left[\bigcap_{i=1}^{n} H_i \right]$$

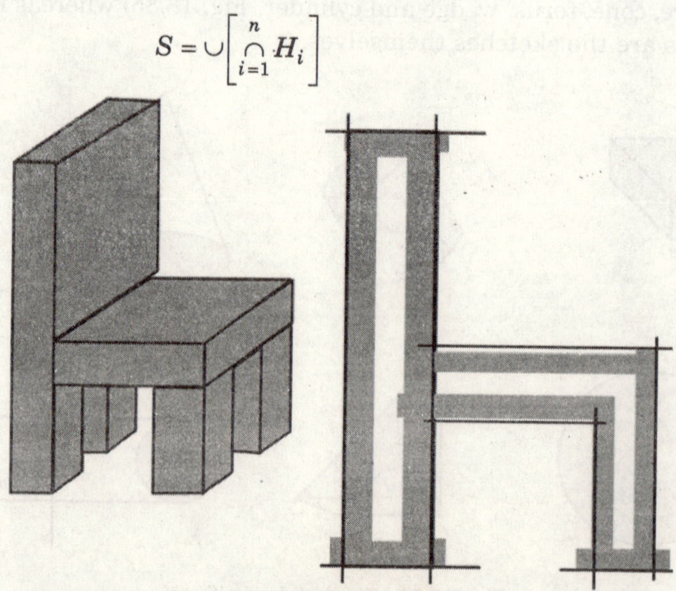

Fig. 18.36: A chair modeled by the half-space methodology

Boundary representation is one of the two most popular and widely used schemes (the other being CSG) and is based on the concept that a solid is made of a set of faces, which are subsets of closed and orientable surfaces. A closed surface is one that is continuous without breaks. An orientable surface is one where it is possible to distinguish two sides by using the direction of the surface normal to point inside or outside the solid model. Each face is bounded by edges and each edge is bounded by vertices. B-Rep solids could be either simple polyhedral objects having straight edges or curved polyhedral objects. For simple polyhedral objects, the B-Rep entities are vertex (*V*), edge (*E*), face (*F*), loop (*L*) which is a closed set of edges contained within a face often arising due to a blind or through hole through that face, genus (G) which is the total count of through holes (which themselves are termed as handles) and body (B) which indicates the closed surface holding certain volume. If a solid has internal pockets then the bodies it consists will be one plus as many number as the internal pockets. Such a solid is termed as disjoint solid [1, 2].

The Euler-Poincare law combines all these variables into an equation which if satisfied then that combination of these B-Rep entities results in a valid solid.

$$F - E + V - L = 2(B - G)$$

18.11. FINITE ELEMENT ANALYSIS

FEM is a powerful technique to solve partial differential equations and has become a widely accepted tool for mechanical systems *analysis* with an aim to examine stresses, strain, forces, kinematics, vibrations, failure, fatigue, fracture, creep etc. Almost all types of analyses are possible and it is an approximate numerical techniques based on the variational calculus. FEM is closely linked with computer usage; FEM would not have been so popular without it! Beware computer has no brain of its own! There is much to learn before one jumps to the computer! Why?

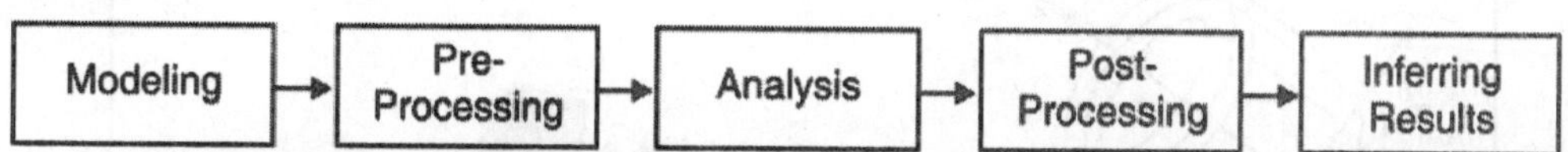

Fig. 18.37: The first and last stages above need much higher skill and appreciation of the physics of the problem than the middle three which have been more or less mechanized by the CAD systems available in the market

It is because more than actually running the solver, the preparation of the given problem into an equivalent finite element model and interpreting the numerical values or colorful pictures of contour plots, deformation animations etc. require a high level of skill and understanding of the mechanical system at hand. FEM comes after the product design is conceptualized. Conceptual design includes trial material, operating conditions, shape and dimensions. Then FEM should be used to check the sufficiency of selection of material, dimensions, operating conditions etc. To be able to model and infer from the results are more important than doing the analysis itself because nowadays there are a good number of commercial packages (Fig. 18.37).

Finite Element Procedure

1. Select the **domain**
2. Choose an **element** type and **material(s)**
3. Create the mesh of **finite elements:** called as **discretization** operation; spatial and temporal
4. **Approximate** the solution within an element; **piece-wise** application
5. Develop **element matrices** and equations
6. Generate the **global finite element matrices equation**
7. **Solve** for the unknowns at the nodes
8. **Interpret** the results –>**Repeat** if needed

Fig. 18.38: Steps in the general procedure of finite element analysis

The general finite element procedure consists of standard steps (Fig. 18.38) and perhaps that is the reason why FEM has become an industry standard for mechanical systems analysis. It is not surprising, from the same stand point, to note that the use of FEM has been smoothly extended to the modern and micro and nano electromechanical systems analysis.

Approximation of a general real 3-D mechanical system to a more amenable approximate 3-D, 2-D or 1-D system is the key to successful use of FEM. One needs a lot of practice and experience for judgment on these matters. An example is heat transfer in a cylinder is shown in Fig. 18.39. If it is known to the designer that the heat flow is predominantly in the radial direction only, as it happens in I^2R heating of electric cables in sheathing, it saves a lot of time and money to analyze the system by a much simpler 1-D radial heat flow than true 3-D analysis.

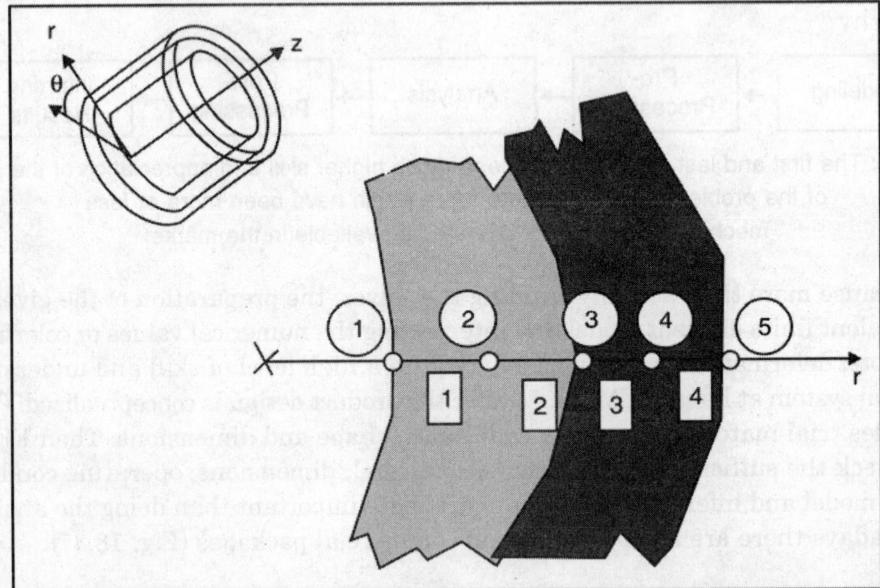

Fig. 18.39: Engineering ingenuity can be put to use to suitably reduce real 3-D systems to equivalent 1-D or 2-D systems for FEA purposes, resulting in a lot of saving in time and money

You can understand the procedure by comparing with FDM because there, largely, are a lot of similarities. Of course, FEM is much different from FEM in its capabilities and far more advanced. How FEM is different from solving the governing differential equations directly and traditional variational calculus? Instead of solving the governing differential (equilibrium) equations directly, the finite element method solves an integral form of these equations. Therefore, we need to convert the governing differential equations into an integral form first. In the traditional variational calculus methods, the solution is approximated over the entire domain. In FEM, the approximation is over each element. That is why, FEM is called the piece-wise application of variational method.

Properties of variational symbol, δ are crucial for appreciating the mathematical development of the finite element equations from the partial differential equations. For example look at the uniaxial bar subjected to an axial tensile load shown in Fig. 18.40.

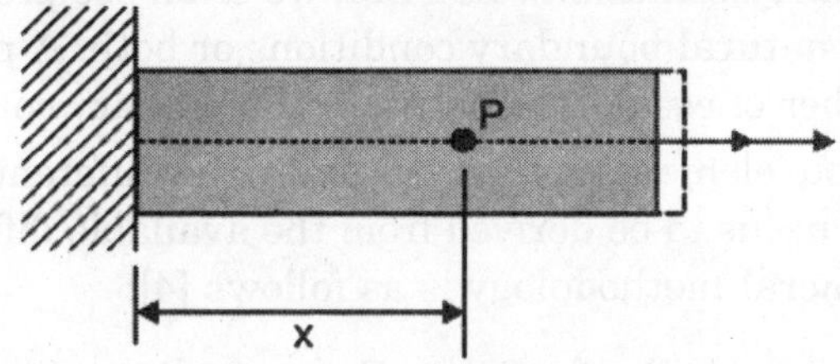

Fig. 18.40: An axially loaded bar

A functional is a function of functions. The functional for the axially loaded bar is the total potential energy and it and its derivatives can be written as follows [3]:

$$F = F(x, u, u')$$

$$\delta F = \frac{\partial F}{\partial u}\delta u + \frac{\partial F}{\partial u'}\delta u'$$

$$dF = \frac{\partial F}{\partial x}dx + \frac{\partial F}{\partial u}du + \frac{\partial F}{\partial u'}du'$$

Displacement, u, and strain, u', change at point P from undeformed to deformed position. Hence the functional, potential energy, changes. But, for small displacements, location of P still is taken to be the same, at x. Hence, δ is treated as independent of x. The other properties of the δ are as follows:

$$\delta(F_1 \pm F_2) = \delta F_1 + \delta F_2$$

$$\delta(F_1 F_2) = F_2 \delta F_1 + F_1 \delta F_2$$

$$\delta(F_1 / F_2) = \frac{F_2 \delta F_1 - F_1 \delta F_2}{F_2^2}$$

$$\delta\left[(F_1)^n\right] = n(F_1)^{n-1}\delta F_1$$

$$\frac{d}{dx}(\delta u) = \frac{d}{dx}(\alpha v) = \alpha\frac{dv}{dx} = \alpha v' = \delta u' = \delta\left(\frac{du}{dx}\right)$$

$$\delta\int_a^b u(x)dx = \int_a^b \delta u(x)dx$$

Development of Integral Equations from the given partial differential equations representing the physics of the mechanical system can be developed in two ways in FEM. Rayleigh-Ritz Finite Element Method requires a functional and methodology is to minimize the functional. Weighed residual method (or Galerkin method) requires no functional and the methodology is to minimize the weighed residual of the partial differential equations resulting when the approximate solution is substituted. The weights themselves can be obtained in many ways and whichever way we construct them, they must satisfy the homogeneous form of the essential boundary conditions of the problem. Boundary conditions themselves can be of two types. The specification of the unknown variable itself (for example say the displacement in elasticity analysis) is termed as essential boundary condition (or Dirichlet's boundary condition) and the specification of the first or higher order differential coefficients of the unknown variable is known as the natural boundary condition (or Neumann boundary condition). In FEM, we often see problems with only essential boundary conditions or only natural boundary conditions or both. A problem is said to be well-posed problem if equal number of essential and natural boundary conditions are specified.

In the Rayleigh-Ritz finite element method, normally the differential equation of the system is available. A functional (Π) needs to be derived from the available differential equation(s). How to derive the functional? General methodology is as follows [4]:

$$L(\phi) - f = 0 \;\; on \;\; \Omega$$

$$\phi = \Phi \;\; on \;\; \Gamma$$

$$\int_{\Omega} \delta\phi\left[L(\phi) - f\right] d\Omega = \delta \int_{\Omega} \phi\left[L^*(\phi) - f^*\right] d\Omega = 0$$

$$\delta\left[I(\phi)\right] = \delta\Pi = 0$$

$$\Pi = I(\phi) = \int_{\Omega} \phi\left[L^*(\phi) - f^*\right] d\Omega$$

As an example, look at the bar having the shape of a frustum of a cone and is subjected to an axial tensile load (Fig. 18.41).

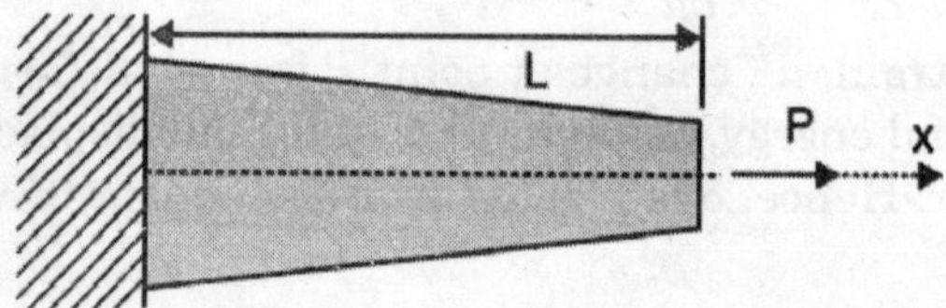

Fig. 18.41: The example bar for which the total potential energy is to be found

$$\frac{d}{dx}\left(AE\frac{du}{dx}\right) = 0 \;\; \text{subject to}$$

because $\sigma = P/A = E\varepsilon = E\dfrac{du}{dx} \Rightarrow$

$$P = AE\frac{du}{dx} \Rightarrow \frac{d}{dx}\left(AE\frac{du}{dx}\right) = \frac{dP}{dx} = \omega = 0$$

$$\int_0^L (\delta u)\frac{d}{dx}(AEu')\,dx = 0 \Rightarrow \int_0^L [(\delta u)d(AEu')] = 0$$

We need integration by parts:

$$\int_0^L (U)\left(\frac{dV}{dx}\right)dx = (U)\int_0^L\left(\frac{dV}{dx}\right)dx - \int_0^L\left\{\left(\frac{dU}{dx}\right)\left\{\int_0^L\left(\frac{dV}{dx}\right)dx\right\}\right\}dx$$

$$\Rightarrow \int_0^L UdV = \underbrace{(UV)\Big|_0^L}_{\therefore \text{ it is a definite integral}} - \int_0^L (dU)V \Rightarrow \int_0^L VdU = (UV)\Big|_0^L - \int_0^L UdV$$

Applying $\left[\int_0^L VdU = (UV)\Big|_0^L - \int_0^L UdV\right]$ to $\int_0^L [(\delta u)d(AEu')] = 0$

where $\quad U = AE(u'),\ V = (\delta u)$

$$\left[[AEu'(\delta u)]\Big|_0^L - \int_0^L (AEu')d(\delta u) = 0,\right.$$

$$d(\delta u) = \frac{d(\delta u)}{dx}dx = \delta\left(\frac{du}{dx}\right)dx = \delta(u')dx$$

$$P\delta u_L - \int_0^L (AEu')(\delta u')\,dx = 0 \text{ where } u_L = \text{displacement at } x = L$$

Now bring δ out of all terms. Since, $u'(\delta u') = \frac{1}{2}\delta\left[(u')^2\right]$,

$$\delta(Pu_L) - \int_0^L \delta\left(\frac{AE}{2}(u')^2\,dx\right) = 0 \Rightarrow \delta\left[\int_0^L\left(\frac{AE}{2}(u')^2\,dx\right) - (Pu_L)\right] = 0$$

$$\Pi = \left[\frac{1}{2}\int_0^L\left(AE(u')^2\,dx\right) - (Pu_L)\right]$$

This functional Π in "elasticity" is nothing but the total potential energy. Similar functionals can be derived in many other problems such as heat transfer, torsion, dynamics, electromagnetic field problem etc. Those functionals will have corresponding physical meanings. The minimization of functional implies in elasticity problem the minimization of the total potential energy for equilibrium.

Potential Energy (Π) = Strain Energy (S) + Work Potential (WP)

For example, for the 1-D continuum that we discussed just now,

$$\Pi = \left[\underbrace{\frac{1}{2}\int_0^L\left(AE(u')^2\,dx\right)}_{\text{Strain Energy }(S)} - \underbrace{(Pu_L)}_{\substack{\text{Work Potential}\\ \text{(WP)}}}\right]$$

For linear elastic materials, the strain energy per unit volume being

$$\frac{1}{2}\{\sigma\}^{T}\{\varepsilon\},$$

where $(\sigma\}$ is the stress tensor and $\{\varepsilon\}$ is the strain tensor,

$$S = \frac{1}{2}\int_V \{\sigma\}^T \{\varepsilon\}\, dV$$

$$WP = \int_V \{u\}^T \{f\}\, dV - \int_S \{u\}^T \{T\}\, dS - \sum_i \{u\}_i^T \{P\}_i$$

Hence,

$$\Pi = \frac{1}{2}\int_V \{\sigma\}^T \{\varepsilon\}\, dV - \int_V \{u\}^T \{f\}\, dV - \int_S \{u\}^T \{T\}\, dS - \sum_i \{u\}_i^T \{P\}_i$$

The principle of minimum potential energy says that for conservative systems, of all the kinematically admissible displacement fields, those corresponding to equilibrium extremize the total potential energy. If the extremum condition is a minimum, the equilibrium state is stable. Conservative system is path independent and entropy is zero. Kinematically admissible displacement fields are those that satisfy the single-valued nature of displacements (compatibility) and the boundary conditions.

From analysis point of view, most mechanical systems continuous but some systems can be simplified to discrete systems thus circumventing the step of meshing (Fig. 18.42).

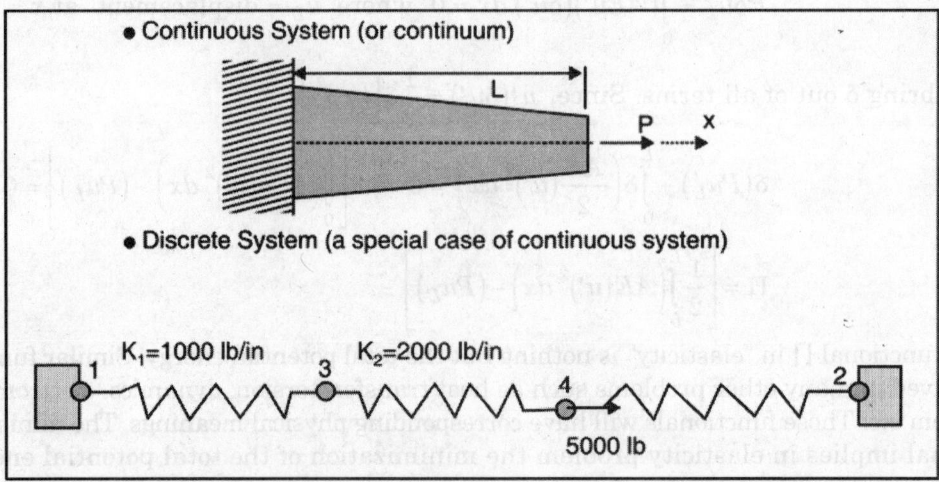

Fig. 18.42: Continuum systems and discrete systems

For continua, Π can be used to find the approximate solution. For an assumed displacement, field is constructed, such as

$$u = \sum a_i \phi_i (x, y, z) \quad i = 1 \text{ to } l$$

$$v = \sum a_j \phi_j (x, y, z) \quad j = (l+1) \text{ to } m$$

$$w = \sum a_k \phi_k (x, y, z) \quad k = (m+1) \text{ to } n$$

The functions φ_i are usually taken as polynomials. Displacements u, v, and w must be kinematically admissible. That is they must satisfy the specified boundary conditions. Introduce the stress-strain relations and strain-displacement relations. Then, in place of displacements, substitute the approximate solution. Then we obtain:

$$\Pi = \Pi(a_1, a_2, a_3, \ldots, a_r)$$

Minimize Π with respect to each unknown constant in the approximation.

$$\frac{\partial \Pi}{\partial a_i} = 0 \quad i = 1, 2, \ldots, r$$

This gives "r" number of simultaneous algebraic equations in "r" number of "a_i", which can be solved very easily. Substituting the "a_i" back in the solution approximation, we have the approximate solution with us.

Why cast the equations in displacements (as the unknown nodal variable) always? In a problem in which the displacements are unknown, compatibility conditions are automatically satisfied (by the displacement solution found). Hence, it is advisable to make displacement rather than strain or stress as nodal unknown variable.

Weighted residual method essentially based on the observation that the governing equation of the mechanical system is not exactly satisfied but leaves a residual when an approximate solution, instead of an exact solution, is substituted. Weighted residual methods employ this residual and use the logic that if there exists an approximation solution such that the residual due to that solution when multiplied by suitable known weight functions and integrated over the domain will be equal to zero, then that solution can be found out.

Obtaining finite element solution is to solve the global FE equation, which is nothing but a set of simultaneous algebraic equations. For linear problems and steady state conditions, direction iteration method with Gaussian elimination is sufficient for moderately sized meshes. Frontal solution, Cholesky method, Skyline method are special methods of solution that enable optimum storage of elements in coefficient matrix. In dynamic and transient problems, spacial approximation is done by FEM whereas temporal approximation is done by FDM. In non-linear problems, an iterative method of solution such as Newton-Raphson method is used. The solution is incremental in nature because for incremental load applied, incremental response of the system is sought. Non-linearity can arise due to different reasons: geometric non-linearity, governing equations being non-linear, constitutive equations being non-linear, for example, non-linear elastic material.

Three sources of error can be there in finite element solutions. Domain approximation error is due to the fact that meshing is a geometric approximation. Approximation error results due to iterative numerical solution approximation based on a specific convergence criterion. Quadrature error and finite arithmetic error results due to numerical integration and numerical computation on computers. Two measures of error are normally used, Energy norm and L_2 norm.

Based on the suitability to the application, the geometry and solution approximation can be suitably made in three different ways, super-parametric, sub-parametric and iso-parametric formulations. If m is the degree of geometric approximation and n is the degree of solution approximation then, If $m < n$, then it is subparametric formulation, if $m > n$, then it is superparametric formulation and if $m = n$, then it is isoparametric formulation. Isoparametric formulation is more often used because it offers saving on computations.

In each of 1-*D*, 2-*D* and 3-*D* problems, natural/local coordinates allow each element in the Cartesian space to be mapped into the natural/local coordinate space. Such mapping simplifies the integration of elemental stiffness matrices because a single "for" loop can be used to do the numerical integration for all elements. This becomes possible because in the natural coordinate space, the limits of integration are always –1 to + 1, in each direction.

$$v = N_1 v_1 + N_2 v_2$$
$$x = N_1 x_1 + N_2 x_2$$

where

$$N_1 = \frac{1-\xi}{2} \qquad N_2 = \frac{1+\xi}{2}$$

ξ = natural coordinate

The shape functions are normally derived by inspection. Most analysis work can safely rely on Lagrangian shape functions which are the most widely used. Shape functions are polynomials in the natural coordinates. They express the approximation of solution and geometry. Shape functions act as weighing functions on nodal values of the variable in the proposed approximate solution (Fig. 18.43).

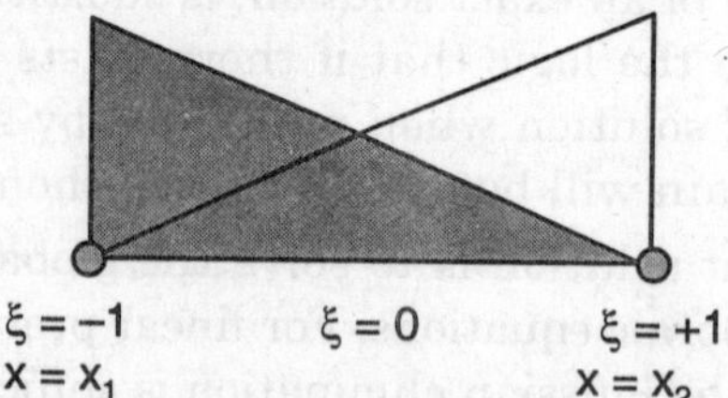

Fig. 18.43: One-dimensional example of shape function in natural coordinates

Element stiffness (or coefficient) matrices often involve integration of certain terms except in 1-*D* element and CST. The integration which is originally in Cartesian coordinates is converted to natural coordinates using the concept of Jacobian. Numerical integration by Gaussian quadrature involves simple summation of function values at Gauss points weighed by the weight functions.

$$\iiint f(u,v,w)\ dxdydz$$

$$\Rightarrow$$

$$\int_{-1}^{+1}\int_{-1}^{+1}\int_{-1}^{+1} f(u,v,w)\ |J|\,dudvdw$$

$$\approx \sum_i \sum_j \sum_k \left[w_i w_j w_k\, f(u_i, v_j, w_k) \right]$$

Geometric transformation means the coordinate system is fixed and the model is moved in that same coordinate system from one location to the other location, by rigid-body motion. It is possible to propose one single transformation matrix [*T*] that enables all the rigid-body transformations – translation, scaling, reflection, rotation, clipping and windowing. Such [T] should have certain specific properties, it should be as general as possible, must apply to all rigid body transformations including, clipping and windowing, both 2-*D* and 3-*D* transformations should be possible, homogeneous form is preferable, because only homogeneous form of transformation, matrix can include translation in the general representation of **P* = P[*T*]**.

$$\vec{P}^* = \vec{P} + \vec{d}$$

$$\vec{P}^* = \begin{Bmatrix} x^* \\ y^* \\ z^* \\ 1 \end{Bmatrix}^T = \begin{Bmatrix} x \\ y \\ z \\ 1 \end{Bmatrix}^T [T] = \begin{Bmatrix} x \\ y \\ z \\ 1 \end{Bmatrix}^T \begin{bmatrix} 1 & 0 & 0 & 0 \\ 0 & 1 & 0 & 0 \\ 0 & 0 & 1 & 0 \\ x_d & y_d & z_d & 1 \end{bmatrix}$$

Example:

$$P = [1 \;\; 1 \;\; 1 \;\; 1]; \; \vec{d} = [2 \;\; 1 \;\; 1 \;\; 1]$$

$$\vec{P}^* = \begin{Bmatrix} x^* \\ y^* \\ z^* \\ 1 \end{Bmatrix}^T = \begin{Bmatrix} x \\ y \\ z \\ 1 \end{Bmatrix}^T [T] = \begin{Bmatrix} 1 \\ 1 \\ 1 \\ 1 \end{Bmatrix}^T \begin{bmatrix} 1 & 0 & 0 & 0 \\ 0 & 1 & 0 & 0 \\ 0 & 0 & 1 & 0 \\ 2 & 1 & 1 & 1 \end{bmatrix} = \begin{Bmatrix} 3 \\ 2 \\ 2 \\ 1 \end{Bmatrix}^T$$

Increase and decrease in size with/without change of shape is achieved by scaling. Scaling is the implementation of the "zoom" operation available in the CAD systems. Scaling involves apparent translation for all the points on the part except that on the origin. Equal amount of scaling in all three Cartesian coordinate direction is called uniform scaling whereas different amounts of scaling in three direction is termed as non-uniform scaling.

Scaling about the origin of the WCS:

$$[P]^* = [P][S]$$

$$[S] = \begin{bmatrix} s_x & 0 & 0 \\ 0 & s_y & 0 \\ 0 & 0 & s_z \end{bmatrix}; \; s_i > 0$$

$$\begin{Bmatrix} x^* \\ y^* \\ z^* \end{Bmatrix}^T = \begin{Bmatrix} x \\ y \\ z \end{Bmatrix}^T \begin{bmatrix} s_x & 0 & 0 \\ 0 & s_y & 0 \\ 0 & 0 & s_z \end{bmatrix} \Rightarrow x^* = s_x x; \quad y^* = s_y y; \quad z^* = s_z z$$

If $[S] = [I]$

Reflection is useful in building parts of symmetry in which only half the part needs to be built and rest is built by reflection.

$$[P]^* = [P][M]$$

$$[M] = \begin{bmatrix} \pm 1 & 0 & 0 \\ 0 & \pm 1 & 0 \\ 0 & 0 & \pm 1 \end{bmatrix} = \begin{bmatrix} m_{11} & 0 & 0 \\ 0 & m_{22} & 0 \\ 0 & 0 & m_{33} \end{bmatrix};$$

Reflection may be done generally through (*a*) either a plane, (*b*) or a line, and (*c*) or a point. Particular cases of these are, 1) Through one of the principal planes ($X = 0$, $Y = 0$ or $Z = 0$), 2) Through one of the principal axes (X, Y or Z), and 3) Through the origin of WCS.

Two types of aspects of rotational transformation is possible. First is rotation of coordinate system $x - y$ to $x^* - y^*$ by positive θ and finding the coordinate values of the same point with respect to the new coordinate system. Second is rotation of the point with respect to the same coordinate system by positive θ. The two are in opposite directions. Hence the minus sign for *sine* component exchanges.

Consider 2-*D* rotation of a point in the *X*-*Y* plane.

$$x^* = x\cos\theta + y\sin\theta$$

$$y^* = -x\sin\theta + y\cos\theta$$

$$\begin{Bmatrix} x^* \\ y^* \\ 1 \end{Bmatrix} = \begin{bmatrix} \cos\theta & \sin\theta & 0 \\ -\sin\theta & \cos\theta & 0 \\ 0 & 0 & 1 \end{bmatrix} \begin{Bmatrix} x \\ y \\ 1 \end{Bmatrix}$$

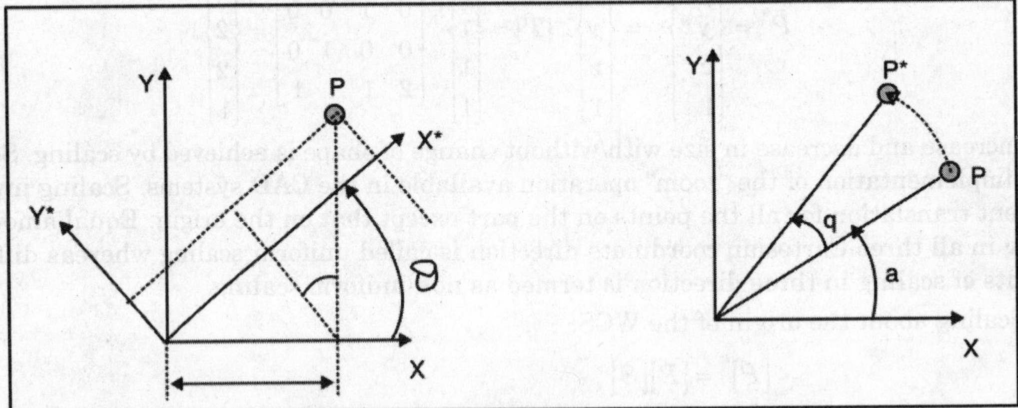

Fig. 18.44: The transformation of a point

Consider 2-*D*.

X-*Y* plane rotation only.

$$x^* = r\cos(\theta + \alpha) = r\cos\theta\cos\alpha - r\sin\theta\sin\alpha$$

$$= x\cos\theta - y\sin\theta$$

$$y^* = r\sin(\theta + \alpha) = r\sin\theta\cos\alpha + r\cos\theta\sin\alpha$$

$$= x\sin\theta + y\cos\theta$$

$$\begin{Bmatrix} x^* \\ y^* \\ 1 \end{Bmatrix} = \begin{bmatrix} \cos\theta & -\sin\theta & 0 \\ \sin\theta & \cos\theta & 0 \\ 0 & 0 & 1 \end{bmatrix} \begin{Bmatrix} x \\ y \\ 1 \end{Bmatrix}$$

Remember, rotation is always done about an axis, not a point. In the special case of planar rotation, the rotational axis appears to be a point. Different problems in rotational transformation are rotation about each of the principal axes (of WCS), two-dimensional rotation about an arbitrary axis that does not pass through the origin of MCS and is parallel to any one of the *X*, *Y* or *Z* principal axes, three dimensional rotation about an arbitrary axis that is inclined to all the principal axes of MCS and passes through the origin of WCS, the most general case of 3-*D* rotation about an arbitrary axis that is inclined to all the three principal axes of WCS and does NOT pass through the origin of WCS.

Transformation matrix for rotation about each of the principal axes need to be developed first. Once these are ready, using them other cases of rotation can be easily treated.

Consider 2-*D*.

X-*Y* plane rotation only.

$$x^* = r\cos(\theta + \alpha) = r\cos\theta\cos\alpha - r\sin\theta\sin\alpha$$

$$= x\cos\theta - y\sin\theta$$

$$y^* = r\sin(\theta + \alpha) = r\sin\theta\cos\alpha + r\cos\theta\sin\alpha$$

$$= x\sin\theta + y\cos\theta$$

$$\begin{Bmatrix} x^* \\ y^* \\ 1 \end{Bmatrix} = \begin{bmatrix} \cos\theta & -\sin\theta & 0 \\ \sin\theta & \cos\theta & 0 \\ 0 & 0 & 1 \end{bmatrix} \begin{Bmatrix} x \\ y \\ 1 \end{Bmatrix}$$

$$[P]^* = [R_z][P]$$

Similarly,

$$[R_x] = \begin{bmatrix} 1 & 0 & 0 \\ 0 & \cos\theta & -\sin\theta \\ 0 & \sin\theta & \cos\theta \end{bmatrix}; \ [R_y] = \begin{bmatrix} \cos\theta & 0 & \sin\theta \\ 0 & 1 & 0 \\ -\sin\theta & 0 & \cos\theta \end{bmatrix}$$

It is essential to be able to represent all geometric transformations as matrix multiplications rather than additions or subtractions for ***concatenating*** different transformations that are done in sequence. As we have seen, translation does not readily does not allow it. Representing points in homogeneous coordinates is the effective answer to this problem. Homogeneous coordinates have enabled geometric transformations to be embedded in the graphics processor/accelerator (hardware) thus enormously speeding up the execution of display commands. Homogeneous coordinates are also useful in obtaining perspective views of geometric models, projective geometry, mechanism analysis and design and robotics. Homogeneous coordinates remove many anomalous situations such as representing points at infinity and the non-intersection of parallel lines, otherwise encountered with simple Cartesian coordinates. Rational parametric curves and surfaces can be more simply represented using homogeneous coordinates. An *n*-dimensional space is mapped into $(n + 1)$-dimensional space where the $(n + 1)$th dimension being the scaling factor h. For the purposes of geometric modeling, $h = 1$ is taken to avoid unnecessary divisions.

Perspective Projection Creates an artistic effect that adds some realism to perspective views. The size of an entity is inversely proportional to its distance from the center of projection–the closer the entity to the center, the larger its size is. This projection is not popular with engineers and draftsmen because actual dimensions and angles of objects and therefore shapes cannot be preserved. Therefore, measurements cannot be taken from perspective views directly. It is basically for better visualizations of 3-*D* objects on 2-*D* screen. This projection cannot preserve parallelism more polular in architecture and fine arts.

REVIEW QUESTIONS

1. Who invented Bezier curve?
2. What are the applications of Bezier curve, past and present?
3. If only one control point is given, what will be the Bezier curve?
4. If there are only two control points P_0 and P_1, what will be the nature of Bezier curve?
5. Prove that the Bezier curve with Bernstein polynomials in general does not pass through any of the control points except the first and last.
6. Prove that the curve is always contained within the convex hull of the control points and that it never oscillates wildly away from the control points.
7. How do you obtain the closed Bezier curve? What is the effect on the degree of the curve?
8. What is meant by multiplicity of control points and what is the consequence of it? Will multiple control points at a coordinate location change the degree of the curve?
9. Why it is necessary to blend multiple Bezier curve segments to model a part represented by a set of data points, instead of building a single curve?
10. Does Bezier curve provide C^2 continuity at the blend point?
11. Is it possible to build two parallel Bezier curves?
12. Can a circle be exactly represented by a Bezier curve?
13. What are the advantages of Bezier curve? Is it possible to build interpolating Bezier curve?
14. How the degree of Bezier curve is decided? Is the curve ever symmetric?
15. What is the advantage of tangency of Bezier curve to the first and last control polygonal segments?
16. What is meant by the curve "well behaving"?
17. What are the limitations of Bezier Curve?
18. What is invariance of Bezier curve?

PRACTICE WORK-1

- Apply the following to your CAD system:
- Two ways to execute commands
  - Menubar approach
  - Toolbar approach
- System level customization
- Document level customization
- Modes of CAD system
- Coordinate systems and sketch planes

PRACTICE WORK-2

- Practice the following on your CAD system
  - Sketching (2D operations)

- Modeling (3D operations)
- 2 ½ dimensional models (or 2 ½ D models)
- 3D models
- Different views, view orientation, view modes, view manipulations
- Productivity tools: enable quicker action such as:
- Geometric modifiers
- Geometric arrays
- Grids

PRACTICE WORK-3

- Practice the following on your CAD system
  - Management Tools
  - Layers
  - Colors
  - Storing and retrieving CAD models
  - Use FTP and Telnet sessions

REFERENCES

1. Zeid I., "*Mastering CAD/CAM*", Tata McGraw-Hill, New Delhi, 2006.
2. Mortenson M., "*Geometric Modeling*", John Wiley, 1997.
3. Reddy J.N., "*Introduction to Finite Element Method*", Tata McGraw-Hill, New Delhi.
4. Chandrupatla T., and Belegundu D., "*Finite Elements in Engineering*", Prentice Hall, Englewood Cliffs, NJ.

CAD OF DIE-CAST PARTS

19.1. INTRODUCTION

Parting direction and parting line determination in case of die-casting parts is a complex issue which depends on shape and topological complexity of the part. Manual determination of parting direction and parting line is time consuming and cumbersome. An automated parting direction and parting line determination can increase the overall efficiency in the mold design and manufacturing. Moreover, it is required for automated cost estimation of die cast parts. This work is an attempt to generate the optimal parting direction and parting line from CAD model of a die cast part. The algorithm developed here takes STEP AP214 file as input. The system determines optimum parting direction with minimizing the number of undercuts and undercut volume as criteria. Parting line is determined subsequently by finding maximum edge loop in the parting direction. These results are further used to determine mold costs. Initial results obtained by testing this system for many die-cast parts, are encouraging.

Die casting is an important near net shape manufacturing process used extensively for producing metallic parts. Figure (19.1) shows schematic of die-casting process. Like many other near net-shape manufacturing processes, die-casting process also utilizes, primarily two mold halves. Product quality and productivity directly affect the way molds are designed and manufactured. Research in the area of die-casting process has focused mainly on simulation of flow patterns in the casting and molding processes. This research has led to the development of models for determination of pressure, temperature, shrink factor, gating and runner system etc.

However process planning automation, particularly selection of optimum parting direction and parting line has received very little attention. This is an important step in process planning, as it dictates the number and shape of the cores and affects all subsequent steps in design and manufacturing of molds. The mold manufacturing and the processing costs are also largely affected by this decision. Guidelines for selection of parting direction and parting lines can be found in handbooks and manuals [1, 2]. Pye [3] suggested that parting line should be along the maximum dimension of the part when viewed in the parting direction. Similarly Ravi and Srinivasan [4] have proposed a nine point criteria for computer aided parting line and parting surface design. These criteria when used assist the mold designer to decide most suitable parting line. The proposed criteria include maximum projected area in the parting direction, minimum number of undercuts, minimum draw distance etc. The user can give different weightage to each of these criteria and choose the optimum parting line.

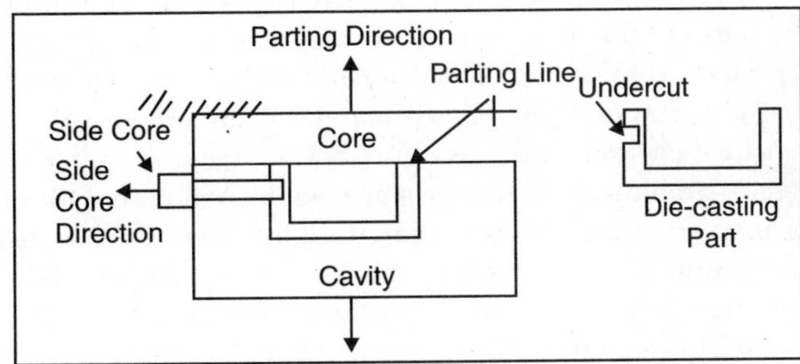

Fig. 19.1: Schematic diagram of the die-casting process

The criteria for selection of parting direction, parting line and parting surface given by the researchers in the past provided a platform for their logical determination. However, it required a large amount of data as an input in terms of part geometry and shape for implementation taking much time and efforts of a die-casting expert. A completely automated approach to parting line determination will not only reduce the manufacturing time but will also help in estimating the cost of mold at an early stage of component design. The same has been attempted here. The algorithm developed here is along the lines of Fu et al. [5, 6, 7]. Unlike, their work the system proposed here takes product data directly from a STEP file and identifies undercut features which are then used for parting line determination. The major purpose of automating the parting line determination in our work is to use the same for automated cost estimation of die cast molds. Two representative example parts are taken for which results of parting line determination and mold cost estimation results are presented.

Selection of parting direction is a prerequisite for the determination of the parting line. Different methodologies have been suggested in the past for the selection of parting direction. Manual examination of the part shape and geometry has always been the preferred method. Chen [8] used the minimum bounding box method to find a suitable parting direction. It was assumed that the suitable parting direction may be one of the three perpendicular directions to the bounding surfaces. The three possible parting directions were evaluated by application of fuzzy logic based on the three criteria out of nine suggested by Ravi and Srinivasan [4]. The criteria for evaluation were minimum number of undercuts, minimum draw distance and maximum projected area. This method though simple to implement may not provide with the optimal solution.

The algorithms based on applications of surface visibility and demoldability has been used to find the optimal parting direction. Chen et al. [9, 10], Fu et al. [5], Nee et al. [11], Weinstein [12] and Yin et al. [13] used algorithms based on the surface visibility. The terms *visibility* and *demoldability* have been defined by Chen et al. [9]. In simple terms a surface *S* on an object is said to be *'completely visible'* in a viewing direction **d** if, for every point *p* on *S*, the ray starting from infinity to *p* in the direction **d** does not intersect the interior of that object. Also, a surface *S* on an object is said to be *'partially visible'* in a viewing direction **d** if, there exists at least one point *p* on *S* such that the ray from infinity to *p* in the direction **d** does not intersect the interior of the object. A part is said to be *'demoldable'* in direction **d** if every point on the part is either visible from direction **d** or **–d.** They used the technique of local interference and visibility map. The methodology used was that by taking the regularized difference between the convex-hull of the part with part itself, we can get all possible undercut features. The set of pockets so obtained were considered as separate objects and visibility maps were drawn for the surfaces of the part which are common with these objects. The set of opposite directions which covered maximum number of visibility maps was selected as the optimal parting direction. Similar methodology was used by Yin et al. [13] to find the optimal parting direction, but instead of the visibility map used by Chen [9], they used the concept of freedom cone. Mochizuki and Yuhara [14] have also presented BRDM (Boolean Regularized Difference Operation) approach to determine undercut features and then optimal parting direction. Weinstein and Manoochehri [12] used the mutual visibility criterion to determine concave and convex regions of the part. Allowable draw range for these regions was evaluated and subsequently the optimal draw range was found. Fu et al. [6] and Nee et al. [11] presented a set of criteria that were based on the normal direction of the faces and axis direction of the symmetrical features.

For the determination of the parting line, the parting direction must be selected first. Pye [3] suggested that parting line should be along the maximum dimension of the part when viewed in the parting direction. But there may be many alternatives available for the maximum edge loop in the parting direction, the problem thereafter becomes, evaluation of different alternatives available based on the selection criteria. Weinstein and Manoochehri [12] presented an algorithm which used the multi-objective function to evaluate the parting line alternatives available to get the optimal one. Wong et al. [15] used a method of slicing a 3D CAD model of the part. The parting line was formed by joining the extreme points of the slice planes. The best parting line was selected from the available alternatives by using the criteria of flatness, draw, projected area, undercut volume and undercut length. Nee et al. [11] used the classification of the part surfaces into core molded, cavity molded and local tool molded surfaces. This classification was then used to find the parting line edge loop. The work presented here addresses the problem of extracting the die-casting specific features from the STEP part model and getting the optimal parting direction and parting line determined by the system. The automated parting line determination is then used for cost estimation of die cast molds.

19.2. FEATURE RECOGNITION METHODOLOGY

A *'feature'* is described as any geometric or topological form that is used in design/manufacturing activity, or it can be an information element representing a *'region of interest'* within the part. Feature recognition is the key for the automation of any process planning activity including for die casting process. Convex hull decomposition, syntactic pattern recognition, graph based and rule based methods are the techniques which have been commonly used for feature recognition (Shah and Mantyla [16]. Mixed or hybrid methods which are a combination of two or more methods have also been used. Our system uses combination of graph based and rule based methods for feature recognition from the part STEP file. A STEP file stores the part data in a neutral format specified under ISO-10303, which needs to be processed for use in any particular system. A translator has been developed which reads geometric data (in B-rep format) from a STEP file. This data forms an input for recognition of die casting features. Extended attributed face adjacency graph (EAFAG) of the part is then prepared with the help of B-rep data. The features are then recognized with the help of knowledge base incorporated in the system, which is in the form of *if-then-else* rules. For example a cylindrical protrusion is identified with the following set of rules. For example *if a surface is having an inner loop, and this loop is circular in shape and it is attached with a solid cylinder, then a cylindrical protrusion may be present.* Many such set of rules are applied to recognize a feature. Figure 19.2 describes the schematic of feature recognition process.

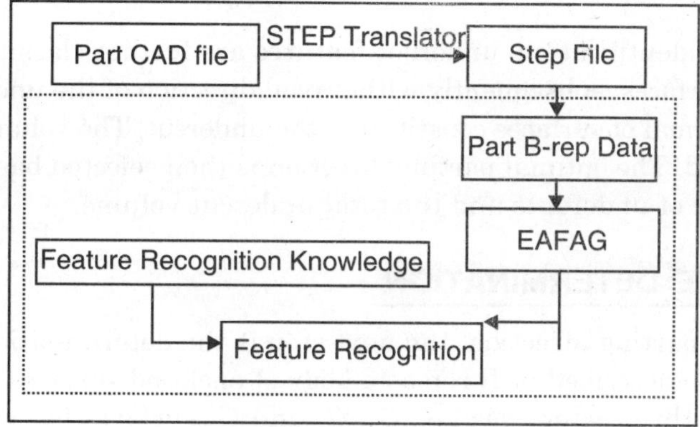

Fig. 19.2: Feature recognition methodology

19.3. UNDERCUT FEATURES AND WITHDRAWAL DIRECTION

Undercut features can be defined as concave or convex portions of the molding (Fu et al. [7]). If the undercut feature cannot be molded by the core or the cavity they will be the real undercuts, as they will need a side core or side cavity in the mold structure. The boss, cylinder, cone and sphere can also be the undercut features. In the automated generation of parting direction, all the undercut features should be recognized and extracted.

The withdrawal direction is the direction in which the undercut feature can be withdrawn after molding. The undercut withdrawal direction will depend upon the type of undercut feature and the mutual relationship of the surfaces constituting the undercut.

The direction and volume of an undercut feature need to be identified for automated generation of parting direction. Nee et al. [11] have defined the various types of undercut features based on their geometrical and topological relationships.

The type of inner-loop present in any surface gives the indication of the type of undercut present. Some of the undercut features and the inner-loop indicating their presence are given in Fig. 19.3.

Type of Inner Edge Loop	Type of Undercut Feature
Three Edges	Triangular Undercut/ Protrusion
Four Edges	Four Edged Undercut/ Protrusion
More Than Four Edges	More than Four Edged Undercut/protrusion
Circular Edge	Cylindrical or Conical Protrusion/Hole Spherical Protrusion/Hole

Fig. 19.3: Inner edge loop and the type of undercut present

After preliminary identification, undercut features are further classified according to the relationship between surfaces; subsequently withdrawal direction of the undercut is determined based on the surface normal of surfaces constituting the undercut. The volume for each undercut feature is also calculated. The optimal parting direction is then selected based on the criteria of minimizing the number of undercuts and the total undercut volume.

19.4. PARTING LINE DETERMINATION

After determination of parting direction, the next step is the determination of parting line. A part or a molding can be described in B-rep as a body of enclosed surfaces. The surfaces of the part are classified into three categories i.e., G_1, G_2 and G_3 surfaces based on the orientation with the parting line [11]. If p and $-p$ are the set of parting directions and p is the parting direction of the core, then G_1 surfaces need to be molded by the core and G_3 surfaces by the cavity. Some G_2 surfaces can be molded by the core or the cavity or by the side core or side cavity. If L_i is the surface normal of any surface, then the scalar product $L_i \cdot p$ will help in deciding what type of surface it is. This classification and the criteria used are explained with the help of Fig. 19.4. In most cases the parting lines are the common boundary of the G_1 and G_3 surfaces. The system generates the maximum contour of the edge loop. The parting line is selected based on the maximum projected area of the edge loop.

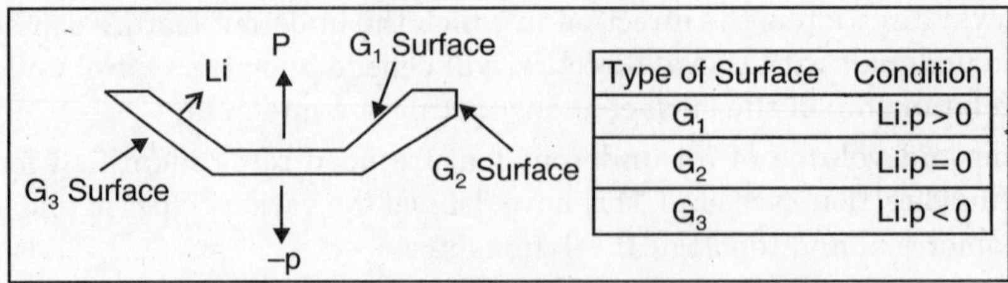

Fig. 19.4: Classification of molding surfaces of the part

19.5. RESULTS AND CONCLUSIONS

The system for automated parting direction determination and parting line generation for die cast parts is developed in a MATLAB environment. The system takes input data of the part from a STEP file. It prepares the EAFAG for the part and extracts the undercut features with the help of the rule base. The optimal parting direction is determined taking the minimum number of undercuts and total undercut volume as criteria. Parting line is determined using the criteria of maximum projected area. The system gives optimal parting direction and parting line as an output. The system developed here has been tested on a number of parts. Figure 19.5 shows two such representative parts and parting direction and parting lines obtained by the system developed here. Our initial results show that results obtained are quite encouraging and match with those obtained by expert process planners of die-casting.

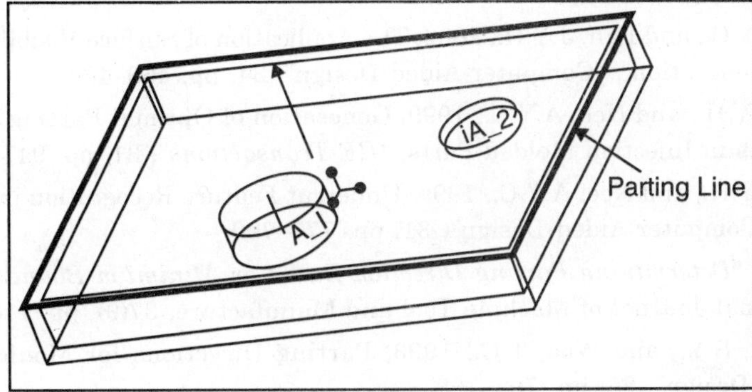

Fig. 19.5 (a): Example part 1 and with the results from the system

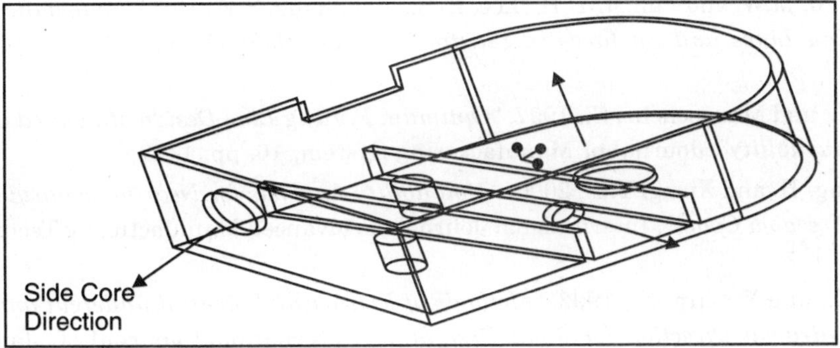

Fig. 19.5 (b): Example part 2 and with the results from the system

The results of parting direction and parting line determination are used for mold cost estimation. Cost of a mold depends on many factors which include geometric complexity of the mold halves, manufacturing process used for mold manufacture, mold material, manufacturing cost of die halves namely core and cavity. Number of side pulls, parting line complexity, threaded cores, feed system etc. have also to be considered for mold cost estimation. We have developed a comprehensive system for die-cast part cost estimation including mold cost estimation. The details of this cost estimation procedure is not a part of this chapter and can be found in [17]. The cost of molds for two representative parts of aluminum A360 alloy shown in Figs. 19.5 (a) and 19.5(b) were found to be Rs. 334144.00 and Rs. 410688.00 respectively.

REFERENCES

1. Bralla, J.G., "*Design for Manufacturability Handbook*", 1999 (McGraw-Hill: New York).
2. American Society of Metals, "*Metals Handbook—Forging and Casting*", Vol. 5, ASM Metals Park, OH, 1970.
3. Pye, R.G.W., 1978, "*Injection Mould Design: A Design Manual for the Thermoplastic Industry*", The Plastic and Rubber Institute.
4. Ravi, B., and Srinivasan, M.N., 1990, Decision Criteria for Computer Aided Parting Surface Design, "*Computer-Aided Design*", 22, pp. 11–18.
5. Fu, M.W., Nee, A.Y.C., and Fuh, J.Y.H., 2002, The Application of Surface Visibility and Moldability to Parting Line Generation, "Computer-Aided Design", 34, pp. 469–480.
6. Fu, M.W., Fuh, J.Y.H., and Nee, A.Y.C., 1999, Generation of Optimal Parting Direction based on Undercut Features in Injection Molded Parts, "*IIE Transactions*", 31, pp. 947–955.
7. Fu, M.W., Fuh, J.Y.H., and Nee, A.Y.C., 1999, Undercut Feature Recognition in An Injection Mold Design System, "Computer-Aided Design", 31, pp. 777–790.
8. Chen, Y.H., 1997, "*Determining Parting Direction Based on Minimum Bounding Box and Fuzzy Logic*", International Journal of Machine Tool and Manufacture, 37(9), pp. 1189–1199.
9. Chen, L.L., Chou, S.Y., and Woo, T.C., 1993, Parting Directions for Mould and Die Design, "Computer-Aided Design", 25, pp. 762–768.
10. Chen, L.L., and Chou, S.Y., 1995, "*Partial Visibility for Selecting a Parting Direction in Mould and Die Design*", Journal of Manufacturing Systems, 14(5), pp. 319–330.
11. Nee, A.Y.C., Fu, M.W. and Fuh, J. Y. H., Lee, K. S., and Zhang, Y.F., 1998, "*Automatic Determination of 3-D Parting Lines and Surfaces in Plastic Injection Mold Design*", Annals of CIRP, 47(1), pp. 95–98.
12. Weinstein M., and Manoochehri S., 1997, "*Optimum Parting Line Design of Molded and Cast Parts for Manufacturability*", Journal of Manufacturing System, 16, pp. 1–12.
13. Yin, Z.P., Ding, H. and Xiong, Y.L., 2000, "*Moldability Analysis for Near-net-shaped Manufactured Parts using Freedom Cones*", International Journal of Advanced Manufacturing Technology, 16, pp. 169–175.
14. Mochizuki, T., and Yuhara, N., 1992, "*Methods of Extracting Potential undercut and Determining Optimal withdrawal Direction for Mold Designing*", International Journal of Japan Society of Precision Engineering, 26, pp. 68–73.
15. Wong T., Tan, S.T., and Sze, W.S., 1998, Parting Line Formation by Slicing a 3D CAD Model, "*Engineering with Computers*", 14, pp. 330–343.
16. Shah, J.J., and Mantyla, M., Parametric and Feature Based CAD/CAM: Concepts, "*Techniques and Applications*", 1995. (John Wiley and Sons Inc.: New York.)
17. Madan, J, Rao, P.V.M., and Kundra, T.K., *An Early Cost Estimation System for Die Cast Parts*, Paper communicated to International Journal of Production Research.

CAE OF TURNING CENTRE

20.1. INTRODUCTION

At present the machine tool manufacturers are facing competition nationally as well as internationally due to worldwide globalization of business. It is required to produce quality product at the minimized rate and to be supplied to the customer at the right time. Moreover the quality of the job produced on these machine tools depends directly on the quality and performance of machine tools produced and some other conditions. To develop good products, design engineers need to study how their designs will behave in real-world conditions. To take care of this condition, often analysis of the proposed design becomes very useful and reliable tool for the design engineer. The performance of a machine tool is ultimately assessed by its ability to produce work piece of the required physical features in the minimum time and at small operating cost. The extent to which the behavior of the various elements of the machine contributes to its overall performance. However, the results of research with the experience of the user form the basis for the design procedures currently available. The aim of this work is to understand and explain the analysis of machine tool structure–a prerequisite of the sound systematic design. This work deals with the application of high-end software–Wild Fire 2.0 to solve the problem lying with analytical analysis method. Here includes the optimization of the Z-Slide of LX-200 with certain permissible deformation which is defined by manufacturer.

20.2. STRUCTURAL ANALYSIS OF LX-200

At present the machine tool manufacturers are facing competition internationally due to worldwide globalization of business. It is required to produce quality product at the minimized rate and to be supplied to the customer at the right time. Moreover the quality of the job produced on these machine tools depends directly on the quality and performance of machine tools produced and some other conditions. To develop good products, design engineers need to study how their designs will behave in real-world conditions. To take care of this condition, often analysis of the proposed design becomes very useful and reliable tool for the design engineer [1, 2].

Introductions to LX-200

LX-200 is CNC Turning Centre with slant bed mono block structure for high rigidity. The machine has simultaneous X, Z, and C axis control through SINUMERIC 802D. The machine facilitates operations like turning, boring facing, threading rigid tapping profiling etc. with stationary tools. While with live tool option, face milling, key way milling, cam milling, profile milling drilling, rigid tapping etc. operations are possible.

The machine bed is in a slanted mono block design. It is specially reinforced with cement concrete. It is able to absorb torsion and vibration, which insures highest precision, metal cutting capacity and tool life. It provides high level rigidity to enable machining of hard parts.

High rigidity, great dynamic and static load carrying capacity, outstanding running smoothness and total enclosure of the carriage are the main features of the cylindrical roller type guide ways. These features result in higher machining rates while enhancing geometrical accuracy and surface quality of the machined work pieces. Improved vibration behavior extends tool life. The weight on the guide way is well balanced and distributed to enhance stability and load capacity.

The spindle is housed in super precision double row cylindrical and a pair of the angular contact bearings in front and a double row cylindrical bearing at the rear. This construction gives very high precision and stiffness in both axial and radial directions and allows heavy material removal rates. The assembly of head stock is carried out in a controlled dust free environment. The bearings are grease lubricated for life.

The twelve stations turret have six driven tools and six stationary tools. This facilitates the turning; milling and drilling operations on him same set ups and set-up times. The turret has VDI type clamping for quick adjustment and rigid gripping of the tool holders. By synchronizing the rotation of tap and its feed, taping speed can be increased as compared to a conventional floating tap holder.

The compact digital wizard with complex functions, integrated digital drives and multiple processors for various functions is designed for high performance machines [3].

For a Turning Center, usual operation will be Turing, drilling, and various types of milling operations, i.e., end milling, face milling, etc.

- So when analyzing the structure of a LX-200 it becomes essential to consider cutting forces in these operating conditions, for different operations. Present work considers turning operation.

Designation of Axis for LX-200

Standard right hand rule works to define all two axis of the machine tool. According to it, the Fig. 20.1 below shows axis designation.

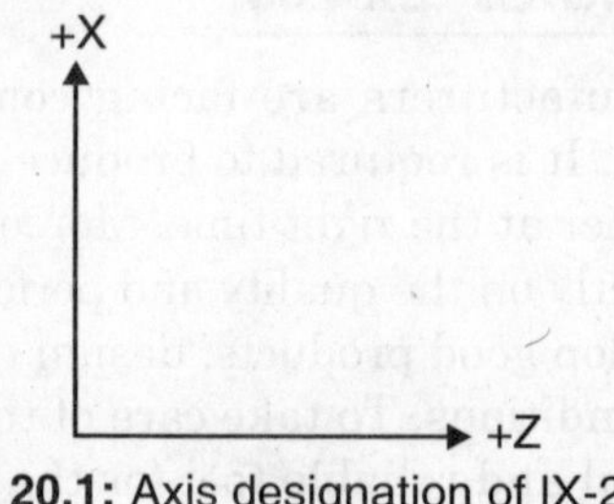

Fig. 20.1: Axis designation of IX-200

As it is explained, for analyzing the structure loads and constraints are required to specify. Loads should be calculated based on the worst condition that has to be bear by the structure during operation. Again one can still give some design safety by overloading the structure. Constraints also require attention, as they restrict and control the behavior of component during actual operation and during analysis, itself. Here for our case of LX-200 cutting forces are calculated for turning operation. Then calculated cutting forces are applied to the models of components of machine tool with constraints and results of analysis are obtained.

20.3. CUTTING FORCE CALCULATION: TURNING OPERATION

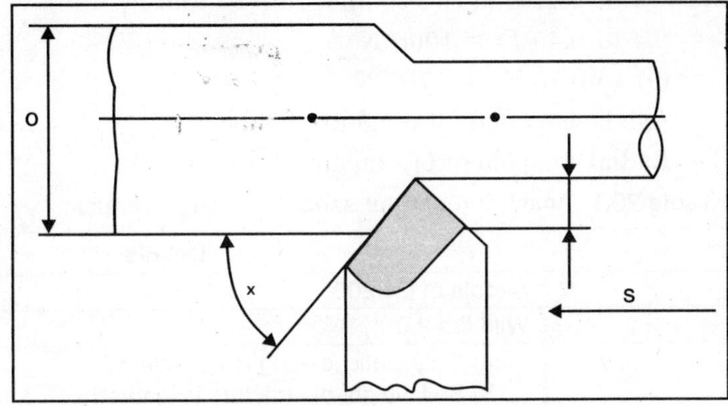

Fig. 20.2: Cutting force during turning operation

Cutting forces are calculated based on several assumptions, regarding cutting conditions and values of cutting parameters, which will generate maximum cutting force. Assumptions are listed below:

- D = Diameter of work piece–50 mm
- N = Number of revolutions–1500 rpm
- s = Feed per revolution–0.5 mm/rev
- d = Depth of cut–2 mm
- P = Motor power–7.5 kW
- Ps = Specific cutting force-kg/mm^2
- Vc = Cutting speed, m/min
- F = Cutting force, kg.
- η = Efficiency of the machine–0.7 to 0.85

$$Vc = \pi \text{ D N}/1000$$
$$= \pi \times 50 \times 1500 / 1000$$
$$= 235.61 \text{m/min (Vc is taken as 250 m/min)}$$
$$P = Vc * f * d * Ps / 6000 * \eta$$
$$Ps = P * 6000 * \eta / Vc * f * d$$
$$= 7.5 * 6000 * 0.75 / 250 * 0.5 * 2$$
$$= 135 \text{ kg/mm}^2$$

Maximum Ps for steel is taken as 250Kg/mm^2

Cutting force F = Ps*f *d

= 250*0.5*2

= 250 kg. = 2500 N [4]

From CMTI Hand Book Page No. 653 Table No. 274:

Take Approach angle: 45°

Rack angle: +10

Nose radius: 0.2–0.4 mm

Now resolving the cutting forces in two components P_X and P_Y

P_X = 0.4 to 0.45 F = 1000 N

P_Y = 0.7 to 0.75 F = 1750 N

where P_X = Axial component of tangential force

P_Y = Radial component of tangential force [5]

Table 20.1: Analysis report for saddle: Turning operation

Particulars	Details
Model description	Saddle of LX-200
Software used	Wild fire 2.0
Assumptions	a) Only static loading is considered b) Self weight of structure is ignored c) Units of measurement for result parameters is: mm Ns i.e. millimeter, Newton, second
Material used	FE 30, Material properties are as follows: E = 1.2 x 10^5 N/mm^2 μ = Poisson's ratio = 0.25 ρ = Density = 7.85 x 10^{-9} tonne/mm^3
Type of element used	3D solid elements
Loads	a) Weight of turret, tool disc and other assy. b) Weight of motor and mounting. b) Cutting force during turning operation.
Constraints	Bottom face of saddle is fixed on the slant bed.
Solution type	Standard design study, static analysis

Results of Analysis:

Parameters	Results
Displacement plot	Max. Displacement value: 0.899 μm

20.4. FOR TURNING OPERATION

As the saddle is directly fitted on the slant bed which is 45° slanted angle with the base of the machine. During the turning operation two components of the load will be transmitted on the saddle. Also the load of the turret, tool disc and other assy. and motor assy is acting on the saddle so far the analysis concern we have to consider all the loads.

Loads and Constraints

Weight of turret, tool disc and other: 2000 N (In –Y direction)

Weight of motor and mounting: 80 N (In –Y direction)

Cutting force during turning operation: P_X = 1500 N (In – X direction)

P_Y = 2500 N (In – Y direction)

As the bottom face of the saddle is attached with the Slant bed, so that surface is taken as the fixed surface, thus restricting all the degree of freedom. Model of the Saddle with applied load and constraint is shown in Fig. 20.4. While interpretation of the results are carried out subsequently are carried out subsequently [6].

Fig. 20.3: Model of saddle

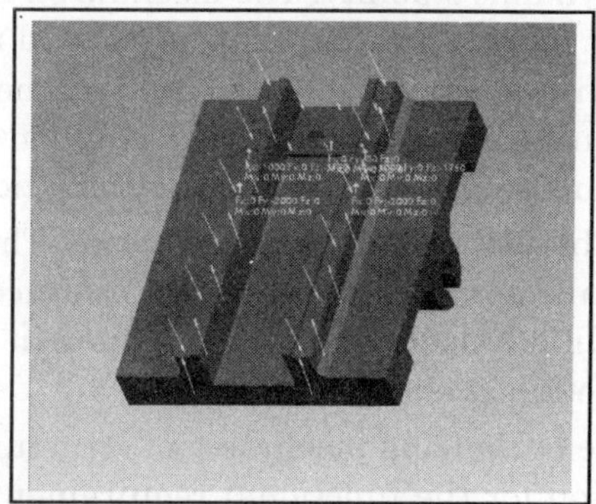

Fig. 20.4: Saddle with load and constraint

Fig. 20.5: Displacement result window for turning operation

Interpretation of Results

As shown from the Fig. 20.5 that the maximum displacement is 0.899 μm which is within the permissible limit. So the performance of the saddle is satisfactory for the turning operation.

20.5. OPTIMIZATION

Optimization, as in general sense, a task that is goal oriented while simultaneously not violating some constraints, which is specified by problem directly or indirectly. Same is the concept for structural optimization, at the same time not compromising with their functional capabilities. The techniques for the optimization of functional properties of machine tool structures aim at the minimization of product cost satisfying the functional requirements, which include the following.

- The minimum weight design exhibiting maximum static stiffness at tool point and dynamic stability. The machine structural elements should exhibit maximum stiffness at tool point rather than as isolated members. Machine elements which have maximum stiffness to weight ratio need not necessarily exhibit minimum tool point deformation unless due consideration is given to its shear center position, thus stressing the importance of functional optimization.
- Thermal stability and accuracies. The functional optimization of precision machine tools calls for thermal stability and accuracy in addition to vibration free machining. Unbalanced heat distribution in the machine tool causes undesirable thermal deformations.

While optimizing machine tool structural elements, importance should be given to their optimization for the functional requirements of maximum tool point stiffness, stability and accuracy and stress level in joints. Thus by reducing production costs and not compromising with functional properties is the goal of achieving optimum result.

Almost all FEA software gives facility for optimization of design in some or other way. Here, optimization is approached with Pro/Mechanica for a Saddle. As it is already stated, to reduce production cost, straight way is to reduce material mass, same time not compromising with its functional properties. Again at this stage, the skill and expertise of design engineer comes into picture, as it is required to decide where to make change? Because what FEA software can do is nothing but the mechanical calculation work, which is rather tedious and long procedure [7].

Optimization in Pro/Mechanica

We are interested in minimizing the weight of the structure – Saddle, without exceeding a specified value of displacement and stress. This calls for an optimization study. These are the most computationally intensive studies. Before proceeding with an optimization, one should usually perform a couple of standard analyses to determine the behavior of the solution. For example, one should have some idea as to the convergence behavior of the model. It also helps if we have a pretty good idea of where the optimum solution lies so that we can reduce the size of the optimization search space; here sensitivity study helps a lot.

It is very important to make sure that part/model will regenerate for all values of the design parameters within the search space. For a more complicated part, changing dimensions of one feature might interfere with the regeneration of another. This is yet another issue that the part designer must keep in mind.

What happens during Optimization?

We are familiar with simple numerical optimization algorithms such as the method of steepest descent. The algorithm in Pro/Mechanica is considerably more complex than this, although the basic idea is the same. The algorithm is considerably more efficient than simple steepest descent, and also must contend with the limits (known as constraints in optimization theory) in the search space. Pro/M evaluates the current design and tries to decide in what direction to move in the search space in order to either remove a constraint violation (like exceeding the allowed stress) or improve on the goal (in our case to reduce the mass). According to the documentation, you can select from two optimization algorithms the sequential quadratic programming (SQP) algorithm and the gradient projection (GDP) algorithm. The default is the SQP, which is generally faster for problems with multiple design variables.

If the initial design point is feasible (that is, no constraints are violated), the algorithm moves the design point in a direction to better satisfy the goal until/unless a constraint boundary is met in the search space. Then it moves in a direction tangent to the constraint surface, all the while seeking out the minimum value of the objective function. If the initial design point is infeasible (i.e., constraints are violated), then one or more correction steps are taken to reach the (nearest?) constraint boundary. Thus, if the first design is infeasible, the design at the end of these first iteration steps is not guaranteed to be feasible. The GDP has the advantage that, if started with a feasible design, it tends to produce a series of intermediate designs that are always feasible, even if it is unable to locate the global optimum design (either due to the objective function or limits set by you). In contrast, the SQP algorithm does not guarantee that intermediate designs are feasible but only that the optimum (if found) is feasible. The advantage of SQP is its generally increased speed over GDP.

Optimization Applied to Saddle

It is interesting rather obvious that as thickness is reducing, displacement in Y direction is increasing and total mass of the structure is almost directly proportional to the thickness. After studying the results of sensitivity study, our search space for finding optimum result will confined to thickness value of 125 to 135. At the same time, constraint for optimized result search is required; which should be obeyed while finding optimum results. In the case of machine tool structure, it will be some limiting value of maximum displacement that is allowable. For the case of saddle (Turning centreLX-200) allowed value for displacement is taken as 1.5 μm. Again it is required to give some convergence value for desired parameter, which will work together with no. of passes specified during optimization for finding the optimized result. Every time, after executing one pass, convergence value is checked for its validity, if it gets satisfied then process of optimized result search will stop. Results of optimization study are shown in Fig. 20.6.

Interpretation of Results

Results of optimization process are shown here. The whole story of process carried out during optimized result search can be seen in the summary of the analysis, which is very useful tool for review. (Not shown here)

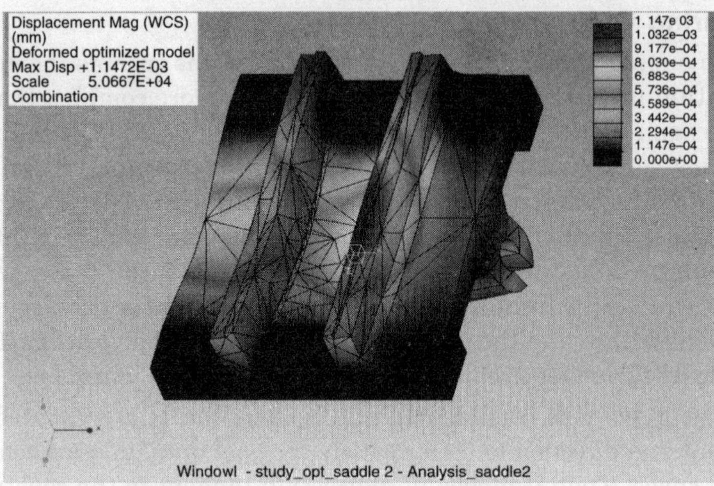

Fig. 20.6: Displacement and stress result after optimization

Studying the results, it is clearly seen that for initial value of design parameter, thickness 125, total mass is minimum and maximum displacement magnitude is also minimum and fall within the specified value. For optimization function total mass is also get reduced by 122 kg for thickness 133 mm to 106 kg for thickness 125 mm, this means that 16 kg reduction in total mass and still not with in the permissible limits of 1.14 μm.

REFERENCES

1. Cowlay A., Fawcett M.A., Analysis of Machine Tool Structure by Computing Techniques, "*Advances in Machine Tool Design and Research*",Vol. 3, No. 2, 1967. pp. 119–137.
2. Badawi M.W., Mohsin M.E., A Comparison of Some Design Details for Machine Tool Structural Elements, "*Advances in Machine Tool Design and Research*", Vol. 5, No. 3, 1963. pp. 355–367.
3. Macpower company catalogue.
4. Widia, cutting tool selection Catalogue, page no. 24.
5. CMTI, "*CMTI Hand book of Machine Tools Design*", Fourth edition, Tata McGraw-Hill Publishing Co. Ltd. New Delhi, 1982.
6. Wild fire 2.0 modeling and analysis software.
7. S.S. Rao, "*Engineering Optimization Theory and Practice*", Third edition, New Age International, New Delhi, 2001. Literature from Pro-Mechanica.

RAPID PROTOTYPING

21.1. INTRODUCTION

Rapid prototyping (RP) technology has been developing very fast in the last 10 years. Some ideas concerning this technology are presented in the chapter for predicting the evolution in the future. The RP issues in the 21st century including part orientation, support structure generation, slicing, path planning, and process parameter selection are discussed. STL files, solid models, and image and reverse engineering data are all considered as possible input to the RP system. Extension of its application in more areas will be beneficial for the advancement of RP.

Rapid prototyping (RP) is a new forming process and principle, which is one of the important breakthroughs of recent technological progress in the manufacturing industry. From the emergence of the first RP system (SLA) in 1988, 5435 RP systems with about 20 kinds of processes were in operation around the world at the end of 1999. At the end of 2000, system manufacturers had sold a total of 6755 systems around the world in 58 countries [1]. In the early days of RP, the automotive and aerospace industries dominated the RP application. But this is no longer the case as RP has spread into many other industries. It shows how fast RP technology is developing. Digital design (CAD) and digital manufacturing (RP) have emerged to facilitate and accelerate product creation. Many kinds of technology require two conditions in order to develop quickly. On the one hand, it needs relevant fundamental theories and technologies to support it. On the other hand, it needs improved capabilities and new application fields. RP technology is no exception. The direction of research and development of RP technology will be analyzed with consideration of these two factors.

21.2. DESIGN INPUT

STL Files

Currently the most widely accepted input to existing RP process planning systems is an STL *file.* An .STL file is constructed from a solid model of a part by first approximating all the surfaces of the part with a mesh of triangular facets and then writing out the nodes and outward normal of each triangular facet to a text or binary file. To represent the surfaces of the object more accurately, more facets are used. The .STL file was adopted as the de facto industry standard input to RP systems because it is simple to construct, robust tessellation algorithms exist, it allows RP software developers to avoid having to interface with a large number of CAD file formats, and the file is easy to manipulate. However, the .STL file format has several drawbacks [2]. For these reasons, there have been several attempts to perform steps on the solid model of a part, rather than first convert the part to an .STL file.

Solid Models

A wide variety of solid modeling systems, representations, and formats exist both commercially and in research laboratories. While most research groups have begun work based on solid models of parts using the resources available to them, two have suggested developing a standard format for the transfer of design and RP data to and from the system. In Carleberg [3], the author points out the need for the product model to be used in selecting process parameters and in planning tool paths. A STEP-based file format is proposed [4] so that information such as process parameters, scanning patterns and material data can be passed on to the RP system.

Other Inputs

In this chapter we also consider other forms of design input such as medical and satellite images, reverse engineering data, and mathematical functions. A common format for these diverse types of data does not exist, but techniques are described for converting image and point data into solid models.

21.3. RAPID PROTOTYPING ISSUES

STL File Repair

Quite often .STL files are flawed, preventing further efforts to be completed successfully. Examples of problems that occur in STL files are missing facets [5, 6, 7], (causing surfaces to be described incompletely), adjacent facets with opposite surface normal, degenerate facets, and facets, which describe a zero-volume object. Problems of this nature result from faulty surface or solid models, which are used to generate the STL file or numerical inaccuracies within the solid or surface model. Efforts have been made to convert surface models into solid models so that the probability of generating a faulty STL file is diminished.

In a surface model [8] is analyzed to determine a topological structure for the part. The surface model can contain both trimmed and untrimmed parametric surfaces, but no *a priori* surface adjacency information is needed. The algorithm developed in this chapter determines which surfaces are adjacent and merges their common edges to form a boundary representation solid model of the part. Missing surfaces and improper intersections are highlighted so that the designer can correct them. Once a solid model representation of the part has been found, an STL file with fewer flaws is generated.

Part Orientation and Placement

The direction in which a part is built can affect its surface finish, the time to build the part, the strength of the part in different directions and the amount of support structure needed.

Originally the designer and/or the builder selected the part orientation manually. They would study the part, select an orientation, and then rotate the part either within the CAD software before the STL file was generated. Recently, the need to automate this process has been recognized.

The objective is to minimize the number of slices required to build the part (thus minimizing build time) and the surface area that will suffer from the stairstep effect. These objective are attained [9] when the software rotates the part in increments about user supplied axes and for each rotation slices the part and calculates the error due to the stairstep effect. The objective in Allen [10] is to minimize the area of the part, which needs to be supported during part build. The orientation that results in the smallest supported area is considered optimum. The method described in Frank [11] to minimize the stairstep affect, build time, and number of supports required. In Thompson et al. [12] four objectives are pursued: minimize the build height and number of supports required and maximizes surface accuracy and strength of the manufactured part. An optimization routine is used to find the best orientation. A feature based approach to part orientation is taken in Suh et al. [13]. Several .STL files are read into the software and the bounding box for each part is found [14]. Then a genetic algorithm is employed to pack as many boxes into the build envelope and use as much surface area as possible. The algorithm described in Beascoechea et al. [15] takes solid models of the parts as input. First the minimal bounding box for each object is computed. Next heuristics and a genetic algorithm are used to determine a schedule for placing the boxes into the build envelope. Finally, once a schedule has been developed, the optimal position for each box is found.

Support Structure Generation

In general, as parts are built up layer-by-layer, support structures are needed for a variety of reasons. They are used to support overhangs, maintain stability of the part so it does not tip over, support large, flat walls, prevent excessive shrinkage, support components that later become connected to the part, and support slanted walls. Part orientation and support structure determination is a coupled problem because some orientations of a part will require support structures while others will not. These support structures are either modeled along with the part using CAD software or are generated automatically by external support structure software Jacobs [16]. On the other hand, the software supplied by Stratasys for the FDM system will generate support structures from the 2D contours formed by slicing.

In addition, different types of supports are used for different situations and different RP apparatus. Examples of different types of supports structures are shown in Fig. 21.1. The FDM process in general uses only thin-walled zig-zag supports built up from the base of the modeling envelope or existing part surfaces. One support structure generation system, described in Allen et al. [10] does not require an .STL file as input. Three cases requiring support structures are detected: overhangs, floating objects during build that later become attached to the part, and instability during build. Methods for generating support structures from contours for the FDM system have been reported [2]. All the contours in the structure are merged to form a containment contour around the outside of the part. Then each part contour is offset and supports are generated using a zig-zag pattern between the offsets and the containment contour.

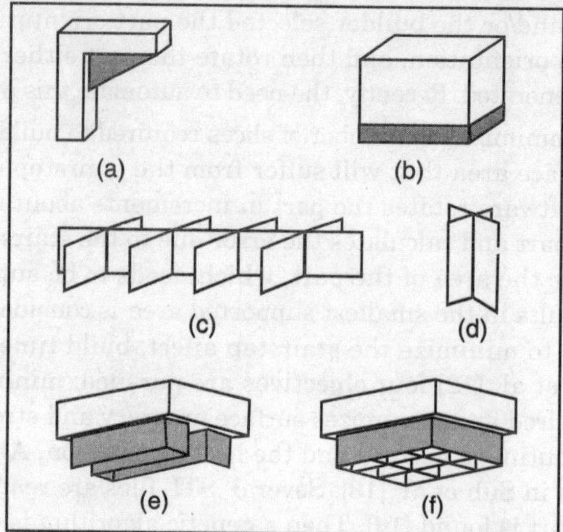

Fig. 21.1: Support structures (a) gusset support (b) base support (c) web support (d) column support (e) thin-walled zig-zag support (for FDM process) (f) perimeter and hatch support

Part Slicing

Once the part is in the proper orientation and support structures have been added it is sliced. Slicing involves the intersection of the part model with a plane whose normal lies in the build direction. If slicing is done on a faceted model, a single slice contains one or more planar polygonal contours representing the cross-section of the part at the height the slice is taken. A series of slices spaced either uniformly or non-uniformly is generated and used to perform path planning. A description [17] by marching algorithm using topological information assumes that the topology of the facets is known, the intersection between the slicing plane and an edge is found. Then the other two edges of the facet the first edge belongs to are searched for an intersection. This insures that the contours are formed sequentially and it reduces the amount of search time to form the slice.

Uniform slicing of objects represented by a constructive solid geometry (CSG) tree has also been considered. In Farouki et al. [18], primitives used to compose the part are limited to those bounded by natural quadric surfaces (spheres, cones, and parallelepipeds), while in Guduri et al. [19, 20] primitives bounded by higher order surfaces are permitted. In addition, adaptive slicing is better suited for parts with features such as horizontal faces and thin protrusions, which are often omitted with uniform slicing which is shown in Fig. 21.2. To ease the discussion of adaptive slicing techniques, Fig. 21.3 illustrates some of the terms that are used. In most adaptive slicing routines, the user specifies a maximum allowable cusp height *d*, which is the maximum distance between the manufactured surface and the desired surface.

The angle the desired surface makes with the build direction is θ. In addition, the particular RP machine that is used in general has a minimum and maximum allowable slice thickness, *lmin* and *lmax* respectively. The goal of the routine is to find the slice thickness *l* for a particular layer. An adaptive slicing technique for faceted parts is described in Dolenc et al. [21]. To minimize the staircase effect on slanted faces, the layer thickness is computed so that it does not violate the maximum allowed cusp height. It is given by:

$$l = d/\sin\theta = d/N_z \quad ...(21.1)$$

Peak features such as horizontal surfaces, pointed edges, and pointed ends of a polyhedral object are detected in Thompson et al. [12] by scanning the faces and examining their vertices, edges and normals. In Frank et al. [11] peak features are detected by examining all the z-values of the vertices of each face in an object. Regions between each peak feature are adaptively sliced using Eqn. 21.1.

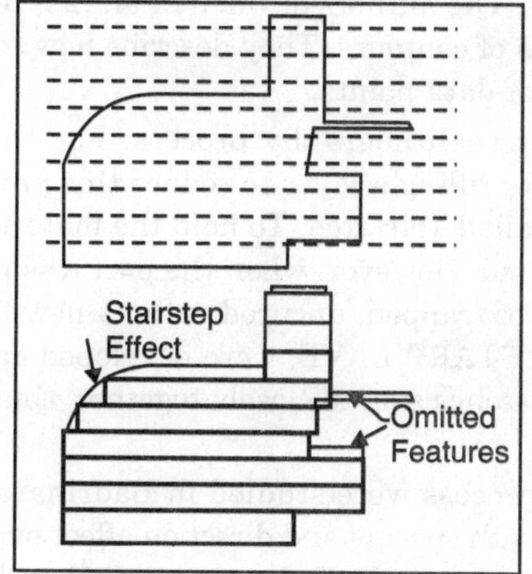

Fig. 21.2: Problems with uniform slicing

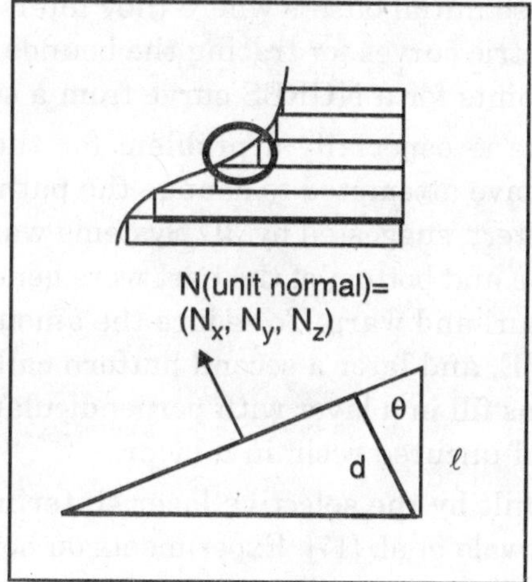

Fig. 21.3: Illustration of cusp height

In Sabourin et al. [22], a polyhedral object is first uniformly sliced, resulting in thick slabs. Then each slab is resliced uniformly using eqn. 21.1. This allows slicing to be done on a parallel processor. Slicing techniques have been well studied and robust algorithms have been developed for both uniform and adaptive slicing. Most commercial systems employ uniform slicing algorithms, but as systems become more advanced it is hoped that the available adaptive slicing routines will be adopted.

Path Planning

The path-planning step determines the tool paths for the RP apparatus that will be used to build the part. Different processes require different types of tool paths. Paths that the RP tool follows for adding a layer of material will affect the time to build the part, surface accuracy of the part, part distortion, and stiffness and strength of the final part. In Chen et al. [23], the authors investigate how different path patterns affect the time to build a part using the stereolithography process. The directions of laser travel, number of path segments, and number of laser jumps (positioning of the laser for a new path) are investigated. They develop a mathematical model for predicting builds time based on the tool paths and process parameters selected. Crawford [24] recognized that in order to maximize the geometric accuracy and achieve desired mechanical properties, scanning and build patterns must be optimized for each part. Therefore an accurate model of the part must be used for path planning.

One approach, described in Chang [25], is to superimpose a mesh of voxels whose height is the desired slice thickness over an STL model of a part. Each voxel is tested to see if it is inside or outside the part. Each row (or column) of voxels is used as a tool path. Where a voxel that is inside the part neighbors a voxel outside the part, the tool starts or stops adding material accordingly. The accuracy of the part will depend on the size of the voxels used. Another method for determining scan paths is reported by Rock et al. [17]. Here parallel rays are fired at an .STL model of a part and where they intersect the part surface, the tool will be turned on or off. A technique for determining the offset of a polygonal contour is given in Yang et al. [26]. This involves first finding an initial offset of each line segment making up the contours and then trimming these initial offsets where they intersect. The authors of Chen et al. [23] recommend using parametric curves for tracing the boundaries of contours. They describe how to calculate the control points for a NURBS curve from a set of data points.

Warpage is especially a problem for the stereolithography process. For this reason, researchers have attempted to change the paths for filling a layer to reduce this problem. The initial fill pattern suggested by 3D Systems was called TriHatch. To hold the material in, skin fills on the top and bottom of the part were generated. However, when the part was post cured, it tended to curl and warp. To reduce the amount of trapped, uncured resin, a new fill pattern called WEAVE, and later a second pattern called STARWEAVE, were developed Jacobs [16]. These patterns fill in a layer with perpendicular hatches spaced closely together, thus reducing the amount of uncured resin in a layer.

Parts built by the selective laser sintering process were studied in Badrinarayan et al. [27] and Agarwala et al. [17]. Experiments on how path spacing and direction affect part strength showed that shorter scan vectors result in stronger parts. In Fodran et al. [28] and Kulkarni et al. [29], parts built by the FDM process were studied. In the first article raster and contour paths were compared based on part strength. In addition, parts impregnated with five different epoxies and composite supplements in differing amounts were compared. The second article investigates how strength differs for raster, contour, and spiral paths and how varying raster angles affects strength. In addition, the authors develop an analytic model of parts based on laminate analysis and a mathematical model for determining the material to void ratio in parts. The effect of path planning on surface accuracy, warpage and material properties is still an open research area.

Process Planning using Image and Reverse Engineering Data

Often design data is stored in some form other than a solid model. For instance, medical data used to construct a prosthetic device may be stored in the form of a stack of planar images such as a series of CT or MRI scans. Reverse engineering data that is used to reconstruct an object may be clouds of three-dimensional points. Different strategies have been employed for RP process planning from these types of data. Medical images are attractive input to an RP process planning system because they are planar slices of an object, which theoretically can be used as slices for manufacturing. However, these images are usually spaced too far apart for RP systems. CT and MRI scans are normally taken at intervals of 0.25–5 cm (0.1–2 in.) whereas slices of RP processes are generally less than 0.25 cm (0.1 in.). The method for generating intermediate slices is to generate a solid model of the object from the given images. Image processing techniques are used to extract planar contours representing the boundary of the object. Contours are grouped into branches and surfaces are fit through individual branches in such a way that they join other surfaces smoothly at common boundaries. Joining all surfaces encloses a solid region. This solid model can then be optimally oriented and sliced. This method allows the build direction to be different from the direction in which the medical scans were taken. Point data from coordinate measuring machines can be used in two ways to perform process planning for RP.

21.4. CLOSURE

As RP processes move away from accepting .STL files as input and adopt solid models, .STL repair algorithms will become obsolete. However, solid models will have to be defined completely and unambiguously to proceed smoothly. Many methods exist for determining an optimal part build orientation for .STL files. No methods have been developed specifically for solid models, though methods for .STL files can be employed if the solid model is temporarily faceted. Different criteria for evaluating a possible orientation have been suggested; however some, especially the measurement of part strength for a given part orientation, should be investigated more fully so that accurate measurements can be made. Many robust methods exist for automatically generating support structures for faceted parts. Little has been done for parts represented by a solid model. Slicing of both polyhedral and non-polyhedral objects has been well covered by the literature and adequate techniques has been developed. Experiments have been performed to determine the effect that tool paths have on the geometry and material properties of finished parts. However, more experiments need to be performed and the results must be incorporated into automatic path planners. Finally, RP processes are developed for metals and used for functional parts and production runs, a set of new process planning tasks will have to be dealt with that will include process speed-ups, hardware issues, etc.

REFERENCES

1. T. Wohlers, Wohlers Report, Rapid Prototyping and Tooling State of the Industry, Annual Worldwide Progress Report, (Wohlers Associates, Inc., U.S.A, 2001).
2. Chalasani, Kumar, Jones, Larry, and Roscoe, Larry: Proc. "*Solid Freeform Fabrication Symposium*", 1995, University of Texas, Austin, 1995.
3. Carleberg, Per: Proc. "*Solid Freeform Fabrication Symposium*", 1994, University of Texas, Austin, 1994.
4. Owen, Jon. STEP: An Introduction, "*Information Geometers*", Winchester, U.K., 1993.
5. Bøhn, Jan H., and M.J., Wozny: Proc., "*Solid Freeform Fabrication Symposium*", 1992, University of Texas, Austin, 1992.
6. Mäkelä, I., and A., Dolenc: Proc. "*Solid Freeform Fabrication Symposium*", 1993, University of Texas, Austin, 1993.
7. Rock, Stephen J., and Wozny, J., Michael: Proc. "*Solid Freeform Fabrication Symposium*", 1991, University of Texas, Austin, 1991.
8. Sheng, Xuejun and Meier, Ingo R., (1995), Generating Topological Structure for Surface Models, "*IEEE Computer Graphics and Applications*", **15**, 35.
9. Bablani, Minoo, and Bagchi, Amit. (1995), "*Quantification of Errors in Rapid Prototyping Processes, and Determination of Preferred Orientation of Parts*", Transactions of the North American Manufacturing Research Institution of the SME, **23**, 319.

10. Allen, Seth and Dutta, Deba. (1995), "*Determination and Evaluation of Support Structures in Layered Manufacturing*", Journal of Design and Manufacturing, **5**, 153.
11. Frank, Dietmar and Fadel, Georges: Proc. Fifth International Conference on Rapid Prototyping, University of Dayton, 1994.
12. Thompson, David C., and Crawford, H., Richard: Proc. "*Solid Freeform Fabrication Symposium*", 1995, University of Texas, Austin, 1995.
13. Suh, Yong S., and Wozny, J., Michael: Proc. "*Solid Freeform Fabrication Symposium*", 1995, University of Texas, Austin, 1995.
14. Wodziak, John R., Fadel, Georges M., and Kirschman, Chuck: Proc. Fifth International Conference on Rapid Prototyping, University of Dayton, 1994.
15. Beascoechea, Alejandro and Wozny, J. Michael: Proc. "*Solid Freeform Fabrication Symposium*", 1995, University of Texas, Austin, 1995.
16. Jacobs, Paul F., "*Rapid Prototyping and Manufacturing*", (McGraw-Hill, New York, 1992).
17. Agarwala, Mukesh K., Bourell, David L., Wu, Benny, and B.J. Joseph: Proc. "*Solid Freeform Fabrication Symposium*", 1993, University of Texas, Austin, 1993.
18. Farouki, Rida T., and Konig, Thomas. (1996), "*Computational Methods for Rapid Prototyping of Analytic Solid Models*", Rapid Prototyping Journal, **2**, 41.
19. Guduri, Sashidhar, Crawford, Richard H., and Beaman, J., Joseph: Proc., "*Solid Freeform Fabrication Symposium*", 1992, University of Texas, Austin, 1992.
20. Guduri, Sashidhar, Crawford, Richard H., and Beaman, J., Joseph: Proc., "*Solid Freeform Fabrication Symposium*", 1993, University of Texas, Austin, 1992.
21. Dolenc, A., and Mäkelä, I., (1994), "*Slicing Procedures for Layered Manufacturing Techniques*", Computer-Aided Design, **26**, 119.
22. Sabourin, Emmanuel, Houser, Scott A., and Bøhn, Jan Helge. (1996), "*Adaptive Slicing Using Stepwise Uniform Refinement*", Rapid Prototyping Journal, **2**, 20.
23. Chen, Calvin C., and Sullivan, Paul. A., (1996), "*Predicting Total Buildtime and the Resultant Cure Depth of the 3D Stereolithography Process*", Rapid Prototyping Journal, **2**, 27.
24. Crawford, H., Richard: Proc., "*Solid Freeform Fabrication Symposium*", 1993, University of Texas, Austin, 1993.
25. Chang, Wei-Ren. "*CAD/CAM for the Selective Laser Sintering Process*", M.S., Thesis, University of Texas, Austin, TX, 1989.
26. Yang, Daniel C.H., Jou, Yungsen, Kong, Tom, and Chuang, Jui-Jen: Proc., Sixth International Conference on Rapid Prototyping, University of Dayton, 1995.
27. Badrinarayan, B., and J.W., Barlow: Proc., "*Solid Freeform Fabrication Symposium*", 1995, University of Texas, Austin, 1995.
28. Fodran, Eric, Koch, Martin, and Menon, Unny: Proc., "*Solid Freeform Fabrication Symposium*", 1996, University of Texas, Austin, 1996.
29. Kulkarni, Prashant and Dutta, Debasish. (1996), "*Deposition Strategies and Resulting Part Stiffness in Layered Manufacturing*", ASME Journal of Manufacturing Science and Engineering, in review.

CAD OF INJECTION MOULDING

22.1. INTRODUCTION

By leveraging various designs for assembly and design for manufacturing methods, Manufacturing Industries, Tool Rooms, Tool Designers, and Product Designers can apply solid modeling, or 3D design, to increase profit margins and decrease the time to deliver the products, tools to customer. In today's demanding markets, customers are required the improved designs with shorter lead times. Depending on customer demand, every manufacturer wants change in products and accordingly changes in processes and methods, which could require less manufacturing cost with increased profit. Only the product, tool and process designer can convert this requirement into reality. Plastics have become a part of society, which accepts and often takes for granted the major role of these fantastic materials and easy method of manufacturing them. Injection mould design has become a routine activity in plastic industry and tool rooms. An intelligent CAD system and 3D modeling technique helps in many ways for converting designers Visualisation to Realisation. Also these are helping tools to cut product introduction time. It extends help in reducing pre-market phase of product life cycle (PLC). This chapter describes methodology for utilization of CAD tools, to reduce pre-market phase of product life cycle of injection moulds and plastic product.

The injection moulding is one of the most widely used production processes for producing plastic parts with high production rate. The process consists of injecting molten plastic material from a hot chamber into a closed mould, allowing the plastic to cool and solidify and ejecting the finished product from the mould [1]. For each new plastic product, the injection-moulding machine requires a new injection mould (IM). Design and manufacture of injection mould is a time consuming and expensive process and traditionally requires highly skilled tool and mould makers. Design of mould includes several factors such as part geometry, mould material, parting line and number of cavities per mould. With the advances in computer technology and artificial intelligence, efforts have been directed to reduce the cost and lead-time in the design and manufacture of an injection mould (IM). Mould design affects the productivity, mould maintenance cost, manufacturability of mould, and the quality of the moulded part. As manufacturing and supply technology is continuously changing, which providing new tools and methods for manufacturing and design of products, processes. Various CAD tools, offers the ability to create solid models, that can be used to calculate optimized design parameters, finite element analysis, interference and kinematics analysis [1]. These calculations results in a much more robust design and ultimately shortens the design and development cycle time. It is supporting to reduce incubation period necessary for plastic product in the product life cycle. These new design CAD tools (created solid model) is not only helps in design stage but also helps in manufacturing stage of machine parts and tooling.

Every product goes through a cycle from birth, into an initial growth stage into a relatively stable period and finally into a declining state that eventually ends in the death of the product. This product life cycle for any product is divided mainly in two phases, one known as pre-market phase and second market phase. Pre-market phase includes research and development and market research. As this is the non-profit zone in PLC, G. every product developer thinks of reducing time period of this pre-market phase so that they can introduce quality product to customer before than competitor, Dieter [2]. This chapter describes the methodology, by which injection mould designer and developer can reduce pre-market phase of PLC while introducing new quality plastic product in market as per customer requirements.

22.2. INJECTION MOULD (IM) DESIGN METHODOLOGY

Basically injection mould (IM) design includes mould assembly design with details and detailed design drawing of all parts to be manufactured [3]. The total design and development (manufacturing) cycle of injection mould (IM) is shown in Fig. 22.1, it describes that design and development process starts with the part/component design information such as detailed drawing, material and no of quantity required. After detailed study of design of desired component the actual design process starts [4, 5].

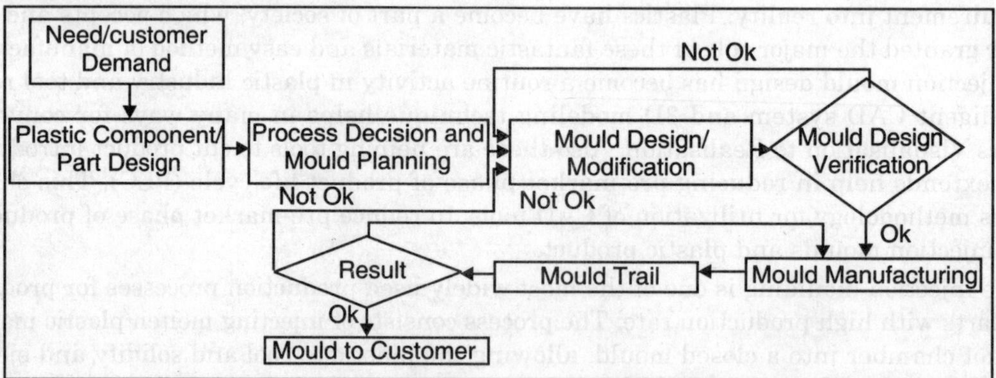

Fig. 22.1: Design and development cycle of injection mould (IM)

Once the design of required mould components, like mold base (various plates ejection system, locating ring, pillars), ejector pins, cores and cavities (plates) is ready, then those passed to manufacturing unit/tool room for manufacturing. During design process specific methodology is adopted for making various decisions (see Fig. 22.2.). The total process is divided in different

11 steps, which provides design solutions step by step with correct decisions, depends on requirement from customer about component. These all steps fulfill complete design requirements, which are generally to be considered by designer during injection mould design. During design, maximum time is consumed in core/cavity planning and ejection system planning. This time span depends on component complexity and various features like ribs, slots, undercuts, projections holes and snap fittings. Where step 2, 5, 6, 8, 11 are more critical from the point of view of accuracy and which takes maximum time for decision. These are specific areas in design process, which may become reason for design corrections and modifications, which consumes maximum time and time to market increases.

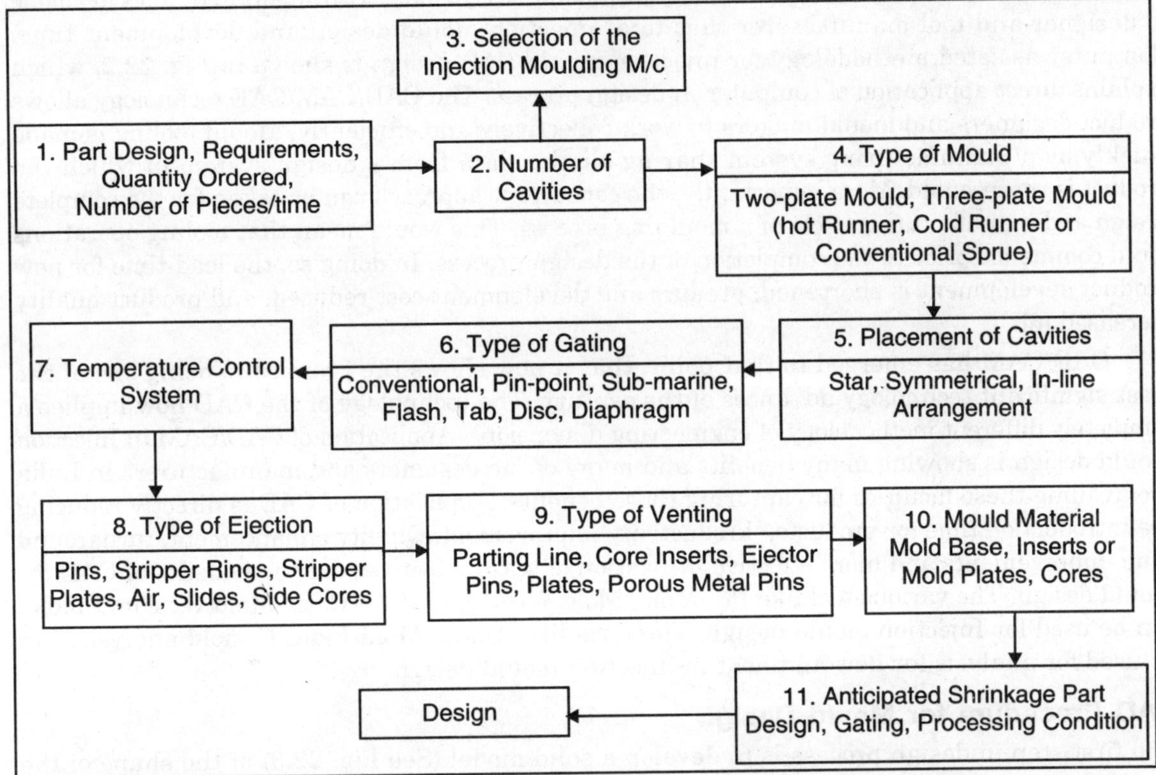

Fig. 22.2: Methodology for designing of injection mould

Following are various activities where maximum time gets consumed in injection mould (IM) Design Development.

- Decision on no. of impressions.
- Cores/Cavities design.
- Decision on Gate location.
- Developing assembly drawing and detailed drawing of parts.
- Developing CNC programming.
- Core/Cavity section manufacturing.
- Developing electrodes for EDM work.
- Assembly.

Due to maximum time consumption in above mentioned many activities of design and development of injection mould (IM), the total process time goes on increasing. This time duration is from 30 days to 4 months and even more than that depending on part complexity and different features included in part design. Again the time consumes in redesign and some design/ manufacturing modification, which increases the lead-time in tool development.

22.3. COMPUTER AIDED INJECTION MOULD DESIGN

Today the market is totally driven by customer and every customer wants products with shorter lead-time moreover with precisely defined quality. Now a days many CAD/CAM/CAE tools available to designer and tool manufacturer that total injection mould design and development time. Computer assisted methodology for injection mould (IM) design is shown in Fig. 22.2, which explains direct application of computer in design process. The CAD/CAM/CAE technology allows product designers and mould makers to work collectively and efficiently. Mould tooling is made quickly by a manufacturing system sharing similar data from a design system in which the product is represented. More importantly, the concurrent approach can be taken for the complete design and manufacturing cycle of a moulding process. This would mean that tooling operations could commence prior to the completion of the design process. In doing so, the lead-time for new product development is shortened, product and development cost reduced, and product quality increased [6].

CAD/CAM has emerged to that point, that it now shows the promise to being one of the most significant technology advances of the century. The technology of the CAD now implies a completely different methodology of engineering dimensions. Application of CAD/CAM in Injection Mould design is showing many benefits and many of the designers and manufacturers in India are availing these facilities for capturing those benefits. Application of CAD is directly reducing pre-introduction time for products. Productivity improvement, Quality enhancement, turnaround time improvements and more effective utilization of scare resources are the benefits of CAD for Mould design. The various tool like Pro-Mold, Mold Wizard, C-Mold, Catia-Mold version, I-mold, can be used for Injection mould design. The tools like Ansys, Mold-Flow, C-mold analysis, can be used for analysis for flow and heat in injection mould design.

CAD Procedure for Mould Design

The first step in design process is to develop a solid model (See Fig. 22.3) of the shape of the plastic part based on functional requirement with any CAD tool. At this stage, designer will take into account basic moulding feature such as parting line and draft angle. In cases where the product model is not "mouldable", the mould designer will make necessary changes to take into account manufacturing considerations such as shrinkage, machine allowances and mould durability. If the moulded part must be machined or ground in a secondary operation, machining allowances is added. Parts with undercuts require a more complex mould with internal undercuts and sliders. Using moulding simulation, designer get early feedback on how wall thickness and gate locations will affect the production of desired part. Predictions of weld lines, fill pattern and temperature distributions allow the mould designers to lay out and optimise the gate and runner systems, as well as to predict clamp tonnage, shot size, and cycle time requirements, all before the mould geometry is finalised. The CAD user has a choice of fully integrated, world-class mould design tools. The specialised mould design tools give mould designer direct control over key parameters such as pull direction, parting line and parting surface, as well as insert and electrode creation. Special attention is paid to the task of splitting the part into core and cavity as described in Fig. 22.4. For parts with undercuts, sliders and/or lifter are also needed, which can be extracted from solid model of desired part [7, 8, 9].

Once the mould has been laid out the mould base is sized and assembled. In some cases, a standard mould base libraries are available in software such as HASCO or DME will be used. These libraries provide with A and B plates in correct size and thickness, additional plates if appropriate, guide pins, bushings, locating rings and more.

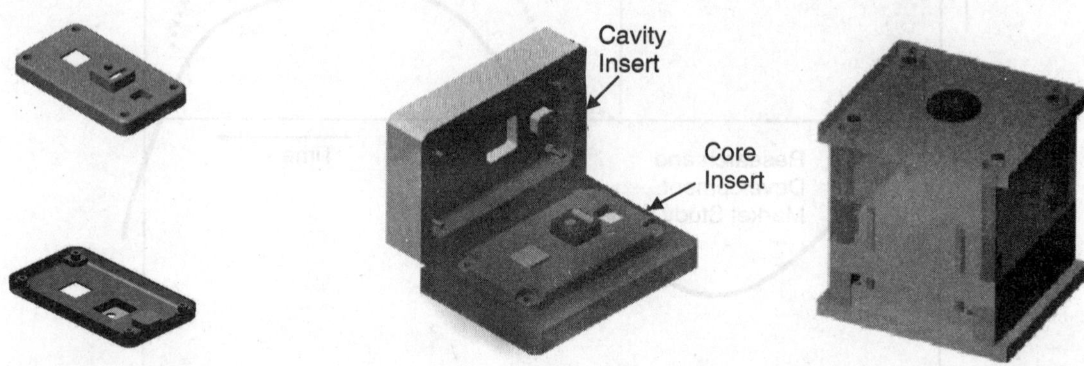

Fig. 22.3: Three dimensional model of desired part

Fig. 22.4: Core and cavity sections

Fig. 22.5: Standard mould base

22.4. GAINS FROM USE OF CAD

Quick response to market demand is the prime directive for any designer and mould shop. They are paid per job, not by the hour, so time directly affects the bottom line, and customers don't like to wait–they want part production to start as soon as possible. With the conventional method a mould set delivered to customer in 16 to 18 weeks must now be delivered in 3 to 6 weeks. This is possible only due to application of various CAD/CAM tools available, which helps to designer and manufacturer a lot. Designer can develop complete design of mould for given part within short period of time (4 hrs to 3 day) in one seating and forward it to production unit for further work. For the above-described desired part (see Fig. 22.3), total actual 25 hours elapsed in designing complete mould up to assembly modeling stage. (Case study conducted at Amritha Tools and crafts ltd. Kushaiguda, Hyderabad.).

CAD benefited in reduction pre-market phase of PLC as shown in Fig. 22.6. In this case basically the product life cycle is divided into the premarket and market phases. A lot of includes activities are involved prior to bring any product to the market. This includes mainly research and development activities and marketing studies. The investment (negative profits) needed to create any as in this case injection mould, is shown along with profit. Due to competition in market all manufacturers spending major part of investment on R and D and trying to bring the product in market earliest than competitor. All the CAD tools are advantageous to accelerate the design process in injection mould design, whereas CAE tools supports to verify the designs to bring required quality and accuracy in design. Similarly CAM tools helps to convert CAD data for related manufacturing so as the product can be launches early in market. With the application of these tools, depends on customer demand, the changes in tools and product have become more easier and less time consuming.

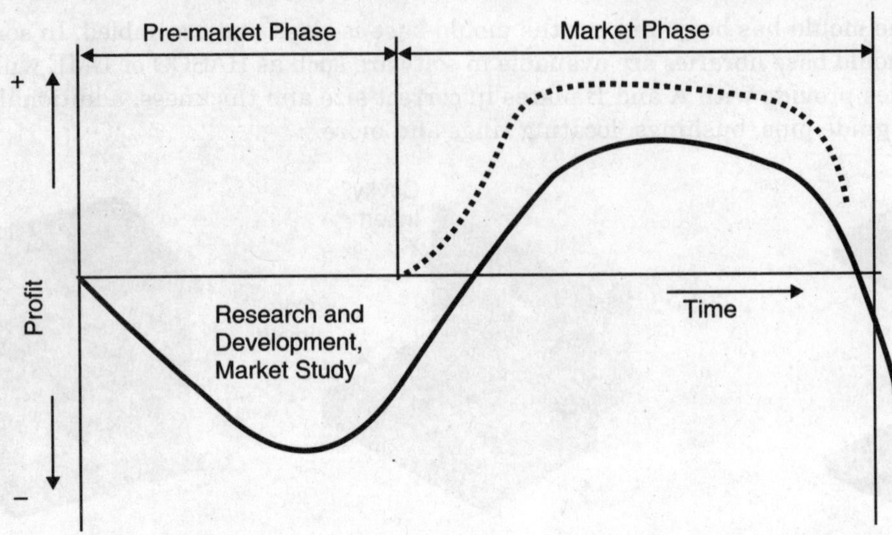

Fig. 22.6: Extended product life cycle for injection mould and plastic product

22.5. CLOSURE

By leveraging various designs for assembly and design for manufacturing methods, manufacturing industry can apply solid modeling, or 3D design, to increase profit margins and decrease the time to market of its product. By utilising 3D CAD packages for developing injection mould (IM) designer could shorten the total design and development time. CAD solid modelers can offer many advantages over traditional method (2D work, design board work), and possibly the greatest advantage is the designer/manufacturing engineer can see the component at actual on screen before it is ever prototyped. This can save both time and money because the design using these advanced tools is more accurate and effective. Also it leads to reduce non-profit activity time in the form of per-market phase in PLC.

However, the modeling or design time required to layout a product increases because of the complexity of the software. The 3D model of desired part through injection mould can be utilised for many purposes i.e. core/cavity development, mold filling analysis for decision of gate location, assembly of mould and virtual prototype. The computer aided injection mould design procedure described in this chapter can be used to develop injection mould (IM) more effectively with shortened development time. Tool designer/mould developer can implement proposed method for converting their *Visualisation to Realisation.*

REFERENCES

1. X.W. Xu, "*Plastic Forming Processes and the Applications of CAD/CAM Technology*", IPENZ Transactions, Vol. 26, No. 1/Emch, (1999), pp. 33–42.

2. G.E., Dieter, "*Engineering Design, A Materials and Processing Approach*", McGraw-Hill, Inc, pp. 76–77.
3. D.V., Rosato, "*Injection Molding Handbook*"., Indian Edition, CBS Publishers. (1987), pp. 315–343.
4. R.G.W., Pye, "*Injection Mould Design*", (1989), pp. 255–296, 513–530.
5. Shad Dowlatshahi, "*An Application of Design of Experiments for Optimization of Plastic Injection Moulding Process*". Journal of Manufacturing Technology Management, Vol. 15, Number **6**, (2004), pp. 445–454.
6. Song Bin, lin Zengzhi, Fu Qinrong and Lu wen Feng, "*A Design Methodology for Compressing Lead Time from Plastic Part Design to Mould Making*". International Journal of CAD/CAM, Vol. 3, No. **2**, (2003), pp. 43–50.
7. Sandeep Patil, Dr. U. M., Choudhary, (2004), "*Feature based Knowledge Management Methodology for Injection Mould Design*", ISAMAP2k4, (IIT Kharagpur, Vol.-I), pp. 283–290.
8. Weigang Hu, Syed Masood, (2002), "*An Intelligent Cavity Layout Design System for Injection Moulds*", International Journal of CAD/CAM Vol. 2, No. 1, pp. 69–73.
9. Yong – Jeong Huh, (2001), "*A Knowledge Based Design System for Injection Moulding*", Journal of Microelectronics and Packing society, Vol. 8 No.3, pp. 11–17.

CAD FOR WELDING PROCESS

23.1. INTRODUCTION

Modeling of welding is technically and computationally demanding, and simplification and idealization of the material behaviour, process parameters and geometry is inevitable. Numerical modeling is a powerful tool for residual stress prediction, but validation with reference to experimental results is essential. In this chapter, a comprehensive methodology for the analysis of residual stresses in a butt welded plate is discussed using the finite element method. The non-linearities in the welded zone due to the variation of material properties and heat transfer coefficients with respect to temperature and also, due to the inclusion of a radiation boundary condition and solid phase transformation effects are considered in the transient thermal and thermo-elasto-plastic analysis. Limitations on the usability of some commercial finite element codes, especially for including thermal effects due to phase transformation and transformation plasticity, are highlighted.

Notation

β_1 – Convective heat transfer coefficient
β_2 – Radiation heat transfer coefficient
Δt – Time increment
ε – Emissivity
λ – Thermal conductivity
$[\lambda]$ – Effective conductivity matrix
$\{\partial\sigma\}$ – Stress increment for an increment in temperature
$\{\partial\varepsilon^e\}$ – Elastic strain increment
$\{\partial\varepsilon\}$ – Total strain increment
ρ – Density
C – Specific heat of material
σ_{sb} – Stefan-Boltzmann constant
T_a – Ambient temperature

In a fusion welding, a weldment is locally heated by the welding heat source. Due to the non-uniform temperature distribution during the thermal cycle, incompatible strains lead to thermal stresses. These incompatible strains due to dimensional changes associated with solidification of the weld metal, metallurgical transformations, and plastic deformation, are the sources of residual stresses and distortion.

Welding induced residual stresses and distortion can play a very important role in the reliable design of welded joints and welded structures. However, the welding process itself is a very complex phenomenon, which has not been fully understood, so that the distribution and magnitude of residual stresses is not readily available from the literature. It is, therefore, necessary to develop a model for the welding process, which would produce not only qualitative but also quantitative information on residual stresses. In this chapter, a mathematical formulation for a butt welding process, through transient thermal and thermo-elasto-plastic analysis is presented.

23.2. MATHEMATICAL FORMULATION FOR BUTT WELDING

Detailed formulations used for finite element analysis of transient temperature and stress distributions during the butt welding process are given in the following sections.

Thermal Analysis

The temperature field in the molten weld pool is assumed to be governed by the same equation as that applied to the solid metal region. The governing partial differential equation for transient heat conduction is considered with internal heat generation ($\dot{Q}$), density (ρ), thermal conductivity (λ) and specific heat of material (C) as functions of temperature and the thermal equilibrium equation is given by the following three-dimensional non-steady state equations [1].

$$\{\lambda(T)(\partial T/\partial X)^2 + (\partial T/\partial Y)^2 + (\partial T + \partial Z)^2\} + \dot{Q} = \rho(T)C(T)(\partial T/\partial t) \quad ...(23.1)$$

Two Dimensionality of the Welding Problem

Although the problem of finding residual stress distribution is three dimensional in nature, a complete three dimensional analysis for temperature and stress distributions is cost prohibitive and time consuming. Simplified two dimensional models are used to obtain results in the cross section (plane strain and axisymmetric) or on the surface (plane stress) depending on the type of geometry. Assuming that the welding speed is high enough, the heat flow in the longitudinal direction (along the X-direction in Fig. 23.1) can be neglected when compared to the heat flow transverse to the weld (e.g. butt welding of thick plates). The three-dimensional transient temperature field can thus be calculated from the two dimensional [2].

Quasi steady state formulation using the transformation for weld travel speed (υ), and time (t).

$$T(x, y, z, t) = T(0, y, z, t-x/\upsilon) \quad ...(23.2)$$

Under this assumption, all the cross-sections normal to the weld line are covered by the same temperature history for a given heat input and weld speed, differing only in time at which a particular temperature distribution occurs. The problem is therefore reduced to finding the two dimensional unsteady field at any section normal to the weld line. Under these assumptions, the thermal equilibrium equation applicable to two dimensional unsteady heat conduction is given by

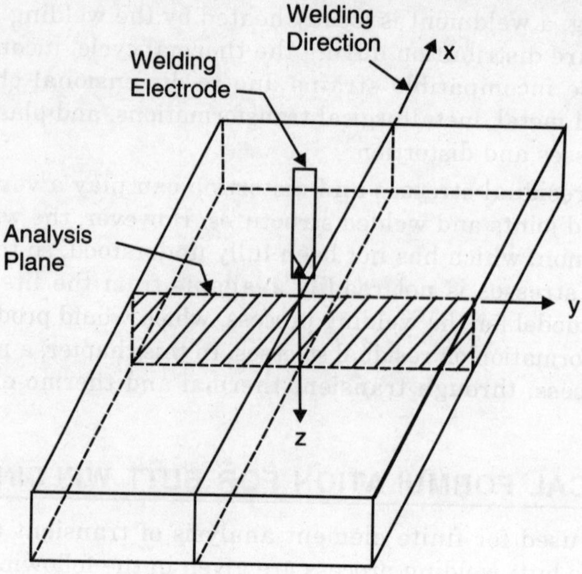

Fig. 23.1: Butt welded plate

$$\lambda(T)(\partial^2 T/\partial^2 Y) + (\partial^2 T/\partial^2 Z) + \dot{Q} = \rho(T)C(T)(\partial T/\partial t) \quad \text{...(23.3)}$$

Boundary Conditions

Surface heat flow that can take place in the model during the butt welding process is specified by the following boundary conditions:

(i) Surface heat input condition

q_e is the rate of heat input for unit surface area (applicable for welding to model the energy input from weld electrode).

(ii) Surface heat loss due to convection and radiation (q_l)

$$q_l = q_l^c + q_l^r$$

where,

q_l^c is heat loss due to convection and

q_l^r is heat loss due to radiation.

Heat loss from the surfaces of butt welded plate at temperatures (T) near the melting point to ambient temperature (T_a) is predominantly due to radiation and is given by

$$\begin{aligned} q_l^c &= \varepsilon\sigma_{sb} A(T^4 - T_a^{\ 4}) \\ &= \varepsilon\sigma_{sb} A(T^2 - T_a^{\ 2})(T + T_a)(T - T_a) \\ q_l^c &= \beta_2(T - T_a) \\ q_l &= \beta_1(T - T_a) + \beta_2(T - T_a) \end{aligned} \quad \text{...(23.4)}$$

Variational Formulation

A variation formulation allows the surface boundary conditions stated above to be treated together with the equilibrium equation. The general weak form is given by [3]

$$\int_y \int_z (f(y,z)W(y,z)dydz) = 0 \qquad ...(23.5)$$

Using weak form, Eqn. (2) given as is

$$\int_y \int_z \lambda(T)(\partial^2 T/\partial^2 Y) + (\partial^2 T/\partial^2 Z)W(y,z)dydz$$

$$+ \int_y \int_z [\dot{Q} - \rho(T)C(T)(\partial T/\partial t)]W(y,z)dydz = 0 \qquad ...(23.6)$$

Applying the bilinear method for Eqn. (23.4)

$$B\,(T, W) = l\,(T)$$

Bilinear method functional $B(T, W)$ is symmetry, a quadratic function exists and this function is given by

$$I\,(T) = 1/2\,B\,(T, T) - l\,(T)$$

The functional Π, for the variation formulation, applicable for the two dimensional heat transfer problem, together with the boundary conditions is given by

$$\Pi = \int_{vol} 1/2\lambda[(\partial T/\partial y)^2 + (\partial T/\partial z)^2]dvol - \int_{vol} T(\ddot{Q} - \rho c)(\partial T/\partial Y)dvol$$
$$- \int_{se} Tq_{se}ds_{se} + \int_{sl} Tq_{sl}ds_{sl}. \qquad ...(23.7)$$

Considering the minimization of Π with reference to temperature, the thermal equilibrium equation is given by

$$\int_{vol} \delta T'^T \lambda T' dvol = \int vol(\delta T^T(\dot{Q} - \dot{T}\rho C)dvol + \int_{se} \delta T^T q_e ds_e - \int_{sl} \delta T^T q_l ds_l \qquad ...(23.8)$$

Incremental Solution Procedure

A time incremental solution is essential to take care of the changes in the ambient temperature and boundary conditions with time. Euler's backward implicit method is used for the analysis, as its unconditional stability allows for larger time steps. Assuming there is no heat generation (i.e. $\dot{Q} = 0$) and substituting the value of q_1 from Eqn. 23.3 and using the equilibrium Eqn. 23.5 corresponding to time $(t + \Delta t)$ is given by

$$\int_{vol} \delta T'^{T^{t+\Delta t}} T' dvol = \int_{vol} \delta T^{T^{t+\Delta t}} \dot{T}^{t+\Delta t}(\rho C)dvol + \int_{se} \delta T^{T^{t+\Delta t}} q_e ds_e$$

$$- \int_{sl} \delta T^{T\;t+\Delta t}(\beta_1 + \beta_2)^{t+\Delta t}(T - T_a)ds_l \qquad ...(23.9)$$

Considering that the variation of parameters is linear between two time steps t and $(t + \Delta t)$, the relationship between parameters at time t and $t + \Delta t$ is given as

$$^{t+\Delta t}T = {}^{t}T + T$$

$$^{t+\Delta t}T = {}^{t}T' + T'$$

$$^{t+\Delta t}\lambda = {}^{t}\lambda + \lambda$$

$$^{t+\Delta t}(\beta_1 + \beta_2) = {}^{t}(\beta_1 + \beta_2) + (\beta_1 + \beta_2)$$

Where T is the temperature added during the time interval, Δt and T', λ, β_1, and β_2 are the increments of the appropriate parameters during Δt. Assuming that

$$^{t+\Delta t}\lambda\ {}^{t+\Delta t}T = {}^{t}\lambda({}^{t}T' - T') \quad \text{...(23.10)}$$

$$^{t+\Delta t}(\beta_1 + \beta_2)\ {}^{t+\Delta t}(T - T_a) = {}^{t}(\beta_1 + \beta_2) + {}^{t+\Delta t}({}^{t}T + T - T_a) \quad \text{...(23.11)}$$

and substituting Eqns. 23.7 and 23.8 into Eqn. 23.6, the resulting equation is given as

$$\int_{vol} \delta T'^{T} \lambda T' dvol + \int_{sl} \delta T^{T}\ {}^{t}(\beta_1 + \beta_2) T ds_l = -\int_{vol} \delta T'^{T}\ {}^{t}\lambda T' dvol$$

$$-\int \delta T^{T}\ {}^{t+\Delta t}q_e ds_e - \int_{sl} \delta T^{T} (\beta_1 + \beta_2)({}^{t}T - {}^{t+\Delta t}T_a) ds_l$$

$$\int_{vol} \delta T'^{T}\ {}^{t}\lambda\ {}^{t}T dvol. \quad \text{...(23.12)}$$

Equation 23.10 is to be solved to obtain the temperature increment during time step Δt, for the corresponding increments in heat input and surface heat losses.

Iteration Scheme

Considering β_1, λ, ρ, C and β_2 are the functions of temperature and the Eqn. 23.9 is non-linear, an iterative solution procedure is necessitated. Referring to ${}^{t+\Delta t}T = {}^{t}T + T$, the temperature increment T during time increment Δt is obtained as a sum of temperature increments (ΔT), over various iterations.

$$T = \sum_{i=1}^{n} \Delta T^{(i)} \quad \text{...(23.13)}$$

Where n is the number of iterations and (i) is the counter. The equilibrium equation for iteration (i) between time steps t and $(t + \Delta t)$ is written as

$$\int_{vol} \delta T'^{T}\ {}^{t}\lambda \Delta T'^{(i)} dvol + \int_{sl} \delta T^{T}\ {}^{t}(\beta_1 + \beta_2) \Delta T^{(i)} ds_l = \int_{vol} \delta T^{T}\ {}^{t+\Delta t}\dot{T}^{(i)}\ {}^{t+\Delta t}(\rho C)^{(i-1)} dvol$$

$$+ \int_{se} \delta T^{T}\ {}^{t+\Delta t}q_e ds_e - \int_{sl} \delta T^{T}\ {}^{t+\Delta t}(\beta_1 + \beta_2)^{(i-1)} X({}^{t+\Delta t}T^{(i-1)} - {}^{t+\Delta t}T_a) ds_l$$

$$-\int_{vol} \delta T'^{T}\ {}^{t+\Delta t}\lambda^{(i-1)}\ {}^{t+\Delta t}T'^{(i-1)} dvol \quad \text{...(23.14)}$$

where $\quad {}^{t+\Delta t}T^{(i)} = {}^{t+\Delta t}T^{(i-1)} + \Delta T^{(i)}$

The convection and radiation heat transfer coefficients are given as

$$^{t+\Delta t}\beta_1^{(i-1)},\ ^{t+\Delta t}\beta_2^{(i-1)}$$

$^{t+\Delta t}\lambda^{(i-1)}$ is the convective heat transfer coefficient corresponding to temperature $^{t+\Delta t}T^{(i-1)}$ is given as

$$^{t+\Delta t}q_e,\ ^{t+\Delta t}T_a$$

$^{t+\Delta t}\dot{T}$ is approximated using a time integration scheme as

$$^{t+\Delta t}\dot{T} = (^{t+\Delta t}T - {}^{t}T)/\Delta t$$

and for the iteration scheme, it is written as

$$^{(i-1)}\dot{T}^{(i)} = (^{(i+\Delta t)}T^{(i-1)} + \Delta T^{(i)} - {}^{t}T)/\Delta t$$

Finite Element Formulation

Using isoparametric formulations coordinates Y and Z (Fig. 23.1), temperature T and its derivatives at any point of the element are obtained by a set of shape functions $[H]$ (Fig. 23.2) and its derivatives $[B]$ as [4]

$$y = H(y, z)\,Y \qquad \text{...(23.15a)}$$

$$z = H(y, z)\,Z \qquad \text{...(23.15b)}$$

$$T(y, z) = H(y, z)\,\upsilon \qquad \text{...(23.15c)}$$

$$\dot{T}(y, z) = H(y, z)\,\dot{\upsilon} \qquad \text{...(23.15d)}$$

$$T,(y, z) = B(y, z)\,\upsilon \qquad \text{...(23.15e)}$$

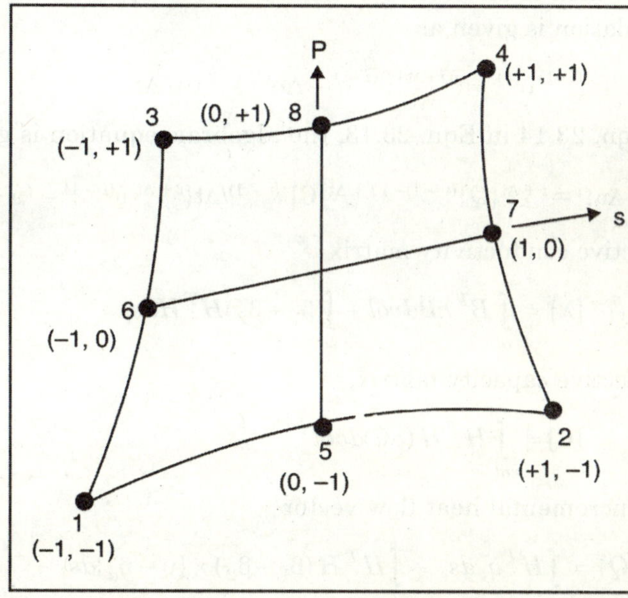

Fig. 23.2: Isoparametric element

Where the matrices [H] and [B] define the variation of the temperature and temperature gradients within the element and υ represents the nodal temperature vector.

The corner nodes are $i = 1, 2, 3, 4$. Corresponding to these nodes the shape functions are

$$h = 1/4(1+ ss_i)\ (1 + rr_i + (ss_i + rr_i - 1))$$

For the mid-side nodes $i = 5,8$ $(s_i = 0)$,

$$h= 1/2(1 - s^2)(1 + rr_i)$$

For the mid-side nodes $i = 6, 7$ $(r_i = 0)$

$$h = 1/2(1 - r^2)(1 + ss_i)$$

Using these relations 23.15a, 23.15b, 23.15c in Eqn. 23.10, the incremental finite element equilibrium equation applicable to individual elements in the domain is written as

$$\int_{vol} \delta\upsilon^T\,{}^t\lambda B\Delta\upsilon^{(i)}dvol + \int_{sl} \delta\upsilon^T\,{}^t(\beta_1+\beta_2)\Delta\upsilon^{(i)}ds_l$$

$$= -\int_{vol} \delta\upsilon^T H^T H\,{}^{t+\Delta t}\dot{\upsilon}^{(i)}\,{}^{t+\Delta t}(\rho C)^{(i-1)}dvol + \int_{se} \delta\upsilon^T H^T\,{}^{t+\Delta t}q_e ds_e$$

$$-\int_{sl} \delta\upsilon^T H^T H \times{}^{t+\Delta t}(\beta_1+\beta_2)^{(i-1)}({}^{t+\Delta t}\upsilon^{(i-1)} - {}^{t+\Delta t}\upsilon_a)ds_l$$

$$-\int_{vol} \delta\upsilon^T B^T\,{}^{t+\Delta t}\lambda^{(i-1)}B\,{}^{t+\Delta t}\upsilon^{(i-1)}dvol \quad ...(23.16)$$

Since $\delta\upsilon^T \neq 0$ which is an arbitrary variation in temperature, the iteration scheme in finite element formulation is given as

$$^{(i-1)}\dot{\upsilon}^{(i)} = (^{(i+\Delta t)}\upsilon^{(i-1)} + \Delta\upsilon^{(i)} - {}^t\upsilon)/\Delta t \quad ...(23.17)$$

Substituting Eqn. 23.14 in Eqn. 23.13, the algebraic equation is given by

$$[[\lambda] + ({}^t[C]/\Delta t)]\{\Delta\upsilon\}^i = {}^{t+\Delta t}\{Q\}^{(i-1)-)}\,{}^{t+\Delta t}[C]^{(i-1)}/\Delta t\{{}^{t+\Delta t}\acute{\upsilon}^{(i-1)} - {}^t\upsilon + {}^{t+\Delta t}\{Q\}^{(i-1)} \quad ...(23.18)$$

where [λ] is the effective conductivity matrix

$$[\lambda] = \int_{vol} B^T\lambda B dvol + \int_{s}(\beta_1+\beta_2)H^T H ds_l \quad ...(23.19)$$

If [C] is the effective capacity matrix,

$$[C] = \int_{vol} H^T H(\rho C)dvol \quad ...(23.20)$$

$\{Q\}$ is the net incremental heat flow vector

$$\{Q\} = \int_{se} H^T q_e ds_e - \int_{sl} H^T H(\beta_1+\beta_2)\times\{\upsilon - \upsilon_a\}ds_l + \int_{vol} B^t\lambda b\upsilon\, dvol \quad ...(23.21)$$

The computer code has been developed for the above mentioned analysis in order to obtain the temperature and residual stress values [5, 6].

23.3. THERMAL ANALYSIS

Calculation of the transient temperature distribution depends on the attainment of quasi-stationary condition, which is developed when the arc is moving at a constant speed and end effects are neglected.

Modeling of Arc

The welding parameters employed in the study are:

Current = 1170 A

Voltage = 33 V

Speed = 5 mm/sec.

Modeling heat input from the arc is the most critical task, for it directly influences temperature profile and hence cooling rates and the size of the melt zone and heat affected zone. The heat input from the arc to the plate is assumed to have a uniform distribution. The arc radius is taken as 10 mm. The arc is applied over a rectangular area of length 10 mm along welding direction and 5 mm in the transverse direction. The heat is applied over the square of side 5 mm for duration of 1 sec.

Assumptions for Thermal Analysis

Natural convective heat transfer occurs on all surfaces of the plates except the plane symmetry. But the area beneath the nozzle of the torch experiences forced convection due to the flow of the shielding gas. Also radiation heat transfer occurs from high temperature areas. Applying radiation boundary is a difficult task in Ansys and to simplify the problem. The overall heat transfer coefficient accounts effects of both convection and radiation and has a value of 15W/mm^2°C. The major assumptions employed while carrying out the thermal analysis are as follows:

- Cartesian co-ordinate system is employed
- Two-dimensional, transient heat conduction is assumed within the weld plates
- The heat flux distribution is assumed to be of Gaussian type, the heat flux is assumed to be of uniform over the entire area
- The material is isotropic, but the material properties, viz., thermal conductivity and specific heat are assumed to the functions of temperature. The material density is assumed to be constant
- The plate is assumed to be stationary and the heat source is moving at the specified rate with respect to time
- Both convection and radiation heat loss is present over the entire surface including the welding zone
- Convective heat transfer coefficient (h) and emissivity of the weld plate are assumed to be constant
- Heat flux is applied normal to the surface and over the surface flux
- Slag formation is not considered
- The addition of filler metals is not considered
- Forced convection effects are not considered
- The effect of gas diffusion in the weld pool and weld cracking are not considered.

Boundary Conditions for Thermal Analysis

The following are the boundary conditions employed during the thermal analysis.

For time $t = 0$, entire solid is assumed to be at uniform temperature of 30°C.

The moving heat source is modeled by applying and deleting heat flux over the corresponding areas with respect to the welding speed.

At time $t > 0$, over the entire surface of the solid, convection heat loss is applied by specifying the constant convective heat transfer co-efficient (h) and surrounding temperature ($T\alpha$).

Thermal Properties Steel

In this analysis requires accurate values of thermal conductivity and specific heat capacity, material density and latent heat of fusion and phase transformation up to the melting point. Low temperature data is readily available, but high temperature data must be data collection. Above the melting point, the complex convective heat transfer mechanisms in the weld pool are very difficult to model even with accurate thermal properties. From the specific heat value, the enthalpy is calculated for different temperatures. The material density was assumed to remain constant at the room temperature value. This is a reasonable assumption as density has only a minor effect on the heat transfer analysis. For the current analysis the temperature dependent thermal conductivity, specific heat capacity values and latent heat capacity value were taken from reference. The values are shown in Fig. 23.3.

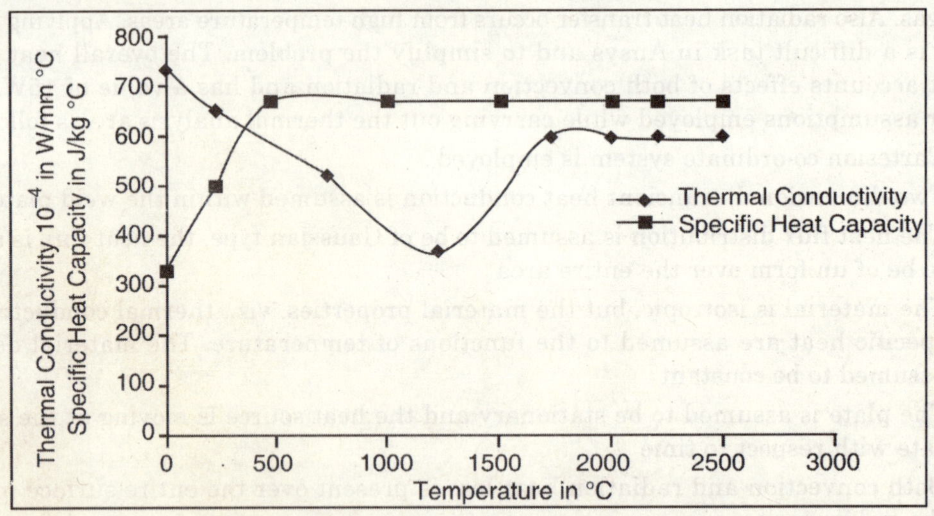

Fig. 23.3: Thermal properties steel

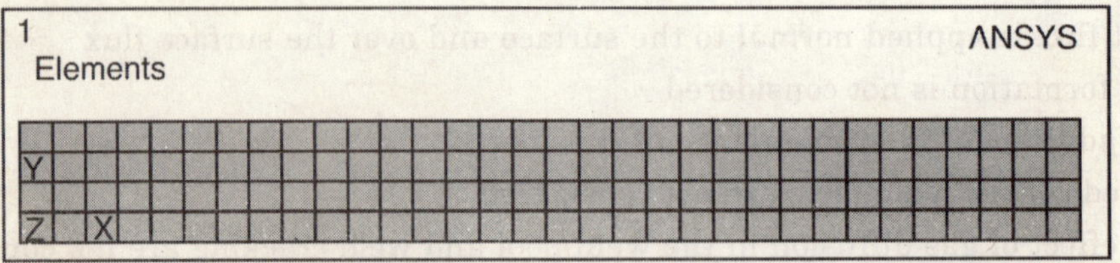

Fig. 23.4: Two dimensional view of the plate employed in the study

Solving the Thermal Problem

The arc is modeled as moving heat source and takes 6.6 seconds to cover the entire length of the plate. The arc heat flux moves along the joint at the rate of the welding speed and for any instant of time, the nodes, which receive heat flux, are identified and the nodal heat input values are calculated and applied to the corresponding nodes. For each time interval, the elemental [C] and [*k*] matrices are computed and assembled to form the global matrices. The arc heat flux is then calculated for the appropriate nodes and the surface heat losses are computed. After formulating the equation, it is solved for various time intervals using Galerkin's time marching scheme. The Jacobian conjugate gradient solver is used as the equation solver.

23.4 MESH GENERATION

Mesh generation plays a crucial role in accuracy of numerical results. The dimensions of the plate are as follows Length = 80 mm, Width = 80 mm, Thickness = 8 mm. The nodes and elements in the plate model are generated directly and nodes are numbered in such a way that the bandwidth is minimum. This is done to achieve further economy in solution. The total number of elements and nodes are 380 and 651 respectively. The analysis uses PLANE-77 element for thermal analysis.

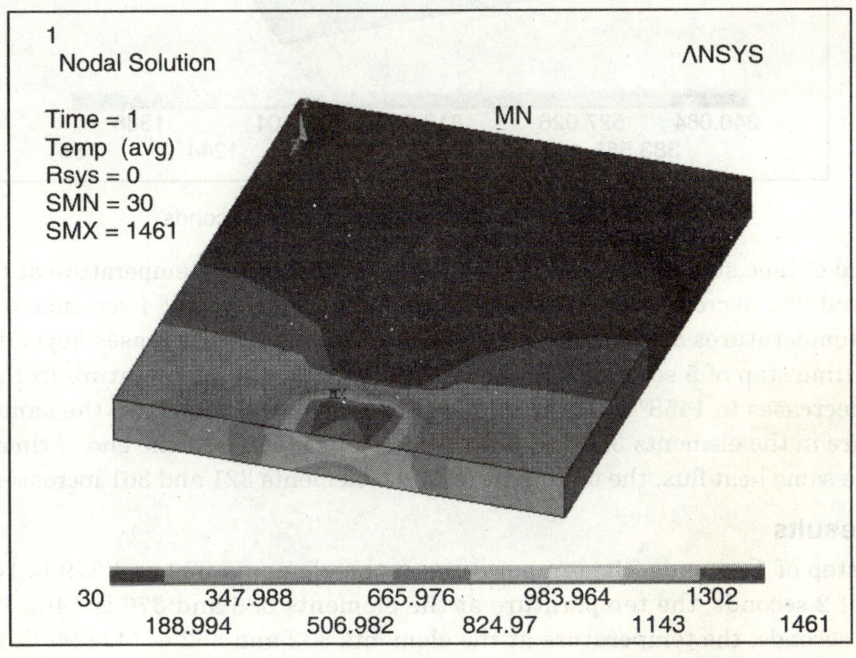

Fig. 23.5: Temperature distribution for 1 seconds

Thermal analysis of a welding process is carried out by having typical values for heat flux (q) as $6*10^9$ w/m^2 , welding speed (w) as 0.005 m/s, Time, t as 6.6 sec., plate thickness (z) as 0.08 m, ambient temperature, T_0 as 30° C and heat transfer co-efficient, h as 20 W/m^2 K. The temperature distribution in the weld interface has been drawn with time along abscissa and temperature along the ordinate. At the time of start of welding, the temperature in the elements 339 and 379 is 1461°C as shown in Fig. 23.5.

Above procedure is following as bellow time steps. At the end of time step of 2 seconds, for the applied heat distribution, the temperature in the adjacent elements increases to 336 and 376 is 1464°C. At any instant, as soon as the temperature in the plate drops below 1464°C by an infinitesimal amount, the liquid metal starts to solidify.

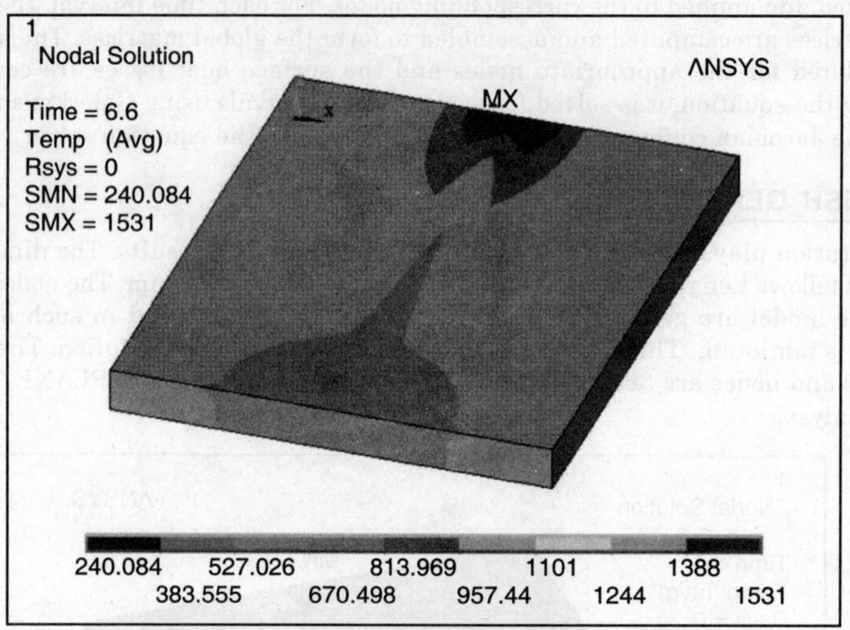

Fig. 23.6: Temperature distribution for 6.6 seconds

At the end of time step of 3 seconds, for the same heat flux, the temperature at the adjacent elements 33 and 352 decreases beyond to 1462°C. At the time step of 4 seconds, for the same heat flux the temperatures at the adjacent elements 330 and 370 decreases beyond to 1460°C. At the end of time step of 5 seconds, for the same heat flux, the temperature in the elements 327 and 367 decreases to 1458°C. At the end of time step of 6 seconds, for the same heat flux, the temperature in the elements 324 and 364 decreases to 1456°C. At the end of time step of 6.6 seconds, for the same heat flux, the temperature in the elements 321 and 361 increases to 1531°C.

Analytical Results

For the time step of 1 seconds, the temperature at the elements 339 and 379 is 1435.6°C. At the end step of 2 seconds, the temperature at the elements 336 and 376 is 1460.8°C. For the time step of 3 seconds, the temperature at the elements 332 and 352 is 1471.99°C. At the time step of 4 seconds, the temperature at the elements 330 and 370 is 1478.3°C. For the time step of 5 seconds, the temperature at the elements 327 and 367 is 1482.42°C. At the time step of 6 seconds, the temperature at the elements 324 and 364 is 1485.28°C. For the time step of 6.6 seconds, the temperature at the elements 321 and 361 is 1486.5°C. The coordinates are obtained from element and node list which was obtained from ANSYS 8.1.software. So both the results are comparable.

Comparison of Software and Analytical Results

The software results and analytical results were compared and it is shown in Fig. (23.7). The variation of these software and analytical results in percentage error were shown in Table 23.1.

Table 23.1: Comparisons of software results and analytical results

S.No.	Time	Peak Temperature from ANSYS in (°C) (A)	Peak Temperature from C in (°C) (B)	Percentage error (%) (A–B)/A
1	1	1461	1435.6	0.017
2	2	1464	1460.8	0.002
3	3	1462	1471.99	0.006
4	4	1460	1478.30	0.0125
5	5	1458	1482.42	0.0167
6	6	1456	1485.28	0.0201
7	6.6	1531	1486.5	0.0290

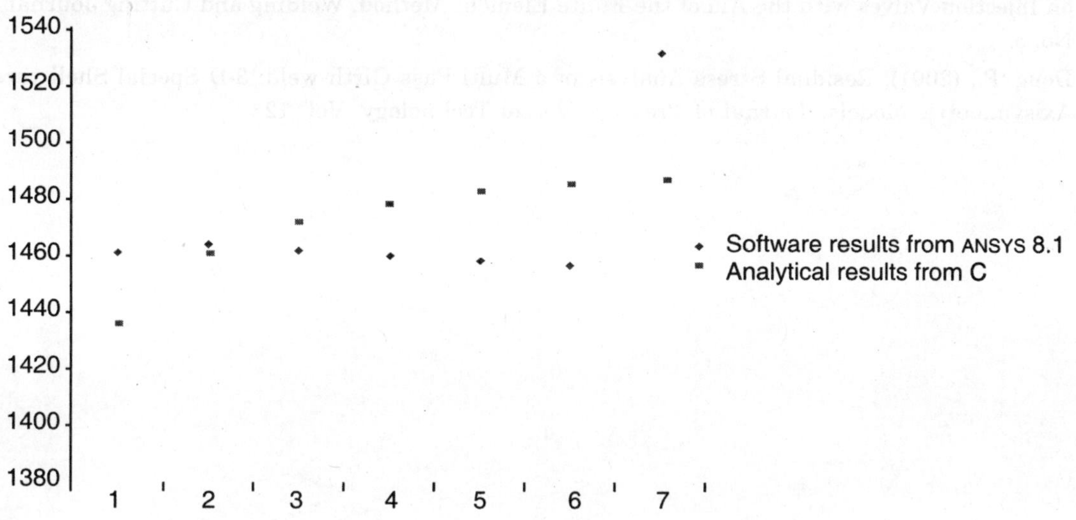

Fig. 23.7: Comparision of software and analytical results

23.5 CLOSURE

A mathematical formulation for a butt welding process is developed for transient thermal and thermo-elasto-plastic stress analysis considering various non-linearity due to (*i*) Material properties variation and heat transfer coefficients variation with temperature, (*ii*) Radiation boundary conditions and (*iii*) Solid phase transformations. The effect of temperature distribution and residual stress were analyzed and the results were plotted. Then the thermal analysis carried out by the software ANSYS 8.1., is compared with analytical result. The major conclusions drawn from the present investigations are: The peak temperature obtained from both the analysis is more than the melting point of the selected material. Based on this peak temperature, the welding has been successfully carried out at the particular time. From the thermal analysis the deviation among the two analysis is less than 5%. So, the model has been validated.

REFERENCES

1. Donald.R. pitts, Leightone.sison, *"Heat Transfer Schaum's Outlines Series"*. McGraw-Hill Book Company (1983).
2. Krutz G.W. and Segerlind J., (1978), "*Finite Element Analysis of Welded Structures*", Welding Journal.
3. O.C., Zienkiewitcz, "*The Finite Element Methods*", McGraw-Hill, New York (1987).
4. S. Kandasamy, K. Thilagavathi, and K. Gunavathi, "*Numerical Method*", S. Chand Publishers (1997).
5. Dipl-Ing Michael Rethmeier et al. (2000); Minimizing the Temperature Load and the Distortion on Injection Valves with the Aid of the Finite Element Method, Welding and Cutting Journal 52, No: 8.
6. Dong. P., (2001); Residual Stress Analysis of a Multi-Pass Girth weld: 3-D Special Shell versus Axisymmetric Models, Journal of Pressure Vessel Technology, Vol. 123.

NURBS Surfaces

24.1. INTRODUCTION

There is no proper general NURBS representation of the offset curve (surface) to a NURBS curve (surface) in the literature. This chapter presents a new algorithm to determine a sequence of approximations to the offset curve (surface) of a given curve (surface) with the properties that the global error of each approximation from the true offset can be bounded and the sequence converges to the true offset.

Offset surfaces are very important in manufacturing, and computation so finding approximations of offset curves and surfaces have undergone extensive research. Creating an offset to a curve is an intuitive operation and has been mathematically known for more than a hundred years [1, 2]. The offset operator is closed for arcs and lines, i.e., an offset of an arc and a line are an arc and a line, respectively. This is not so for Bezier and NURBS curves, in general, so approximations are usually derived.

Two methods for finding approximations to offset curves are commonly used. The first approximates the curve using piecewise lines and arcs and then finds the representation of the exact offset to the approximation. The second method attempts to approximate the offset by directly transforming the curve representation, in particular the control points [1, 3, 4]. To improve the accuracy of the approximation in the second method, the original curve is subdivided [2, 4] or manually refined when the error is above pre-specified tolerance level and the same technique is applied to each of the subdivided pieces. The original curve is usually subdivided in the middle of its parametric domain, although that is usually not the optimal location. Curve inflection points have also been considered as splitting points for planar offsets [4].

Both approaches do not bound the offset error globally. But in this chapter we will take you through a robust method which constructs a NURBS surface that accurately approximate the data points (forming the surface) and also accomplish this in a practical amount of time.

Without delving too much into the fundamentals of NURBS representation we would directly discuss the algorithm in Sec. 24.2 before moving onto the results in Sec. 24.3 and finally some conclusions in Sec. 24.4.

24.2. ALGORITHM — THE INITIAL APPROXIMATION OF DATA BY A NURBS SURFACE

The initialization stage involves plotting the polynomial curve using the set of data points and the NURB curve using the control points selected by the user. Having done this, we estimate the error between these curves by finding distances between the corresponding points on either of the curves. This is done as follows:

- A fixed number of points are laid on both the curves
- The corresponding points laid on either of the curves is calculated and mapped by using the distance, velocity equation (distance = velocity * time) and with the fact that $t = 0$ and $t = 1$ at the start and end of both the curves respectively
- The distance (error) between these mapped points on the curves is calculated.

To continue with the process of fitting a NURBS surface to contours of data, we must specify the control points, weights, and knot vectors for the initial NURBS surface. We start by assigning u and v parameter values corresponding to each data point. We base this on the convention of summing arc lengths along the two parameter dimensions.

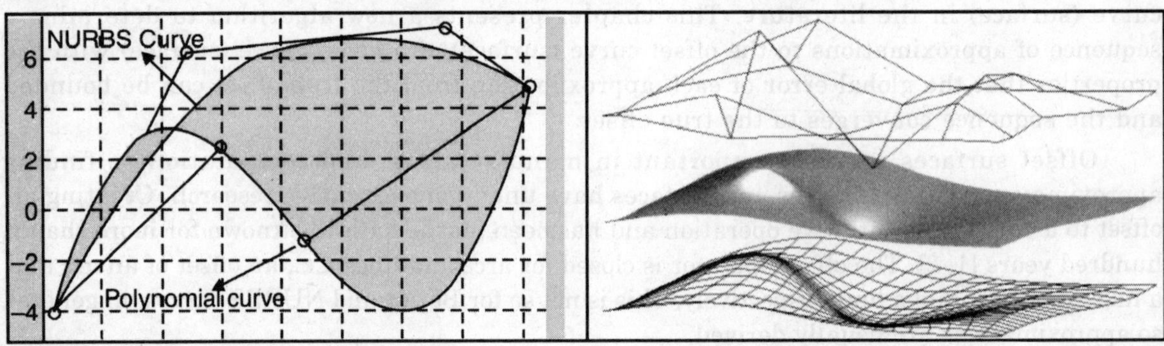

Fig. 24.1: Estimates the error between corresponding points of Polynomial and NURBS curve (Striped Lines)

Fig. 24.2: Example of a NURBS surface. The NURBS control points are drawn as a mesh above the surface. Below the surface is the dense mesh of points that would be required to render a similar polynomial surface

Let $d_{i,j}$ be the jth data point in contour i. The distance between successive points in the v direction along a contour is

$$\Delta_v d_{i,j} = \left\| d_{i,j} - d_{i,j-1} \right\|$$

The total arc length along a contour is found by summing $\Delta_v d_{i,j}$ along the contour. The v parameter value, $v_{i,j}$ for each data point along the contour is defined to be

$$v_{i,j} = \begin{cases} \dfrac{\sum_{k=1}^{j} \Delta v d_{j,k}}{\sum_{k=1}^{n_i} \Delta v d_{i,j}}, & j > 0 \\ 0, & j = 0 \end{cases}$$

where n_i is the number of data points in contour i. This results in v values that range from 0 to 1. The values of $u_{i,j}$ are assigned similarly.

Now the initial control points can be placed. Given that the NURBS surface is defined to have n rows and m columns of control points, we can specify u and v values corresponding to each control point assuming that they will be distributed evenly over the ranges of u and v. Thus, the u and v values for the control point in row i and column j, are

$$u(c_{i,j}) = \frac{i}{n-1} \quad \text{and} \quad v(c_{i,j}) = \frac{j}{m-1},$$
$$\text{for } i = 0,\ldots,n-1 \text{ and } j = 0,\ldots,m-1.$$

Now we find the four data points, d_1, d_2, d_3 and d_4 and their corresponding parameter values, (u_1, v_1), (u_2, v_2), (u_3, v_3), and (u_4, v_4), that form the vertices of the plane containing the parameters of the control point. The coordinates of the control point are then found by linearly interpolating the data points:

$$c_{i,j} = (1 - v_f)[u_f d_3 + (1 - u_f) d_1] + v_f [u_f d_4 + (1 - u_f) d_2]$$

After the $n \times m$ grid of control points is initialized, we add duplicate control points along the $v = 0$ and $v = 1$ ends of the grid to create a smooth join along the v direction, which is along each contour. This specializes the NURBS surface for the types of cylinders we are dealing with in approximating contour data. A total of four control points, two on either side of the join, are added for each row of the control point grid.

All components of the weight matrix, W, are initialized to 1. The knot vector in the u direction is initialized to the standard non-uniform knot vector for third-degree NURBS curves,

$$u_f = \frac{u(c_{i,j}) - u_1}{u_3 - u_1} \left(= \frac{u(c_{i,j}) - u_2}{u_4 - u_2} \right)$$

$$v_f = \frac{v(c_{i,j}) - v_1}{v_2 - v_1} \left(= \frac{v(c_{i,j}) - v_3}{v_4 - v_3} \right)$$

The knot vector in the v direction is initialized to the uniform knot vector.

24.3. NURBS OPTIMIZATION

We define a NURBS surface to be optimized if the mean-squared error, E, between the data points and the corresponding NURBS surface points is a local minimum. E depends on the control points, C, the weights, W, the knot vectors, U and V, and the set of data points D. Let $\hat{u}$ and $\hat{v}$ be the parameter values for the point on the NURBS surface that corresponds to data point d. This notation is also applied to other variables. Then, we can define E to be

$$E(C,W,U,V,D) = \frac{1}{2|D|}\sum_{d\in D} \| d - P(\hat{u}, \hat{v}) \|^2$$

Local minima in E can be found by performing a gradient-descent procedure to optimize C and W. (The knot vectors U and V can also be optimized in this way, but we have not yet explored this.)

First, we drop the arguments to E for clarity. The gradient of E with respect to each coordinate of the control points, C_i, for $i = 1,\ldots,3$ for the x, y, and z coordinates, is

$$\nabla_{C_i} E = \frac{-1}{|D|}\sum_{d\in D} \hat{R}(d_i - P_i(\hat{u}, \hat{v})) \bullet \hat{W},$$

$$R = \frac{N_u^T U^T \hat{V} \hat{N}_V}{U N_u W N_v^T V^t}$$

and the $\bullet$ operator means component-wise multiplication. Similarly, the gradient of E with respect to the weights, W, is

$$\nabla_w E = \frac{-1}{|D|}\sum_{d\in D} \hat{R}\sum_{i=1}^{3}(d_i - P_i(\hat{u}, \hat{v}))(\hat{C}_i - P_i(\hat{u}, \hat{v}))$$

Gradient descent based on the above gradients will arrive at a local minimum of E. This procedure involves iterations of the following update equations:

$$C \leftarrow C - \alpha_C \nabla_C E$$
$$W \leftarrow W - \alpha_W \nabla_W E$$

Figure 24.3 is a schematic representation of the flow chart of the algorithm.

24.4. CLOSURE

As far as the prediction of control points using analytical approach vis-à-vis neural networks approach is concerned, we have invested considerable effort in developing an intelligent approach to the initialization of the control points for a NURBS surface, given the data to be approximated. As the previous figures show, the initial NURBS surface approximates the data well in some places, but not in others. For those areas where the initialization is a poor fit to the data, the gradient-descent procedure requires many iterations to improve the fit. Our current development is in 2-D, but is easily extended to the full 3-D case.

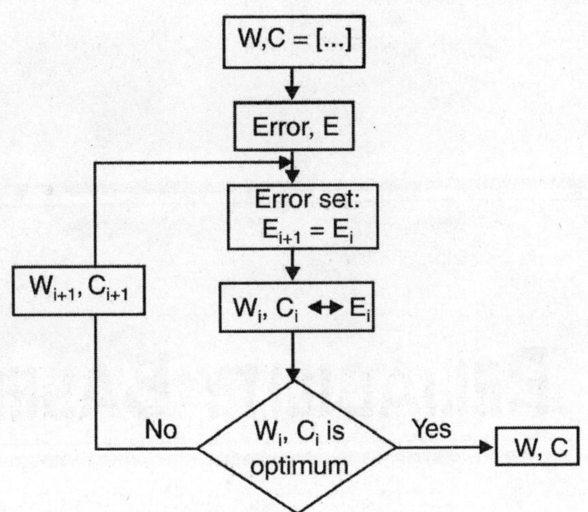

Fig. 24.3: General flow chart of our approach

REFERENCES

1. B. Cobb, "*Design of Sculptured Surfaces Using the B-spline Representation*". Ph.D. thesis, University of Utah, Computer Science Department, June 1984.
2. B. Pham. Offset Approximation of Uniform B-splines. "*Computer Aided Design*", Vol. 20, No. 8, pp. 471–474, October 1988.
3. J. Hoschek and N. Wissel. Optimal approximate conversion of spline curves and spline approximation of offset curves. "*Computer Aided Design*", Vol. 20, No. 8, pp. 475–483, October 1988.
4. R.T. Farouki and V.T. Rajan. Algorithms for Polynomials in Bernstein form. "*Computer Aided Geometric Design*", 5, pp. 1–26, 1988.

RELIABILITY BASED DESIGN

25.1. INTRODUCTION

The work illustrates simulation approach for parametric design and optimizes the performance of a 2-DOF R-R planar manipulator. Using dynamic and kinematic models of a manipulator different performance measures for the manipulator are obtained for different combination of parameters with effect of noise incorporated to imitate the real time performance of the manipulator. A novel approach has been used to model, the otherwise difficult to model, noise effects. The data generated during simulation for various parameter combinations are utilized to analyze the statistical significance of kinematic and dynamic parameters on performance of manipulator using ANOVA technique. Optimal parameter combinations, which gave optimum performance measures, for a point in workspace are obtained and compared.

In recent times, attention of researcher has been focused on the optimal design of robotic manipulators. In the past, little work has been done in this area because of difficulties associated with geometrical constraints and complex model of manipulator. Mostly robots are required to perform several routine tasks with a higher precision than their human counterparts. When a robot is commanded, to move its end-effector to a specified position, the actual position reached may be quite different from the desired one. This variation in the actual and desired position of the end-effector is termed as positional error of a robot and the average precision with which the manipulator moves its arm to the commanded position is termed as its positional accuracy. Parametric design is a concept to select a combination of control factors of a product/process in such a way that its performance will be insensitive to noise factors. Objective of the work reported here is to find out the optimal combination of parameters to minimize positional error without any constraints such as maximum velocity and acceleration of links. As it is very tedious, time consuming and uneconomical to conduct physical experiments on the actual manipulator by varying the values of control factors and finding the optimum setting of factors which will result in maximum positional accuracy, the approach adopted is to formulate a mathematical model relating all factors to the positional accuracy, conduct experiments on the model and obtain optimum combination of parameters. In the past studies for optimal performance were conducted only with one prototype with fixed parameters and if the performance variation was found to be large then it was reduced by adoption of suitable control strategy.

A variety of approaches have been tried by different researchers to simulate the performance of manipulator, among them Monte-Carlo simulation techniques were used by some researchers, such as Benhabib et al. [1] to explore the probabilistic effects of the error sources. A number of models have been proposed for analysis and determination of geometric error. The model developed by Chen and Chao [2] and Driel and Swayze [3] used several geometric parameters to characterize the robot completely. Bhatti and Rao [4] presented a concept of reliability as a statistical measure of manipulator performance using geometric parameters. Manoochehri and Seireg [5] have developed a computer programme for form synthesis and optimal design of robot manipulator. Azadivar [6] cited that it is important that robotic manipulators should be designed to perform in some optimal manner to give optimal accuracy, minimum vibrations, minimum energy usage, etc. Imura et al. [7] and Deol and Pai [8] indicated that designing an optimal manipulator is very complicated and not well defined and sometimes depends on definition and criterion of optimality. Luh et al. [9] pointed out that for various reasons the numerical values of the kinematic parameters for a robot may not be correct. This may be due to manufacturing tolerances, deviation such as link and joint compliance, or time-dependent effects such as gear wear and component damage. Imura et al. [7] suggested that robot calibration, which, addresses static errors as well as dynamic errors. Offodile and Uguwu [10] investigated how various process variables such as speed of the tool and payload affect the robot repeatability and indicated experimentally that speed at which robot traveled in work envelope has noticeable effects on its repeatability. Parkinson et al. [11] and Chen et al. [12] indicated that the early design phase of a product has greatest impact on life cycle and its quality of performance. Therefore, significant improvement in quality can be realized by optimizing product design. An extensive review of the methodologies available to obtain a robust design that have less performance variation due to the variations of control factors and noise factors is given by Rout and Mittal [13]. An attempt was made by Rout and Mittal [14] for finding the best setting parameters for reduction of performance variation using Taguchi method by increasing positional accuracy of a manipulator and illustrated the ways to incorporate effects of noise in performance.

25.2. IDENTIFICATION OF SOURCES OF NOISE IN MANIPULATOR

The main function of the manipulator is to move its end effector to a desired point of path accurately. However, the discrepancy exists between the actual and desired point reached because of noise factors. The noise factors are those factors that are difficult, expensive or hard to control during production or operation. The noise factors which contribute to the performance variation of manipulator are enumerated as Environmental condition in which operates, Errors in manufacture and assembly, Fluctuations in electricity supply affecting functioning of the joint actuators, Friction at the joint, Joint compliance between the angular encoder and the actual angular output and Bearing clearances. There can be several other noise factors, which have influence over performance variation but are not discussed, in the present work.

25.3. KINEMATIC AND DYNAMIC MODEL OF 2-DOF R-R PLANAR MANIPULATOR

Considering the 2-DOF R-R planar manipulator having link lengths l_1, l_2, link masses m_1, m_2 and joint torques τ_1, τ_2 respectively, and assuming links as slender members with mass concentrated at center of gravity, the kinematic model can easily be obtained. Let $P(x, y)$ be the target position of arm end-point in the Cartesian workspace of the manipulator, the point of interest, for which the performance in terms of positional accuracy has to be modeled and optimized [15]. The coordinates of P in Cartesian coordinate for joint angles θ_1 and θ_2 are given by

$$x = l_1C_1 + l_2C_{12} \qquad \text{...(25.1)}$$

$$y = l_1S_1 + l_2S_{12} \qquad \text{...(25.2)}$$

where, $C_i = \cos\theta_i, S_i = \sin\theta_i$, $C_{ij} = \cos(\theta_i + \theta_j)$ and $S_{ij} = \sin(\theta_i + \theta_j)$ with $i, j = 1, 2$ for the two links, link1 and link2, respectively.

From Eqns. (25.1) and (25.2) joint variables θ_1 and θ_2 can be obtained as:

$$\theta_1 = \tan^{-1}\left[\frac{y(l_1 + l_2C_2) - xl_2S_2}{x(l_1 + l_2C_2) + yl_2S_2}\right] \qquad \text{...(25.3)}$$

$$\theta_2 = \cos^{-1}\left[\frac{x^2 + y^2 - l_1^2 - l_2^2}{2l_1l_2}\right] \qquad \text{...(25.4)}$$

Differentiating Eqns. (25.3), and (25.4) and solving for joint velocities $\dot{\theta}_1$ and $\dot{\theta}_2$ gives

$$\dot{\theta}_1 = \frac{\dot{x}C_{12} + \dot{y}S_{12}}{l_1S_2} \qquad \text{...(25.5)}$$

and

$$\dot{\theta}_2 = -\frac{\dot{x}x + \dot{y}y}{l_1l_2S_2} \qquad \text{...(25.6)}$$

where $(\dot{x}, \dot{y})$ represent end-effector velocity and the dynamic model of 2-DOF R-R planar manipulator is

$$\tau_1 = \left[\left(\frac{m_1}{3} + m_2\right)l_1^2 + \frac{m_2}{3}l_2^2 + m_2l_1l_2C_2\right]\ddot{\theta}_1 + m_2\left(\frac{l_2^2}{3} + \frac{l_1}{2}l_2C_2\right)\ddot{\theta}_2 - m_2l_1\,l_2S_2\dot{\theta}_1\dot{\theta}_2$$
$$-\frac{m_2}{2}l_1l_2S_2\dot{\theta}_2^2 + \left(\frac{m_1}{2} + m_2\right)gl_1C_1 + \frac{m_2}{2}gl_2C_{12} \qquad \text{...(25.7)}$$

$$\tau_2 = \left[\frac{m_2}{3}l_2^2 + \frac{m_2}{2}l_1l_2C_2\right]\ddot{\theta}_1 + \frac{m_2}{3}l_2^2\ddot{\theta}_2 + \frac{m_2}{2}l_1l_2S_2\dot{\theta}_1^2 + \frac{m_2}{2}gl_2C_{12} \qquad \text{...(25.8)}$$

Equations 25.7 and 25.8 are solved for $\ddot{\theta}_1$ and $\ddot{\theta}_2$ and it is observed that apart from the variables $\theta_1, \theta_2, \dot{\theta}_1$, and $\dot{\theta}_2$, $\ddot{\theta}_1$ and $\ddot{\theta}_2$ depends on following six independent parameters:

Length of link 1, l_1; Length of link 2, l_2; Mass of link 1, m_1; Mass of link 2, m_2; Torque applied at joint 1, τ_1; Torque applied at joint 2, τ_2. These six parameters are classified as control factors because designer has a control over them, can choose as per requirements. These form the six input parameters for the parametric design optimization study.

25.4. PERFORMANCE MEASURES FOR MANIPULATOR

To assess the performance variability of manipulator following measures have been taken into consideration in this work.

(a) Mean Positional Error: Mean positional error $\bar{\varepsilon}$ is defines as

$$\bar{\varepsilon} = \frac{1}{n}\sum_{i=1}^{n}\varepsilon_i \qquad \text{...(25.9)}$$

where, n is number of experiments or replications and ε_i is positional error for ith experiment and is expressed by distance between the actual points reached by the end-effector (x_i, y_i) and desired point (x, y) respectively and calculated as

$$\varepsilon_i = \sqrt{\left((x_i - x)^2 + (y_i - y)^2\right)} \qquad \text{...(25.10)}$$

(b) Reliability: The reliability (R) of a manipulator is defined as the probability of the end-effector reaching a point or in a close vicinity of it within specified range. If the end effector reaches any point within the specified region, it is considered as a successful experiment (5). Hence, the reliability R is given as,

$$R = \frac{\text{Number of successful experiments}}{\text{Total number experiments}} \qquad \text{...(25.11)}$$

The specified range around a target point is called the permissible error region and its shape and size depends on the intended use of the manipulator. The reliability as performance measure has been used to evaluate the overall performance of all control factors combinations. The results obtained are used for validating the outcome of factorial design to manipulator parameter design.

25.5. SIMULATION METHODOLOGY

The effect of noise is assumed to follow normal distribution with mean zero and standard deviation. After randomly generating noise for each control factor for specified standard deviation, it is incorporated as the effect of noise into all control factors. Subsequently Eqns. 25.3, 25.4, 25.5, 25.6, 25.7 and 25.8 are utilized to obtain accelerations of link 1 and link 2 for the manipulator. To compute the point actually reached by end-effector with the presence of noise factors, a search technique was used as discussed in following section. After conducting similar type of experiments for several times various performance measures are estimated using the output of experiments. The performance measures comsputed are positional error and reliability.

Search Technique

The method discussed above is utilized to compute the angular displacements, angular velocities and angular accelerations of two links with effect of noise included that is $\theta_1^n, \theta_2^n, \theta_1^n, \theta_2^n, \theta_1^n, \theta_2^n$, for link 1 and link 2, respectively. Treating $\ddot{\theta}_1^n$, $\ddot{\theta}_2^n$ as output from the experiment, search for the angular displacements of links θ_1^e and θ_2^e is carried out. These angular displacements are always different from θ_2^n, θ_1^n computed from kinematic model, because while performing search it uses angular accelerations of links. While pursuing the search nominal values of control factor level values are utilized in search to obtain angular displacements of links θ_1^e and θ_2^e. Finally, the coordinates of the point $P(x_i, y_i)$ actually reached by end-effector of manipulator are computed from the computed values of angular displacements using Eqns. 25.1 and 25.2. Later, all the performance measures discussed in Sec. 25.5 are calculated for the manipulator in each experiment.

Simulation

The numerical values used to simulate the performance of 2-DOF RR manipulator and parameters used for experimental run are given below:

Number of levels for each control factor = 2, Values of the six control factors for two levels listed in Table 25.1.

Number of combinations in full factorial design = 2^6

Number of replications for each combination = 8 (for factorial design), 1000 (for reliability).

Coordinates of target point in workspace, x=0.40 m, y = 0.30 m. The control factors, with mean values as level values in Table 25.1 and standard deviations are assumed as below:

$\sigma_{l_1} = 0.0001$ m, $\sigma_{l_2} = 0.0001$ m, $\sigma_{m_1} = 0.01$ kg, $\sigma_{m_2} = 0.01$ kg, $\sigma_{\tau_1} = 0.10$ Nm, $\sigma_{\tau_2} = 0.10$ Nm

Tolerance range selected around target point for computation of reliability $\Delta x = \pm\, 0.0005$ m, $\Delta y = \pm\, 0.0005$ m.

Design matrix indicating different control parameter combinations (2^6) and simulated results are provided in Table 25.4.

Using above numerical values, simulation is been carried out. The simulations are run for 8 replications for factorial design and statistical analysis of results is provided in Table 25.2. After comparing performance measures the optimal results are provided in Table 25.3. The relationship between positional error and the control factors are provided in Appendix 1. To validate the results obtained in factorial design the simulation is run for 1000 cycles and results are provided in Table 25.4. Discussions of results are provided in following section.

25.6. DISCUSSIONS AND ANALYSIS

Although the statistical analysis technique shown here are standard approaches and commonly used in statistical quality control, but their application to robotics seem to be minimal. Factorial designs are widely used in experiments involving several factors where it is necessary to study the combined effect of factors on a response [16]. It allows the effect of a factor to be estimated at several levels of the other factors yielding conclusions that are valid over a range of experimental conditions and more importantly part of this study is Analysis of variance (ANOVA). In ANOVA, statistical test like F-test compares the variance among the treatment means versus the variance of individuals within the specific combination of parameters. High values of F indicate that one or more of the means differ from another. ANOVA technique helps in determining statistically significant contributors to performance variation of manipulator though all control factors participate in simulating the performance.

Target at x = 0.40 m, y = 0.30 m in Workspace

It is important to specify, that for ANOVA, responses utilized for analysis are ε_i. From analysis it was observed that control factors l_1, l_2, m_2, and interacting factors $m_1\tau_1$, $m_2\tau_1$, $l_1l_2m_1\tau_1\tau_2$ are significant terms by comparing computed F-statistic value with tabulated F-statistic value for given factor or interaction factor combination. These results of analysis are displayed in Table 25.2. It indicates all control factors have statistically significant role to play in performance measuer apart from the interactions. So, none of the control factors go out of contention for parameter design. Observing this other performance measures, mean positional error and reliability, are used to find the suitable control factor combinations for which performance is optimal. It can be observed that mean positional error is maximum (2.200282 cm) at combination number 42. Mean positional error is lowest (0.0735486 cm) at combination no. 29. But maximum reliability is observed at 16th and 33rd combination number of control factors.

25.7. CLOSURE

The work presented gives an insight to the use of simulation method for modeling and optimizing the performance of robot manipulators. It illustrates a novel approach to simulate different performance measure of manipulator. Subsequently use of ANOVA technique to determine statistically significant kinematic and dynamic parameters based on the simulated performance measure. The approach discussed will help in determining significant factors responsible for performance variation and selection of the optimum factors rather than spending effort in controlling performance of manipulator.

Table 25.1: Values for control factors at two levels

Control Factors	Low Level	High Level
l_1 (m)	0.40	0.50
l_2 (m)	0.25	0.35
m_1 (kg)	5.5	6.5
m_2 (kg)	4.0	5.0
τ_1 (Nm)	–500	–800
τ_2 (Nm)	–100	–105

Table 25.2: Analysis of variance for factorial design

Source	Sum of Squares	DF	Mean Square	F-Value	Remark
l_1	11.52	1	11.52	3.67	Significant
l_2	47.33	1	47.33	58.72	Significant
m_1	3.154×10^{-5}	1	3.154×10^{-5}	3.913×10^{-5}	
m_2	30.19	1	30.19	37.46	Significant
τ_1	1.02	1	1.02	1.26	
τ_2	2.56	1	2.56	3.18	
l_1l_2	1.13	1	1.13	1.040	
$m_1\tau_1$	3.53	1	3.53	4.38	Significant
$m_2\tau_1$	4.03	1	4.03	5.00	Significant
$l_1l_2m_1\tau_1\tau_2$	3.56	1	3.56	4.41	Significant
Residual	385.27	478	0.81		
Corrected Total	506.35	511			

Table 25.3: Optimal level values of control factors for optimal performance

Performance measure	Values of optimal performance measure	Optimal level values of control factor					
		l_1 m	l_2 m	m_1 kg	m_2 kg	τ_1 Nm	τ_2 Nm
Reliability	0.073	0.50	0.35	5.5	4	–500	–100
Positional error	0.00739	0.40	0.35	6.5	5	–500	–105

Table 25.4: Control factor combinations (Design Matrix) with simulated performance measures

Serial No.	l_1 (m)	l_2 (m)	m_1 (kg)	m_2 (kg)	τ_1 (Nm)	τ_2 (Nm)	Mean positional error in cm $\bar{\varepsilon}$	Reliability(R)
1	0.40	0.25	5.5	4	–500	–100	1.562474	0.032
2	0.40	0.25	5.5	4	–500	–105	1.50228	0.001
3	0.40	0.25	5.5	4	–800	–100	1.402717	0.011
4	0.40	0.25	5.5	4	–800	–105	1.396989	0.021
5	0.40	0.25	5.5	5	–500	–100	1.177036	0.052
6	0.40	0.25	5.5	5	–500	–105	1.30733	0.053
7	0.40	0.25	5.5	5	–800	–100	1.080913	0.054
8	0.40	0.25	5.5	5	–800	–105	1.190189	0.042
9	0.40	0.25	6.5	4	–500	–100	1.573897	0.006
10	0.40	0.25	6.5	4	–500	–105	1.808581	0.011
11	0.40	0.25	6.5	4	–800	–100	1.458038	0.021
12	0.40	0.25	6.5	4	–800	–105	1.646157	0.032
13	0.40	0.25	6.5	5	–500	–100	1.394521	0.033
14	0.40	0.25	6.5	5	–500	–105	1.412347	0.042

(Contd...)

15	0.40	0.25	6.5	5	–800	–100	1.066661	0.023
16	0.40	0.25	6.5	5	–800	–105	1.418603	0.041
17	0.40	0.35	5.5	4	–500	–100	1.140686	0.035
18	0.40	0.35	5.5	4	–500	–105	1.125238	0.042
19	0.40	0.35	5.5	4	–800	–100	1.130304	0.055
20	0.40	0.35	5.5	4	–800	–105	1.115178	0.061
21	0.40	0.35	5.5	5	–500	–100	0.812012	0.019
22	0.40	0.35	5.5	5	–500	–105	0.909786	0.059
23	0.40	0.35	5.5	5	–800	–100	0.800313	0.054
24	0.40	0.35	5.5	5	–800	–105	0.898607	0.062
25	0.40	0.35	6.5	4	–500	–100	1.060104	0.032
26	0.40	0.35	6.5	4	–500	–105	1.108661	0.038
27	0.40	0.35	6.5	4	–800	–100	0.976335	0.052
28	0.40	0.35	6.5	4	–800	–105	1.223492	0.046
29	0.40	0.35	6.5	5	–500	–100	0.735486	0.029
30	0.40	0.35	6.5	5	–500	–105	0.907035	0.043
31	0.40	0.35	6.5	5	–800	–100	0.791805	0.051
32	0.40	0.35	6.5	5	–800	–105	0.806899	0.061
33	0.50	0.25	5.5	4	–500	–100	1.708904	0.040
34	0.50	0.25	5.5	4	–500	–105	2.023467	0.011
35	0.50	0.25	5.5	4	–800	–100	2.197954	0.004
36	0.50	0.25	5.5	4	–800	–105	1.952398	0.022
37	0.50	0.25	5.5	5	–500	–100	1.512813	0.032
38	0.50	0.25	5.5	5	–500	–105	1.419334	0.031
39	0.50	0.25	5.5	5	–800	–100	1.83867	0.011
40	0.50	0.25	5.5	5	–800	–105	1.629347	0.001
41	0.50	0.25	6.5	4	–500	–100	2.025348	0.021
42	0.50	0.25	6.5	4	–500	–105	2.052949	0.007
43	0.50	0.25	6.5	4	–800	–100	1.827379	0.011
44	0.50	0.25	6.5	4	–800	–105	2.200282	0.021
45	0.50	0.25	6.5	5	–500	–100	1.761378	0.032
46	0.50	0.25	6.5	5	–500	–105	1.587568	0.033
47	0.50	0.25	6.5	5	–800	–100	1.621176	0.004
48	0.50	0.25	6.5	5	–800	–105	1.638508	0.020
49	0.50	0.35	5.5	4	–500	–100	0.981427	0.073
50	0.50	0.35	5.5	4	–500	–105	1.263425	0.021
51	0.50	0.35	5.5	4	–800	–100	1.167455	0.019
52	0.50	0.35	5.5	4	–800	–105	1.12552	0.021
53	0.50	0.35	5.5	5	–500	–100	0.854406	0.061
54	0.50	0.35	5.5	5	–500	–105	0.846924	0.050
55	0.50	0.35	5.5	5	–800	–100	0.925765	0.051
56	0.50	0.35	5.5	5	–800	–105	0.902866	0.063
57	0.50	0.35	6.5	4	–500	–100	1.13169	0.022
58	0.50	0.35	6.5	4	–500	–105	1.237134	0.041
59	0.50	0.35	6.5	4	–800	–100	1.236303	0.020
60	0.50	0.35	6.5	4	–800	–105	1.149676	0.049
61	0.50	0.35	6.5	5	–500	–100	1.017523	0.061
62	0.50	0.35	6.5	5	–500	–105	0.997188	0.041
63	0.50	0.35	6.5	5	–800	–100	1.002658	0.039
64	0.50	0.35	6.5	5	–800	–105	1.018837	0.011

REFERENCES

1. B., Benhabib, R.G. Fenton, and A.A., Goldenberg, "*Computer Aided Joint Error Analysis of Robot Joints*", IEEE Journal of Robotics and Automation, Vol. RA–3, 1987, 317–323.
2. J., Chen and L.M., Chao, "*Positioning Error Analysis for Robot Manipulators with all Rotary Joints*", IEEE Journal of Robotics and Automation, 1987, pp. 535–545.
3. M.R., Driels and W.E., Swayze, "*Automated Partial Pose Measurement System for Manipulator Calibration Experiments*", IEEE Transactions on Robotics and Automation, 1993, Vol. 10, pp. 430–440.
4. P.K., Bhatti and S.S., Rao, Reliability Analysis of Robot Manipulators, ASME Journal of Mechanisms, "*Transmissions and Automation in Design*", 1988, Vol. 110, pp. 175–181.
5. S., Manoochehri and A.A., Seireg, "*A Computer Based Methodology for the Form Synthesis and Optimal Design of Robot Manipulators*", Journal of Mechanical Design, 1990, Vol. 112, pp. 501–507.
6. S., Azadivar, The Effect of Joint Position Errors of Industrial Robots on their Performance in Manufacturing Operations, "*IEEE Journal of Robotics and Automation*", 1987, Vol. 2, pp. 109–115.
7. J.I., Imura, Y., Yokohji, and T., Youshokawa, "*Robust Control of Robot Manipulator Based on Joint Torque Sensor Information*", International Journal of Robotic Research, 1994, Vol. 13, pp. 432–442.
8. K., Deol, and D.K., Pai, "*Performance Measure for Robot Manipulators: A Unified Approach*", International Journal of Robotic Research, 1996, Vol. 15, pp. 92–111.
9. J.Y.S., Luh, M.W., Walker and R.P.C., Paul, Resolved Acceleration Control of Mechanical Manipulators, "*IEEE Transaction on Automatic Control*", 1980, Vol. AC-25, pp. 468–474.
10. O.F., Offodile and K., Ugwu, Evaluating the Effect of Speed and Payload on Robot Repeatability, "*Robotics and Computer Integrated Manufacturing*", 1991, Vol. 8, pp. 27–38.
11. A., Parkinson, S., Sorenson, and N., Pourhassan, A General Approach for Robust Optimal Design, "*ASME Journal of Mechanical Design*", 1993, Vol. 115, pp. 74–80.
12. W., Chen, J.K., Allen, K.L., Tsui and F., Mistree, "*A Procedure for Robust Design: Minimizing Variation Caused by Noise Factors and Control Factors*", Transactions of the ASME Journal of Mechanical Design, 1996, Vol. 118, pp. 478–485.
13. B.K., Rout and R.K., Mittal, A Review on Minimizing Variability in Product Performance with the Help of Computational Tools by Controlling Active and Noise Factor for Robust Design, Proceedings of National Conference on Current Applications of "*Computers in Design Engineering*", at J.N.V. University, Jodhpur, India, 2001, 13–23.
14. B.K. Rout and R.K. Mittal, Parametric Robot Manipulator Design Optimization Using Taguchi Method, Proceeding of 48th Congress Indian Society for "*Theoretical and Applied Mechanics*", 2003, held at IIT Guahati, India, 15–27.
15. R.K. Mittal and I.J. Nagrath, "*Robotics and Control*", 2003, Tata McGraw Hill Publication, New Delhi.
16. D.C., Montgomery, "*Design and Analysis of Experiments*", 2001, John Wiley & Sons, New York.

APPENDIX 1

Equation A1 for positional in terms of coded control factors:

$$\varepsilon_1 = 1.32 + 0.15\, l_1 - 0.30\, l_2 - 2.482\times10^{-4}\, m_1 - 0.24\, m_2 + 0.045\, \tau_1 - 0.071\, \tau_2 - 0.047 l_1 l_2 + 9.893\times10^{-3} l_1 m_1 - 0.037\, l_1 \tau_1 + 0.013\, l_1 \tau_2 + 0.077\, l_2 m_1 - 0.016\, l_2 \tau_1 + 0.036\, l_2 \tau_2 - 0.083\, m_1 \tau_1 + 5.710\times10^{-3} m_1 \tau_2 - 0.089\, m_2 \tau_1 - 0.046\, \tau_1 \tau_2 - 0.031\, l_1 l_2 m_1 - 8.274\times10^{-4} l_1 l_2 \tau_1 - 0.017\, l_1 l_2 \tau_2 - 0.042\, l_1 m_1 \tau_1 + 0.034\, l_1 m_1 \tau_2 + 0.021\, l_1 \tau_1 \tau_2 + 0.056\, l_2 m_1 \tau_1 - 0.010\, l_2 m_1 \tau_2 + 0.058\, l_2 \tau_1 \tau_2 + 0.071\, m_1 \tau_1 \tau_2 + 0.025\, l_1 l_2 m_1 \tau_1 - 0.024\, l_1 l_2 m_1 \tau_2 - 4.077\times10^{-3} l_1 l_2 \tau_1 \tau_2 + 0.053\, l_1 m_1 \tau_1 \tau_2 - 0.014\, l_2 m_1 \tau_1 \tau_2 - 0.083\, l_1 l_2 m_1 \tau_1 \tau_2 \quad \text{(A1)}$$

SECTION - III

MANUFACTURING SYSTEMS

26

CASE STUDY: CAE SET UP FOR CRASH SIMULATION

26.1. INTRODUCTION

Finite Element Analysis (FEA) in automotive design has become a standard tool in the recent decades. Besides gaining time in automotive product development because of fewer prototypes, CAE engineer receives an overview of the product behavior through FEA simulation. Now, there is a need to reduce the simulation development cycle time and increase the quality and consistency of the process. This requires a robust automated workflow of the process. An attempt is made here in this direction to reduce the preprocessing time and to eliminate the repetitive work, for occupant/crash simulation using Altair HyperWorks. Altair has developed a HyperWorks Process Manager Template to automate the numerical simulation as per the Federal Motor Vehicle Safety Standards (FMVSS) regulation standards. An easy to use process automation template has been developed with user friendly graphical user interface.

Finite Element Analysis (FEA) has gone a long way in automotive design and has become a standard tool. This tool has integrated very well into the design process. Better understanding of this mathematical tool and availability of computational facilities at a relatively lower cost helps design engineer to perform multiple numerical simulations. Now, the results of numerical simulation have reached a level of accuracy that is comparable to physical testing. CAE also provides tools to help improve the maturity of the design by simulating the design and avoiding or minimizing the need to build and test prototypes. Companies worldwide recognize the need to integrate computer-aided engineering (CAE) into their product development process as quickly as possible. In the automotive industry FEA has been integrated so deeply in the design process to the extent that CAE drives the design process. Applications of FEA in automotive design include many disciplines like internal and external aerodynamics, static and dynamic structural analysis, crashworthiness design and occupant safety. By simulating important characteristics, the design can be debugged and precious development time reduced. CAE is not just a validation tool. CAE will lead the design process.

Current trends in virtual product development:

- Increased use of CAE simulation techniques – reduces development time
- CAE toolset standardization and CAE process automation
- Batch CAE processing
- Robust design
- Structural optimization
- Engineering data management.

A CAE task involves a series of steps that an analyst performs:

- Import design geometry
- Build CAE model
- Set up loads and boundary conditions
- Run analyses
- Interrogate results and recommend a better design.

In the design departments of major automotive manufacturers hundreds of engineers perform thousands of Finite Element Analysis in the course of product development. During the optimization of a component or the complete assembly very similar configurations have to be analyzed many a times. In this scenario a strong demand to automate the FEA process arises [1].

26.2. PROCESS AUTOMATION OF CAE PROCESS

Companies invest time to develop CAE methods that provide the best results. These best practices can be captured using Process automation and distributed throughout an organization. Process automation lets companies implement standard CAE procedures for a variety of CAE problems.

CAE process automation automates the routine FE modeling, job submission, and analysis results processing tasks. It significantly improves engineers' productivity, and quality of work. It allows CAE engineers to respond much faster to analysis requests, and satisfy program-timing requirements. It provides a common basis for continuous improvements. Also, it allows CAE engineers to have more time and use better FE models to develop design recommendations. CAE process automation has a predefined workflow, and uses customized modules to perform routine functions.

Process Automation when applied to a CAE process has the following advantages:

- Standardize CAE processes
- Minimize CAE analysis set-up time
- Rerun previously-conducted CAE processes
- Reduces time and mouse clicks needed to execute series of operations
- Provide integration with CAD, PDM systems, databases, intranets, and job submission systems
- Streamline CAE activities
- Establish and distribute "Best Practice" processes corporate wide.

Many users of FEA have made efforts in the last years to automate their processes either by developing process automation software in house or together with partners from software industry.

26.3. CAE AUTOMATION TOOLS AT ALTAIR

Batch Mesher

- Significant reduction in time and labor to complete meshes by eliminating manual meshing.
- Distribute meshing jobs on work load management tool, Altair ® PBS Pro ®.
- Store meshes back in data management tools.

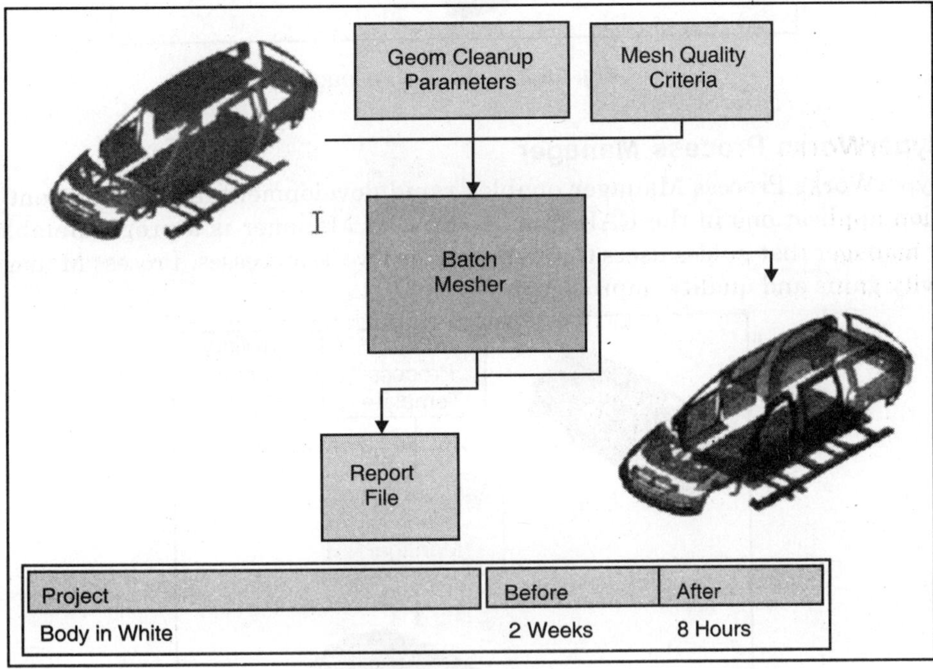

Fig. 26.1: Batch mesher

Assembly Manager

- Imports Bill of Materials (BOM) and connection information.
- Imports and positions meshed parts and assemblies.
- Exports fully assembled models.
- Store meshes back in data management tools.

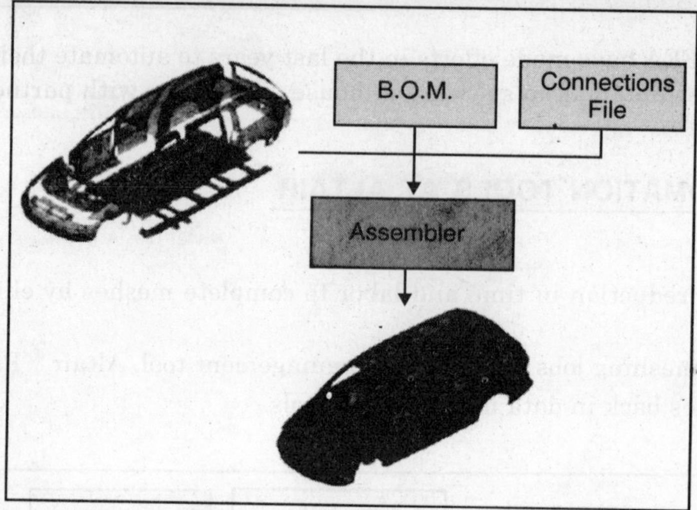

Fig. 26.2: Assembly manager

Altair HyperWorks Process Manager

Altair HyperWorks Process Manager enables rapid development and deployment of process automation applications in the CAE domain. Process Manager is a programmable personal workflow manager that guides users through standard work processes. Process Manager delivers productivity gains and quality improvements.

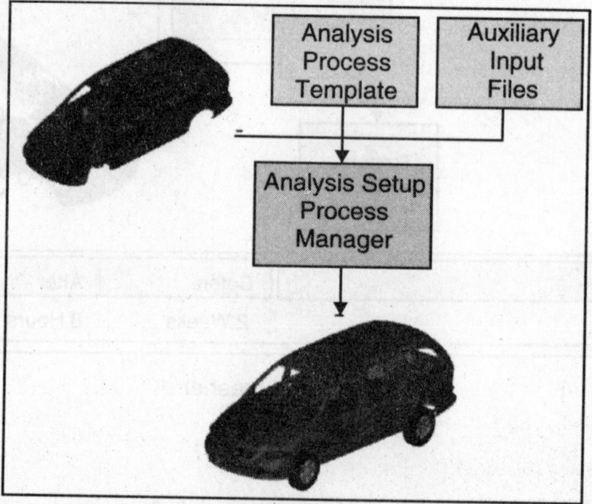

Fig. 26.3: Altair hyperWorks process manager

User interface features

- Process Manager client is integrated with the HyperWorks desktop
- The "Process Tree" displays a series of steps the user executes
- A simple checkbox is used to mark completed steps

- The interface provides the ability to re-execute an individual or series of steps automatically
- Each process can be run in an interactive or automatic mode
- Break-points can be set to stop at any step for user-input if/when appropriate
- The state of the project can be saved to a persistent file
- Process studio is available for authoring Process Manager templates.

26.4. CASE STUDY OF PROCESS AUTOMATION: CAE MODEL SET UP FOR CRASH SIMULATION USING ALTAIR HYPERWORKS

Virtual Impact Template

Virtual Impact Template was developed to automate the crash test analysis process. Current Federal Motor Vehicle Safety Standards (FMVSS) have been incorporated into the virtual analysis process. Individual corporate standards can also be included in this process. This tool provides a step-by-step approach to conducting a simulated crash test using automated wizards. The wizards will guide the user through the steps of selecting which event(s) to run, which Finite Element (FE) models to use in the events and the entry of other parameters such as key locations, measurement device locations, etc.

Using HyperWorks Process Manager, commercially available software, a customized process automation template is developed. The customized interface guides the CAE Analyst through the process while automating the steps in importing the meshed model, creation of component assemblies, importing and position of dummy based on the selected reference position, seat belt routing, creation of accelerometers, contacts and control cards, exporting the deck, and finally post-processing of results. The automated solution results in significant productivity gains while improving repeatability and reliability of the analysis [2].

Features of virtual impact template

- Basic file management through directory structure creation. Wherever possible, reusable data will be stored into the current working project directory for use with other test modes that require the same input data.
- Configuration files which helps in standardization of the CAE process.
- Selection tool (i.e., desired test mode, vehicle model).
- Automated import of FE model.
- Automated model check with report generation.
- Automatically remove any existing initial or undesirable boundary conditions.
- Tools to simplify the grouping of key components into groups.
- Tools to simplify the creation of accelerometer and section force output requests.
- Automated positioning of moving barrier, dummy models, and impactors.
- Automated creation of contacts and initial conditions.

Benefits of using virtual impact template

- Time and effort savings
- Automatic storage of analysis runs and results
- Standardized process and establish best practices
- Increased reliability and repeatability
- Enables consistent result output
- Reduces the time spent on report generation and formatting.

Crash Simulation using Altair HyperWorks Process Manager

Process methodology: The CAE model set up for frontal impact analysis is typically considered as complex among others since it involves the vehicle model, dummy positioning, seat belt routing and barrier positioning along with other typical steps. The complexity of the analysis can be overcome with quick turnaround times by employing process automation. Much of the work can be automated through the Tcl/Tk programming interface of HyperWorks Process Manager. In Fig. 26.4, the various steps involved in setting up a CAE model as per FMVSS208 are captured as tasks in HyperWorks Process Manager Tree [3].

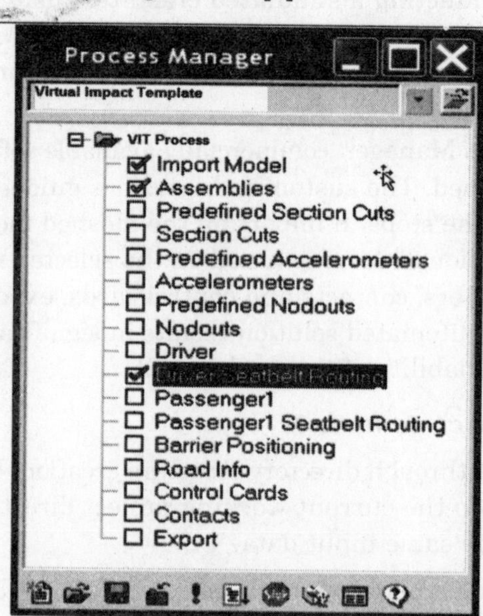

Fig. 26.4: Overview of the frontal crash impact (FMVSS208) simulation process

Configuration files (xml): Configuration files which helps in standardization of the CAE process is basically an xml file where the different standards and defaults can be set by the user. Some of the salient features of the configuration files have been listed below:

- The uniform units to be followed throughout the process can be defined in the configuration file.
- Assemblies can be predefined in the configuration file and they get created automatically when assemblies task is made active for the very first time in the process.

- A predefined section cut can also be defined from the configuration file. The assembly on which the section cut has to be made, the location of the cut or the start and end location will also come from the configuration file. Similarly, predefined accelerometers or nodouts can also be defined.
- Barrier parameters like barrier types, gap and offset % are all configurable from the xml file.

Sample GUI snap shots of Virtual Impact Template

The dummy positioning and the seat belt routing are among the tasks that consume a lot of time and effort of a crash analyst. So from the Virtual Impact Template only these two tasks have been picked and their features have been highlighted as below:

Importing and positioning the dummy

- The dummy is imported to the H-point as provided by the OEMs. User can fine tune the dummy if necessary. Optionally the contact definition can be created between the dummy and the seat, steering wheel and instrument panel during the time of import through a simple check button as shown below (Fig. 26.5).

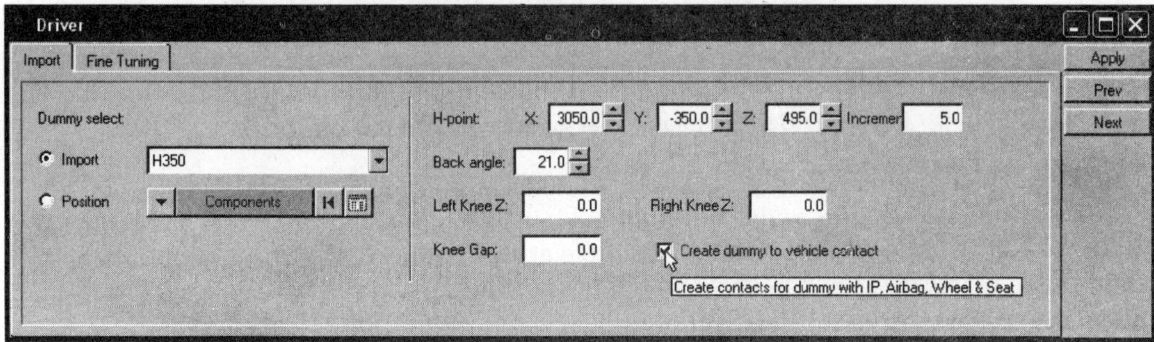

Fig. 26.5: Dummy import panel

- The user can move the knee of the dummy to represent the actual driving position. The tool internally identifies the knee assembly and performs the required translation or rotation to achieve the desired position (Fig. 26.6).

Fig. 26.6: Dummy fine tuning panel

Seat belt routing: In this step the seat belt is routed from the anchorage point in the common anchorage zone over the dummy torso to the retractor point and then to the lower anchorage point over the dummy lap.

- If the xml file has been configured with the OEM specific or Regulations specific seat belts the list box shown below gets automatically updated when the tool is loaded.
- Both 3 point and 4 point belt routing is supported. Once the user selects a particular configuration, the picture on the right gets automatically updated as shown in Fig. 26.7. This serves as a guideline for the analyst.

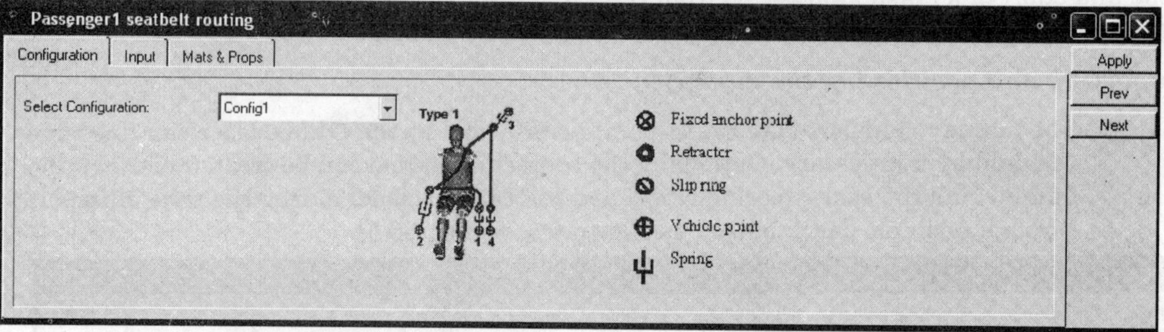

Fig. 26.7: Configuration panel for the seat belt routing

- From the input page of this task (Fig. 26.8) the user can create 1D/2D belts, specify the element size, gap, width and the end length. Once the user selects the required nodes on vehicle and the dummy, a simple apply button creates the appropriate belt.
- The user has the option of an interactive belt creation where he can pick and pull the auto created belt to the desired orientation.

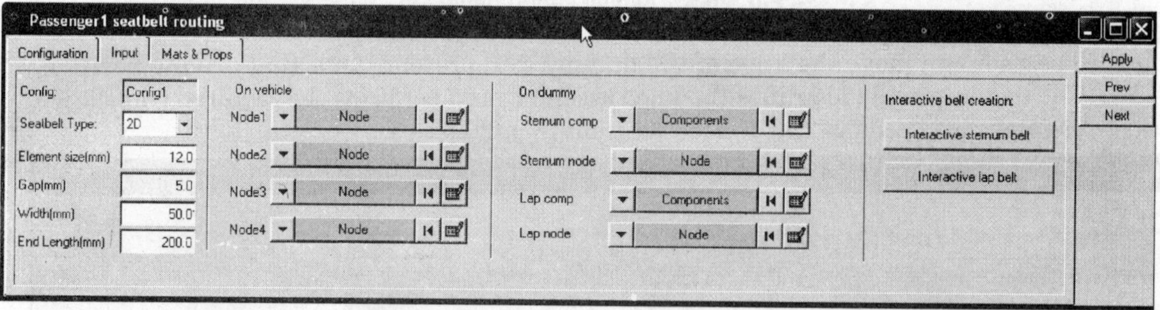

Fig. 26.8: Input panel for the seat belt routing

- From the Mats and Props page of this task (Fig. 26.9) the belt material properties like the loading and unloading curves, damping coefficient etc. can be specified. These properties get automatically associated with the seat belt during its creation.

This process can be repeated as many times as possible by changing the anchorage point location within the acceptable anchorage zone.

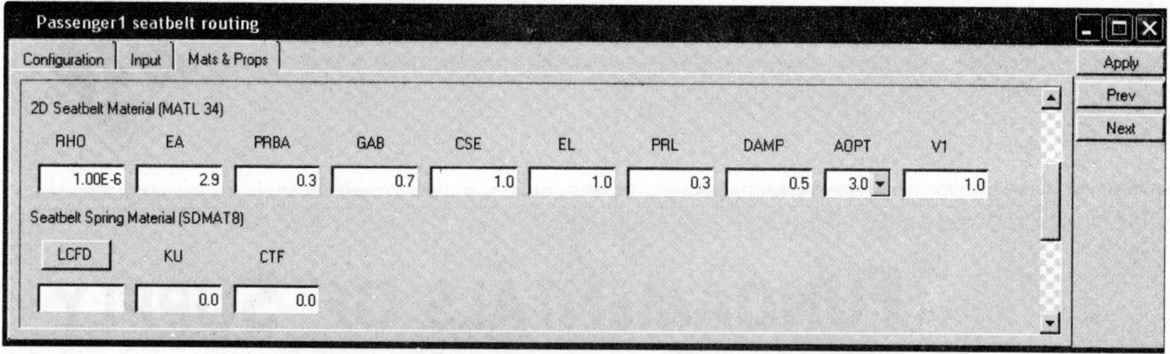

Fig. 26.9: Material and properties panel for the seat belt

26.5. CLOSURE

Table 26.1: Results for crash simulation test using virtual impact template

Project	Before	After	Reproducible quality
Frontal Impact FMVSS 208	40 hours	8 hours	100 %
Head Impact FMVSS 201	4–6 hours	15 min	100 %
Batch meshing of Body in White (BIW)	2 weeks	8 hours	100 %

This HyperWorks Process Manager template helps the designers to automatically set up the CAE model as per FMVSSS208 regulation. It also enables Best Practices through Standardising CAE processes and minimizing CAE analysis set-up time to check for the compliance with respect to FMVSS208 regulation.

BENEFITS SUMMARY

This Process Manager template delivers productivity gains and quality improvements through:

- Standard process flow
- Reduced manual input and duplicate data entry
- Checks for input validity
- Maximized resource usage.

REFERENCES

1. Richard A., Wilson, "*A Review of Vehicle Impact Testing*": How it began and what is being done, SAE Technical papers, Document Number 700403, pp. 834–850.
2. Federal Motor Vehicle Safety Standards, Crashworthiness, Occupant crash protection, Standard 208 (FMVSS208 regulation), part 571.
3. Altair HyperWorks 7.0 suite manuals.

FUNDAMENTALS OF SUPPLY CHAIN MANAGEMENT

27.1. INTRODUCTION

Intense competition in today's economy, the shrinking life cycles of products, superior product quality, excellent customer service, expanded variety, and high expectations of customers have forced business enterprises to focus their attention on correctly controlling their supply chains. Coupled with the continuing advancement in rapid information exchange, these have motivated the ceaseless development of supply chain knowledge and techniques to manage it. In a typical supply chain, raw materials are procured, products are manufactured at one or more factories, shipped to warehouses for intermediate storage and then shipped to retailers or customers. So the supply chain consists of suppliers, manufacturing centers, warehouses, distribution centers and retail outlets as well as raw materials, work in progress inventory and finished products that flow between facilities. The effective management of this chain so that product is produced and distributed at the right time, at right quantities and to the right locations in order to minimize the total system wide cost while satisfying service level requirements is known as supply chain management. Therefore, in today's markets, no business can be successful without mastering the issues, problems and possibilities in managing supply chains. That is why SCM has become an important topic of discussion among managers and academicians alike.

27.2. EVOLUTION OF SCM

The literature suggests that SCM has its roots in the evolutionary path followed through materials management and physical distribution after World War II, functional logistics (different mangers for all functions), and integrated logistics (one manger for all functions) (Ganeshan *et al.* [1]. Forrester [2] justified the first step beyond functional logistics by using a system analysis approach to describe the forces that determine growth, fluctuation, and decline. He developed a complete company model that describes the flows of information, materials, manpower, capital, equipment and money. Bowersox [3] discussed the evolution of integrated logistics and touches upon what will become known as the supply chain. He states that related responsibilities seldom terminate when product ownership transfer occurs and that firms are linked together in cooperative vertical marketing systems providing total supply chain performance. Langley [4] suggested four stages of development: (*i*) cost control, (*ii*) profit center orientation recognizing the positive impact on sales, (*iii*) view of logistics as key to product differentiation, and (*iv*) as a principal strategic advantage. Masters and Pohlen [5] described the evolution of logistics into three phases: (*i*) functional management (1960–70) – functions such as purchasing, shipping and distribution are each managed separately, (*ii*) internal integration (1980s)–the management of supply chain functions of a single facility are unified and become the responsibility of a single individual and (*iii*) external integration (1990s)–the management of supply chain functions throughout the chain are unified requiring cooperation and co-ordination between links in supply chain. Lalonde [6] described the evolution of integrated logistics in three phases (*i*) physical distribution – the distribution of the goods is all that needs to be managed by a logistics manger, (*ii*) internal linkages – it is important for the logistics manager to control both internal supply functions and physical distribution and (*iii*) external linkages–logistics management requires cooperation in management with upstream and downstream entities in order to maximize the benefits of the total logistics system.

27.3. DEFINITIONS OF SCM

Tsay et al. [7] defined a supply chain as two or more parties linked by a flow of goods, information and funds. Many definitions of the concept of supply chain (SC) have been provided in the literature. Mabert and Venkataraman [8] discussed different views on supply chains i.e. (*i*) the "relational" activities between a buyer and seller, (*ii*) including all "upstream" suppliers, and (*iii*) a "value chain" approach, in which all activities required to bring a product to the market-place are considered part of the supply chain". Stevens defined a supply chain as a connected series of activities which is concerned with planning, coordinating and controlling materials, parts, and finished goods from supplier to customer. It is concerned with two distinct flows (material and information) through the organization" [9]. SCM has been defined in various ways (see Table 27.1), many of which define the same issue in different words. Some commonalties can be distinguished and are given below:

- Embracing an integral perspective across the supply chain
- Enhancing customer responsiveness
- Adopting flow management of materials and information
- Optimal co-ordination and configuration of the supply chain process
- Minimizing the total system wide costs along the supply chain
- Rejection of intra- and inter-organizational boundaries
- Achieving partnership arrangements

Table 27.1: Definitions of supply chain management

Authors	Definition
Jones and Riley [10]	SCM is an integrative approach to deal with the total flow of materials from suppliers through end users.
Goldratt and Cox [11]	SCM is defined as an enabler, a tool to achieve a company's goal.
Houlihan [12]	SCM covers the flow of goods from supplier through manufacturer and distributor to end-user.
Ellram and Cooper [13]	SCM is an integrative philosophy to manage the total flow of a distribution channel from the supplier to the ultimate user.
Ellram [14]	SCM is defined as an integrative approach to dealing with the planning and control of the materials flow from suppliers to end-users.
Novack and Simco [15]	SCM covers the flow of goods from the supplier through the manufacturer and distributor to the end user.
Scott and Westbrook [16]	SCM is used to refer to the chain linking each element of the production and supply process from raw materials through to the end customer.
Lee and Billington [17]	SCM is the networks of manufacturing and distribution sites that procure raw materials, transform them into intermediate and finished products, and distribute the finished products to customers.
Cooper and Ellram [18]	SCM is an approach whereby the entire network from which suppliers through the ultimate customer, is analyzed and managed in order to achieve the "best" outcome for the whole system.
Davis [19]	SCM is an approach [that] will enable a manufacturing operation to better manage its supply chain, ultimately improving customer satisfaction levels while reducing overall costs.
Lamming [20]	SCM is a networking approach to value chain optimization.
Martha and Lisa [21]	SCM is an integrating philosophy to manage the total flow of a distribution channel from supplier to the ultimate customer.
Turner [22]	SCM is a technique that looks at all the links in the chain from raw material suppliers through various levels of manufacturing to warehousing and distribution to final customer.
Berry et al. [23]	SCM aims at building trust, exchanging information on market needs, developing new products, and reducing the supplier base to a particular original equipment manufacturer [OEM] so as to release management resources for developing meaningful, long-term relationship.
Johannson [24]	SCM requires all participants of the supply chain to be properly informed. With SCM, the linkage and information flows between

	various members of the supply chain are critical to overall performance.
Metz [25]	SCM is defined as the coordination or integration of a series of activities/processes, which procure, produce, and deliver products or services to customers.
Lalonde and James [6]	SCM is the delivery of enhanced customer and economic value through synchronized management of the flow of physical goods and associated information through sourcing to consumption.
Berry [26]	SCM is a system whose constituent parts include material supplies, production facilities, distribution services and customer linked together by feed forward flow of materials and feedback flow of information.
Slack et al. [27]	SCM is the concept of strategically coordinating supply chains.
Walton and Miller [28]	SCM is the strategic integration of trading partners.
Giunipero and Brand [29]	SCM is a strategic tool used to enhance overall customer satisfaction that is intended to improve a firm's competitiveness and profitability.
Johnson and Wood [30]	SCM is a process of strategically managing the movement and storage of materials, parts, and finished inventory from suppliers through the firm and on to the customers.
Lamming [31]	SCM is the extent of vision required on the part of managers with respect to their value-adding system, and the influence they must seek to have over it.
Thomas and Griffin [32]	SCM is the management of material and information flows in and between facilities such as vendors, manufacturing and assembly plants and distribution centers.
Harland [33]	SCM is defined as managing business activities and relationships [i] internally within an organization, [ii] with immediate suppliers, [iii] with first and second-tier suppliers and customers along the supply chain and [iv] with the entire supply chain.
Bechtel and Yayaram [34]	SCM focuses attention on the interactions of channel members to produce an end product/service that will provide value for the end user.
Caldwell and Down [35]	SCM is defined as managing value across organizational boundaries, which mandates new forms and new ways of managing inter-corporate relationships.
Cooper et al. [36]	SCM is the integration of all key business processes across the supply chain.
Cooke [37]	SCM is the coordination and integration of all activities associated with moving goods from the raw materials to the end users, for sustainable competitive advantage. This includes systems

Contd...

Table 27.1: Contd...

Authors	Definition
	management, sourcing, production scheduling, order processing, inventory management, transportation, warehousing, and customer service.
Kopczak [38]	SCM is the set of entities including suppliers, logistics services providers, manufacturers, distributors and resellers, through which materials, products and information flow.
Lee and Ng [39]	SCM is the network of entities that starts with the supplier's supplier and ends with the customer's customer involved in the production and delivery of goods and services.
Rich and Hines [40]	SCM is centered on organizational restructuring and extends to the development of a company-wide collaborative culture but also embraces a strong sense of the integration of all activities, which control the timing, and synchronization of material flows.
Towill [41]	SCM is an integrating process based on flawless delivery of basic and customized services.
Alber and William [42]	SCM is used to deliver products and services from raw materials to end customers through engineered flow of information, physical distribution, and cash by global network.
Christopher [43]	SCM is the management of upstream and downstream relationships with suppliers and customers to deliver superior customer value at less cost to the supply chain as a whole.
Brown [44]	SCM is a complex, dynamic network or system of interconnected and interdependent individuals, groups, companies, organizations, and relationships whose goal is to satisfy and add value to their particular customer.
Lambert et al. [45]	SCM is the integration of key business processes from end user through original suppliers that provides products, services, and information that add value for customers and other stakeholders.
Mabert and Venkataramanan [8]	SCM is the network of facilities and activities that performs functions of product development, procurement of material from vendors, the movement of materials between facilities, the manufacturing of products, the distribution of finished goods to customers, and after-market support for sustainment.
Ross [46]	SCM is a continuously evolving management philosophy that seeks to unify the collective productive competencies and resources of business functions found both within the enterprise and outside in the firm's allied business partners located along interesting supply channels into a highly competitive, customer enriching supply system focused on developing innovative solutions and synchronizing the flow of market place, products, services and information to create unique, individualized sources of customer value.

Contd...

Tan et al. [47]	SCM encompasses materials/supply management from the supply of basic raw materials to final product [and possibly recycling and re-use]. SCM focuses on how firms utilize their suppliers' processes, technology and capability to enhance competitive advantage. It is a management philosophy that extends traditional intra-enterprise activities by bringing trading partners together with the common goal of optimization and efficiency.
Tyndall et al. [48]	SCM is the coordinated flow of materials and products across the enterprise and with trading partners. It also includes the management of information flow, cash flow and process/work flow.
Hoek [49]	SCM is characterized by control based on networking and integration of processes across functional, geographical, and organizational interfaces.
Akkermans et al. [50]	SCM increases customer service and profitability through coordination/integration of multiple echelons, processes, and functions like suppliers, purchasing, manufacturing, distribution, marketing/sales, and customers.
Beamon [51]	SCM is an integrated process where raw materials are transformed into final products then delivered to customers.
Hicks [52]	SCM is systematic effort to provide integrated management to meet customer needs and expectation from the suppliers of raw materials through manufacturing to end-customers.
Ayers [53]	SCM is more than the physical movement of goods from earth to earth. It is also information, money movement, and the creation and deployment of intellectual capital.
Ballou et al. [54]	SCM involves all activities associated with the transformation and flow of goods and services, including their information flows, from sources of raw materials to end-users. For coordination to continue there is a need for metrics that can identify and capture chain wide benefits and costs, information sharing mechanism to distribute this data among chain members, and an allocation mechanism for redistributing the rewards of collaboration.
David et al. [55]	SCM is a set of approaches to efficiently integrate suppliers, manufacturers, warehouses, and stores, so that merchandise is produced and distributed at right quantities, to the right locations, and at the right time, in order to minimize system-wide costs while satisfying service level requirements.
Hopp and Spearman [56]	SCM is defined as the overall system wide coordination of inventory stocks and flows.
Chopra and Meindl [57]	SCM involves the management of flows [i.e. information, product and funds] between and among the stages in supply chain to maximize total profitability.

Contd...

Stanley and Gregory [58]	SCM is the collaborative effort of multiple channel members to design, implement and manage seamless value added processes to meet the real needs of the end customer. The development and integration of people and technological resources as well as the coordinated management of materials, information, and financial flows underlie successful supply chain integration.
John [59]	SCM is defined as the systematic, strategic coordination of the traditional business functions within a particular company and across businesses within the supply chain, for the purpose of improving the long-term performance of the individual companies and supply chain as whole.
Norina and Bailey [60]	SCM is defined as an integration of each supply chain member's organizational activities in order to achieve particular objectives through achieving system-wide objectives.
Sheikh [61]	SCM is defined as the planning, design, and control of the flow of information, materials, and money along the supply chain in order to meet customer requirements in an efficient manner, now and in the future.
Charles and Bauer [62]	SCM refers to the methods, systems and leadership that continuously improve an organization's integrated processes for product and service design, sales forecasting, purchasing, inventory management, manufacturing or production, order management, logistics, distribution and customer satisfaction.
Michiel et al. [63]	SCM is a systems approach to managing the entire flow of information, materials, and services from raw materials suppliers through factories and warehouses to the end customer.
Monczcka et al. [64]	SCM is the integration of all activities associated with the flow and transformation of goods from the raw materials stage [extraction], through to end users, as well as the associated information flows through improved supply chain relationships to achieve a sustainable competitive advantage.
Gaither and Frazier [65]	SCM involve all management functions related to the flow of materials from company's direct suppliers to its direct customers, including purchasing, warehousing, inspection, production, materials handling, and shipping and distribution.
Coyle et al. [66]	SCM can be viewed as a pipeline or conduit for the efficient and effective flow of products/materials, services, information, and financials from the supplier's suppliers through the various intermediate organizations/companies out to the customer's customers or the system of connected logistics networks between original vendors and ultimate final customer.
Handfield and Nichols [67]	SCM is the integration and management of supply chain organizations and activities through cooperative organizational relationships, effective business processes, and high levels of information sharing to create high-performing value systems that provide member organizations a sustainable competitive advantage.

27.4. SUPPLY CHAIN MANAGEMENT ANTECEDENTS

Mentzer et al. [68] defined supply chain orientation (SCO) as "the recognition by a company of the systemic, strategic implications of the activities and processes involved in managing the various flows in a supply chain". He also identified eight antecedents to SCM (i.e. trust, commitment, interdependence, organizational compatibility, vision, key processes, leader, and top management support.). The first two, trust and commitment are widely recognized as necessary preconditions to any successful long-term relationship [69]. The third antecedent identified by Mentzer et al. [68] is interdependence. Companies that do not recognize some level of dependence on other companies in their supply chain have little motivation to work with those companies. The next two antecedents are organizational compatibility and vision. The overriding theme is that firms in a supply chain should have similar operating philosophies and corporate cultures and visions [36], as well as decision-making style [70]. A shared vision among all member firms of a supply chain provides each firm with the directions, objectives and strategies in how to manage the supply chain [68]. To work together successfully, the member firms must have similar operating philosophies and corporate cultures. The sixth antecedent is identifying key processes. Member firms should agree on the key processes of the supply chain [45] from which to begin to integrate. The identification of the key processes is what is crucial as those key processes become successfully managed at a supply chain level, and then other processes can be included. As with any initiative, SCM requires a supply chain leader [68, 71]. One firm will play the key role in coordinating and overseeing the entire supply chain. Several factors may determine the supply chain leader, for example, size or economic power [72]. Top management support is necessary and vital to sustain any long-term initiative and is the last antecedent. Resources originate from top management. Lack of support translates into lack of resources, which result in a failed initiative.

27.5. SUPPLY CHAIN DRIVERS

The supply chain drivers play a significant role in supply chain management's ability to support a firm/organization's competitive strategy to maintain the competitive advantage. Chopra and Meindl [57] identified three logistical (facilities, transportation and inventory) and three cross functional drivers (sourcing, pricing and information) for supply chain management. These drivers interact with each other to determine the supply chain's performance in terms of responsiveness and efficiency. If we think of inventory as what is being passed along the supply chain and transportation as how it is passed along, then facilities are the where of the supply chain. The major components of logistical drivers decisions are: location and capacity for *facilities*; cycle inventory, safety inventory, seasonal inventory and level of product availability for *inventory* and design of transportation network and choice of transportation mode for *transportation*. The major components of cross functional drivers decisions are: push versus pull, coordination and information sharing, forecasting and aggregate planning and enabling technologies for *information*; in-house or outsource, supplier selection and procurement for *sourcing* and pricing and economies of scale, everyday low pricing versus high-low pricing and fixed price versus menu pricing for *pricing*.

27.6. SUPPLY CHAIN MODELS

Considering a broad spectrum of the supply chain concept, there may be various classification schemes to categorize supply chain models. The taxonomy of supply chain based on mathematical models can be classified as: deterministic and stochastic [73, 74, 75, 76]. As noted by Budnick, et al. [75] Silver [77] and Zipkin [78], some supply chain models based on inventory theory and simulation contain both deterministic and stochastic elements and consequently should be treated as hybrids. Therefore, a hybrid model to the category is added. Another category called 'IT-driven models' was added to the taxonomy to reflect the current advances in IT for improving supply chain efficiency. These categories are somewhat different from the taxonomy developed by Beamon [73] who did not report the evolution of IT-driven models. To elaborate, the supply chain models are classified into four major categories: deterministic (non-probabilistic); stochastic (probabilistic); hybrid and IT-driven.

Deterministic Models

One of the earliest efforts to create an integrated supply chain models dates back to Geoffrion and Graves [79] and Glover et al. [80]. Geoffrion and Graves [79] introduced a multi-commodity, logistics network design model for optimizing finished product flows from plants, to the distribution centers to the final customers. Glover et al. [80] developed a computer-based production, distribution, and inventory (PDI) planning system that integrated three supply chain segments comprised of supply, storage/location, and customer demand planning. The core of the PDI system was a network model and diagram that increased the decision maker's insights into supply chain connectivity. The model however was confined to a single-period and single-objective problem. Geoffrion and Powers [81] gave a review of evolution of distribution strategies over the past twenty years, describing how the descendants of the above model can accommodate more echelons and cross commodity detail. The international plant location problem model presented by Hodder and Dincer [82] is a large scale non-linear MIP, and it is very difficult to solve. The model is non-linear due to the inclusion of the financial variables to allocate the costs to plants. By developing an approximation procedure, the authors transform the model into a more tractable, but still non-linear, formulation. Cohen and Lee [83] developed a mixed-integer non-linear, value-added chain model that coordinated the supply chain process comprised of sourcing, centralized production planning, and inter-plant transhipment. The model incorporated capacity, demand, and production constraints, but failed to capture risk factors inherent in a global setting. Cohen and Moon [84] presented a mixed integer, multi-commodity model to find inbound raw material flows, assignment of product lines and specification of production volumes and outbound finished product flows in a production distribution network. The model contains binary variables for assigning products to plants and for determining the part of the concave curve of production costs to be applied. Research results show that global optimal solutions are frequently difficult to obtain unless a special structure exists. A variant of the generalized Benders decompositions technique is applied. Arntzen et al. [85] presented a mixed-integer programming model, called global supply chain model (GSCM) which evaluated global supply chain alternatives involving multiple products and multiple stages (echelons). More specifically, GSCM took into account the interdependence of production, inventory and delivery processes to minimize activity days and costs associated with production, inventory, material handling, and transportation. Ashayeri and Rongen [86] refined a grid model and the multi-criteria solution method called ELECTRE to formulate the distribution centre (DC) repositioning strategy based upon the analyses of material flows, DC locations, and throughput times. Although the proposed model and solution method were simple to use, they were confined to single-period and un-capacitated problems. Another multiple objective approach was proposed by Min and Melachrinoudis [87] to configure multi-echelon supply chain networks connecting material flows among suppliers, manufacturers, break-bulk terminals, and customers. Their analytic hierarchy process-based model also considers contingency planning associated with supply chain reconfiguration. It, however, did not consider multiple periods, capacity constraints and risk factors.

More recently, Melachrinoudis and Min [88] extended their previous work by designing a multi-objective, multi-period mixed integer programming model that determined the optimal relocation site and phase-out schedule of a combined manufacturing and distribution facility from supply chain perspectives. A similar problem was solved by Melachrinoudis et al. (2000) using a novel methodology called physical programming which allows a decision maker to express multi-criteria preferences not in the traditional form of weights but in terms of range of different degrees of desirability. In addition to the selection of the new site, the model determines the optimal schedule for relocating capacity from the existing site to the new site, the production levels in the two sites during the transitional period, and the shipments to the customers in each period of the multi-period planning and horizon. In an effort to integrate inventory, transportation, and location functions of a supply chain, Nozick and Turnquist [89] proposed an approximate inventory cost function and then embedded it into a fixed-charge location model. The fixed-charge facility location model was designed to consider a trade-off between demand coverage and cost associated with the location of automobile DCs. Although the model deals with multiple objective (service-cost trade-off) issues, it is confined to a single period, single echelon problem with no capacity constraint.

Stochastic Models

In an increasingly competitive environment, there are many uncertain or random elements in the supply chain such as customer demands, lead times, and production fluctuation. The stochastic models take into account these uncertain and random elements. One of the pioneering works dealing with the stochastic nature of the integrated supply chain is credited to Milder, who developed a dynamic programming model based on optimal control theory for selecting an optimal combination of transportation modes, commodity flows, and re-routing of carriers from customers and suppliers over a multi-period planning horizon. Tapiero and Solilman [90] utilized optimal control theory to solve multi-commodity transportation, multi-regional production and inventory planning problems over time with uncertain demand. Despite its merit, the model combining linear and parametric programs created severe computational difficulty. Lee and Billington [91] attempted to integrate the material flow of marketing, manufacturing, and material ordering policy, the customer service level for each product, and postponement strategies. Similarly Lee and Feitzinger [92] and Swaminathan and Tayur [93] presented stochastic models to formulate postponement (delayed product differentiation) strategies. In particular, Swaminathan and Tayur [93] provided stochastic programming models and effective computational procedures to study inventories of common components, the use of vanilla boxes for postponement and the effect of assembly task sequencing on operational performance. They also utilized the inherent structure of problems to develop computationally efficient algorithms based on sub-gradient methods. In robust optimization, the uncertainty about the problem data is treated as deterministic, unknown but bounded (e.g. via intervals of confidence for the data). A robust solution is one that tolerates changes in the problem data up to a given bound known as a priori. Ahmed and Salinities [94] developed a robust optimization framework for the problem of supply chain planning in the process industries. Since the standard stochastic programming formulation of the problem does not address the variability of the uncertain recourse costs across the uncertain parameter scenarios, they extended the stochastic programming formulation to account for robustness of the recourse costs through the use of an appropriate variability criterion. To overcome the difficulty associated with solving the robust models that include non-separable terms, they developed a heuristic procedure for the restricted recourse formulation. This method iteratively enforces recourse robustness while solving the standard stochastic program in each step. Their models can provide the decision maker with a tool to analyze the trade-off associated with the expected profit and its variability. In the meantime, several attempts have been to quantify the effects of imbalance between supply and demand in the supply chain. These attempts include Fisher et al. [95] who developed a stochastic program that aimed to minimize under production and overproduction costs as a result of imbalance between supply and uncertain demand in the supply chain. Similarly, Lee et al. [96] investigated the bullwhip effect that might arise when order variances distorted customer demand and consequently created imbalance between supply and demand in the supply chain. Metters (93) followed up on Lee et al. [96] by developing a dynamic programming model that aimed to minimize the expected costs of production, inventory holding, and excess demand penalty, subject to production obeying capacity constraints.

Hybrid Models

The single highest cost in a supply chain is inventory which accounts for nearly half of the total logistics costs [97]. Due to the heavy influence of inventory on the supply chain cost, the literature dealing with inventory theoretic models is relatively rich. Baumol and Vinod [98] are credited with the introduction of one of the most classic inventory theoretic models. Although they did not differentiate between truckload (TL) and less-than-truckload (LTL) freight rates, their model allowed a decision maker to make trade-offs among direct shipping cost, in-transit carrying cost, ordering cost and a consignee's inventory carrying cost. Since in-transit carrying cost is a surrogate measure of transit time (speed of delivery), their model can be used for analysing trade-offs between cost (inventory level) and time (transit time). Das [99] modified the inventory theoretic model proposed by Baumol and Vinod [100] by considering a more general estimate of the demand variability. His model can also be applied to inter-modal solutions. Karmarkar and Patel [101] used a decomposition approach to solve a single product, single period, multiple location inventory problem with stochastic demands and transhipment between locations. To consider interactions between inventory management and transportation modal choice, Constable and Whybark [102] further extended the inventory theoretic model. They added expected backorder cost to the original inventory theoretic model suggested by Baumol and Vinod [100]. Herron [103] employed the inventory theoretic model to not only relate the level of customer service to the inventory level, but to also determine the frequency of expedite shipment in case of stock-outs. Blumenfeld et al. [104] developed a model similar to Herron [103] to make a trade-off between freight expedition and safety stock holding cost. Some variants of the inventory theoretic model were used on a location-inventory problem where demand is uncertain and re-distribution of inventories is permissible between order cycles [105]. Schwarz [106] developed an inventory theoretic model based on Clark and Scarf [107] to determine the size of the total inventory investment and the location of inventory stocking points simultaneously. Their model, however, is limited to a single commodity problem. Another shortcoming of the model is severe computational complexity that may prohibit its application to a real-world problem. Singh and Vrat [108] developed a more practical model that was designed to determine the optimal location of repair part stocking pints and the allocation of repair part inventory to that location. Despite the popularity of the inventory theoretic models in the supply chain modelling literature, some alternative models proposed by Bookbinder, McAuley, and Schulte [109], Cachon [110], Karabakal et al. [111] and Newhart et al. [112]. To elaborate, Bookbinder et al. [110] employed both spreadsheet-based simulation and linear programming models to evaluate inventory/production alternatives and then select the best alternative, while making a trade-off between transportation cost and inventory investment. However, like past studies reviewed earlier, Bookbinder et al. [109] did not estimate the impact of backhauls on transportation cost. Newhart et al. [112] used both spreadsheet-based inventory and mathematical programming models to determine the most cost efficient methods of production and inventory with consolidation. More recently, Cachon [110] utilized a game theory to take into account an infinite horizon, stochastic demand inventory problem between one supplier and one retailer. In this game theory, Cachon [110] considered the possibility of 'double marginalization', buy-back contracts and quantity discounts to develop the optimal joint inventory policy. Karabakal et al. [111] used a combination of simulation and mixed-integer programming models to determine the number and location of automobile distribution and processing centers as well as set of market areas covered buy by each distribution and processing centre, while evaluating customer performance measures, such as ability of the supply chains to deliver a customer's preferred vehicles within short time windows. Given that a simulation model is well-suited for evaluating dynamic decision rules under 'what if scenarios', a few attempts have been made to develop simulation models to improve supply chain dynamics involving demand amplification with the presence of many supply chain players. These include Towil et al. [41] and Winker, Towil and Naim [113]. The main purpose of both studies is to create best decision rule that will allow decision maker to reduce lead times, compress the distribution channel and coordinate information flow throughout the supply chain. One of the most intriguing features of their models is the use of influence diagrams that visualise the cause-and-effect relationship between the decision rule and the improvement in the supply chain performances. As suggested by Swaminathan et al. [114], reengineering the supply chain because of business dynamics is becoming a necessity, but it is not an easy task. Although software simulation tools can ease the burden of analysis, it is still a major endeavor. The use of simulation as a vehicle for understanding issues of organizational decision-making has gained considerable attention and momentum in recent years (Feign et al. [115], Kumar et al. [116] and Maloni, and Benton [117]. Towill et al. [41] used simulation techniques to evaluate effects of various supply chain strategies on demand amplification. Tzafestas and Kapsiotis [118] utilized a combined analytical/simulation model to analyze supply chains. Swaminathan et al. [119] utilized a simulation to study the effect of sharing supplier's available-to-promise information.

Petrovic et al. [120] developed a fuzzy generative supply chain model to determine target order-up-to levels of inventories along a supply chain under uncertain demand and external supply of raw materials. The results of the fuzzy model were then used as input data for an evaluative simulation model that aimed to calculate replenishment quantities during a finite planning horizon. The simulation model also provides the user with the assessment of supply chain performance. These fuzzy and simulation models, however, are confined to a single production problem with no capacity constraints. Petrovic [121] extended these models by incorporating the element of uncertain lead times during the replenishment process into the fuzzy model framework. Both these simulation studies are useful for understanding supply chain dynamics under uncertainty, but are still limited to periodic review systems.

IT Driven Models

One of the critical drivers of supply chain success is enhanced visibility through an information sharing mechanisms linking supply chain partners. Considering the significant role of IT in the supply chain success, IT-driven models are high in demand. Nevertheless, IT-driven modelling efforts are still in their infancy. IT-driven models that are worth noting include Al-Mashari and Zari [122], Camm et al. [123], Johnston et al. [124], Min and Melachrinoudis [88], Richmond and Peters [125] and Talluri [126]. More specifically Camm et al. [124] combined an integer programming model involving the location of distribution centres and sourcing of multiple products with GIS to develop a flexible decision support system (DSS). However, their model based DSS did not include capacity constraints. Johnston et al. [123] developed a stand-alone GIS model for managing and integrating multi-facility warehousing and production systems. The model aimed to find near-optimal storage locations for stock items in a multi-facility warehousing environment. Min and Melachrinoudis [88] blended GIS into a mixed integer programming model based DSS which was designed to develop warehouse restructuring strategy. Their DSS was intended to determine which warehouses to retain and which warehouses to phase-out in such a way that the restructured supply chain network minimised total logistics costs, while meeting capacity, demand and delivery requirements. Through descriptive rather than normative (analytical), Richmond and Peters [125] presented a conceptual warehouse management systems (WMS) model comprised of three phases: (i) WMS investment strategy, (ii) WMS implementation and planning and (iii) WMS execution. Phase one focus on operation strategy and software selection. Phase two concerned with the detailed design and conference room pilot as well as the detailed specifications of material handling equipment solution. Phase three got involved in the construction of material handling equipment solutions, organisational structure, worker training, system testing, conversion and support. Developed a SAP R/3-based ERP architecture and a conceptual diagram in an effort to create value-oriented supply chains that enable a high level of integration and communication among all supply chain processes. Their ERP architecture consisted of three layers: presentation based on graphical user interface, application and database. Talluri [126] proposed a goal programming model for an effective acquisition and justification and maintenance, costs, flexibility, execution accuracy and compatibility.

27.7. CLOSURE

The rapid development of the Internet has led to many changes in ways business are run. More and more companies are moving online for procuring raw materials, interacting with customers, and delivering directly to them by bypassing existing channels for product flow. Stores are turning into showrooms; computers are becoming the shopping stores; checkout counter replaces shipment delivery boy. In a world where the customer has a lot of options and loyalty changes with the click of a button, achieving customer satisfaction will take much more than old-fashioned manufacturing shop floors and supply chains. The battleground of the next decade will be supply chain vs. supply chain. Therefore, everyone should understand the concept of supply chain management. Now present trends are outsourcing, cross-boundary working, flattened hierarchies, teams, empowerment and mass customization. Therefore supply chain should be flexible to adopt all these.

REVIEW QUESTIONS

1. Define the terms "supply chain" and "supply chain management".
2. What do you understand by supply chain management antecedents? Discuss the eight SCM antecedents.
3. Explain the statement "supply chain is dynamic in nature" with proper justification.
4. How can supply chain drivers improve the supply chain performance? Explain it through a manufacturing supply chain.
5. Mention and explain the different supply chain models.
6. Discuss this statement "The strength of the supply chain is equal to its weakest link" with proper justification.
7. Mention the model one should choose in uncertain environment in supply chain. Give your answer with proper justification.
8. What is difference between value chain and supply chain?
9. Mention and explain the different components of each logistical supply chain driver.
10. Mention and explain the different components of each cross functional supply chain driver.
11. Mention the difference between supply chain management and logistics management.

REFERENCES

1. Ganeshan, R., Magazine, J.E. and Paul, S. (1999), A Taxonomic Review of Supply Chain Management Research: The Quantitative Models for *"Supply Chain Management"*, Kluwer Academic Publishers.
2. Forrester, J.W. (1958), "*Industrial Dynamics, A Major Break Through for Decision Makers*", *Harvard Business Review*, Vol. 36, pp.37–66.
3. Bowersox, D.J. (1969), Reading in Physical Distribution Management, "*The Logistics of Marketing*", Macmillan, New York.
4. Langley, C.J. (1992), The Evolution of the Logistics Concept: "*The Strategic Issues*", Pitman Publishing.
5. Masters, J.M and Pohlen, T.L. (1994), "*Evolution of the Logistics Profession*", The Logistics Handbook, Free Press, New York.
6. Lalonde, J. and James, M.M. (1994), "*Emerging Logistics Strategies: Blueprints for the Next Century*", International Journal of Physical Distribution and Logistics Management, Vol. 24, No.7, pp.35–47.
7. Tsay, A., Nahmias, S., and Agrawal, N., (1998), "*Modelling Supply Chain Contracts: A Review*", Kluwer Academic Publishers, Boston, MA.
8. Mabert, V.A. and Venkataramanan, M.A. (1998), Special research focus on supply chain linkages: challenges for design and management in the 21st century, "*Decision Sciences*", Vol. 29, No.3, pp.537–552.

9. Stevens, G.C., (1989), "*Integrating the Supply Chain*", International Journal of Physical Distribution and Materials Management, Vol. 19, pp. 3–8.

10. Jones, T.C. and Riley, W.D. (1985), "*Using Inventory for Competitive Advantage through Supply Chain Management*", International Journal of Physical Distribution and Materials Management, Vol. 15, No. 5, pp. 16–26.

11. Goldratt, M. E., and Cox, J. (1986), The Goal, North River Press, New York.

12. Houlihan, J. (1988), "International Supply Chains: A new approach", Management Decision, Vol. 26, no. 3, pp. 13–19.

13. Ellram, L.M. and Cooper, M.C. (1990), "*Supply Chain Management Partnerships and the Shipper-Third Party Relationship*", International Journal of Logistics Management, Vol.1, No.2, pp.1–10.

14. Ellram, L.M. (1991), "*Supply Chain Management: The Industrial Organization Perspective*", International Journal of Physical Distribution and Logistics Management, Vol. 21, No.1, pp. 13–22.

15. Novack, R.S. and Simco, W. (1991), "*The Industrial Procurement Process: A Supply Chain Perspective*", Journal of Business Logistics, Vol. 12, No.1, pp.145–167.

16. Scott, C., and Westbrook, R., (1991), "*New Strategic Tools for Supply Chain Management*", International Journal of Physical Distribution and Logistics Management, Vol. 21, No. 1, pp. 23–33.

17. Lee, H. L. and Billington, C. (1992), "*Managing Supply Chain Inventory: Pitfalls and Opportunities*", Sloan Management Review, Vol. 33, No. 3, pp. 65–73.

18. Cooper, M.C. and Ellram, L.M. (1993), "*Characteristics of Supply Chain Management and the Implications for Purchasing and Logistics Strategy*", International Journal of Logistics Management, Vol. 4, No. 2, pp. 13–24.

19. Davis, T. (1993), "*Effective Supply Chain Management*", Sloan Management Review, pp. 35–46.

20. Lamming, R., (1993), "*Beyond Partnership: Strategies for Innovation and Lean Supply*", Prentice Hall, New York, USA.

21. Martha, C. and Lisa, M.E. (1993), "*Characteristics of Supply Chain Management and the Implications for Purchasing and Logistics Strategy*", The International Journal of Logistics Management, Vol. 4, No. 2.

22. Turner, R. (1993), Integrated supply chain management: what's wrong with picture? "*Industrial Engineering*", Vol. 25, No. 12, pp. 52-55.

23. Berry, D., Towill, D.R., and Wadsley, N. (1994), "*Supply Chain Management in the Electronics Product Industry*", International Journal of Physical Distribution and Logistics Management, Vol. 24, No. 10, pp. 20–32.

24. Johannson, L. (1994), How can a team approach add value to your supply? "*Total Quality Environmental Management*", Vol. 3, No. 4, pp. 521–530.

25. Metz, P. (1994), "*Supply Chain Management*", Report on the state of the art; MIT Supply Chain Management Team, Centre for Transportation Studies, MIT.

26. Berry, D. (1995), "*The Analysis, Modeling and Simulation of a Re-engineered PC Supply Chain*", PhD Thesis, University of Wisconsin, USA.

27. Slack, N., Chambers, S.C., Harrison, A. and Johnston, R. (1995), "*Operations Management*", Pitman, UK.

28. Walton, W. L., and Miller, G.L., (1995), "*Moving towards LIS Theory Development*": A Framework of Technology Adoption within Channels, Journal of Business Logistics, Vol. 16, No. 2, pp. 117–223.
29. Giunipero, L.C. and Brand, R.R. (1996), "*Purchasing Role in Supply Châin Management*", International Journal of Logistics Management, Vol. 7, No. 1, pp. 29–37.
30. Johnson, C. and Wood, D.F. (1996), "*Contemporary Logistics*", Prentice Hall, NJ.
31. Lamming, R. (1996), "*Squaring Lean Supply with Supply Chain Management*", International Journal of Operations and Production Management, Vol. 16, No. 2, pp. 183–196.
32. Thomas, J., and Griffin, P.M., (1996), "*Coordinated Supply Chain Management*", European Journal of Operations Research, Vol. 94, pp. 1–15.
33. Harland, C.M. (1996), "*Supply Chain Management: Relationships, Chains and Networks*", British Academy of Management, Vol. 7, pp. S63–S80.
34. Bechtel, C. and Yayaram, J. (1997), "*Supply Chain Management: A Strategic Perspective*", International Journal of Logistics Management, Vol. 8, No.1, pp. 15–34.
35. Caldwell, N.S. and Down, S. (1997), Faking it, Supply Chain Relationships, "*Management Research News*", Vol. 20, No. 2, pp. 53–61.
36. Cooper, M.C., Ellram L.M., Gardner J.T., and HANKS A.M. (1997), "*Meshing Multiple-alliances*", Journal of Business Logistics, Vol. 18, No. 1, pp. 67–89.
37. Cooke, A. (1997), "*The Solid-gold Supply Chain*", Logistics Management and Distribution Report.
38. Kopczak, L.R. (1997), Logistics Partnership and Supply Chain Restructuring: Survey Results from the USA Computer Industry, "*Production and Operations Management*", Vol. 6, No. 3, pp. 226–247.
39. Lee, H.L. and NG, S.M. (1997), Introduction to the Special Issue on Global Supply Chain Management, "*Production and Operations Management*", Vol. 6, No. 3, pp.191–192.
40. Rich, N., and Hines, P., (1997), "*Supply Chain Management and Time-based Competition*", The Role of the Supplier Association, International Journal of Physical Distribution and Logistics Management, Vol. 27, pp. 210–225.
41. Towill, D.R., Naim M.M. and Wikner, J. (1992), Industrial Dynamics Simulation Models in the Design of Supply Chains, International Journal of "*Physical Distribution and Logistics Management*", Vol. 22, No. 5, pp. 3-13.
42. Alber, K.L., and William, T.W., (1998), "*Supply Chain Management: Principles and Techniques for the Practioner*", Research paper Series, APICS Educational and Research Foundation, Falls Chruch, VA.
43. Christopher, M. (1998), "*Logistics and Supply Chain Management Strategies for Reducing Cost and Improving Service*", Second Ed., Published, Financial Times Professional Limited.
44. Brown, G.C. (1987), Design and Operation of Multi-commodity Production/distribution System using Primal Goal Decomposition, "*Management Science*", Vol. 33, No. 11. pp.1469–1480.
45. Lambert, M., Martha, C.C., and Pagh, J.D. (1998), Supply Chain Management Implementation Issues and Research Opportunities, International Journal of "*Logistics Management*", Vol. 9, No. 2, pp.1–19.
46. Ross, D.F., (1998), Competing through Supply Chain Management Creating Market – Winning Strategies through Supply Chain Partnerships, Chapman and Hall, USA.
47. Tan, K.C., Kannan, V.R., and Handfield, R.B., (1998), "*Supply Chain Management*": Supplier Performance and Firm Performance", International Journal of Purchasing and Material Management, Vol. 34, No. 3, pp. 2–9.

48. Tyndall, G., Christopher, G. P., Wolfgang, and John, K. (1998), *"Supercharging Supply Chains"*, John Wiley and Sons, New York.
49. Hoek, R.I.V. (1998), Measuring the Un-measurable–Measuring and Improving Performance in the Supply Chain, *"Supply Chain Management Review"*, Vol. 31, No. 4, pp. 187–192.
50. Akkermans, H., Paul, B. and Bart, V., (1999), Virtuous and Vicious Cycles on the Road towards International Supply Chain Management, International Journal of *"Operation and Production Management"*, Vol. 9, pp. 565–581.
51. Beamon, M.B. (1999), *"Measuring Supply Chain Performance"*, International Journal of Operation and Production Management, Vol. 19, No. 3, pp. 275–292.
52. Hicks, D.A. (1999), *"The State of Supply Chain Strategy"*, *IIE Solutions*, pp. 24–29.
53. Ayers, J., (2000), *"A Primer on Supply Chain Management"*. The Executive's Journal, Vol. 16, No. 2, pp. 6–15.
54. Ballou, R.H., Stephen, M.G., and Mukherjee, A. (2000), "New Managerial Challenges from Supply Chain Opportunities", *"Industrial Marketing Management"*, Vol. 29, No. 1, pp.7–18.
55. Norina, L. V. and Bailey, W. C. (2001), *"Quantitative Approach to Supply Chain Relationships"*, 11th Annual Food and Agribusiness Symposium, Australia.
56. Hopp, W.J. and Spearman, M.L. (2000), *"Factory Physics"*, McGraw-Hill, Boston, MA.
57. Chopra, S. and Meindl, P. (2001), *"Supply Chain Management: Strategy, Planning and Operation"*, First Edition, Pearson Education Asia.
58. Stanley, F., and Gregory M., (2001), *"Achieving World Class Supply Chain Alignment: Benefits, Barriers and Bridges"*, Center for Advanced Purchasing Studies.
59. John, T. M. (2001), *"Supply Chain Management"*, Sage Publications, California.
60. Norina, L.V., and Bailey, W.C., (2001), *"Quantitative Approach to Supply Chain Relationships"*, 11th Annual Food and Agribusiness Symposium, Australia.
61. Sheikh, K., (2001), *"Manufacturing Resource Planning (MRP II) with Introduction to ERP, SCM, and CRM"*, First Edition, Tata McGraw-Hill, New Delhi.
62. Charles, P. and Bauer, M. (2002), *"E-supply Chain Using the Internet to Revolutionize Your Business"*, Berrett-Koehler publishers, San Francisco.
63. Michiel, R.L., Harold, E.F., Flynn, A.E. and Fraser, J.P. (2002), *"Purchasing and Supply Chain Management"*, Twelfth Edition, McGraw-Hill, New York.
64. Monczcka, R., Trent, R. and Handfield, R. (2002), *"Purchasing and Supply Chain Management"*, Second Edition, Thomson Learning.
65. Gaither N. and Frazier G. (2002), *"Operations Management"*, Thomson Asia, Singapore.
66. Coyle, J., Bardi, E.J. and Langley, J.C. (2003), *"The Management of Business Logistics, a Supply Chain Perspective"*, Seventh Edition, Thomson Learning.
67. Handfield, R.B. and Nichols, E.L. (2003), *"Supply Chain Redesign: Transforming Supply Chains into Integrated Value System"*, Pearson Education Asia.
68. Mentzer, J. T. et al. (2001), *"Defining Supply Chain Management"*, *Journal of Business Logistics*, Vol. 22, No. 2, pp.1–25
69. Morgan, R., and S., Hunt. (1994), *"The Commitment-trust Theory of Relationship Marketing"*, Journal of Marketing, Vol. 58, pp. 20–38.

70. Spekman, R.E., Spear, J., and Kamauff, J. (2001), "*Supply Chain Competency: Learning as a Key Component*", Working paper, University of Virginia.

71. Houlihan, J. (1988), International Supply Chains: A New Approach, "*Management Decision*", Vol. 26, No. 3, pp. 13–19.

72. Bowersox, D.J. and Closs, D. J. (1996), "*Logistical Management: The Integrated Supply Chain Process*", McGraw-Hill, Inc.

73. Beamon, B.M. (1998), "*Supply Chain Design and Analysis: Models and Methods*", International Journal of Production Economics, Vol. 55, pp. 281–294.

74. Bradley, S.P., Hax, A.C., and Magnanti, T. L., (1977), "*Applied Mathematical Programming*", Addison-Wesley, Reading, MA.

75. Budnick, F.S., Mcleavey, D. and Mojena, R. (1988), "*Principles of Operations Research for Management*", Irwin, Homewood.

76. Mentzer, J.T., and Schuster, A.D. (1983), "*Computer Modeling in Logistics: Existing Models and Future Outlook*", Journal of Business Logistics, Vol. 3, No. 2, pp. 1–55.

77. Silver, E.D. (1981), Operations Research in Inventory Management: A Review and Critique, "*Operations Research*", Vol. 29, pp. 628–645.

78. Zipkin, P.H. (2000), "*Foundations of Inventory Management*", McGraw-Hill, New York.

79. Geoffion, A.M. and Graves, G.W. (1974), Multi-commodity Distribution Design by Benders Decomposition, "Management Science", Vol. 20, pp. 822–844.

80. Glover, F., Jones, G., Karney, D., Klingman, D. and Mote, J. (1979), An Integrated Production, Distribution and Inventory Planning System, "*Interfaces*", Vol. 9, No. 5, pp. 21–35.

81. Geoffrion, A.M. and Powers, R.F. (1995), Twenty Years of Strategic Distribution System Design: An Evolutionary Perspective, "*Interfaces*", Vol. 25, No. 5, pp.105–127.

82. Hodder, J.E. and Dincer, M.C. (1986), A Multifactor Model for International Plant Location and Financing under Uncertainty, "*Computers and Operations Research*", Vol. 13, pp. 5, pp. 601–609.

83. Cohen, M.A. and Lee, H.L. (1989), Resource Deployment Analysis of Global Manufacturing and Distribution Networks, Journal of "*Manufacturing and Operations Management*", Vol. 2, pp. 81–104.

84. Cohen, M.A. and Moon, S. (1991), An Integrated Plant Loading Model with Economies of Scale and Scope, European Journal of "*Operations Research*", Vol. 50, pp. 266-279.

85. Arntzen, B.C., Brown, G.G., Harrison, T.P., and Trafton, L.L., (1995), Global Supply Chain Management at Digital Equipment Corporation, "*Interfaces*", Vol. 25, pp. 69–93.

86. Ashayeri, J., and Rongen, J.M.J., (1997), Central Distribution in Europe: A Multi-criteria Approach to Location Selection, The International Journal of "*Logistics Management*", Vol. 9, No. 1, pp. 97–106.

87. Min, H. and Melachrinoudis, E., (1999), The Relocation of a Hybrid Manufacturing/distribution Facility from "*Supply Chain Perspectives: A Case Study*", Omega, Vol. 27, No. 1, 75–85.

88. Min, H., and Melachrinoudis, E., (2001), Restructuring a Warehouse Network: Strategies and Models, "*Handbook of Industrial Engineering*".

89. Nozick, L.K. and Turnquist, M.A. (2001), Inventory, Transportation, Service Quality and the Location of Distribution Centres, European Journal of "*Operational Research*", Vol.129, pp. 362–371.

90. Tapiero, C.S. and Soliman, M.A. (1972), Multi-commodities transportation schedules ovver time networks, "*AIIE Transactions*", Vol. 2, pp. 311-327.

91. Lee, H.L. and Billington, C. (1996), Material Management in Decentralized Supply Chains, "*Operations Research*", Vol. 41, No. 5, pp. 835–847.

92. Lee, H.L. and Feitzinger, E. (1995), "*Product Configuration and Postponement for Supply Chain Efficiency*", Institute of Industrial Engineers, Fourth Industrial Engineering Research Conference Proceedings, pp. 43–48.

93. Swaminathan, J.M., and Tour, S.R., (1999), Stochastic Programming Models for Managing Product Variety: Quantitative Models for "*Supply Chain Management*", Lower Press.

94. Ahmed, S., and Salinities, N.V., (1998), Robust Process Planning under Uncertainty, "*Industrial and Engineering Chemistry Research*", Vol. 37, pp. 1883–1892.

95. Fisher, M., Hammond, J., Obermeyer, W., and Raman, A. (1997), Configuring a supply chain to reduce the cost of demand uncertainty, *"Production and Operations Management"*, Vol. 6, No. 3, pp. 211–225.

96. Lee, H. L., Padmanabhan, V. and Whang, S. (1997), Information distortion in a supply chain: the bullwhip effect, "*Management Science*", Vol. 43, No.4, pp. 546-558.

97. Metters, R. (1997), Quantifying the Bullwhip Effect in Supply Chains, "*Journal of Operations Management*", Vol. 15, pp. 89-100.

98. Baumol, W.J., and Vinod, H.D. (1970), An inventory theoretic model of freight transport demand, "*Management Science*", Vol. 16, No. 7, pp. 413–421.

99. Das, C. (1974), Choice of transport service: An inventory theoretic approach, *"The Logistics and Transportation Review"*, Vol.10, No. 2, pp.181–187.

100. Baumol, W.J., and Vinod, H.D. (1970), An inventory theoretic model of freight transport demand, "*Management Science*", Vol. 16, No. 7, pp. 413–421.

101. Karmarkar, U.S. and Patel, N.R. (1977), The one-period, n-location distribution problem, *"Naval Research Logistics*", Quarterly, Vol. 24, Vol. 4, pp. 559–575.

102. Constable, G.K. and Whybark, D.C. (1978), The interaction of transportation and inventory decisions, "*Decision Sciences*", Vol. 9, pp. 688–699.

103. Herron, D.P. (1983), Management science in industrial logistics, "*Applications of Management Science"*, Vol.3, pp. 49–85.

104. Blumenfeld, D.E., Hall, R.W., and Jordan, W.C. (1985), Trade-off between freight expediting and safety stock inventory costs, "*Journal of Business Logistics*", Vol. 6, No. 1, pp. 79–100.

105. Das, C. (1975), Supply and redistribution rules for two-location inventory systems: One-period analysis, "*Management Science*", Vol. 21, No. 7, pp.765–776.

106. Schwarz, L.B., (1981), Physical distribution: The analysis of inventory and location, *"AIIE Transactions"*, Vol. 13, No. 2, pp. 138–150.

107. Clark, A.J. and Scarf, H. (1960), Optimal Policies for a Multi-echelon Inventory Problem, "*Management Science*", Vol. 6, No. 4, pp. 475–490.

108. Singh, N., and Vrat, P., (1984), "A location-allocation model for a two-echelon repair-inventory system", "*IIE Transactions*", Vol. 16, No. 3, pp. 222–228

109. Bookbinder, J.H., Mcauley, P.T. and Schulte, J. (1989), Inventory and transportation planning in the distribution of fine papers, *"Journal of the Operational Society"*, Vol. 40, No. 2, pp. 155–166.

110. Cachon, G.P. (1999), "*Competitive Supply Chain Inventory Management*", Kluwer Academic Publishers, Boston, MA.

111. Karabakal, N., Gunal, A. and Ritchie, W. (2000), Supply-Chain Analysis at Volkswagen of America, "*Interfaces*", Vol. 30, No. 4, pp. 46–55.
112. Newhart D.D., Stott, K.L. and Vasko, F.J., (1993), Consolidating product sizes to minimize inventory levels for a multi-stage production and distribution systems, "*Journal of the Operational Research Society*", Vol. 44, No. 7, pp. 637–644.
113. Wikner, J., Towill, D.R., and Naim, M., (1991), Smoothing supply chain dynamics, "*International Journal of Production Economics*", Vol. 22, No. 3, pp. 231–248.
114. Swaminathan, J.M., Smith, S.F., and Sadden, N.M., (1998), Modelling supply chain dynamics: A multivalent approach, "*Decision Sciences*", Vol. 29, No. 3, pp. 607–632.
115. Feign, G.E., AN, C., Connors, D. and Crawford, I. (1996), "*Shape Up, Ship Put*", *OR/MS Today*, Vol. 23, No. 2, pp.24–30.
116. Kumar, A., Own, P.S. and Perpetual, M.J. (1993), "Organizational simulation and information systems design: an operations level example", *Management Science*, Vol. 39, No. 2, pp. 218–240.
117. Maloni, M.J. and Benton, W.C. (1997), "Supply Chain Partnership: Opportunity for Operation Research", *European Journal of Operations Research*, Vol. 10, pp.419–429.
118. Tzafestas, S. and Kapsiotis, G. (1994), "Coordinated control of manufacturing/supply chains using multi-level techniques", *Computer Integrated Manufacturing Systems*, Vol. 7, No. 3, pp. 206–212.
119. Swaminathan, J.M., Sadden, N.M., and Smith, S.F., (1995), "*Information Exchange in Supply Chains*", Technical Report, Carnegie Mellon University, Pittsburgh, PA.
120. Petrovic, D., ROY, R., and Petrovic, R. (1998), "Modelling and simulation of a supply chain in an uncertain environment", *European Journal of Operational Research*, Vol.109, pp. 299–309.
121. Petrovic, D., (2001), "Simulation of supply chain behaviour and performance in an uncertain environment", *International Journal of Production Economics*, Vol.71, pp. 429–438.
122. AL-Mashari, M., and Zairi, M., (2000), "Supply chain re-engineering using enterprise resource planning (ERP) software of a SAP R/3 implementation case", *International Journal of Physical Distribution and Logistics Management*, Vol. 30, No. 3/4, pp. 296–313.
123. Camm, J.D., Chorman, T.E., Dill, F.A., Evans, J.R., Sweeney, D.J. and Wegryn, G.W. (1997), "Blending OR/MS judgment, and GIS: Restructuring P&G's supply chain, "*Interfaces*", Vol. 27, No. 1, pp. 128–142.
124. Johnston, D.A., Taylor, G.D. and Visweswaramurthy, G. (1999), Highly constrained multi-facility warehouse management system using a GIS system, "*Integrated Manufacturing Systems*", Vol. 10, No. 4, pp. 221–232.
125. Richmond, B., and Peters, E., (1998), Warehouse Management Systems Technologies: Technology-Enabled Warehouse Excellence: Strategic Supply Chain Alignment: Best Practice in "*Supply Chain Management*", Gower Publishing Limited, Hampshire.
126. Talluri, S., (2000), "An IT/IS acquisition and justification model for supply chain management", *International Journal of Physical Distribution and Logistics Management*, Vol. 30, No. 3/4, pp. 221–237.

PRODUCTIVITY

28.1. INTRODUCTION

The global economy has created a highly competitive environment for companies with significant pressure on timely delivery and lower profit margins. This can be achieved by a shortened process cycle and stream lined supply chain. A case study was undertaken at M/s Precision Electricals. By implementing the time study methods to the conventional method of manufacturing the product, the finished product were produced of the same quality standard but in a shorter period of time. The number of pieces assembled increased by 25% without increasing the number of workers and by only rearranging the work schedule with the help of time study methods and statistical analysis.

Time study is a structured process of directly observing and measuring the work in order to establish the time required for completion of the work by a qualified worker working at a defined level of performance [1].

The basic procedure of systematic work measurement includes:

1. Analysis (of the work into small, easily measurable components or elements)
2. Measurement (of those components)
3. Synthesis (from those measured components to arrive at a time for the complete job)

First we undertake preliminary observation of the work (a pilot study) to identify small elements, which can be clearly recognized on subsequent occasion and are convenient, in terms of the time required for the completion of the job, for measurement.

The number of cycles that are observed depends on the variability in the work and the accuracy required. Since time study is essentially a sampling technique in which the value of time required for the job is based on the observed times for a sample under observation. Hence total numbers of observations are taken over a range of conditions and on a range of workers to arrive at a basic time.

Once a basic time for each element has been determined, allowances have been added to derive the standard time.

The case study was conducted at the M/s Precision Electricals to improve the productivity through the time study methods. M/s Precision Electricals is a manufacturer of electrical accessories. It manufactures the electrical accessories under the brand name of precision. The company produces a wide range of switches, sockets, switch-sockets, regulators, lamp holders, ceiling roses etc. In the case study conducted I have taken up the switch-sockets for the time study analysis. Figure 28.1 shows the front and the backside image of the switch-socket under study. Figure 28.2 represents the pictorial view of the stores department and the assembly department.

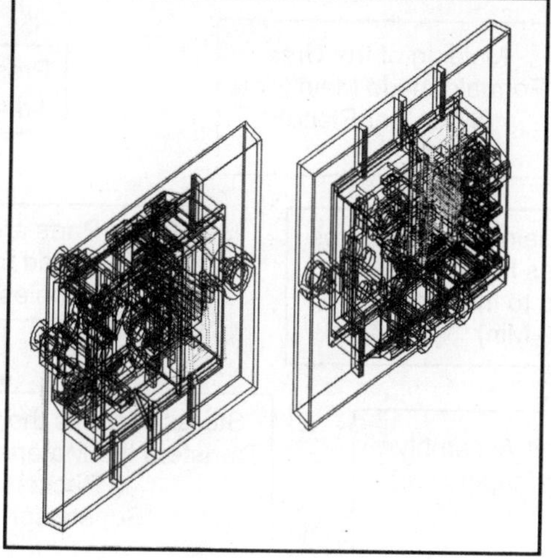

Fig. 28.1: The switch-socket

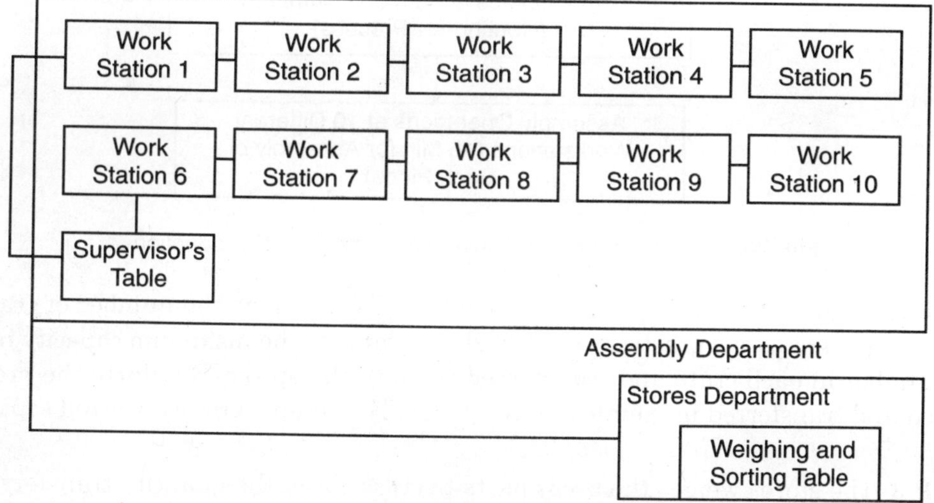

Fig. 28.2: The pietorial view of the stores department and the assembly department

28.2. THE CONVENTIONAL METHOD OF PRODUCTION

Figure 28.3 shows the conventional method of production and the steps involved in the production of the switch-socket are as described below.

STEP 1–Initialy the molding process and the press tool operations are started together. Here it was observed that each piece of the back plate or the main plate required 5 min for the moulding. During the moulding process the machine operator remained idle. His work started only after the moulding was done i.e., of unloading the moulded piece and putting the powder for the next moulding.

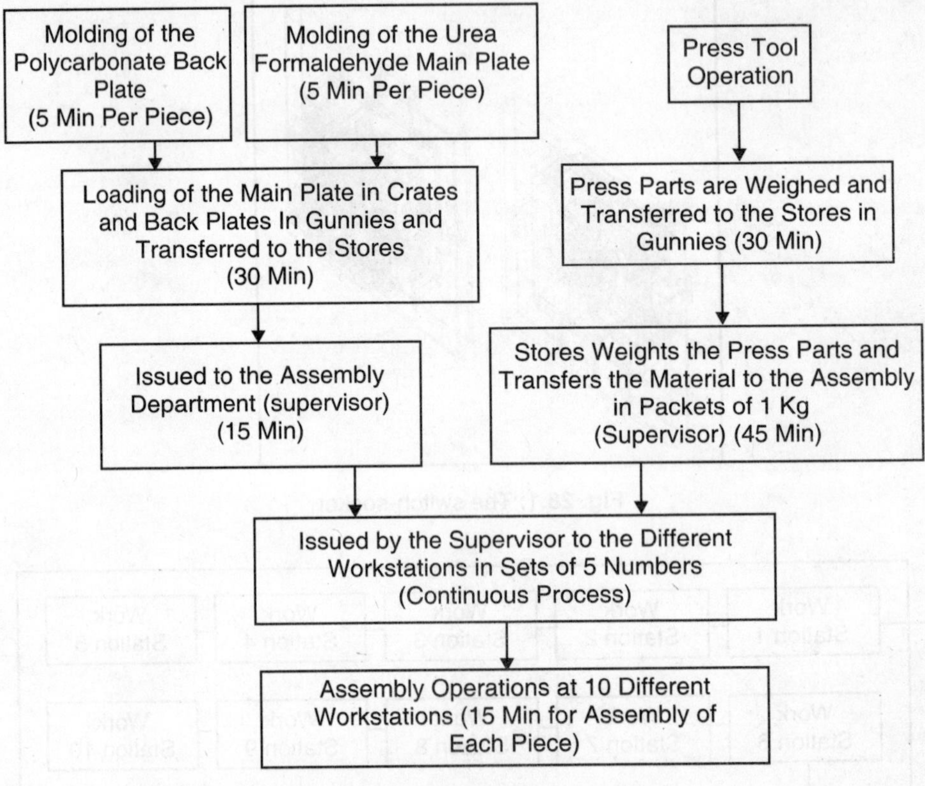

Fig. 28.3: Flow chart of the conventional method of production

STEP 2–Loading the main plates in the crates. To minimize the number of crates to be transferred to the stores the practice was to load the crates to the maximum capacity indicating the total number in each crate and transferred them to the stores. Similarly the press parts were weiged and transferred in gunnies to the stores. The scrap were transferred separately to the stores.

STEP 3–The stores weiges the press parts to cross check the quantity transferred to the stores from the press department and in the process, makes plastic packets of 1 kg of each type of the press parts.

STEP 4–After receiving the requisition from the production department the stores department transfers the main plates, back plates and the press parts along with accessories like screws and nuts to the production department. The main plates and the back plates are transferred in crates as has been received from the molding department and the press parts and the accessories in terms of weight.

STEP 5–After receiving all the material from the stores department the supervisor in the production department makes sets of 5 numbers and distributes them to the 10 Workstations. After the work starts at all the workstations the supervisor again repeates the process. The process of distributing the material at different workstations is the continuous job for the supervisor.

STEP 6–Finally it is the job at the workstations. Assembly of each piece requires minimum 15 minutes at each workstation.

28.3. SCOPE OF IMPROVEMENT IN THE PRODUCTION LINE

During the study it was seen that the job, which the supervisor was performing, was restricted to only issuing of the material to the workstations. The supervisor physically counts each component used in the assembly and issues them to the workstations in sets of 5 numbers. So it was decided that if the supervisor's job of counting and distributing the subassemblies was reduced then the supervisor (being the main link between the management and the workers) could be handed over with some data-collecting job required for improving quality of the product.

Also it was observed that the molding operator remains idle during the molding process. The molders job is basically restricted to the loading and unloading of the main plate and back plate. Hence it was decided to utilize the molder's idle time for cleaning of the flash, runner and burrs on the molded pieces. Hence a file for cleaning the urea formaldehyde plates and a cutter for cleaning of the polycarbonate plates were issued to the molders. A training session was organized for the molders to give them the correct idea for doing the job. But since the molders could not expertise in fine cleaning of the job, their responsibility was restricted to rough cleaning only and fine cleaning of the main plates and the back plates was done at the workstations.

28.4. REVISED SET UP OF PRODUCTION

An revised set up was arranged with a pilot batch of 500 pieces as shown in Fig. 28.4. The main aim of the experimental set up is whether it is practically possible to free the supervisor and reduce the assembly time.

STEP 1—Initialy the molding process and the press tool operations are started together. Here it is observed that each piece of the back plate or the main plate require 5 min for the moulding. During the moulding process the machine operator does the rough finishing of the previous molding.

STEP 2— Loading the main plates in the crates. The crates are loaded with 50 numbers of the main and back plates each and transferred to the stores. Similarly the press parts are weiged and transferred in gunnies to the stores.

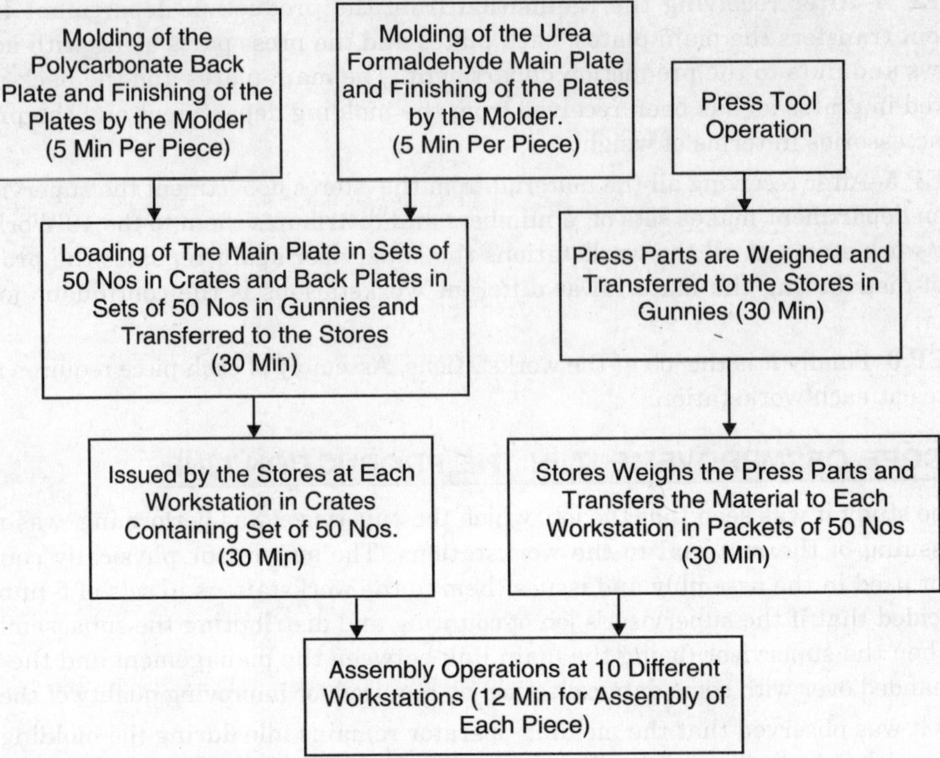

Fig. 28.4: Flow chart of the revised set up of production

STEP 3—The stores weighes the press parts to cross check the quantity transferred to the stores from the press department and in the process makes plastic packets of 50 numbers of each type of the press parts using an Atco weighing machine set of 50 numbers instead of packets of 1kg.

STEP 4—After receiving the requisition from the production department the stores department transfers the main plates, back plates and the press parts along with accessories like screws and nuts to the production department. The main plates and the back plates are transferred in crates as has been received from the molding department and the press parts and the accessories in terms of packets of 50 numbers to each of the 10 workstations.

STEP 5—Finally it is the job at the workstations. Assembly of each piece requires minimum 12 minutes at each workstation, since the main plates and the back plates already had rough finishing done and saved time at the workstations by doing only the fine finishing.

28.5. OBSERVATIONS DURING THE REVISED PROJECT RUN

The time taken by the supervisor to count the main plates and the back plates and distribute them at the different workstations in sets of 5 numbers continuously is reduced as the molding department is sending the main and back plates in sets of 50 numbers.

An Atco weighing machine was purchased which had the provision for giving the number of the pieces loaded on the scale. The stores department is instructed to make packets of 50 numbers of each press part sub-assembly instead of making packets of 1kg. It is observed that the stores department requires the same amount of time to make packets of 50 numbers as it used to take for making packets of 1 kg.

Since the molder was idle during the molding process this time is utilized to do the rough finishing of the main and the back plates. Thus saving time of the assembler, as he has to do only the fine finishing. Hence the time taken for assembly reduces from 15 min to 12 min.

Instead of stocking up all the main plates, back plates and the press parts at the supervisor's place/work table the stores department is asked to deliver the sub-assemblies directly at the 10 workstations as all the press parts, main plates and the back plates are already available in sets of 50 numbers ready for delivery. Since this operation would take 10 times more time as compared to stocking up all the components at the supervisors workplace, hence it is decided that the assembler at the start of the day's production, along with the supervisor will collect the material for the assembly of 50 sets from the stores. This activity takes 30 minutes for the assembler and hence the assembly time is reduced from 8 hrs to 7 hrs and 30 min. But this activity made the supervisor free and he is also available for the assemble job. During the lean period the supervisor's job involved collecting data for the continuous improvement of the quality of the product and increasing the productivity.

28.6. CALCULATIONS

(a) Conventional Method

The production time available is 8 hours = 8 × 60 = 480 minutes.

Assembly of each piece required 15 min.

So the day's production at each workstation is = 480/15 = 32 pieces.

Hence total production at 10 workstations = 32 × 10 = 320 pieces.

(b) Revised Method

The production time available is 7 hrs and 30 min = (7 × 60) + 30 = 450 min.

Assembly of each piece required 12 min.

So the day's production at each workstation is = 450/12 = 37 pieces.

Hence total production at 10 workstations = 37 × 10 = 370 pieces.

Since the supervisor was also available for the assemble job

Hence total production at the end of the day = 370 + 37 = 407 pieces.

i.e., 27% rise in the total production.

Since the supervisor is not always available for the assembly job and other day-to-day activities are handed over to him hence assembled pieces varied from 400 pieces per day to 407 pieces per day that becomes rise of minimum 25% in the total production as shown in Table 28.1.

Table 28.1: Comparison of the total productivity for the conventional and the revised method of production

Production time	Number of workstations	Time required for assembly	Total number of pieces assembled per day	% Increase
8 Hours	10	15 min	320	–
7 Hours and 30 min	10 + Supervisor	12 min	400 to 407	25% to 27%.

28.7. CONCLUSION

It was observed that the stores department required the same amount of time to make packets of 50 numbers as it used to take for making packets of 1 kg. Since the supervisor was not always available for the assembly job and other day-to-day activities were handed over to him hence assembled pieces varied from 400 pieces per day to 407 pieces per day that becomes rise of minimum 25% in the total production. The molder was idle during the molding process this time was utilized to do the rough finishing of the main and the back plates. Thus saving time of the assembler, as he had to do only the fine finishing. Hence the time taken for assembly reduced from 15 min to 12 min.

REFERENCE

1. "*Industrial Engineering and Management*"—by O.P. Khanna. Dhanpat Rai Publications (P) Ltd. Reprint 2001., Chapter 9–Work Study.

SUPPLY CHAIN PERFORMANCE MEASURES

29.1. INTRODUCTION

A decision framework is proposed for selection of supply chain performance measures along the product life cycle, which considers the dynamic nature of the supply chain, changing needs of the customers, product characteristics and the external and internal environments. The process of choosing appropriate supply chain performance measures is difficult due to the complexity of the system, and the existing performance measurement theory fails to provide the necessary support to develop performance measures for any supply chain considering various system wide aspects. The proposed decision framework is able to solve the above problem and is applicable to all types of supply chains. In this paper, an effort has been made to obtain effective supply chain performance measures considering all influencing parameters of the supply chain.

In order to survive in a globally competitive environment and sustain long-term advantages, more and more enterprises have introduced the practice of Supply Chain Management (SCM) [1]. From the beginning of this decade, this subject has been studied, practiced and reported in the literature. According to Mentzer et al. [2], it is still a very popular development area among many companies. Most of them realize that supply chains need to be assessed by their performance. But the process of choosing appropriate effective supply chain performance measures is difficult due to the complexity of the system, and the existing performance measurement theory fails to provide the necessary support to develop performance measures for a supply chain considering, various system wide aspects. According to Beamon [3], there is very little literature available in performance measurement system (PMS) design and performance measures selection for supply chains. Although researchers attempt to build new measures and metrics for SCM, most of current supply chain PMS's in place possess too many defects to meet with SCM requirements. Besides the criticism about non-connection with strategy and a biased focus on financial metrics, there are some in-depth problems of supply chain PMS's: lack of a balanced approach to integrate financial and non-financial measures and lack of system thinking, in which a supply chain must be viewed as one whole entity with a measurement system that spans the entire supply chain.

Furthermore, performance measurements provide an approach to identify success with the potential of impacting management strategies and facilitating the understanding of the supply chain's progress and position. Additionally performance measures assist in directing management attention, revising company goals, and re-engineering business processes [4, 5, 6]. It is very important to measure and evaluate supply chain activity and performance to support better decision making and better communication, to provide performance feedback and to motivate and direct behavior. Performance measurement provides the means by which a company can assess whether its supply chain has improved or degraded. The selection of the right performance measures that vary from one supply chain to another is the first step of supply chain performance measurement. The need for a comprehensive, flexible and adaptable framework for selecting supply chain performance measures has long been felt. But little work has been done in the selection the supply chain performance measures considering a system perspective. One can find a few examples, (Beamon, [7]; Narasimhan and Jayaram [8]; Van Hoek [4]; Holmberg [9]; Gunasekaran et al. [10]; Bullinger et al. [11]; Lai and Cheng [12]; Chan and Qi [13]; Lockamy III and McCormack, [14] which talk about performance measures and a PMS of the supply chain. In the proposed decision framework, changing needs of the customer, product characteristics and the external and internal environments provide the means by which a company can assess whether its supply chain has improved or degraded.

29.2. METHODOLOGY FOR SELECTION OF SUPPLY CHAIN PERFORMANCE MEASURES ALONG THE PRODUCT LIFE CYCLE

In order to reveal a real insight of supply chain performance, the appropriate performance measures that best suit to a supply chain should be selected and evaluated. To achieve this objective, the literature reviews along with discussions and interviews with supply chain experts and practitioners are carried out which become the basis to develop a decision framework for selection of supply chain performance measures. The proposed decision framework is generic in nature. The supply chain performance measures are also classified into three categories i.e., time dependent performance measures, quantity dependent performance measures and quality dependent performance measures. The details about the supply chain management performance measures, proposed decision framework and case situation for selection of supply chain performance measures are discussed in below sections.

29.3. PERFORMANCE MEASURES OF SCM

SCM brings the revolutionary philosophy and approach to manage the business with the sustained competitiveness. It also plays a key strategic factor for increasing organizational effectiveness such as enhanced competitiveness, better customer care and increased profitability. Therefore organizations must work together to help each other in the SCM environment. Each node in the supply chain is a strategic link. Strong links make strong supply chains and weak links hurt every member in the supply chain. According to Cox [15] "companies are instructed to construct ever efficient and responsive supply chains because it will no longer be company with company, but supply chain competing against supply chain". Therefore, there is an emerging requirement to focus on the performance of the extended supply chain or network in which company is a partner. One-dimensional metrics such as capacity utilization, inventory turns or costs tend to give a distorted picture of the supply chain. Likewise, it is often found many companies achieved great improvements in their operational performance, but they do not impact on the end consumers due to overall poor performance in the supply chain. In the early eighties, it happened with General Motor [16]. Performance measures provide the means by which the achievement of the organizational goals and objectives are assessed in quantifiable terms [17]. In theory and practice, we often find the quotation that "You can not manage what you do not measure". But even measuring the operations need not necessarily lead to good results. Managers and employees are warned that "You get what you measure" or "be careful what you measure you might get it" [18, 19]. According to Sahay et al. [20], "each department measures supply chain performance in its own terms. Its members are evaluated on their ability to meet objectives consistent with departmental goals. Not surprisingly, then, they drive operations towards improving performance in their own department, often at the expense of other functional areas. When each sets its performance measures in isolation from other, narrow functional interests often conflict with broader supply chain goals". Traditionally, performance measurement is defined as the process of quantifying effectiveness and efficiency of action [21]. In modern business management, performance measurement goes well beyond just quantification and accounting. It is supposed to contribute much more to business management and performance improvement in the industries. From the management perspective, performance measurement provides necessary information of management feedback for decision makers and process managers. It plays the critical roles of monitoring performance, enhancing motivation and communication, and diagnosing problems [22, 23]. Furthermore, performance measurement provides an approach for identifying success and potentials of management strategies, and facilitating understanding of progress and position. With these, it assists in directing management attention, revising company goals, and re-engineering business processes [19, 5, 6]. Henceforth, accurate performance measurement is significantly helpful for improvement of SCM. Beamon [3] identifies three types of measures: resources, output, and flexibility. Supply chain models have predominately utilized two different performance measures: cost and a combination of cost and customer responsiveness. Costs may include inventory costs and operating costs. Customer responsiveness measures include lead-time, stock out probability, and fill rate. The supply chain performance measures are classified into three categories i.e., time dependent performance measures, quantity dependent performance measures and quality dependent performance measures. Time dependent performance measures are those measures, which may increase or decrease with respect time measurement scale. The time dependent performance measures are order fulfillment cycle time; customer inquiry resolution time; customer inquiry response time; supply chain response time; cash to cash or cash conversion cycle time; supply chain cycle time (source/make cycle time); supply chain inventory days; vendor lead-time for basic order; vendor lead time for special orders and new product introduction time. Quantity dependent performance measures are those measures for which there exists single direct numerical measurement. The time dependent performance measures are not included in it. The quantity dependent performance measures are item fill rate or product fill rate; order fill rate; cycle service level; perfect order fill rate; dollar fill rate; stock out rate; average number of late orders; average value of late orders; asset utilization and total supply chain cost. Quality dependent performance measures are those measures for which there is no single direct numerical measurement, although some aspects of them may be quantified. The quality dependent performance measures are installation, warranty, repairs and service parts; customer complaints, claims and returns; product replacement; product tracking; order and product flexibility; effective risk management; supplier performance; social criteria and function duplication minimization.

29.4. DEVELOPMENT OF A DECISION FRAMEWORK FOR SELECTION OF PERFORMANCE MEASURES IN SUPPLY CHAIN ALONG THE PRODUCT LIFE CYCLE

A decision framework is developed for selection of performance measures in supply chain along the product life cycle, which consists of four elements (i.e., competitive strategy, product life cycle phase, product characteristics and the external and internal environments) and it is shown in Fig. 29.1. Each element is discussed below.

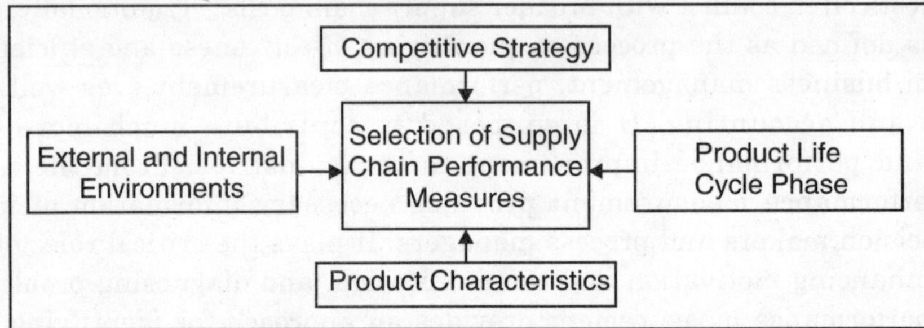

Fig. 29.1: Decision framework for selection of performance measures in supply chain along the product life cycle

Competitive strategy: A company's competitive strategy defines the set of customer needs that it seeks to satisfy through its products and services (i.e., product cost, delivery time, variety and quality). Competitive strategies need not to be same for the similar type of organizations. It depends on supply chain capability, organizational structure and market conditions. For example, Wal-Mart aims to provide high availability of a variety of reasonable quality products at low prices. McMaster Carr sells maintenance, repair, and operations (MRO) products. It offers over 200,000 different products through both catalog and website. Its competitive strategy is built around providing the customer with convenience, availability and responsiveness. With this focus on responsiveness, McMaster does not compete based on low price. Clearly, the competitive strategy at Wal-Mart is different from that at McMaster. For the same product, the competitive strategy may be different. The good example is Dell computer and HP computer. Dell sells PCs directly to customers where as HP sells its PCs through retailers. Dell has stressed customization and variety at a reasonable cost, with customers having to wait approximately one week to get their product. In contrast, a customer can walk into a computer retailer, be helped by a salesperson and leave the same day with an HP computer. The amount of variety and customization available at the retailer is limited. A McMaster Carr customer places greater emphasis on product variety and response time than on cost. A Wal-Mart customer, in contrast, places greater emphasis on cost. A Dell customer, purchasing online, places greater emphasis on product variety and customization. A customer purchasing a PC at a retailer is most concerned with the help in product selection and faster response time. Thus, a firm's competitive strategy will be defined based on the customer's priorities. Competitive strategy targets one or more customers segments and aims to provide products and services that satisfy these customers' needs. Therefore the competitive strategy plays an important role for selection of performance measures.

Table 29.1: Categorization of product characteristics

Category	Characteristics	Category	Characteristics
Based on the value/weight ratio	High Medium Low	Based on demand density	High Medium Low
Based on risk and profit	High risk and high profit High risk and low profit Low risk and high profit Low risk and low profit	Based on market spread	Local National International Global
Based on demand variability	Highly seasonal Moderately seasonal Normal	Based on product life cycle duration	Long Medium Short
Based on the application	Industrial House hold Both	Based on status after the usage	Remanufacturable Decomposable

Product life cycle phase: Every product follows different phases in its entire life cycle. Some of the product characteristics change along its life cycle. In general, these phases are introductory, growth, maturity, decline and withdrawal. To be effective, the supply chain strategy has to be changed along product life cycle phase. As supply chain strategy changes, the performance measures of the supply chain have to be changed along product life cycle phase. In introductory stage, the appropriate marketing tactics include intensive advertising, stimulation of buyer awareness, production often operates inefficiently due to overcapacity, short production runs and uncongealed standards in the product design. Creativity and innovativeness become more important than the efficiency of operations. Most of the performance measures in this stage are designed on the basis of "product success in the market". In the growth stage, the consumer is still tolerant of uneven quality. Changing the image or quality reputation of the product is possible at this point even if flaws existed earlier. This stage is most important in its entire life cycle because it influences the supply chain profit in its rest of the life cycle. In maturity stage, the superior quality of the product has been established by this time, but the market is saturated. In order to maintain profits, costs of production are scrutinized as well as the processes themselves to reduce waste. Every effort is made to produce a zero-defect product. Most of the performance measures should be on this basis only. After some time, the demand for the product decreases that is decline stage. This stage is identified with the decrease in demand rate. The rate of decrease depends on many other factors like new technology, competition and product characteristics. It is very clear that after sometime the product is to be withdrawn from the market, which is known as withdrawal stage. Therefore, while selecting the performance measures, the product life cycle stage has to be taken into account.

Product characteristics: It is essential to know all the possible details of the product for effective selection of supply chain performance. Product can be categorized with reference to many factors. These categories are shown in Table 29.1. The performance measures changes as per product characteristics. The requirements for efficient running of the supply chain depend on all these product characteristics.

Products can be classified on the basis of value/weight ratio. Most of the times, value/weight ratio depends on type of the product and its purpose. Available technology shows the impact on the value/weight ratio of the product. As technology improves, there is a possibility of increase in value/weight ratio. For example, the laptops come under the category of high value/weight ratio and desktops come under the category of medium value/weight ratio. Products can be classified on the basis of demand density. Demand plays major role in existence of any organization. Every organization tries to increase the demand of their product. Demand depends on many characteristics like brand name, product application and etc. In this classification, product category mainly depends on the nature of the application. For example, products like fast moving consumer goods come under the category of high demand products whereas products like computers come under the category of medium demand products. Products can be classified on the basis of risk and profit. Risk involves in everywhere in the supply chain. Intensity of the risk depends on the type of product, brand name, demand for the product and etc. It is observed that very few numbers of organizations prefers to low profit and high risk products. Consumer goods like soaps and toothpastes come under the category of low risk and low profit. Products like computers will have high risk and high profit and automobile products will have medium risk and high profit. Products can be classified on the basis of market spread. A local product is available in a portion of a national market. The term regional product is synonymous with local product. These products may be new products that a company is introducing using a rollout strategy or a product that is distributed exclusively in that region. A national product is one that, in the context of a particular company, is offered in a single national market. International products are offered everywhere or most parts in the world. A truly global product is offered in every world region and in countries at every stage of development. Products can be classified on the basis of demand variability. For example products like refrigerators and air-conditioners have high demand in the summer and school/college bags will have high demand only in the starting of the academic year. Demand to products like consumer goods will have the same demand in all the days. The music systems come under the category of moderately seasonal because they have demand throughout the year and have high in the festival seasons. Demand for the normal category products will be constant but for the highly seasonal products, the demand is high in a particular season. Products can be classified on the basis of product life cycle duration. Duration of the product life cycle depends on the many factors. Major factors are brand name, product application, competitors and new technology. Successful product will have the small growth period compare to other stages. Products can be classified on the basis of the application. This classification is based on the product usage segment. All the products cannot be used for industrial applications or household applications or in both segments. It depends on the product application. Cost and value of the industrial products will be very high compare to other two categories. Products can be classified on the basis of status after the usage. The efficient utilization of resources leads to high profits.

Reduction in losses helps in increase of profit. After the usage, a re-manufacturable product can be used as raw material to the new products, which leads to reduction in required raw material. Some products cannot be re-manufacturable. Mainly re-manufacturability depends on product chemical properties, available technology and environmental factors. Decomposing a product is a typical task. In some countries, it is the responsibility of the organization to take care about the product after usage.

External and internal environmental factors: The effect of government policies/ regulations is not in the hands of supply chain. These will change from region to region. Some times these will affect the productivity and conversion processes itself. Some times due to competition also the supply chain has to take major changes for their existence. Especially, this situation rises when the organization is planning to capture market in new environment. Brand name, quality and cost of the product will affect the competition. The other dangerous factor is new technology. Nobody knows exactly about the effect of new technology on the supply chain. Impact of new technology on variation of supply chain performance measures is very high. Expectations and requirements of product changes globally. In such situations the performance measures change in entire life cycle. This can be controlled with past statistics data.

29.5. DEVELOPMENT OF A CASE SITUATION FOR SELECTION OF PERFORMANCE MEASURES IN THE SUPPLY CHAIN ALONG THE PRODUCT LIFE CYCLE

Company X is a leading world-class manufacturer of audio, video, communications and information technology products for the consumer and professional markets. Its music, motion picture, television, computer entertainment, and online businesses make it as one of the most comprehensive entertainment companies in the global market. Most of the products are highly innovative and qualitative and is discussed in Table 29.2.

Table 29.2: Competitive strategy, product characteristics and external and internal environments of the company X

<table>
<tr><th>Factors</th><th>Details</th><th>Factors</th><th>Details</th></tr>
<tr><td>Competitive strategy</td><td>World class quality
product availability
high responsiveness</td><td rowspan="2">Product characteristics</td><td rowspan="2">High value/weight ratio
Low demand density
High risk and high profit
Global spread
Medium product life cycle duration
Both industrial and house hold application
Re-manufacturable</td></tr>
<tr><td>External and Internal environments</td><td>High competition
Medium customer demand</td></tr>
</table>

The supply chain executives now face the problem to select the right supply chain performance measures to evaluate their performance. A thorough analysis of the problem is required along with the identification of the important supply chain performance measures involved. The selection of supply chain performance measures has been determined through the discussions held with experts taking consideration of the proposed decision framework (see Fig. 29.1). The supply chain performance measures are shown in Table 29.3 for different stages of the product life cycle.

Table 29.3: Performance measures for the case situation along the product life cycle

	TD	QND	QTD
I	Customer enquiry resolution time Customer enquiry response time On time delivery Response time for the returns Supply chain inventory days Purchase order cycle time	Item fill rate Perfect order fill rate Reverse logistics cost Average distance for returns	Product tracking Flexibility in delivery After sales service Customer claims management Flexibility of supplier
G	Customer enquiry response time Customer inquiry resolution time Purchase order cycle time On time delivery Order fulfillment cycle time Response time for returns Supply chain inventory days Total cash flow time	Item fill rate CSL Stock out rate Perfect order fill rate Growth orientation rate Average distance for returns BOM accuracy Average inventory WIP Forecast accuracy Reverse logistics cost Cost of logistics Capacity utilization Asset utilization Set up cost Total supply chain cost	Customization of marketing strategies Customer claims management Flexibility in delivery Shipment accuracy Range of services After sales service Claims management Product tracking Function duplication minimization Flexibility of supplier Social criteria
M	Supply chain inventory days Total cash flow time Purchase order cycle time On time delivery Supply chain response time Supply chain cycle time Order fulfillment cycle time Customer inquiry resolution time Customer inquiry response time	Item fill rate Order fill rate Perfect order fill rate CSL Average number of late orders Stock out rate Average inventory Capacity utilization WIP Asset utilization Cost of reprocessing Cost of logistics Reverse logistics cost Set up cost Market share Total supply chain cost	Range of service Flexibility in delivery Claims management Shipment accuracy Customer claims management After sales service Product tracking Function duplication minimization Maintenance policy Social criteria World wide network
D	On time delivery Supply chain inventory days Order fulfillment cycle time Customer inquiry resolution time Customer inquiry response time	Capacity utilization Reverse logistics cost BOM accuracy Cost of logistics Stock out rate Cost of reprocessing Asset utilization Forecast accuracy Item fill rate Multiple/few suppliers Average inventory Average distance for returns WIP	Innovative product planning Maintenance policy Customer claims management After sales service Claims management Function duplication minimization Product tracking Shipment accuracy Flexibility of supplier Flexibility in delivery Range of services Social criteria Acceptance of risk

W	Order fulfillment cycle time Customer inquiry resolution time Customer inquiry response time	Reverse logistics cost Cost of reprocessing Cost of logistics Supply chain inventory days Average inventory	Maintenance policy Customer claims management After sales service Product tracking Asset utilization Flexibility of supplier

(TD—Time dependent; QND—Quantity dependent; QTD—Quality dependent; I—Introductory stage; G—Growth stage; M—Maturity stage; D—Decline stage; W—Withdrawal stage; CSL—Cycle service level; WIP—Work in process and BOM—Bill of materials)

29.6. CLOSURE

In order to improve the effectiveness of the supply chain, performance measures must be selected which will allow for a more complete and accurate analysis. The proposed decision framework provides the necessary foundation for the selection of performance measures for supply chain along the product life cycle. All influencing possible factors, which effect supply chain performance measures, are considered and explained. One case situation is elucidated in order to reinforce the salient features of the concept. But it can be applicable to different types of industries/supply chain by allowing managers to structure their unique problems, which can reflect their own considerations. The multi-attribute decision models like Analytical Hierarchy Process (AHP) can be used for proper prioritization of the supply chain performance measures once they are selected for a supply chain.

REFERENCES

1. Felix, T.S.C. and Qi, H.J., (2003) Feasibility of performance measurement system for supply chain: a process-based approach and measures, "*Integrated Manufacturing Systems*", Vol. 14, No. 3, pp. 179–190.
2. Mentzer, J., De, W.W., Keebler, J., Min, S., Nix, N., Smith, C., and Zacharia, Z., (2001), Defining supply chain management, "*Journal of Business Logistics*", Vol. 22, No. 2, pp. 1–25.
3. Beamon, B.M., (1998), Supply chain design and analysis: models and methods, "*International Journal of Production Economics*", Vol. 55, pp. 281–94.
4. Van Hoek, R.I., (1998) Measuring the unmeasurable – measuring and improving performance in the supply chain, "*Supply Chain Management*", Vol. 3 No. 4, pp. 187–92.
5. Bourne, M., Mills, J., Wilcox, M., Neely, A., and Platts, K., (2000), Designing, implementing and updating performance measurement systems, "*International Journal of Operations and Production Management*", Vol. 20, No. 7, pp. 754–71.
6. Kuwaiti, M.E., and Kay, J.M., (2000), The role of performance measurement in business process re-engineering, "*International Journal of Operation and Production Management*", Vol. 20, No. 12, pp. 1411–1426.
7. Beamon, B.M. (1999), Measuring supply chain performance, "*International Journal of Operation and Production Management*", Vol. 19, No. 3, pp. 275–92.
8. Narasimhan, R., and Jayaram, J., (1998), Causal linkages in supply chain management: an exploratory study of North American manufacturing firms, "*Decision Science*", Vol. 29, No. 6, pp. 579–605.

9. Holmberg, S. (2000), A system perspective on supply chain measurement, "*International Journal of Physical Distribution and Logistics Management*", Vol. 30, No. 10, pp. 847–68.
10. Gunasekaran, A., Patel, C., and Tirtiroglu, E., (2001), Performance measures and metrics in a supply chain environment, "*International Journal of Operations and Production Management*", Vol. 21, No.1/2, pp. 71–87.
11. Bullinger, H.J., Kuhner, M., and Hoof, A.V. (2002) Analysing supply chain performance using a balanced measurement method, "*International Journal of Production Research*", Vol. 40, No. 15, pp. 3533–3543.
12. Lai, K.H., and Cheng, T.C.E (2003), Supply chain performance in transport logistics: An assessment by service providers, "*International Journal of Logistics: Research and Applications*", Vol. 6, No. 3, pp. 151–164.
13. Chan, F.T.S., and Qi, H.J. (2003), Feasibility of performance measurement system for supply chain: a process-based approach and measures, "*Integrated Manufacturing Systems*", Vol. 14, No. 3, pp. 179–190.
14. Lockamy III, A., and McCormack, K., (2004), Linking SCOR planning practices to supply chain performance: An exploratory study, "*International Journal of Operations and Production Management*", Vol. 24 No. 12, pp. 1192–1218.
15. Cox, A. (1999), Power, value and supply chain management, "*Supply Chain Management: An International Journal*", Vol. 4, pp. 192–198.
16. Hausman, W.H., (2000), Supply chain performance metrics, White Paper Series on Supply Chain Trends and Developments, The Stanford Global Management Forum.
17. Carmen, R., and Conrad, S., (2000), Key performance indicators: Putting customer first, "*Supply Chain Management Review*", pp. 90–95.
18. Lapide, L., (2000), Technology: True measures of supply chain performance, "*Supply Chain Management Review*", pp. 25–28.
19. Hoek, R.I.V. (1998), Measuring the un-measurable–measuring and improving performance in the supply chain, "*Supply Chain Management Review*", Vol. 31, No. 4, pp. 187–92.
20. Sahay, S., Cavale, V., Rajani, R.L., Mohan, R., and Gupta, P., (2001), Supply chain practices in Indian Industries, "*KPMG and Management Development Institute's Joint Survey*", New Delhi.
21. Neely, A., Gregory, M., and Platts, K., (1995), Performance measurement system design: A literature review and research agenda, "*International Journal of Operation and Production Management*", Vol. 15, No. 4, pp. 80–116.
22. Waggoner, D.B., Neely, A.D., and Kennerley, M.P., (1999), The forces that shape organizational performance measurement systems: an interdisciplinary review, "*International Journal of Production Economics*", Vol. 60, pp. 53–60.
23. Rolstandas, A. (1995) Performance measurement: a business process benchmarking approach, Chapman and Hall, New York, NY, USA.

JIT MANUFACTURING SIMULATION

30.1. INTRODUCTION

This chapter discusses the basics of Just-In-Time (JIT) Manufacturing viz. Designing, Planning, Scheduling and Requirements for JIT Manufacturing. Also it focuses on Critical Elements, Integrated Process Control, Marketing, and JIT Manufacturing as a whole. To survive in the competitive global market, any manufacturing organization is required to produce good quality product and to supply these products in time. To achieve cost reduction and improvement of productivity: production must promptly and flexibly adapt to change in markets demand without having slack time. Just-In-Time is a management philosophy of improvement through aggressively discovering and resolving any problems or weakness that impede the organization's effectiveness and efficiency. Just-In-Time seeks to eliminate all waste within the organization, including the waste of under-utilization the talents, skills and potential of its employees through producing only the right amount and combination of parts at the right pace at the right time. This also deals with the cost effect on wastes resulting from storing the products instead of transferring them from one place to another at right time. The facility must deliver finished products demanded by outside buyers at fixed interval points in time. The Just-In-Time goal can be achieved by [1]:

(i) Eliminating wastes,

(ii) Achieving required product quality standards,

(iii) The respect for humanity,

(iv) Applying creative thinking or innovative ideas,

(v) shortening of the set-up time,

(vi) Standardization of operation to attain line-balancing,

(vii) Machine/process layout and multi-function workers for the flexible work-force concept,

(viii) Applying the concept of smoothed productio, and

(ix) Improvement activities by small groups and suggestion system.

In the present paper an attempt has been made to develop a successful simulation model in JIT manufacturing process for improving productivity of an organization.

Just-In-Time (JIT) basically means to produce the necessary units in the necessary quantities at the necessary time for the right place. At the same time it will minimize the wastes due delay of transfer of right items at right time. The JIT goal of producing the necessary items in the necessary quantities at the necessary time is an internal driver of production and operations management. The founder of JIT, Toyota Motor Corporation, Japan has integrated it with computer integrated manufacturing (CIM) technology and a strategic information system (SIS) to further reduce cost, improve product quality and reduce lead-time. JIT also considers, how to adopt production schedule to the demand changes in the market place while satisfying the goals of low cost, high quality and timely delivery.

A common belief that JIT can be applied effectively into a large industry upon mass production only a misconception. In fact, JIT is applicable to any country regardless of location, economic and civil development. Additionally, JIT can be utilized for any size company and in any industry. Also the application of JIT can be utilized in any matter related to time factor which are discussed in the next article.

30.2. ROLE OF JIT

Over the years, the role of JIT has become more crucial in two ways, firstly, the integration of computer integrated manufacturing and strategic information system into the JIT approach to facilitate flexibility in responding to customer demand and secondly, the introduction of continuous improvement activities help in maintaining the integrity of the overall process to boost worker morale. A continuous flow of production or adaptation to demand changes in quality, quantity and variety created by achieving the following key concept, viz., JIT, Automation, Flexible Workforce and creative thinking is going on. Some examples are given below:

(a) Automotive radiators are the essential component of vehicles' cooling system. Without a properly functioning radiator, car or truck engine is going to overheat. A well designed and manufactured radiator will not only improve the effectiveness, but also reduce the space and material for a specific heat transfer. The automatic radiator is predominantly a forced convection heat exchanger. It consists of number of rows of copper tubes and steel fins that serve as extended surfaces. The tubes which carry the hot water from the engine is straight passages with rectangular fins normal to the direction of radiator.

(b) Flexible manufacturing system may be equipped with general-purpose computer capable of controlling the machines and communicating with the other machines in the system. The various operational research problems arising in flexible manufacturing systems range from which type of system to buy to on-line operation control and monitoring policies. Though strongly interdependent flexible manufacturing system planning problems are usually divided into smaller hierarchically structured sub problems. The sub problems are often of a kind similar to problems known from the traditional production systems and thus accessible by standard technique. One of the most difficult problems arising in flexible manufacturing system is the scheduling problem. Therefore the objective should be to find a feasible simulation/scheduling to minimize the maximum completion time for a given set of jobs on a set of machines.

(c) In the job shop scheduling problem, a set of jobs are processed to complete on a set on unrelated machines. Each job requires processing on each machine exactly once. For each job, technology constraints specify a complete, distinct routing which is fixed and known in advance. Processing times are sequence independent, fixed and known in advance. Each machine is continuously available from zero time and operations are processed without preemption. The objective is to minimize the maximum completion time (makespan).

(d) Scheduling/simulation is considered to be a major task for shop floor productivity improvement. Scheduling allocation of resources applying the limiting factors of time and cost to perform a collection of tasks. In most cases, scheduling/simulation problem is taken up only after solving the fundamental planning problem. A set of multiple operation jobs are available for processing at zero time. Each individual operation of job is processed till its completion without break. The main difference of the shop flow scheduling form the basic single machine scheduling is that the inserted ideal time may be advantageous in shop flow scheduling.

The Just-In-Time goal can be achieved by:

(i) Eliminating wastes (i.e., excessive production resources, overproduction, excessive inventory and unnecessary capital investment),
(ii) By achieving required product quality standards,
(iii) To have respect for humanity,
(iv) Applying creative thinking or innovative ideas,
(v) Shortening of the set-up time,
(vi) Standardization of operation to attain line-balancing,
(vii) Process layout and multi-function workers for the flexible work-force concept,
(viii) Applying the concept of smoothed production, and
(ix) Improvement activities by small groups and suggestion system.

30.3. HOW TO ELIMINATE THE WASTES?

The elimination of wastes can be done by the followings:

- Overproduction – make only what is needed now.
- Waiting – coordinate flows between operations and balance load imbalances by flexible workers and equipments.
- Transportation – design facility layouts that reduce or eliminate materials handling and shipping.
- Unneeded production– eliminate all unneeded production steps.
- Work in process (WIP)–eliminate by reducing setup times, increasing production inventories rates and better coordination of production rates between work stations.
- Motion and Efforts – improve productivity and quality by eliminating unnecessary human motions, make necessary motions more efficient, mechanize, then automate.
- Defective products – eliminate defects and inspections. Make perfect products.

Table 30.1: Relation among objectives, ideas and techniques in Just-In-Time

Objective	Ideas	Techniques
Waste Elimination	JIT	- Pull systems with kanbans - Single minute exchange of die (SMED) - Level (smoothing) production - Cellular (flow) manufacturing - Standard operations - Multi process (skill) handling
	Automation	- Industrial engineering (IE) - Jidoka (Automation) - Visual control - Five S movement - Total productive maintenance (TPM)
	Trouble Free	- Poka-Yoke systems - Quality inspection at source - Total quality control

30.4. JIT DEMAND-PULL

In traditional approach we could manufacture a product and have quality control at the end. If the quality didn't line-up. It would be remanufactured or a new piece would be put on. It was after-the-fact control. World class companies moving this point from after-the-fact in to real-time control so that, in fact, once product is manufactured they know it is right.

30.5. GLOBAL STATUS OF JIT IMPLEMENTATION

Japan is the home of Just-In-Time. The basic techniques were developed and are still being developed there. The core philosophy of waste reduction and the variety of approaches Japanese have taken to adopting Just-In-Time to a variety of circumstances in manufacturing industries. Some of the front rubber factories like Sumitomo Rubber Industries have been able to develop a holistic approach in implementing Just-In-Time practices in its Nagwa and Koube factories in Japan. They have developed mechatronic devices also in order to reduce the die changing times to a substantial amounts.

Japan is the founder and home of JIT. JIT was founded by Taiichi Ohno, vice-president of Toyota. The basic techniques of JIT were developed and are still being developed there. The core philosophy of waste reduction and the variety of approaches, Japanese have taken to adopt JIT to a variety of circumstances in manufacturing industries. Some of the rubber factories like Sumitomo Rubber Industries have been able to develop a holistic approach in implementing JIT practices in its Nagwa and Koube factories in Japan. They followed the technique like "no delay"— either due to lot-size production or due to unequal production times of different work places and balancing of conveyance time. They have also developed Mechatronic devices in order to reduce the die changing times to a substantial amounts.

Pattern is very different in UK industries. In one of the rubber machinery like Y-Box extruder mainly employed this technique. Other companies in UK by large accepted the need for the change in manufacturing techniques and process because of global competitiveness.

US response has been very similar to that of UK considering output on the basis of Just-In-Time. UK companies are predominantly stilled on the plateau whereas many of US firms is moving to fulfill the application. Companies in USA adopted Just-In-Time initially as a defense against Japanese's encroachment on their domestic and international markets for the survival. In USA also much greater attention has been paid on the union involvement training and communication regarding Just-In-Time (JIT).

The diversity of European countries has apearheaded a variety of approaches to implement JIT in order to improve manufacturing practice. European companies are more resistant to change than their Japanese counter part due to some of the fundamental problems which are high labour cost, type of work force and organization structure etc.

30.6. SIMULATION FOR JIT PRODUCTION PROCESSES

Purpose

Simulation is a useful tool for assisting companies in exploring the benefits and risks associated with implementing a fully functional pull manufacturing strategy. Since JIT systems must be properly fine-tuned in order to operate efficiently. Simulation is also useful for performing sensitivity analysis. The success of a JIT implementation effort is dependent partly upon using simulation software which is easy to use and possesses great flexibility in modeling manufacturing system requirements [2, 3].

Methodology

Simulation has been used in JIT studies for two main reasons:

To evaluate the relative performance of JIT production with other types of production systems such as, MRP, order point systems and push type systems.

To identify factors detrimental to the success of JIT implementation.

Simulation Language

A major task is building a simulation model is to convert a conceptual model into an actual computer program. Conceptual model for JIT simulation is a discrete one, software packages, generally used are KANBAN, SLAM, SIMAN, GPSS, Q-GERT, DYNAMO etc. (used commercially) and MASS (program in-house).

Design of Experiment

The key components in any simulation study is the design and organization of the experiment which might be include:

- Selecting experimental factors.
- Measures of performance.
- Determining the steady state (equilibrium) condition.
- Determining the length of simulation run.
- Number of replications.

Simulation Model

The assumption of a simulation model affects the realism of the simulation results. These assumptions may reflect well on the characteristics of JIT production. Distribution and random variables-to ensure simulation integrity, most studies often generate data from random number of generators. Simulation studies of any JIT production model can be categorized in the followings:

- Processing (service) time – exponential distribution and normal distribution
- Set up time – uniform distribution
- Lead time – normal distribution
- Time between arrival – constant
- Time between machine repair – exponential distribution
- Time between machine failure – normal distribution

The important aspect of the proposed simulation model is the combination of three word views. These word views are commonly referred to as the event, activity scanning and process orientations. We assume that changes in the state of the system occur at discrete instants in time in accordance with predefined operating rules. Due to the fact that the experimental conditions are fully known, controllable and reproducible for given values of:

- A – the machine age at which the preventive maintenance is necessary
- B – the machine age at which it is necessary to stock parts
- X – the number of parts to maintain in stock (threshold value)

We adopt the network structure model without losing and general approach. The proposed network represents the system of interest by a set of symbols (nodes and branches) interfaced with user written subroutines, if necessary. Our process consists of a time-ordered sequence of events altered with several error, activities and oriental branches. Also this proposed simulation model consists of several networks, each of them describes a specific contribution in the system modeling purpose policy generation (i.e., demand generation, design, planning, control policy generation, state of the machine description, inventory control etc.) which is very helpful for any organization related to time.

Production Control Policy

The production control policy is modeled by using θ defined by [4, 5, 6]

$$\theta = 0, \text{ if } a(t) < B \quad \text{and} \quad \theta = X, \text{ if } a(t) \geq B$$

where, $a(t)$ is the function of machine age.

Machine States

$P[a(t)]$ is the break down probability distribution of the machine. Based on a practical view point, such a probability can be chosen such that:

$$\text{Lim}_{a(t)\to o} P[a(t)] = 0 \text{ and } \text{Lim}_{a(t)\to} P[a(t)] = 1$$

where $a(t)$ is defined here as the number of produced parts since the last intervention on the machine (repair or preventive maintenance). We assume that $P[a(t)]$ is given by

$$P[a(t)] = 1 - \exp[-f\{a(t)\}]$$

Different trends of $P[a(t)]$ (curves C_1 to C_4) are given by the followings:

$K_1\ a(t)$ for curve C_1, $K_1 = 5 \times 10^{-3}$

$f[a(t)] = K_2\ [a(t)]^2$ for curve C_2, $K_2 = 10^{-5}$

$K_3\ [a(t)]^3$ for curve C_3, $K_3 = 10^{-8}$

$K_4\ [a(t)]^4$ for curve C_4, $K_4 = 10^{-11}$

The above mentioned expressions illustrated the impact of a machine age (i.e. production of nos. of parts) on its dynamics manufacturing machines such as machine tools have increased failure rates and corresponds to curve C_3 or C_4 (i.e., milling machining, lathe etc.) the one machine, one product system considered [7, 8, 9].

30.7. CLOSURE

JIT is a revolutionary concept that is a challenge by its simplicity. It introduces no advanced technology or complicated principles but instead of strives to eliminate unnecessary burdens of complexity into the system. It helps a supplier to deliver the expected goods frequently in required small lots. Ideally in a JIT environment, a buyer is expected to synchronize his production capacity with the buyers demand so that the inventory in the supply line is reduced and eventually eliminated.

Computer simulation is recognized as an excellent facility for studying JIT practice. Computer simulation under a JIT philosophy to study the impact of internal and external factors on JIT implementations, and to compare the performance of a JIT approach with other design, planning and control techniques.

Researchers should pay due attention to such issues as model verification, experimental design and statistical output analysis and interpretation. Also they are advised to implement the JIT environment in any organization related to time factor.

REFERENCES

1. Buffa, E.S., (1995), "*Modern Production/Operations Management*", Wiley Eastern Ltd., Japan.
2. Charry, S.N., (2000), "*Production and Operations Management*", Tata McGarw-Hill Publishing Company Limited, New Delhi.
3. Dilwort, J.B., (1998), "*Operations Management*", Tata McGraw-Hill Publishing Company Limited, New Delhi.
4. Philipoom, P.R. et al. (1990), A Mathematical programming approach for determining workcentre lot sizes in a JIT system with signal Kanbans, International Journal of Production Research, Vol. 28 (1), pp.10–15.
5. Amuthakannan R., Kannan S.M., and Pandian D.S., (2005), Modeling approach for predicting and improving the reliability of the real time system, Proc. National Conference on Product Development with Mechatronics System for Global Quality, May 2–3, Madurai, TN.
6. Latkar S.D., and Ingale S.M., (2005), An Optimum Batch Size For a JIT Manufacturing System, Proceedings of National Conference on Advances in Mechanical Engineering, May 13–14, pp. 167–170.
7. Thiel, D., (1996), Analysis of the behavior of production systems using continuous simulation, Int. Journal of Production Research, Vol. 34(11), pp. 3227–3251.
8. Yasuhiro, M., (1965), "*Toyota Production System*" (An integrated approach to JIT), Second edition, pp. 10–14. Japan.
9. Senthil N. S.K, Senthil V.P., and Selladurai V., (2005), Optimizing the make span for job shop scheduling problem using ant colony, Proceedings of National Conference on Advances in Mechanical Engineering, May 13–14, pp. 129–132.

ADVANCES IN SUPPLY CHAIN MANAGEMENT

31.1. INTRODUCTION

The purpose of this chapter is to explore the understanding on Supply Chain Management (SCM) and identify the scope for future research. Chapter is based on review of research papers, articles and views published in the renowned journals and magazines during last decade. The whole journey is divided into three segments – understanding the basics, classifying the issues involved and exploring the directions for future research. It reviews the literature on three broader areas– INBOUND–MANUFACTURING – OUTBOUND of supply chain.

Supply Chain Management (SCM) has gained significance as one of the 21st century manufacturing paradigms for improving organizational competitiveness. SCM has been considered as a competitive strategy for integrating suppliers and customers with the objective of improving responsiveness and flexibility of manufacturing organizations. The term Supply Chain Management (SCM) was first introduced by consultants in the early 1980s. The concept was mainly used to discuss the benefits of integrating a firm's internal business functions; such as purchasing, manufacturing, sales and distribution [1]. The popularity of SCM is attributed to several driving forces, i.e. global sourcing, an emphasis on time and quality based competition, and their respective contributions to greater environmental uncertainty [2]. Logistics has always been a central and essential feature of all economic activity. It is the time related positioning of resources or the strategic management of the total supply chain. The broader spectrum of activities include procurement, manufacturing, distribution and waste disposal, together with associated transport, storage and information technology.

Effective management of logistics activity can bring substantial savings in cost and improves overall competitiveness of industry. For example, logistics influences about 64% of all costs in a manufacturing company. American companies had spent 10% of their GDP on supply chain related activities in 1997. In Europe logistics related costs hover around 55–65% of total turnover. SCM is a way of obtaining vertical integration benefits without its formal ownership costs [3]. Lambert and Cooper [4] and Ellarm and Cooper [5] identify following six issues important in SCM.

- Supply chain information sharing and monitoring systems;
- Co-ordination systems of multiple levels inventory reduction;
- Joint planning systems on different management levels;
- The reduction of supplier bases for better coordination;
- The increase of speed of operations, information and inventory flows; and
- Supply chain costs and benefits systems.

The focused objectives of this chapter are:

- To develop basic understanding on subject area;
- To explore the views reported for broader areas of supply chain; and
- To identify the future directions and scope for research.

Next section provides an issue based classification of SCM. A comprehensive understanding on various issues is presented in tabular format which further provides an opportunity to synthesize and relate the issues of various domains. Section 31.3 develops insights into the future directions of SCM by narrating some of the misperceived issues.

31.2. VIEWS REPORTED

The post-World War-II supply chain was a set of linear, individualized processes that linked manufacturers, warehouses, wholesalers, retailers and consumers together in the form of a human/paper chain [6]. A diagram depicting basic elements of supply chain is presented in Fig. 31.1. SCM is a new name given to traditional concept logistics with special thrust on managing relationship, chain optimization through integration of IT, alliances, vendor evaluation and purchasing strategy, outsourcing and so on. In the last two decades enormous literature is generated, in which, some common concepts like definitions of supply chain, benefits over earlier concept of logistics, stress on core-competence, use of third party logistics, performance parameters of supply chain, network design and simulation, customer centric design approach, global sourcing and international issues, IT infrastructure requirements, B2B and B2C transactions, ownership issues, trust and sharing of information, value addition based profit sharing, etc. are found repetitive and predominant. Here an issues based classification of supply chain literature is presented. Mainly issues are categorized under the basic elements of supply chain like inbound, manufacturing and outbound. Figure 31.2 provides the overall idea on adopted strategy for classifying the literature.

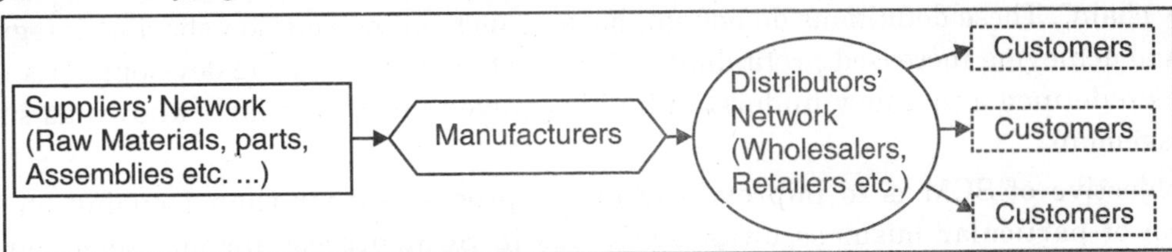

Fig. 31.1: Basic elements of supply chain

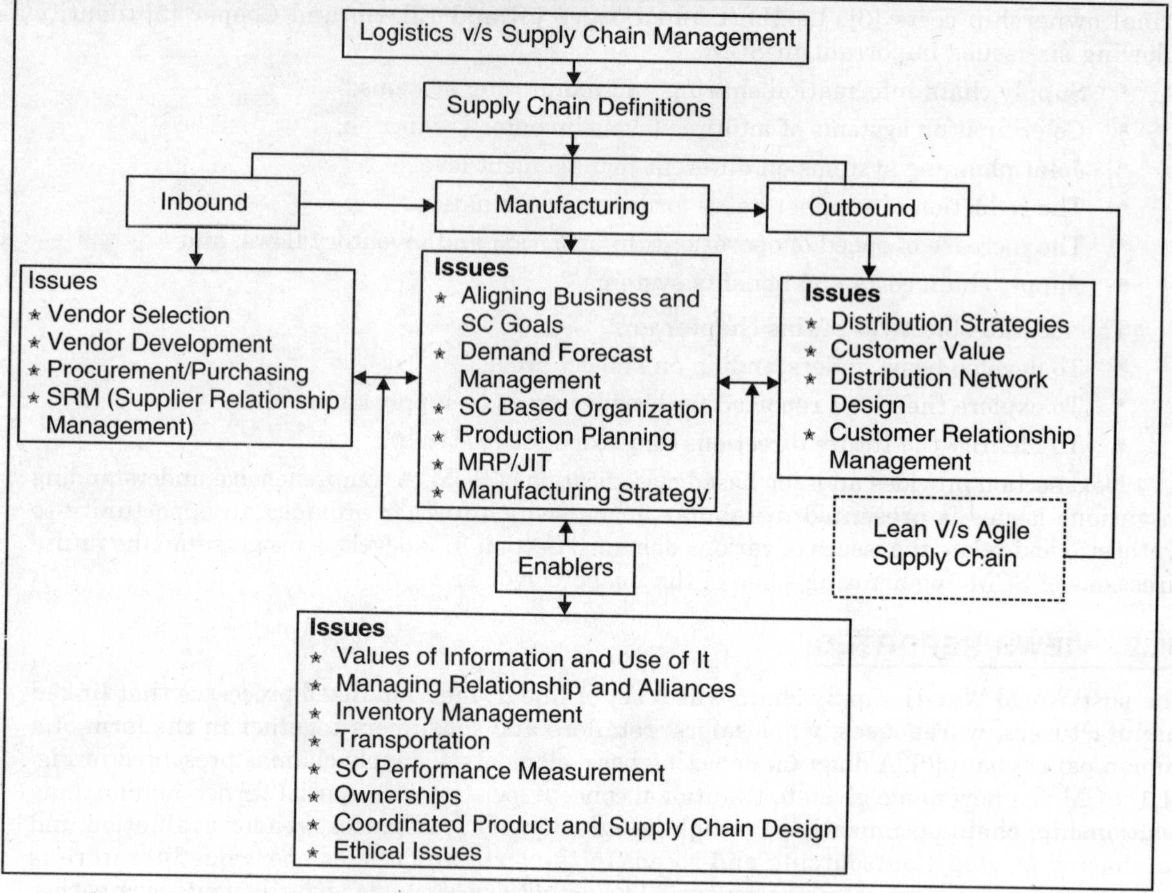

Fig. 31.2: Classification of SCM literature

Logistics v/s Supply Chain Management

Logistics is the flow of material, information, and money between consumers and suppliers [7]. Logistics, defined as the technology for controlling physical flows and information flows [8] is a major contributor to performance development in large corporations, and in this context it is more commonly known as logistics efficiency. Further, Institute of Logistics [9] defines the term, as "Logistics is the time related positioning of resources or, the strategic management of total supply chain". These definitions do not emphasize much on information sharing; integration of system and achieving increased profitability of complete chain. With the developments in IT, concept was redefined and a new understanding is evolved more popularly known as supply chain management.

The objective of SCM is to improve the entire process rather than focusing on local optimization of particular business units. Activities in SCM process include selection and development of vendors, purchase, set up and maintenance of production technologies, pricing, order processing and internal order communication, cost control, production, order fulfillment and service, distribution and delivery, and payment administration.

Just as the supply chain's definition can have different focuses, so too can the extent of the supply chain. It has now evolved from the original "direct supply chain" to the "extended supply chain" and then to the "ultimate supply chain". These evolutions provide organizations with greater potential to utilize their supply chains for strategic purposes, but at the same time, such extensions also provide increased complexity [10]. A comprehensive summary of select views is presented in Table 31.1.

Table 31.1: Supply chain defined

S. No.	Author(s)	Definition	Perspective
1.	Cooper and Ellarm (18)	An integrating philosophy to manage the total flow of a distribution channel from supplier to the ultimate customer.	Integration of flow
2.	New and Payne (49)	Describe supply chain management as the chain linking each element of the manufacturing and supply process from raw materials through to the end user, encompassing several organizational boundaries.	Linkage of various elements
3.	Harland (29)	Describes supply chain management as managing business activities and relationships (1) internally within an organization, (2) with immediate suppliers, (3) with first and second-tier suppliers and customers along the supply chain, and (4) with the entire supply chain.	Management of business activities and relationships
4.	Institute of Logistics (34)	The supply chain is a sequence of events intended to satisfy a customer. It can include procurement, manufacture, distribution and waste disposal, together with associated transport, storage and information technology.	Customer satisfaction
5.	Lambert et al. (39)	Supply chain management is the integration of key business processes from end user through original suppliers that provides products, services and information that add value for customer and other stakeholders.	Integration of processes, value addition to stakeholders
6.	Handfield and Nichols (28)	The supply chain encompasses all activities associated with the upstream and down-stream flow and transformation of goods and information from the raw materials stage (extraction) through to the end user. Supply chain management is the integration of these activities through improved supply chain relationships, to obtain a sustainable competitive advantage.	Improved supply chain relationships and competitive advantage
7.	Simchi-Levi et al. (56)	Supply chain management is a set of approaches utilized to efficiently integrate suppliers, manufacturers, warehouses, and stores, so that merchandise is produced and distributed at the right quantities, to the right locations, and at the right time, in order to minimize system wide costs while satisfying service level requirements.	Minimization of system wide cost and satisfying service level
8.	Chopra and Meindl (12)	A supply chain consists of all parties involved, directly or indirectly, in fulfilling a customer request.	Customer
9.	Heizer and Render (33)	Defined as the integration of activities to procure materials, transforms them into intermediate goods and final products, and delivers to customers.	Integration of activities

10.	Mentzer et al. (45)	Supply chain management is defined as the systemic, strategic coordination of the traditional business functions within a particular company and across businesses within the supply chain, for the purposes of improving the long-term performance of the individual companies and the supply chain as a whole.	Performance improvement of a company, and supply chain as a whole
11.	SCOR model (54)	Supply chain management is about managing the flow of products and services and the associated information, across the whole business system to maximize the value to the end customer.	Flow of information
12.	London and Kenley (42)	SCM is a melding of logistics (i.e., of distribution and production), procurement, industrial organization economics, marketing and strategy, which emerged as a distinct area of research in the mid-1980s.	Melding of logistics
13.	Leenders et al. (41)	Supply chain management is a system approach to managing the entire flow of information, materials, and services from raw materials suppliers through factories and warehouse to the end customer.	System approach
14.	Frazelle (21)	Supply chain logistics is the flow of material, information, and money between corporations (intervorkstation, interfacility, intercorporate, and interchain).	Flow of material, information and money
15.	Burt (6)	Supply chain management is the collaborative effort of multiple channel members to design, implement, and manage seamless value-added processes to meet the real needs of the end customer.	Integration of people and technological resources

Issues Involved in SCM–Literature Review

In research, various views and thoughts are reported on each issue. Researchers have supplemented as well contradicted the views for varying context of SCM. Few significant contributions are summarized in Table 31.2.

Table 31.2: Views reported on SCM issues

S. No.	Issue	Reference	View
	INBOUND		
1.	Vendor selection	Dzever et al. (17)	Vendor rating and rationalization are dependent on product conception, geographical proximity, commercial capability, contractual relationship, image and technical service.
		Sarkis and Talluri (53)	Supplier selection decisions must include strategic and operational factors as well as tangible and intangible factors in the analysis.
		Ghodsypour and O'Brien (25)	Selecting the right suppliers significantly reduces the purchasing costs and improves corporate competitiveness
		Humphreys et al. (30)	Reports on three different methods of supplier evaluation —categorical; cost ratio; and weighted point.
2.	Vendor development	Dzever et al. (17)	Vendor development focuses on mutual dependence of buyer-supplier with the objective of achieving long-term customer satisfaction as a result of demand aggregation in the demand chain.

3.	Procurement and purchasing	Fung (23)	Purchasing management is concerned with the acquisition of suppliers' goods and resources, in order to contribute to the administrative and strategic objectives of the organization.
		Kalakota and Robinson (36)	Drivers for the popularity of procurement are trends in global sourcing, emphasis on time to market, product quality based competition, customer uncertainty and the need to improve bottom-line costs. Purchase of goods and services represents the single largest cost item for any given enterprise.
		Fung (23)	Within the context of interactive buyer-seller relationship, purchasing is more than buying as marketing is more than selling.
		Monczka et al. (48)	The traditional purchasing process leaves much to be desired, as it is neither efficient nor effective. For most large manufacturing companies, purchased materials and services can represent 55–75 per cent of their cost of goods sold.
		Thompson (59)	Applying a strategic approach to a poorly purchased good or service generates a saving of between 20 per cent and 40 per cent in expenditure.
4.	Supplier Relationship Management (SRM)	Humphreys et al. (30)	The real productivity, design and quality improvements are not obtainable unless the suppliers in the collaborative relationship innovate to the best of their abilities, in conjunction with the buyer organization.
		Goffin et al. (27)	Supplier management is one of the key issues of supply chain management because the cost of raw materials and component parts constitutes the main cost of a product and most of the firms have to spend considerable amount of their sales revenues on purchasing. Close relationships with suppliers restrict price rises to below market norms.
		Thompson (59) Fung (23)	In practice, purchasing managers have to respond creatively to internal customer's need on the one hand and to maintain a mutually profitable relationship with suppliers on the other.
		Willis (65)	Having a successful relationship with a supplier results in trust and the ability to be customer driven, customer intimate and customer focused.
	MANUFACTURING		
5.	Aligning business and SC goals	Simchi levi et al. (56)	Supply chain goals are conflicting in nature and hence a proper integration of supply chain issues with business strategies and goals is necessary. Supply chain partners modularity specifies the extent to which individual companies participating in the creation of an agile supply chain will have to align operations through the redesign of management practices and interfaces for the flow of goods and information.

6.	Demand forecast management	Yu et al. (66)	Variability of an upstream member's demand is greater than that of the downstream member.
		Lee et al. (40)	The causes of the bullwhip effect have been identified as: demand forecasting, order batching, price fluctuation and rationing game.
7.	SC based organization	Flynn et al. (20)	The traditional organizational structure needs to change to accommodate the necessary information flows and problems created from intensifying the interdependence of activities.
		Christopher (13)	The problem with the traditional vertically designed organization structure is that it is inwardly focused and concentrates primarily on the use of resources rather than upon the creation of outputs.
8.	MRP/JIT	Arogyaswamy and Simmons (4)	With JIT, inventories are eliminated and interdependence among the operations is intensified; the tighter linkages. make operations more responsive to one another.
9.	Manufacturing strategy	Christopher (13)	Identifies three issues for the development of manufacturing strategies–leanness means reduction in wasteful activities, agility means how the firm responds to changes in marketplace requirements–particularly for volume and variety and third key word in manufacturing today is flexibility which is the ability of to produce any variant in any quantity, without significant cost penalty.
10.	Production planning	Mason et al. (44)	ERP has extended the reach of the planning system to include the entire enterprise, from marketing to product development, and to achieve total organizational excellence through integration.
	OUTBOUND		
11.	Distribution strategies	Chilerhouse and Towill (10)	The overriding objective of supply chain is the generation of profit and so to satisfy the revenue source that is customer. However, this cannot be accomplished by one "all-embracing" strategy because customer requirements are very diverse and range from market place to market place.
		Christopher (13)	The critical task that underlies successful distribution today is demand management which is the process of anticipating and fulfilling orders against defined customer service goals.
12.	Customer value and customer relationship management	Eloranta (19)	Suggests that the most important aspects of customer service are delivery time promised and performance, and that customer service should be the primary objective of production management efforts.
		Towill (60)	In all cases supply chain must be integrated and operate in a seamless manner, but also must be tailored to specific customer requirements.
		Zhou et al. (67)	Logistics quality is often measured by the logistics manager's ability to distribute products on spec, on time, and on budget. Thus, the main drive to improve logistics productivity is the enhancement of customer services and asset utilization through a significant reduction in order cycle time (lead time) and logistics costs.
		Reiner (51)	Due to the direct relation between customer satisfaction and performance measures as well as indicators, customer satisfaction is relevant for evaluating process effectiveness.

13.	Distribution network design	Mourtis and Evers (47)	The decision on issues of distribution network like location and size of distribution centers, the logistic activities to be performed at these centres, the capacities required to fulfill these activities, their allocation tc specific product groups etc. are all closely interrelated, making it difficult to develop a sound distribution strategy.
		Amiri (2)	A common objective in designing such a distribution network is to determine the least cost system design such that the demands of all customers are satisfied without exceeding the capacities of the warehouses and plants.
	ENABLERS		
14.	Value of Information and use of IT	Spalding (57)	Investment in information technology (IT) for managing supply chains has been an effective way of obtaining competitive advantage.
		Short (55)	The Internet changes the way companies do business. The changes are permanent in the transition from the industrial economy to the network economy. SCM has been enabled by convergence, which refers to the integration of computer and communication technology.
		Chou et al. (11)	The Internet has created a medium to connect fragmented buyers and sellers, to lower transaction costs, and to make price transparent in the supply chain.
		Rusell and Hoag (52)	The most significant challenges in implementing IT are not technical in nature, but human.
		Mohamed (46)	IT implementations have faltered because of a lack of user awareness, project management, and industry or firm culture. Such root causes evoke the notion that human endeavor in organizations relies on human communication
		Jharkharia and Shankar (35)	Planning and logistical integration throughout the SCN requires centralized co-ordination of key data (drawings, specifications, instructions, order forecasts, production plans and delivery schedules).
		Agrawal and Shankar (1)	As the IT enablement of a supply chain is a strategic and capital-intensive issue, many researchers have highlighted the importance of mutual trust for long-term relationships and the confidentiality of information among partners.
15.	Managing relationship and alliances	Chandra and Kumar (7)	A successful strategic alliance or a partnership must be based on extreme trust, loyalty, positive sum game (a win-win relationship), cross functional teams, sharing common goals and cooperation that includes willingness to assist and positive negotiations based on fairness.
		Yu et al. (66)	By coordinating different parties along the logistics network, or establishing business partnerships, supply chain management creates a win-win situation for all members.
		Harreld (32)	Concept known as "collaborative commerce" has altered the traditional relationship between suppliers and manufacturers from one of "haggling and hedging bets on product orders" to a mutually beneficial model that holds promise to improve the competitive position of all parties involved.
		Stein and Sweat (58)	Managing the organizational culture and the relationships involved along the supply chain are as important as the software itself.

		Tyndall et al. (63)	Supply chain partnerships move from open-market negotiations, to cooperation, to coordination and finally to collaboration. This collaboration leads to a high degree of interdependence along the supply chain.
		Tompkins (61)	The ability to function as a team is an important part of the concept of a highly integrated workflow encompassing total operations, known as "supply chain synthesis".
		Hoek (31)	In supply chains control is no longer entirely based on ownership and functional hierarchies but rather on networking and cross-company relations.
16.	Inventory management	Chen et al. (9)	Stated that order variability could result in overstocking and inefficient use of resources.
		Fransoo and Wouters (22)	Discussed measurement issues of the bullwhip effect. According to the authors, bullwhip effect in a supply chain channel may be measured based on the relative value of the coefficient of variation of orders created and the coefficient of variation of demand or orders received by the channel.
		Olhager (50)	With global markets, increasing global competition and shorter product life cycles, the choices and shifts between make-to-order (MTO) and make-to-stock (MTS) policies must be made faster and at a strategic level.
		Giannoccaro et al. (26)	Inventory management is to a greater extent relevant when a whole supply chain (SC), namely a network of procurement, transformation, and delivering firms, is considered.
		Lee et al. (40)	Lack of a coordinated inventory management throughout the SC often causes the bullwhip effect.
17.	SC performance measurement	Hoek (31)	In measuring performance in the supply chain, where control is no longer based on ownership only, but rather on networking across interfaces, the measurement system may reflect a system of measuring the unmeasurable.
		Bechtel and Jayaram (5)	State that measurement in the supply chain may use integrated measures that are cross-functional and can be applied to the entire process (for example, time from cash to cash) in order to avoid optimization at one point in the chain without considering potential consequences at other points in the chain.
		Chan and Qi (8)	There are some in-depth problems of PMSs in supply chain context including: lack of balanced approach to integrating financial and non-financial measures; lack of system thinking, in which a supply chain must be viewed as one whole entity and the measurement system should span the entire supply chain; and loss of supply chain context, and hence this kind of PMS encourages local optimization.
18.	Coordinated product and supply chain design	Korhonen (37)	Distinguishes the product based and the geographical approach. The geographical approach centers on analyzing local or regional networks of material flows such as in eco-industrial parks, while the product-based approach usually builds on the life cycle of products.
		Simchi levi et al. (56)	Addresses the issue from three perspectives–Economic packaging and transportation, Concurrent and parallel processing, and postponement.

19.	Ethical issues	Burt et al. (2004)	Identifies seven area where ethical considerations are important in supply chain – avoid sharp practices, negotiation, samples, treating sales people with respect, sub standard materials and services.
		Trevisan (1986)	Found possibility of unethical practices in the issues like - giving preferential treatment to suppliers, canceling a purchase order for parts in process while also seeking to avoid cancellation charges, lying to or grossly misleading a sales person in a negotiation, making supplier dependent on the supply management organization etc.

Table 31.3: Comparison of lean supply with agile supply: The distinguishing attributes

Distinguishing attribute	**Lean supply**	**Agile supply**
Typical products	Commodities	Fashion goods
Marketplace demand	Predictable	Volatile
Product variety	Low	High
Product life cycle	Long	Short
Customer drivers	Cost	Availability
Profit margin	Low	High
Dominant costs	Physical costs	Marketability costs
Stockout penalties	Long-term contractual	Immediate and volatile
Purchasing policy	Buy materials	Assign capacity
Information enrichment	Highly desirable	Obligatory
Forecasting mechanism	Algorithmic	Consultative

(Source: Mason-Jone, Naylor and Towill (44))

As an outcome, supply chain can be attributed as an effective or responsive supply chain. Efficiency emphasize on cost reduction and making the system lean while responsiveness demands higher speed in satisfying customer demands. Attributes of SCM have made agility – a necessity.

Lean capabilities – as a sole answer – fall fundamentally short because of their focus on optimizing internal supply chain operations, rather than creating the basic capabilities to respond to what matters to the customer. Customer centric view suggests that a key to agility is to let customers define their own preferred size and type of service window. Mason-Jone et al. (44) based on literature review consolidates factors (Table 31.3) that impact the viability and relevance of agility.

Further, Van Hoek and Harisson (64) proposes a framework matching supply characteristics with demand characteristics as shown in Fig. 31.3. In this, volatility, variability and forecastability are embedded in market predictability and time pressure is included as a separate dimension.

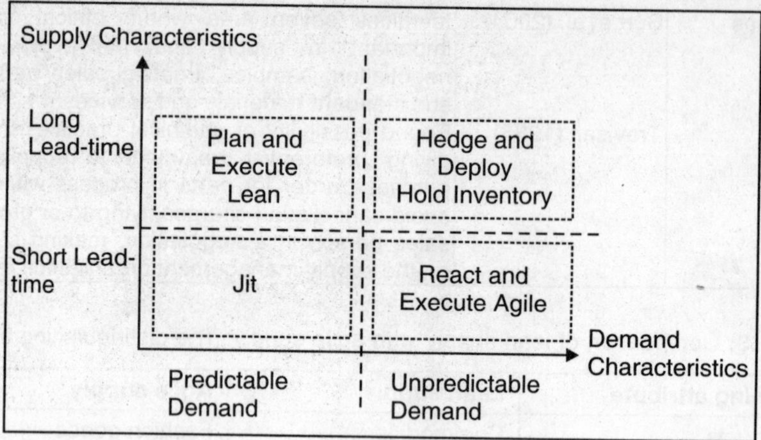

Fig. 31.3: Leanness and agility under demand and supply

31.3. EXPLORING FUTURE

Looking to the broad spectrum of supply chain literature and conflicting issues, following observations are noteworthy to carry out research work on futuristic dimensions. Good relationship between the customer and the supplier contributes to reliable information flows, and reliable demand information flows in turn contribute to high efficiency—these are well-researched issues also in other industry environments. Traditional vertically designed organizational structure needs to be modified for horizontal one to take the full advantage of SCM. This requires the development and deployment of system like ERP is an essential part of it. The development of le-agile supply chain which offers cost effectiveness with improved responsiveness is essential from two aspects. The first is the linking of customer relationship management to the management of the supply chain and second is to manage the supply chain in a dynamic commercial environment that is making network management rather than supply chain management.

Present performance measurement systems are deeply grounded in the accounting disciplines of recording profit. Looking to the complex nature of supply chain activities performance measurement matrix in this direction need to be developed; and outsourcing of non-core functions improves the effectiveness and responsiveness of a particular supply chain. Selection of right kind of service and provider and needed understanding on tangible and intangible aspects is a new direction for research.

On the closer, it is worth to mention that companies will not seek to achieve cost reductions or profit improvement at the expense of their supply chain partners, but rather seek to make the supply chain as a whole more competitive. In short, the contention that it is supply chains, and not single firms, that compete is a central tenet in the field of supply chain management. The whole discussion clearly indicates that supply chain management is not a standalone process. Many a times word is misperceived and only focus was paid to the supply side of the business or to the purchasing function. To summarize, a clear and comprehensive understanding on supply chain management, we address "4 W's" (Fig. 31.4) on it.

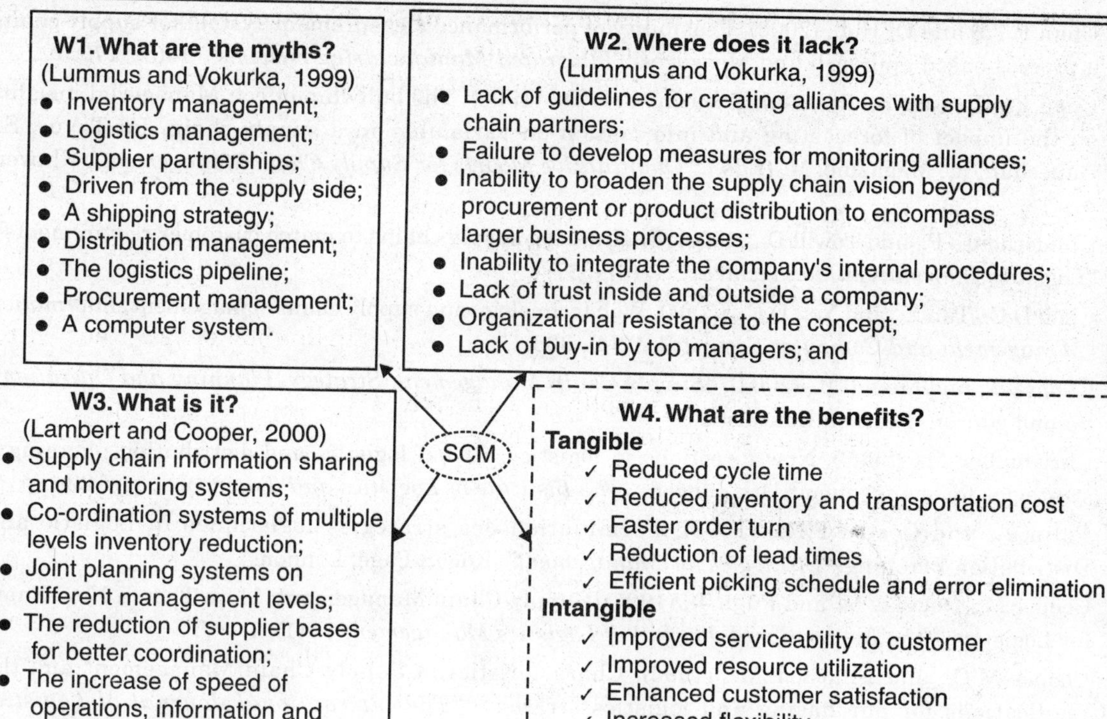

Fig. 31.4: Understanding 4W's on SCM

REFERENCES

1. Agarwal A., and Shankar R., (2003), Online trust building in e-enabled supply chain, "*Supply Chain Management*": An International Journal, 8 (4), 324.
2. Amiri A., (2004), Designing a distribution network in a supply chain system: Formulation and efficient solution procedure, "*European Journal of Operational Research*", accepted 6 September 2004.
3. Arend R.J., and Winser J.D., (2004), Small business and supply chain management: is there a fit? "*Journal of Business Venturing*", accepted on 30 November 2003.
4. Arogyaswamy B., and Simmons R.P., (1991), Thriving on interdependence: the key to JIT implementation, "*Production and Inventory Management Journal*", 32 (3), 56.
5. Bechtel C., and Jayaram J., (1997), Supply chain management: a strategic perspective, "*The International Journal of Logistics Management*", 8(1), 15.
6. Burt D.N., Dobler D.W., and Starling S.L., (2004), "*World Class Supply Chain - The Key to Supply Chain Management*", Seventh edition, Tata McGraw-Hill publication.
7. Chandra C., and Kumar S., (2000), Supply chain management in theory and practice: a passing fad or a fundamental change?" "*Industrial Management and Data Systems*", 100 (3), 100.

8. Chan F.T.S and Qi H.J., (2003), Feasibility of performance measurement system for supply chain: a process-based approach and Measures, *"Integrated Manufacturing Systems"*, 14(3), 179.

9. Chen F., Drezner Z., Ryan J.K., Simchi-Levi D., (1998), The bullwhip effect: Managerial insights on the impact of forecasting and information on variability in a supply chain, In: Tayur, S., Ganeshan, R., Magazine, M. (Eds.), *"Quantitative Models for Supply Chain Management"*. Kluwer, p. 418.

10. Childerhouse P., and Towill D., (2000), Engineering supply chains to match customer requirements", *"Logistics Information Management"*, 13(6), 337.

11. Chou D.C., Tan X., and Yen D.C., (2004), Web technology and supply chain management, *Information "Management and Computer Security"*, 12(4), 338.

12. Chopra S., and Meindl P., (2004), *"Supply Chain Management: Strategy, Planning and Operation"*, Second edition, Pearson Education.

13. Christopher M., (2003), New directions in logistics, Global logistics and distribution planning – strategies for management, 4th edition, *"The Institute of Logistics and Transport"*, p.p. 22.

14. Colin J., and Costes F.N., (1993), Formulation des stratégies logistiques in Logistic and Distribution Planning: *"Strategies for Management"*, Kogan Page, London.

15. Cooper M., Douglas M. and Pugh J., (1997), Supply Chain Management: More Than a New Name for Logistics, *"The International Journal of Logistics Management"*, 8(1), 1.

16. Cooper M.C., and Ellarm L.M., (1993), Characteristics of Supply Chain Management and the implications for purchasing and logistics strategy, *"The International Journal of Logistics Management"*, 4(2), 13.

17. Dzever S., Merdji M., and Saives A.L., (2001), Purchase decision making and buyer-seller relationship development in the French food-processing industry, *"Supply Chain Management: An International Journal"*, 6(5), 216.

18. Ellram L., and Cooper M., (1993), Characteristics of supply chain management and the implications for purchasing and logistics strategy, *"International Journal of Logistics Management"*, 4(2), 1.

19. Eloranta E., (1988), "Managing production for customer service and plant efficiency", *"Industrial Engineering"*, 20(3), 36.

20. Flynn B.B., Sakakibara S., and Schroeder R.G., (1995), Relationship between JIT and TQM: practices and performance, *"Academy of Management Journal"*, 38(5), 1325.

21. Frazelle E., (2004), *"Supply Chain Strategy"*, Tata McGraw-Hill, New Delhi.

22. Fransoo J.C., Wouters M.C.F., (2000), Measuring the bullwhip effect in the supply chain, *"Supply Chain Management"*, 5(2), 78.

23. Fung P., (1999), Managing purchasing in a supply chain context - evolution and resolution, *"Logistics Information Management"*, 12(5), 362.

24. Ganeshan R., (2002), Web-enabling the supply chain: an exploratory case study, New Directions in Supply–Chain and Technology Management, *"Technology, Strategy, and Implementation"*, Amacom, New York, NY.

25. Ghodsypour S.H., and O'Brien C., (2001), The total cost of logistics in supplier selection, under conditions of multiple sourcing, multiple criteria and capacity constraints, *"International Journal of Production Economics"*, 73, 15.

26. Giannoccaro I., Pontrandolfo P., Scozzim B., (2003), A fuzzy echelon approach for inventory management in supply chains, "*European Journal of Operational Research*", 149, 185.
27. Goffin K., Szwejczewski M., and New C., (1997), Managing suppliers: when fewer can mean more, "*Journal of the Operational Research Society*", 50.
28. Handfield R.B., Nichols E.L., (1999), "*Introduction to Supply Chain Management*", Prentice Hall, Upper Saddle River, New Jersey, p. 2.
29. Harland C.M., (1996), "*Supply Chain Management: Relationships, Chains and Networks*", British Academy of Management, 7 (Special Issue), 63.
30. Humphreys P., Mak K.L., and Yeung C.M., (1998), A just-in-time evaluation strategy for International procurement, "*Supply Chain Management*", 3(4), 175.
31. Hoek R.I., (1998), Measuring the unmeasurable – measuring and improving performance in the supply chain, "*Supply Chain Management*", 3(4), 187.
32. Harreld H., (2001), "*Supply-chain Collaboration*", InfoWorld, Vol. 23 No. 52/53, pp. 22–5.
33. Heizer J. and Render B. (2001), "*Operations Management*", 6th Edition, Prentice-Hall, Englewood Cliffs, NJ.
34. Institute of logistics (1998), "*Members' Directory*", p. 8, Institute of Logistics and Transport, Corby.
35. Jharkharia S., and Shankar R., (2004), IT enablement of supply chains: modeling the enablers, "*International Journal of Productivity and Performance Management*", 53 (8), 700.
36. Kalakota R., and Robinson M., (2001), "E-Business 2.0: Roadmap for Success", Boston: Addison-Wesley.
37. Korhonen J., (2002), Two paths to industrial ecology: applying the product-based and geographical approaches, "*Journal of Environmental Planning and Management*", 45(1), 39.
38. Lambert D.M., and Cooper M.C., (2000), Issues in supply chain management, "*Industrial Marketing Management*", 29, 65.
39. Lambert D., Cooper M., and Pagh J., (1998), Supply Chain Management: Implementation issues and research opportunities, "*The International Journal of Logistics Management*", 9(2), 1.
40. Lee H.L., Padmanabhan V., Whang S., (1997), The bullwhip effect in supply chains, "*Sloan Management Review*", 38(3), 93.
41. Leenders M.R., Fearon H.E., Flynn A.E., and Johnson P.F., (2002), "*Purchasing and Supply Management*", Twelfth Edition, McGraw-Hill, Inc., New York, p.20.
42. London K.A., Kenley R. (2001), An individual organization economic supply chain approach for the construction industry: a review, "*Construction Management and Economics*", 19, 777.
43. Lummus R.R., and Vokurka R.J., (1999), Defining supply chain management: a historical perspective and practical guidelines, "*Industrial Management and Data Systems*", 99(1), 11.
44. Mason-Jones R., Naylor J.B., and Towill D. (2000), Engineering the agile supply chain, "*International Journal of Agile Management Systems*", 1.
45. Mentzer J.T., DeWitt W., and Keebler J., (2001), What is supply chain management? in Mentzer, J.T. (Ed.), "*Supply Chain Management*", *Sage Publications, Inc., Thousand Oaks, CA.*
46. Mohamed S., (2003), Web-based technology in support of construction supply chain networks, "*Work Study*", 52(1), 13.
47. Mourtis M., and Evers J.M., (1996), Distribution network design: an integrated planning support framework", "*Logistics Information Management*", 9(1), 45.

48. Monczka R.M., Petersen K.J., Handfield R.B., Ragatz G.L., (1998), Success factors in strategic supplier alliances: the buying company perspective, *"Decision Sciences"*, 29(3), 553.
49. New S.J., Payne P., (1995), Research frameworks in logistics: three models, seven dinners and a survey, *"International Journal of Physical Distribution and Logistics Management"*, 25(10), 60.
50. Olhager J., (2003), Strategic positioning of the order penetration point, *"International Journal of Production Economics"*, 85, 319.
51. Reiner G., (2004), Customer-oriented improvement and evaluation of supply chain processes supported by simulation models, *"International Journal of Production Economics"*, accepted 20 July 2004.
52. Russell D.M., and Hoag A.M., (2003), People and information technology in the supply chain Social and organizational influences on adoption, *"International Journal of Physical Distribution and Logistics Management"*, 34(2), 102.
53. Sarkis J., and Talluri S., (2002), A model for strategic supplier selection, *"Journal of Supply Chain Management"*, 38(1), 18.
54. SCOR Model (2001), The Supply Chain Operations Referencemodel (SCOR), http://www.supply-chain.org/.
55. Short D., (2002), Dean's forum – convergence, Mid-American Journal of Business, 17(1).
56. Simchi-Levi D., Kaminsky P., and Simchi-Levi E., (2000), Designing and managing the supply chain: *"Concepts, Strategies and Case Studies"*, Erwin Mc Graw-Hill, Singapore.
57. Spalding J.O., (1998), Transportation industry takes the right-of-way in the supply chain, ILESolutions, 30(7), 24–28.
58. Stein T., and Sweat J., (1998), Supply chains emerge as both a tool and a philosophy, *"Computer Reseller News"*, November, 102.
59. Thompson M., (1996), Effective purchasing strategy: the untapped source of competitiveness, *"Supply Chain Management"*, 1(3), 6.
60. Towill D.R., (1997), The seamless supply chain: the predator's strategic advantage, *"International Journal of Technology Management"*, 13(1), 37.
61. Tompkins J.A., (1998), Time to rise above supply chain management, *"Transportation and Distribution"*, 39(10), 16.
62. Trevisan R.E., (1986), Developing a statement of ethics: A case study, *"Journal of Purchasing and Materials Management"*, Fall 1986, p.13.
63. Tyndall G., Gopal C., Partsch W., and Kamauff J., (1998), Supercharging Supply Chains: New Ways to Increase Value through *"Global Operational Excellence"*, *Wiley, New York, NY.*
64. Van Hoek R., and Harisson A., (2001) Editorial, Special issue on creating the Agile Supply Chain, *"International Journal of Physical Distribution and Logistics Management"*, 31(4), 231.
65. Willis A.K., (1995), Customer delight and demand management: can they be integrated? Conference proceedings, American publication and inventory control society.
66. Yu Z., Yan H. and Edwin Cheng T.C., (2001), "Benefits of information sharing with supply chain partnerships", *"Industrial Management and Data Systems"*, 101(3), 114.
67. Zhou G., Min H., and Gen M., (2003), A Genetic Algorithm Approach to the Bi-criteria Allocation of customers to warehouses, *"International Journal of Production Economics"*, 86, 35.

CASE STUDY: SUPPLY CHAIN INVENTORY PERFORMANCE MEASURES THROUGH ARENA SIMULATION TOOL

32.1. INTRODUCTION

Inventory is both an asset and a liability. It exists at all stages of the supply chain. It is an operational decision and its planning is controlled by different uncertain parameters in the supply chain. Therefore simulation technique should be used which has the ability to capture the uncertain parameters easily and can do the sensitivity analysis. The supply chain manager can imitate the real life situation through simulation model and able to find the performance measures for set of particular input parameters, which represent a particular situation. In this chapter, a simulation model is developed using ARENA simulation tool for evaluating the supply chain inventory performance measures. One case situation (i.e., a serial supply chain configuration of five manufacturers, five wholesalers and eight retailers for a pool of thirty products) is elucidated in order to reinforce the salient features of the concept.

Simulation is one of the most popular of all quantitative techniques because it can be applied to operational problems that are difficult to model and solve analytically. It is a technique to mimic the operations of various kinds of real world facilities or processes, usually on a computer with appropriate software [1, 2]. Determining the objective of a simulation is a starting point for the process. This is followed by a set of steps to develop the simulation model and eventually the model is run to generate some results [3]. The steps of the simulation process are as follows: definition of the problem, formulation of the model, validation of the model, generation of the results after simulation runs, analysis of the results and decision making based on the analysis. Simulations can be expressed in software using convectional programming languages. However, the most cost effective approach is usually to use a commercial simulation system like ARENA 5.0. These systems include graphical tools for building models, automated routines for testing them under different conditions, and reporting tools for analyzing the results. Simulators incorporate variability quite naturally which is an important advantage because sales, shipments, prices and other aspects vary quite a bit in real world supply chains and this has a major impact on how the supply chains perform. The simulation models handle the variability by using proper distribution. Although simulations are better than mathematical models at exploring the effects of variability, they are not as good at finding optimal solutions. The best way to handle this in simulation is to vary the value of one or more parameters in a systematic way and look for the one that gives you the best fit. According to Sunil and Meindl (2001) "a supply chain is dynamic and involves the constant flow of information, products and funds between different stages. Each stage of the supply chain performs different processes and interacts with other stages of the supply chain". Enterprises have researched under a supply chain management environment on how to satisfy a customer's various needs considering many uncertain variables with stochastic properties in the supply chain. Existing analytical methods could not treat all variables with stochastic properties in the supply chain and therefore these methods are optimally able to represent partial supply chains only. In the analytical method, as the problem size increases, obtaining solutions becomes more difficult. Moreover, even for reasonable sized problems, it is not easy to consider all aspects of the problem in analytical solutions, especially the uncertainty. This is where simulation approach is preferable. Moreover it can also work for the global optimization of planning an entire supply chain with finding local optimum values within each component [4]. Furthermore, existing supply chain networks can also be evaluated in order to assess as to whether the desired performance levels are being achieved. Though product and information flow can have clearly continuous factors in a supply chain, most supply chain problems are usually solved using discrete event simulation modeling method since, real world supply chain systems can be decomposed into a set of logically separate processes which in turn can be simulated as discrete events. Towill et al. [5] conducted simulation study to analyze the effect of system redesign strategies on the performance of a supply chain. They simulate a supply chain with three echelons: factory, distributor, and retailer. The various strategies tested include the effect of integrating information flow throughout the supply chain and removing the distributor echelon. Swaminathan et al. [6] studied the influence of sharing supplier capacity information on the performance of a supply chain. They use simulation for comparing different information sharing scenarios after deriving the optimal inventory policy for the manufacturer under stochastic demand. Inventory exists at all stages of the supply chain. The primary purpose of these inventories is to buffer against uncertainty arising from demand, process [7] and supply. Inventory usually represents from 20% to 60% of the total assets of manufacturing firms. Therefore, inventory management policies prove critical in determining the profits of such firms [1]. Inventory management is indeed a major issue in supply chain management (i.e., an approach that addresses supply chain issues under an integrated perspective) [8, 9]. Therefore, it is essential to evaluate the supply chain inventory performance measures along the supply chain before deciding inventory design parameters along the supply chain in order to know its effect to tackle the particular situation. In this chapter, a simulation model is developed using ARENA simulation tool that captures all the real life scenarios for evaluating the supply chain inventory performance measures. One case situation (i.e., a serial supply chain configuration of five manufacturers, five wholesalers and eight retailers for a pool of thirty products) is elucidated in order reinforce the salient features of the concept.

32.2. DEVELOPMENT OF A CASE SITUATION FOR SCIP MEASURES USING ARENA

A simulation model is developed for a serial supply chain consisting of multiple retailers, distributors/wholesalers and manufacturers. All retailers and wholesalers follow a continuous review inventory policy to control their inventories. Each retailer is assumed to have a group of 10 products taken from a pool of 30 known products. Wholesalers will have any combination out of those 30. The difference here lies in the fact that retailers often have over-lapping products whereas in wholesaler and manufacturer stages, the products are present only at unique venues. Manufacturers produce six products each in continuous production. The starting point of the model is the arrival of customers. The customer arrival rates are given for each of the retailers anticipating the market conditions with the help of corresponding create modules and are normally distributed. Initially the entity type is set as the customer. Each customer is then assigned a set of attributes that govern the number of each product that the customer has arrived to purchase. It is assumed that customers arriving at particular retailers will have a need only for that retailer's products. The number of orders for each product (per customer) is random and is achieved using a discrete probability distribution along with a triangular distribution. Thereafter the mean arrival rate and the number of customers entering the retailer are recorded with the help of record modules. After that, the arriving customers enter the retail sub-models. A sub-model aggregates the logic concerning a particular process thus enabling easy navigating through the model especially as the complexity involved increases. Separate sub-models are used for each of the retailers, wholesalers and manufacturers. The logic involved in each of these three sub-models is explained below.

Retailer Sub-model

Customers arrive at each retail shop at normally distributed rates. Each customer decides how much of each product he wishes to buy, the customer enters the shop. The retail shops are structured as in real life with aisles containing shelves of products. As has been mentioned, each retailer contains a random 10 products and customers arriving are in need only of those 10 at that retailer. The need for a product may also be nil (with a 40% chance as set in the discrete probability) in which case the customer is simulated to merely walk by a particular shelf without picking up a product. Inventory levels in the retailers are maintained at two degrees. There is the main retail inventory, which is the stock that the retailer keeps in store for each product. There is also the secondary shelf inventory from which the customer actually picks the products depending on how many he needs. At each shelf, the number of orders the customer has for that shelf's product is first compared with the current shelf inventory. In the case of there being sufficient shelf inventory, the customer picks his product and that amount is now decremented from the shelf. The fill rate is also calculated at this stage. A normally distributed process time is also defined for the picking process so as to take into account possible real-life delays. If the shelf inventory is not sufficient, it is assumed that the customer will pick up at least the amount that remains and not leave unsatisfied. The attribute that defines the number of orders for that product is now reassigned to this new value. At this time the shelf inventory becomes zero. For arriving customers beyond this period, a shelf stock-out counter is maintained. The aisles are all treated as named stations and the customers traverse from one station to another sequentially. After the entire picking process is complete, the customers arrive at the billing counters area, which contains 3 billing counters. The counter to which they must go to get billed is chosen by the pick station module which will send a particular customer to the billing counter with the least people already in queue and the least enroute to it. In this way, efficiency is maintained. Once the customer is at the counter, a retailer clerk resource processes the order. The billing process constitutes the most complex part of the model. The number of orders of each customer is processed one after the other; at the conclusion of which the customer leaves the store. The processing of each order takes place in the following way. The shelf inventory is first checked as to whether it is empty or not. If it is empty, a pre-decided amount (called Preshelf) is subtracted from the main inventory and added to the shelf inventory of that particular product. The main inventory is then compared with the ROP and a check is also performed as to whether order (if needed) has been placed already. If no order needs to be placed, the next product is processed. If ROP has been reached and an order needs to be placed, a separate module is used to create a duplicate of the entity. One of the duplicates continues the product processing for the remaining products that the customer may have bought. The other duplicate now becomes a retail order entity and stores an attribute corresponding to the product that needs to be ordered and another entity naming the retailer that the order comes from. An order check is assigned after which the entity then proceeds to the wholesaler sub-model for further processing. The inbound at each retailer accepts the orders when the transporters bring them in. The inventory is then incremented by the necessary amount and the order check is reset.

Wholesaler Sub-model

The arriving entities for the wholesale sub-models are the retail orders. The retailers to be supplied by a particular wholesaler are assigned an attribute with the help of which they can easily be identified. The product number corresponds to a particular index of each wholesaler. This index is the basis for all calculations in this sub-model. The queues for waiting retail orders are in the form of queue sets and products of each wholesaler index are made to wait in their corresponding queue. Here too, as in the retail level a decide module is first used to check is the wholesale inventory level for that product is greater than the retail order quantity for the retail order concerned. This retail order value is stored in a 20×5 matrix as it must account for the products present in all retailers. If it is, then the retail order is processed, the wholesale inventory level is decremented and the number of retail orders satisfied is recorded. The resource used to carry out the process is labeled as a wholesaler and the number of resources can be set as required. If the wholesale inventory level is less than the retail order, then it is considered as a stock out and the number of stock-outs is recorded. At this stage, a hold module is used to ensure that the retail order waits at the wholesaler until his reorder arrives. A wait-time has also been calculated for each product in wholesaler. At this point, a separate module is used to ensure that while the entity waits, simultaneous reorder may also take place. At the wholesale level too, a continuous review inventory policy is followed. Thus after each decrement in the wholesale inventory level for a particular product, a check is made to determine if it has fallen below a predetermined ROP. If it has then a check is made again to see if and order has already been placed and is in transit. When this condition is not satisfied an order is placed to the manufacturer and a wholesale lead time attribute is initialized to the current simulation time. Finally, the entity type is now reset to wholesale order and the control flows out of the wholesale sub-model. An attribute is also defined here which identifies the number of the wholesaler from which the order arrives. The wholesale warehouse is divided into two separate types of stations, one for inbound goods and a set of stations for outbound goods. At the outbound stations a transporter picks up the retail order and delivers it to the specific retailer which can be identified by its unique attribute as stated earlier. The transporter is picked from a pool of transporters assigned to each wholesaler on the basis of the product to be transported. At the inbound station, yet another transporter unloads the goods which have been sent by the manufacturer. Thereafter the wholesale inventory level for that product is incremented by the wholesale order quantity. The lead time attribute is recorded to estimate the mean lead time for the wholesaler as well as to determine the mean lead time demand. The inbound and outbound stations are station modules and are labeled with unique station numbers. The distances between the manufacturer outbound station and the wholesale inbound stations and those between the wholesale outbound stations and the retail inbound stations are set in the distance module. The average velocities of the transporters delivering goods from the wholesaler to the retailer and those delivering goods from the manufacturer to the wholesaler are also set before hand in the transporter module.

Manufacturer Sub-model

The logic used to simulate the processes happening at the manufacturer is similar to that of the retail and wholesale sub-models. Since a three stage supply chain is being modeled starting from the retailer and culminating at the manufacturer raw material inventories and their replenishment by suppliers are not considered. It is assumed that each manufacturer produces 6 unique products out of the pool of 30. Since the supply chain is modeled taking (FMCG) products into account a continuous production is assumed to be undertaken by the manufacturer. This is done by using separate create modules in the manufacturer sub-model to simulate continuous production of various products. The agglomerated raw materials are assumed to arrive together and the arrival rate is set in the create module. Consequently, a process module is used to simulate final assembly or production of the product. The production time may be set by the user. The resource used for this process is labeled as manufacturer and the number may be set as required. Thereafter the finished products are batched and then moved to the factory warehouse and thus the factory warehouse inventory level is continuously incremented. The product number of the arriving wholesale order corresponds to a particular index of each manufacturer. This index is the basis for all calculations in this sub-model. The queues for waiting wholesale orders are in the form of queue sets and products of each manufacturer index are made to wait in their corresponding queue. The inventory level check is then performed at the manufacturer. If the wholesaler can be satisfied, a transporter then performs the task of transporting the wholesale order back to the designated wholesaler on the basis of the attribute assigned earlier. Here too, the transporter is chosen from a pool assigned to the manufacturer on the basis of the product. The successful satisfying of the order is recorded. In the case of a stock-out, the order is made to wait at the manufacturer and a corresponding wait-time is calculated.

32.3. INPUT PARAMETERS FOR THE CASE SITUATION

One case is elucidated in order to reinforce the salient features of the concept. The complete sample inputs from the user interface is taken and simulated so as to get a perception of performance of the supply chain configuration of 5*5*8 where first number stands for number of manufacturers, second number stands for number of wholesalers and third number stands for number of retailers. The simulation run length is 30 days. The common inputs to retailers are as follows:

Customer arrival rate: NORM (2,1) min (where NORM: Normal distribution)
Number of orders per product per customer: ANINT (NORM (3,2))
Aisle routing time: NORM (2,1) min
Product picking time per shelf: NORM (2,1) min
Billing time: NORM (5,3) min
Retail re-order placing time: NORM (10,2) min
The retailer specific inputs are given in Table 32.8.

Table 32.1: Input parameters for retailer 1

Product number	Initial inventory level	Re-order point (units)	Order quantity (units)
1	100	25	100
4	100	25	100
9	100	25	100
12	100	25	100
14	100	25	100
18	100	25	100
22	100	25	100
25	100	25	100
28	100	25	100
29	100	25	100

Table 32.2: Input parameters for retailer 2

Product number	Initial inventory level	Re-order point (units)	Order quantity (units)
2	200	40	200
5	200	40	200
8	200	40	200
11	200	40	200
15	200	40	200
17	200	40	200
20	200	40	200
24	200	40	200
26	200	40	200
30	200	40	200

Table 32.3: Input parameters for retailer 3

Product number	Initial inventory level	Re-order point (units)	Order quantity (units)
4	175	30	175
7	175	30	175
10	175	30	175
13	175	30	175
15	175	30	175
19	175	30	175
21	175	30	175
23	175	30	175
27	175	30	175
29	175	30	175

Table 32.4: Input parameters for retailer 4

Product number	Initial inventory level	Re-order point (units)	Order quantity (units)
3	300	40	300
6	300	40	300
8	300	40	300
11	300	40	300
14	300	40	300
17	300	40	300
22	300	40	300
25	300	40	300
28	300	40	300
30	300	40	300

Table 32.5: Input parameters for retailer 5

Product number	Initial inventory level	Re-order point (units)	Order quantity (units)
2	250	50	250
5	250	50	250
9	250	50	250
12	250	50	250
15	250	50	250
16	250	50	250
18	250	50	250
20	250	50	250
24	250	50	250
27	250	50	250

Table 32.6: Input parameters for retailer 6

Product number	Initial inventory level	Re-order point (units)	Order quantity (units)
2	100	30	100
4	100	30	100
9	100	30	100
13	100	30	100
16	100	30	100
19	100	30	100
21	100	30	100
23	100	30	100
26	100	30	100
30	100	30	100

Table 32.7: Input parameters for retailer 7

Product number	Initial inventory level	Re-order point (units)	Order quantity (units)
1	300	45	300
5	300	45	300
7	300	45	300
11	300	45	300
14	300	45	300
17	300	45	300
19	300	45	300
22	300	45	300
24	300	45	300
27	300	45	300

Table 32.8: Input parameters for retailer 8

Product number	Initial inventory level	Re-order point (units)	Order quantity (units)
3	200	40	200
6	200	40	200
7	200	40	200
10	200	40	200
12	200	40	200
16	200	40	200
18	200	40	200
21	200	40	200
25	200	40	200
26	200	40	200

The common inputs to wholesaler are as follows:
Retail order processing time: NORM (10,2) min

Wholesale re-order placing time: NORM (10,2) min
The wholesaler specific inputs are given in Table 32.9.

Table 32.9: Input parameters for wholesalers

Wholesaler number	Product distribution	Initial inventory	ROP
1	1, 3, 5, 6, 15	2000	500
2	2, 4, 8, 10, 12, 13, 18, 20	2000	600
3	7, 9, 11, 14, 16, 17, 19	2500	1000
4	21, 25, 26, 29, 30	2000	750
5	22, 23, 24, 27, 28	3000	1250

The common inputs to manufacturers are as follows:
Production rate per product: NORM (2,0.2) min.
Initial manufacturer inventory per product: 20000.
The manufacturer specific inputs are given in Table 32.10.

Table 32.10: Input parameters for manufacturers

Manufacturer number	Batch quantities	Product distribution
1	100	1, 2, 3, 4, 21, 22
2	200	5, 6, 7, 8, 23, 24
3	200	9, 10, 11, 12, 25, 26
4	200	13, 14, 15, 16, 27, 28
5	300	17, 18, 19, 20, 29, 30

The distance matrix between wholesalers and retailers are given in Table 32.11.

Table 32.11: Input parameters for wholesalers

	Wholesaler 1 (20)	Wholesaler 2 (30)	Wholesaler 3 (40)	Wholesaler 4 (50)	Wholesaler 5 (50)
Retailer 1	20	20	20	20	20
Retailer 2	30	30	30	30	30
Retailer 3	40	40	40	40	40
Retailer 4	50	50	50	50	50
Retailer 5	60	60	60	60	60
Retailer 6	70	70	70	70	70
Retailer 7	40	40	40	40	40
Retailer 8	40	40	40	40	40

(Distances are in km, Quantities in () indicate velocity of transporters in km/h). The distance matrix between manufacturers and wholesalers are given in Table 32.12. Distances are in km, Quantities in () indicate velocity of transporters in km/h.

Table 32.12: Input parameters for wholesalers

	Manufacturer 1 (50)	Manufacturer 2 (60)	Manufacturer 3 (75)	Manufacturer 4 (90)	Manufacturer 5 (100)
Wholesaler 1	40	40	40	40	40
Wholesaler 2	50	50	50	50	50
Wholesaler 3	60	60	60	60	60
Wholesaler 4	50	50	50	50	50
Wholesaler 5	60	60	60	60	60

Table 32.13: Customer order fill rate in percentage

Retailer 1									
P1	P4	P9	P12	P14	P18	P22	P25	P28	P29
85	59	58	57	61	56	60	58	82	82
Retailer 2									
P2	P5	P8	P11	P15	P17	P20	P24	P26	P30
59	59	89	62	80	64	87	61	60	64
Retailer 3									
P4	P7	P10	P13	P15	P19	P21	P23	P27	P29
63	59	87	87	78	62	61	85	62	86
Retailer 4									
P3	P6	P8	P11	P14	P17	P22	P25	P28	P30
85	85	83	61	62	59	61	59	83	61
Retailer 5									
P2	P5	P9	P12	P15	P16	P18	P20	P24	P27
65	63	64	63	84	41	63	90	63	64
Retailer 6									
P2	P4	P9	P13	P16	P19	P21	P23	P26	P30
59	61	61	87	40	60	61	64	59	61
Retailer 7									
P1	P5	P7	P11	P14	P17	P19	P22	P24	P27
90	61	61	63	66	61	63	0	61	62
Retailer 8									
P3	P6	P7	P10	P12	P16	P18	P21	P25	P26
92	92	63	92	62	41	62	63	67	63

(P1 stands for product 1; P2 stands for product 2 and like that)

The performance measures of the supply chain are shown in Tables 32.13 to 32.22. The customer order fill rate in percentage is shown in Table 32.13. The number of retailer's stock-outs, number of retail orders satisfied, number of retail order processed and retail order fill rate in percentage are shown in Tables 32.14 to 32.17. The number of wholesale orders processed, number of wholesale backorders, number of wholesale orders satisfied and number of wholesaler fill rate in percentage are shown in Tables 32.18 to 32.21. The number of manufacturer stock-outs and number of manufacturer backorders are shown in Tables 32.22 to 32.23.

Table 32.14: Number of retailer's stock-outs

Retailer 1									
P1	P4	P9	P12	P14	P18	P22	P25	P28	P29
835	2342	2400	2466	2217	2514	2264	2431	983	986
Retailer 2									
P2	P5	P8	P11	P15	P17	P20	P24	P26	P30
2384	2347	622	2117	1098	2102	736	2243	2276	2040
Retailer 3									
P4	P7	P10	P13	P15	P19	P21	P23	P27	P29
2108	2317	731	708	1228	2167	2179	831	2129	796
Retailer 4									
P3	P6	P8	P11	P14	P17	P22	P25	P28	P30
812	820	940	2174	2167	2342	2224	2240	967	2205
Retailer 5									
P2	P5	P9	P12	P15	P16	P18	P20	P24	P27
1981	2100	1997	2148	857	3393	2109	557	2072	2027
Retailer 6									
P2	P4	P9	P13	P16	P19	P21	P23	P26	P30
2311	2165	2262	685	3413	2213	2207	775	2356	2192
Retailer 7									
P1	P5	P7	P11	P14	P17	P19	P22	P24	P27
556	2223	2174	2047	1891	2198	2046	2062	2225	2164
Retailer 8									
P3	P6	P7	P10	P12	P16	P18	P21	P25	P26
412	416	2102	445	2108	3351	2144	2077	1842	2055

Table 32.15: Number of retail orders satisfied

Wholesaler 1							
P1	P3	P5	P6	P15	–	–	–
188	123	123	123	195	0	0	0
Wholesaler 2							
P2	P4	P8	P10	P12	P13	P18	P20
188	256	119	158	186	223	186	132
Wholesaler 3							
P7	P9	P11	P14	P16	P17	P19	–
140	237	117	166	125	118	189	0
Wholesaler 4							
P21	P25	P26	P29	P30	–	–	–
206	182	198	216	184	0	0	0
Wholesaler 5							
P22	P23	P24	P27	P28	–	–	–
166	222	123	130	179	0	0	0

Table 32.16: Number of retail orders processed

Retailer 1									
P1	P4	P9	P12	P14	P18	P22	P25	P28	P29
138	99	95	94	100	94	100	96	134	136
Retailer 2									
P2	P5	P8	P11	P15	P17	P20	P24	P26	P30
50	50	74	52	67	55	73	51	51	53
Retailer 3									
P4	P7	P10	P13	P15	P19	P21	P23	P27	P29
59	56	82	81	73	58	57	81	57	80
Retailer 4									
P3	P6	P8	P11	P14	P17	P22	P25	P28	P30
46	45	45	33	32	32	33	32	45	33
Retailer 5									
P2	P5	P9	P12	P15	P16	P18	P20	P24	P27
43	42	43	43	56	28	42	59	41	42
Retailer 6									
P2	P4	P9	P13	P16	P19	P21	P23	P26	P30
98	101	102	143	65	99	99	141	97	101
Retailer 7									
P1	P5	P7	P11	P14	P17	P19	P22	P24	P27
50	34	34	35	37	34	35	36	34	34
Retailer 8									
P3	P6	P7	P10	P12	P16	P18	P21	P25	P26
77	78	53	76	52	35	53	53	57	53

Table 32.17: Retail order fill rate in percentage

Wholesaler 1							
P1	P3	P5	P6	P15	–	–	–
100	100	95	100	97	0	0	0
Wholesaler 2							
P2	P4	P8	P10	P12	P13	P18	P20
97	98	100	100	96	100	97	100
Wholesaler 3							
P7	P9	P11	P14	P16	P17	P19	–
98	99	98	98	99	97	98	0
Wholesaler 4							
P21	P25	P26	P29	P30	–	–	–
97	97	97	100	96	0	0	0
Wholesaler 5							
P22	P23	P24	P27	P28	–	–	–
98	100	98	96	100	0	0	0

Table 32.18: Number of wholesale orders processed

Wholesaler 1							
P1	P3	P5	P6	P15	–	–	–
14	14	15	14	20	0	0	0
Wholesaler 2							
P2	P4	P8	P10	P12	P13	P18	P20
15	15	14	15	15	14	15	14
Wholesaler 3							
P7	P9	P11	P14	P16	P17	P19	–
12	12	12	12	8	12	12	0
Wholesaler 4							
P21	P25	P26	P29	P30	–	–	–
15	15	15	14	15	0	0	0
Wholesaler 5							
P22	P23	P24	P27	P28	–	–	–
10	9	10	10	9	0	0	0

Table 32.19: Number of wholesale backorders

Wholesaler 1							
P1	P3	P5	P6	P15	–	–	–
0	0	6	0	6	0	0	0
Wholesaler 2							
P2	P4	P8	P10	P12	P13	P18	P20
6	6	0	0	6	0	6	0
Wholesaler 3							
P7	P9	P11	P14	P16	P17	P19	–
3	3	3	3	0	3	3	0
Wholesaler 4							
P21	P25	P26	P29	P30	–	–	–
6	6	6	0	6	0	0	0
Wholesaler 5							
P22	P23	P24	P27	P28	–	–	–
3	0	3	3	0	0	0	0

Table 32.20: Number of wholesale orders satisfied

Manufacturer 1					
P1	P2	P3	P4	P21	P22
14	14	14	14	14	9
Manufacturer 2					
P5	P6	P7	P8	P23	P24
14	14	11	14	9	9
Manufacturer 3					
P9	P10	P11	P12	P25	P26
11	14	11	14	14	14
Manufacturer 4					
P13	P14	P15	P16	P27	P28
14	11	19	7	9	9
Manufacturer 5					
P17	P18	P19	P20	P29	P30
11	14	11	14	14	14

Table 32.21: Number of wholesaler order fill rate in percentage

Manufacturer 1					
P1	P2	P3	P4	P21	P22
100	88	100	88	88	90
Manufacturer 2					
P5	P6	P7	P8	P23	P24
88	100	92	100	100	90
Manufacturer 3					
P9	P10	P11	P12	P25	P26
92	100	92	88	88	88
Manufacturer 4					
P13	P14	P15	P16	P27	P28
100	92	90	100	90	100
Manufacturer 5					
P17	P18	P19	P20	P29	P30
92	88	92	100	100	88

Table 32.22: Number of manufacturer stock-outs

Manufacturer 1					
P1	P2	P3	P4	P21	P22
0	3	0	3	3	2
Manufacturer 2					
P5	P6	P7	P8	P23	P24
3	0	2	0	0	2
Manufacturer 3					
P9	P10	P11	P12	P25	P26
2	1	2	3	3	3
Manufacturer 4					
P13	P14	P15	P16	P27	P28
0	2	3	1	2	0
Manufacturer 5					
P17	P18	P19	P20	P29	P30
2	3	2	0	0	3

Table 32.23: Number of manufacturer backorders

Manufacturer 1					
P1	P2	P3	P4	P21	P22
0	2	0	2	2	1
Manufacturer 2					
P5	P6	P7	P8	P23	P24
2	0	1	0	0	1
Manufacturer 3					
P9	P10	P11	P12	P25	P26
1	0	1	2	2	2
Manufacturer 4					
P13	P14	P15	P16	P27	P28
0	1	2	0	1	0
Manufacturer 5					
P17	P18	P19	P20	P29	P30
1	2	1	0	0	2

32.4. CLOSURE

This simulation model can be extended for evaluating supply chain inventory performance measures for different configurations and uncertain situations. To tackle the uncertain environment, the sensitivity analysis can be carried out easily. One case situation is elucidated in order reinforce the salient features of the concept. In this chapter, an attempt is made to demonstrate the usefulness of the ARENA simulation tool for the evaluation supply chain inventory performance.

REFERENCES

1. Arnold, J.R.T. (1998), "*Introduction to Materials Management*", Prentice Hall, New Jersey.
2. Chopra, S., and Peter, M., (2001), "*Supply Chain Management Strategy, Planning and Operation*", First Edition, Pearson Education Asia.
3. Christopher, M., (1992), "*Logistic and Supply Chain Management*", Pitman Publication, London.
4. Gentry, J.J., (1992), The role of carriers in buyer-supplier strategic partnerships: a supply chain management approach, "*Journal of Business Logistics*", Vol. 17, pp. 33–35.
5. Kelton, W.D., Sadowski, R.P., and Sadowski, D.A., (2002), "*Simulation with ARENA*", Second edition, McGraw-Hill.
6. Law, A.M., and Kelton, W.D., (2000), "*Simulation Modeling and Analysis*", Third Edition, McGraw-Hill.
7. Lee, H.L., and Billington, C., (1992), Managing supply chain inventory: Pitfalls and Opportunities, "*Sloan Management Review*", Vol. 33, pp. 65–73.
8. Schniederjans, M.J., (1999), "*International Facility Acquisition and Location Analysis*", Quorum Books, London.
9. Strader, T.J., Lin, F., and Shaw, M.J., (1998) Simulation of order fulfillment in divergent assembly supply chains, "*Journal of Artificial Societies and Social Simulation*", Vol 1. No. 2.
10. Towill, D.R., Naim, M.M., and Wikner, J., (1992), Industrial dynamics simulation models in the design of supply chains, "*International Journal of Physical Distribution and Logistics Management*", Vol. 22, 5, pp. 3–13.
11. Swaminathan et al. (1998).

QUALITY FUNCTION DEPLOYMENT

33.1. INTRODUCTION

This chapter explores the potential of QFD as a tool in evaluating the usability of software. Usability engineering has many tools, most of them based on heuristics. QFD, which has found many applications in product design and in engineering, could reduce the uncertainty involved in heuristic tools as it provides semi-empirical methods for evaluation and grading.

A new methodology based on concepts of QFD has been conceived for use in the area of usability evaluation of software. A number of available software have been tested for their usability using this tool. This paper presents the new QFD based tool as a case study.

QFD (Quality Function Deployment) is the systematic translation of the "voice of the customer" to actions of the supplier required to meet the customers' desires, based on a matrix comparing what the customer wants to how the supplier plans to provide it. This basic matrix can been expanded to provide additional insight to the supplier, and cascaded to identify process parameters that must be controlled to meet the customer requirements. QFD must start with an understanding of the customers' needs and wants. These 'whats' become rows in a matrix, against columns representing the suppliers processes ('hows') designed to provide products (or services) meeting the customers' desires.

33.2. USABILITY ENGINEERING

Usability can be defined as "The extent to which a product can be used by specified users to achieve specified goals in a specified context of use with effectiveness, efficiency and satisfaction." (ISO 9241-11 International Organization for Standardization) "The measure of the quality of the user experience when interacting with something whether a website, a traditional software application, or any other device the user can operate in some way or another" [1]. "Usabilty means that the people who use the product can do so quickly and easily to accomplish their own tasks" [2]. Jacob Nielsen, a usability specialist, identifies the following attributes of usability:

Learnability

The system should be easy to learn so that the user can rapidly start doing some work.

Efficiency

The system should be efficient to use so that once the user has used the system, a high level of productivity is possible.

Memorability

The system is easy to remember so that the casual user is able to return to the system after some period of not having used it without having to learn everything all over again.

Errors

The system should have a low error rate, so that users make few errors during it's use and so that if they do make errors they can easily recover from them. Further catastrophic errors should not occur.

The system should be pleasant to use, so that users are subjectively satisfied when using it; they like it. Usefulness is defined in terms of the user's need for the product in the context of the user's goal. Usability comes into picture at every stage of product development, i.e., at various iterative stages of User Centered Design. According to Barnum, the process that incorporates User Centered Design in product development is called as Usability Engineering. Usability engineering is frequently viewed as the umbrella term that describes the set of activities underlining the process of creating user-centered product designs. The three types of Usability evaluation methods are namely, testing, inspection and inquiry.

33.3. USABILITY TESTING

This is the process of learning from users about a product's usability by observing them using the product. A further distinction is to view most usability engineering methods-other than usability testing-as "analytical" and to view usability testing as "empirical," meaning that the information comes from the experiences of the user [3]. According to Rubin [4] Usability Testing means the "process that employees participants who are representatives of the target population to evaluate the degree to which a product meets specific usability criteria. There are two types of testing models:

Testing Within a Lab

It is a formative evaluation of the product, meaning the data gathered can be applied to the continuing development of the product that is still forming. This type of testing brings the user into a controlled environment in which they are instructed to do certain task over a specified period of time and their activities are captured, monitored and sometimes discussed with them later. Generally for this type of testing to gather information protocol analysis and thinking aloud technique could be used.

Field Testing

It means going to the field, evaluating the product i.e., how it is used in industrial setting, offices or at home. This kind of technique is Summative Evaluation meaning the data gathered could be applied to the next version of the product as it is done at the end of the development stage of the product. It offers many advantages over Lab testing, some of them are, you get to study the environment in which the user is working as in the lab it is not the real environment, it offers you the chance to know what other tasks the user has to do while performing the task that you are studying like attending a telephone, attending some client etc. Field testing could be done in one of the two ways, either through Structured Observation or by doing Longitudinal Field Studies.

33.4. USABILITY INSPECTION METHOD-HEURISTIC EVALUATION

Another method for detecting usability problems is heuristic evaluation. In this method an inspector reviews the user interface and judges its compliance with an accepted set of usability rules (heuristics). Nielsen described it as one of the two "discount" usability methods, the other being usability testing with a small number of participants. This method is a discount method because it is cheap and effective. The set of usability heuristics developed by Molich and Nielsen are as follows:

1. Visibility of system status
2. Match between system and the real world
3. User control and freedom
4. Consistency and standards
5. Error prevention
6. Recognition rather than recall
7. Flexibility and efficiency of use
8. Aesthetic and minimalist design
9. Help users recognize, diagnose, and recover from errors
10. Help and documentation

33.5. COGNITIVE WALKTHROUGHS

A Cognitive Walkthrough is a usability inspection method designed to evaluate "ease of learning, particularly by exploration" [5] Cognitive Walkthrough inputs a description of the user interface, a set of task scenarios, assumptions about the knowledge a user will bring to the task, and the specific actions a user must perform to accomplish the task led by a moderator, to examine each step in the correct action sequence by asking a set of pre-defned questions.

33.6. PROBLEMS AND SHORTCOMINGS OF USABILITY TESTING

The main weakness of usability testing is cost. A fully developed state-of-the-art usability lab costs $75,000 to $150,000 depending on the specific needs of the company for testing products [6]. Although many studies have demonstrated huge savings and profits for companies that employ usability testing, some managers are reluctant to provide the money for usability testing. Also, usability tests are inadequate for detecting problems which do not affect user performance, but may affect the user's perception of product quality. For example, consider the issue of consistent text-field labeling. In a usability test, a participant may complete all of the tasks without error. He or she may even manage to do so when some text fields have inconsistent names or abbreviations. However, this inconsistency may decrease the participant's overall impression of the product's quality.

33.7. PROBLEMS AND SHORTCOMINGS OF HEURISTIC EVALUATION

One weakness of this method is that it may produce false positives. An example of a false positive is a minor bug that does not negatively impact user performance or the user's perception of product quality. The danger is that developers may spend valuable time addressing bugs that do not impact user performance or satisfaction to the exclusion of those that do.

The second weakness is that heuristic evaluation does not approximate the conditions under which real users would use the system. Heuristic evaluation does not take place in the context of a real user task; therefore, its results may not be as valid as those derived from a usability test. Moreover, these results may be more difficult to defend than those uncovered in a usability test.

Third, heuristic evaluation provides little information about the magnitude of the problems that are detected. It simply indicates that a particular aspect of the product may have a usability problem. Conversely, a usability test provides you with an appreciation of the problem's severity.

Finally, it is doubtful that the heuristics themselves are helpful in detecting bugs. The evaluators rely on common sense and previous experience. A common bug category detected may not fit into any of the heuristics in the list.

33.8. THE QFD METHODOLOGY

Quality Function Deployment [7] is a method for focusing the effort and limited resources of a project team on what delivers the best value to the most important stakeholders. The concept of QFD was introduced in Japan by Yoji Akao in 1966, and was first applied at the Kobe Shipyard of Mitsubishi Heavy Industries Ltd. Since that time, QFD has become the accepted methodology for development of products and services in Japan and has enabled businesses to successfully develop and introduce products in a fraction of the time required without it [8]. In the early 1980s, QFD was introduced at Xerox, and since then American businesses have exhibited substantially growing interest in using it. QFD "is a method for developing a design quality aimed at satisfying the consumer and then translating the consumer's demand into design targets and major quality assurance points to be used throughout the production phase. . . . (QFD) is a way to assure the design quality while the product is still in the design stage." As a very important side benefit. When appropriately applied, QFD has demonstrated the reduction of development time by one-half to one-third. QFD is a structured process, a visual language, and a set of interlinked engineering and management charts. According to Jackson and Frigon [8], QFD uses the following seven management tools: affinity diagram, interrelationship diagram, tree diagram, matrix diagraph, process decision program chart, arrow diagram, and matrix data analysis chart.

It establishes customer value using the voice of the customer and transforms that value to design, production, and manufacturing process characteristics. The result is a systems engineering process that prioritizes and links the product development cycle so that it assures product quality as defined by the customer/user. Additional power derives from use within a concurrent engineering environment. QFD, as applied under the kaizen philosophy and within total quality control, is by far the most highly developed form of integrated product and process development in existence [9]. There are several approaches to QFD; each of these approaches makes use of matrices to organize and relate pieces of data to each other. Many times these matrices are combined to form a basic tool of QFD called the "House of Quality" [10] shown in Fig. 33.1.

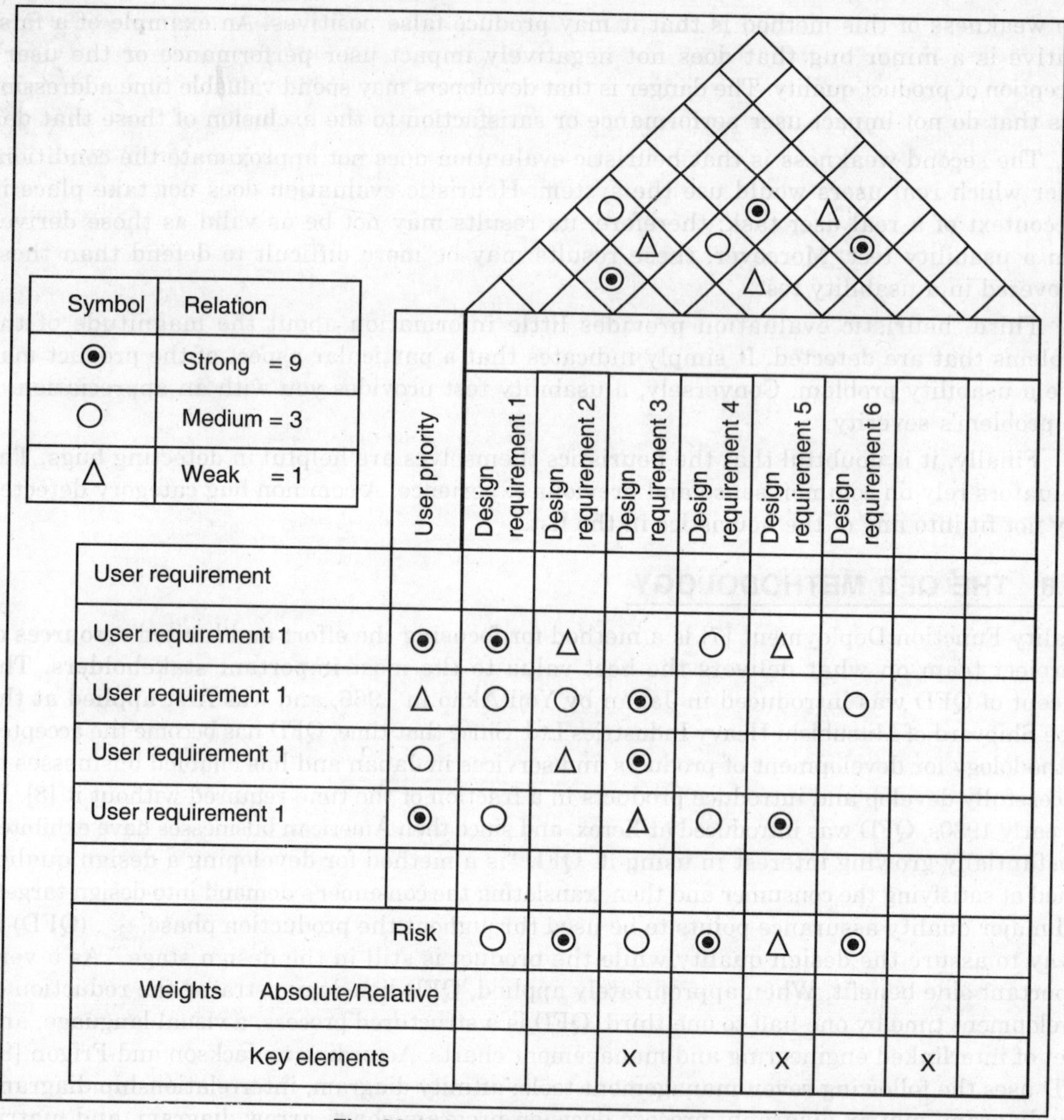

Fig. 33.1: A view of house of quality

The House of Quality starts with a "What-How" Matrix that identifies the wants, desires, and needs of the customer. These customer requirements are shown on the left part of the House of Quality. The ceiling of the House, i.e., the horizontal axis below the roof, shows the design or technical requirements, while the body of the House visually displays the relationships between the customer requirements and design requirements. The symbols used in each box, if any, show whether the relationship between the two corresponding elements is strong or weak, positive or negative. This way the House of Quality quickly reveal patterns and identifies weak points in the design requirements. The Interaction Matrix, also known as the Correlation Matrix, is the "roof" of the House of Quality. It is established to determine the technical interrelationships among the design requirements. This information is valuable as the basis for decisions regarding technical trade-offs [7].

The House of Quality is the first of many matrices in the QFD process of translating customer needs into product requirements and system specifications.

The number of matrices corresponds to the number of phases in the development process. For example, the American Supplier Institute (ASI) follows a four-phase approach, in which a QFD team deploys customer requirements into product characteristics, product characteristics into part characteristics, part characteristics into process characteristics, and finally process characteristics into production characteristics. Other approaches may include more phases, from customer requirements to design requirements, engineering design, product characteristics, manufacturing/purchasing operations, and finally production/quality controls. King [10] and Clausing and Pugh [5] provide more extensive and complete views than the simple "House of Quality" view. Dean [9] views QFD as a system engineering process which transforms the desires of the customer/user into the language required, at all project levels, to implement a product. It also provides the glue necessary, at all project levels, to tie it all together and to manage it. QFD is far more than had previously been disclosed, i.e., the mechanism for deploying quality, reliability, cost, and technology throughout the product, the project to bring forth the product, and the enterprise as a whole. Despite the many views that exist around the application of QFD, in this chapter we confine ourselves to the basic House of Quality, in our attempt to apply QFD for evaluating software usability. This, as the purpose here is not to make a detailed analysis of how QFD could be used in this new context, but to give an overview of the potential that this structured approach has in an unexplored environment, that of usability evaluation. In the following section we discuss the main steps of applying QFD for evaluating software usability and define the basic House of Quality.

33.9. APPLYING QFD FOR EVALUATING SOFTWARE USABILITY

The basic QFD methodology can also be utilized with common usability testing methods to create a hybrid and more accurate methodology which can help to solve the shortcomings of common usability testing and heuristic evaluation methods. In order to satisfy the customer within the schedule and resource constraints that all projects face, it is necessary to concentrate the best efforts on those things of greatest importance to the users of the system. Software design presents interesting challenges for several reasons. Software is an intangible product that is not always conducive to explicit acceptance measures. Design elements are coupled and interdependent, which is different from physical designs that can be deconstructed into independent but functional sub-assemblies and parts. Software is not so easily divisible, creating additional design challenges. Thus an effective evaluation will not only identify problems but will also provide guidance for next iteration of design activities.

33.10. THE CONCEPTUAL FRAMEWORK FOR EVALUATING SOFTWARE USABILITY USING QFD

In the evolutionary course of Software development and design, we propose that the application of QFD methodology can satisfy the increasing need for more usable, user-centric software.

The first step in applying QFD to the usability testing of a software is to identify the "What's" and "How's" that will form the basis of the House of Quality. The "What's", also referred to as user needs or requirements, is a list of what the user wants or what is to be achieved. In the case of a Software, these would be user statements about what they want the Software to function and offer, i.e., what are the desirable software features. On the other hand, the "How's", also referred to as design or technical requirements/attributes, is a list of what one can measure and control in order to ensure that the users' requirements will be satisfied. Typically, the entries on this list are parameters for which a means of measurement and a measurable target value can be established. Again, in the case of a software, these would be the specific software characteristics, visible or not, that satisfy the specific user needs.

33.11. QFD "WHAT'S" FOR USABILITY EVALUATION OF SOFTWARES

The factors that might be influential in improving the usability of Softwares, i.e., what does a user look for in a software are exactly the "what's" of QFD. Below, we state some of the basic "what's" that we have identified:

(a) **Efficiency:** The number of tasks per unit of time that the user can perform using the software. We look for the maximum speed of user task performance. The higher software usability is the faster the user can perform the task and complete the job.

(b) **Functionality:** An equally important "what's" to software availability, is the ease of using a software for both expert and novice users. Functionality is often called the key to successful software. It is what helps users navigate through the software, use the software features, and find the solution they are looking for. Achieving true functionality requires balancing attractive interface design, simple navigation solutions and good content that's worth revisiting. Functional software is one where designers anticipate the needs of users, and where users can easily find what they need.

(c) **User-friendliness:** The software interface should be easy to remember, so that the casual user is able to return to the software after some period of not having used it, without having to learn it all over again.

(d) **Responsiveness:** Users also require that software respond to their specific needs, a requirement usually translated into customization features and personalized services. For example, a benefit that users get while designing any 3-D model or any graphic design using software is availability of various tools and inbuilt features which help user to generate ideas in comparatively lesser time than he/she takes while designing on paper by his/her own imagination.

(e) **Clarity to easily learn:** Attributes of software that bear on the users' effort for learning its application.

33.12. QFD "HOW'S" FOR USABILITY EVALUATION OF SOFTWARES

The degree to which a software satisfies the above user requirements, i.e., the "what's" of QFD, depends on specific software characteristics. We have identified the following key characteristics universally relevant to software.

(a) **Performance:** The term performance refers to properties such as the time it takes for the software to start functioning, the response time experienced by users requesting a specific software service, the number, type and frequency of error messages popping-up, etc. As such, this attribute is associated both with a software's availability but also with a software's responsiveness. The technical characteristics that affect performance include: the programming platform technology used for making the software, the platform capacity, the software's underlying architecture.

(b) **Services offered:** As software development advances, users become more and more demanding regarding the attributes they would expect to find in a software. The additional features they look for include:

(c) **Interactivity:** Defined as the facility for individuals and organizations to communicate directly with one another regardless of time and space. The level of interactivity of a software would be critical in converting software users from interested contacts into interactive customers.

(d) **Personalization:** Of content and services, i.e., dynamic adaptation of a software's content and structure to match the profile of each user.

(e) **Aesthetics:** One of the factors that influence a user's response to a software is the first impression that will be created upon interacting with the software, usually the software's interface. This impression depends on the overall look-and-feel of the software, the colours, the graphic tools page each browser has set up. After completing above tasks, users were asked to tell us what they required most in a browser while browsing over internet. Next section explains what users demanded most.

(f) **Content:** From the user's standpoint content of the software i.e., help documentation and other functional assets need constant updating. Another important element of content is the way it is expressed in writing. Thus, on software, information must be presented in a manner that exploits the medium's possibilities and respects its limits.

(g) **Functional Consistency:** In the interface context, the term functional consistancy covers aspects that influence a interface's consistency of use, consistency of navigability throughout the interface.

(h) **Structure:** No matter how creative, informative and fast a interface is, it is useless if the user cannot find what he is looking for. Under this perspective, it is important that a interface has a clear and consistent structure, organized in the way users expects to find it.

33.13. PROPOSED FUNCTIONING OF THE HOUSE OF QUALITY FOR EVALUATING SOFTWARE USABILITY

Figure 33.2 depicts our attempt to create the basic House of Quality for the usability testing of software, illustrating the interrelationships between the user requirements and the software characteristics discussed in Sections 33.1 and 33.2.

Furthermore, the software's efficiency is directly linked to the software's performance, thus imposing a strong relationship between this customer requirement (Efficiency) and the corresponding design attribute (Performance). Also, the design attributes content and services offered are linked with weak relationships to the efficiency customer requirement. Following a similar reasoning, we proceed to the completion of all interrelationships between user requirements and the design attributes. As shown in Fig. 33.2, the first step of the QFD process identified Performance, Functional Consistancy and Interactivity as the key software characteristics that must meet user requirements. These key elements correspond to the highest relative weights (135, 75 and 135, respectively), which are a combination of the risk, or else importance for the user, and the overall value of the symbols associated with each element. For example, the relative weight of Performance is calculated as 135. In this calculation 9 is multiplied by 15, where 9 stands for the strong risk this attribute has and 15 is the sum of all the symbols in that column (Strong + Medium + Weak + Weak + Weak = 9 + 3 + 1 + 1 + 1), i.e., a measure of the value that this attribute has for all the user requirements.

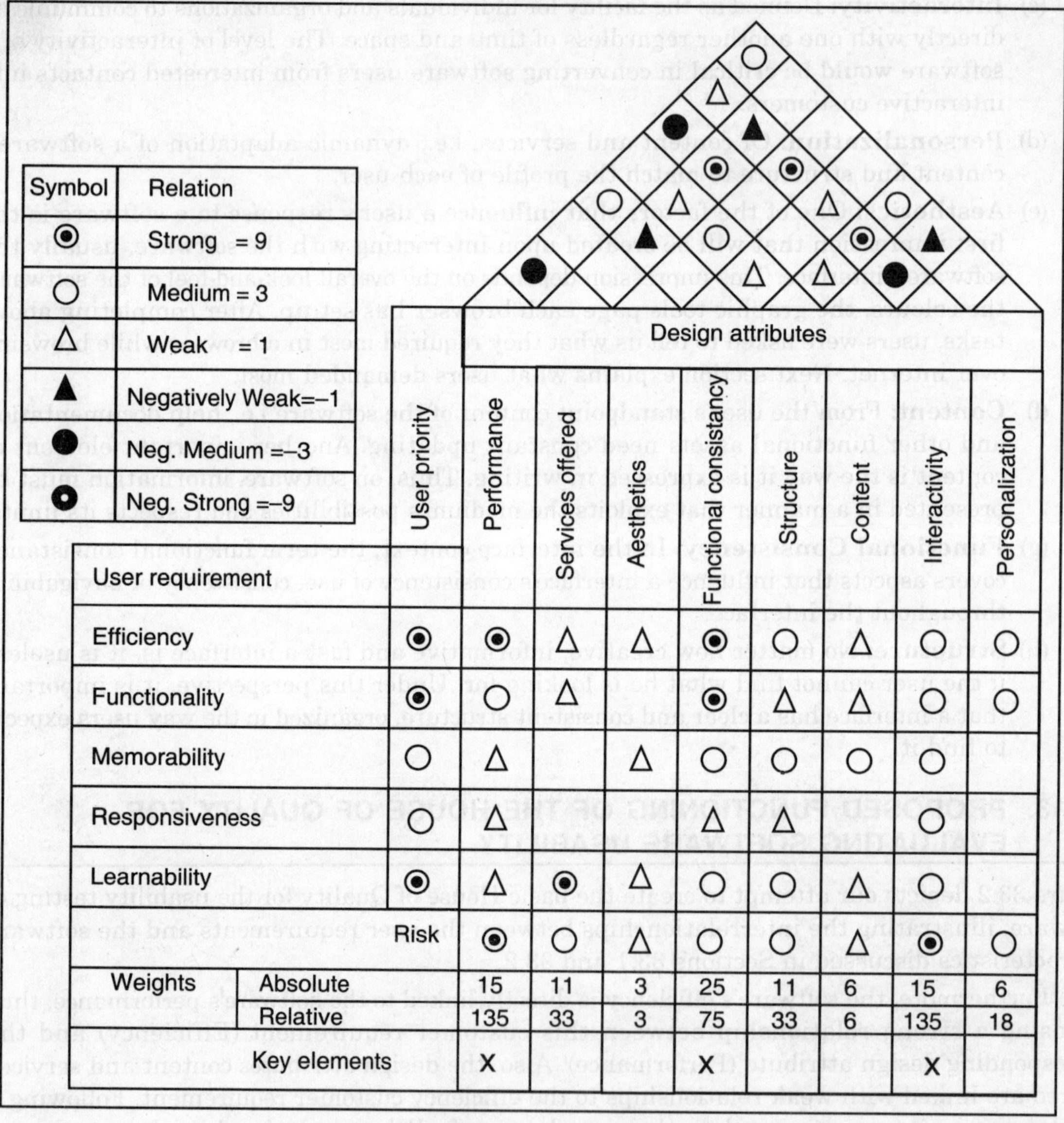

Fig. 33.2: House of quality for usability testing

33.14. APPLYING QFD TO USABILITY EVALUATION OF WEB-BROWSERS

Using QFD approach for usability testing of software, three similar pieces of software were evaluated in this usability test—Opera Web browser (Opera), Microsoft Internet Explorer (Internet Explorer) and Mozilla firefox browser (Mozilla) to ascertain the overall applicability of QFD to intangible products.

The purpose of this test was to:

(a) Determine which web browser best would suit users who were quite knowledgeable about the Internet

(b) Obtain key design features of web browsers by evaluating relative weights.

(c) Compare user productivity with each web browser.

Test Subjects: Twenty users (19–22 years, M = 17 and F = 3) were tested, and all had basic Windows XP experience. All users were also familiar with the basic functions of web browsers such as:

(a) Where to type a URL address (i.e., http://www.webpage.com).

(b) How to click on links in a web page to move from one page to another.

(c) A "hand" appearing over a section in a web page, meant it was "clickable".

(d) How to stop a page while it is loading by hitting the stop button.

Task Scenario: The tasks were designed to simulate skills fundamental to typical browser use. Also, the tasks were developed to allow each user to use the same functions of each web browser including

(a) Visiting different web pages.

(b) Opening a local file for viewing.

(c) Saving web pages in the Favorites/Bookmarks folder.

(d) Downloading and saving pictures and files to the hard drive.

(e) Clearing out the browser's cache/temporary Internet files.

(f) Searching the Internet using the search options.

User Requirements of Web-browsers: In this section we make an attempt to identify the specific user requirements of Web-browsers that have the greatest importance for the end user. These are associated with the general browser characteristics that have the greatest weight. As a result of this attempt we have found that the features "Low memory usage", "High Security" and "Ease of file downloading" have the maximum user priority (Fig. 33.4). Some other commonly required features like "Easy to customize", "Easy to control pop-up windows" and "Easy to switch between multiple browser windows" have medium user priority. While "Shortcuts Availability", "Integrated Web-search options" came out as least required features.

Design Attributes: As shown in Fig. 33.4, the QFD process identified "Performance", "Services Offered" , "Interactivity" and "Functional Consistency" as key design attributes, having strong/ medium level of "Risk" in software development process. In the inter-relationship matix shown in the middle of Fig. 33.4 , we have identified the results on the pre-defined scale of relationship strength between User Requirements and Design Attributes. As a result of this analysis we have found that "Performance" has the strong relationship with "Low memory usage" whereas "Interactivity" has negligible relationship with "Low memory usage". Using a similar approach and findings, we set the inter-relationships between other User requirements and Design Attributes.

	User Requirements	Browsers		Importance as rated by users				
				Low = 1	2	3	4	High = 5
1	Low memory usage	Internet Explorer	X =	0	0	7	9	4
		Mozilla	"	0	4	10	5	1
		Opera	"	3	11	5	1	0
2	Easy to configure/customize	Internet Explorer	X =	1	1	6	9	3
		Mozilla	"	0	3	11	6	0
		Opera	"	0	3	3	10	4
3	Integrated web-search options	Internet Explorer	X =	15	3	2	0	0
		Mozilla	"	0	0	7	12	1
		Opera	"	0	0	8	11	1
4	Easy to switch between multiple browser windows	Internet Explorer	X =	4	12	4	0	0
		Mozila	"	3	2	10	4	1
		Opera	"	0	3	4	11	2
5	Easy to control popup windows	Internet Explorer	X =	1	13	6	0	0
		Mozilla	"	0	0	5	14	1
		Opera	"	0	1	4	12	3
6	Low vulnerability/More security	Internet Explorer	X =	0	4	10	5	1
		Mozilla	"	1	13	5	1	0
		Opera	"	0	5	9	3	3
7	Shortcuts Availability	Internet Explorer	X =	2	12	4	2	0
		Mozilla	"	0	5	11	4	0
		Opera	"	0	1	2	13	4
8	Easy and resumable file downloading	Internet Explorer	X =	12	5	3	0	0
		Mozilla	"	3	13	4	0	0
		Opera	"	0	1	5	14	0

Fig. 33.: User rating based on their relative importance scale gathered from 20 test users testing three types of browsers to evaluate usability

X = No. of Users ranked corresponding User Importance level Low User Importance = 1 High User Importace = 5	Importacne scale:

Competitive Evaluation: In the competitive evaluation section of Fig. 33.4 we have shown the ranking of user requirements based on their relative importance scale. (Highest Importance = 5 and Lowest Importance =1). Figure 33.3 shows how twenty users ranked different browsers based on importance scale. As a result of this evaluation , we have indentified what active and quite regular users of internet ranks different browsers on different user requirements.

Test Results: According to our methodology of calculating weights, we have found that "Mozilla" is most vulnerable for most of internet users where "Opera" suits best for switching between multiple browser windows. Similiar results are shown in Fig. 33.4. The key design features for a web-browser that we have identified using QFD are as follows:

(a) Performance (b) Services offered

(c) Functional Consistency (d) Interactivity

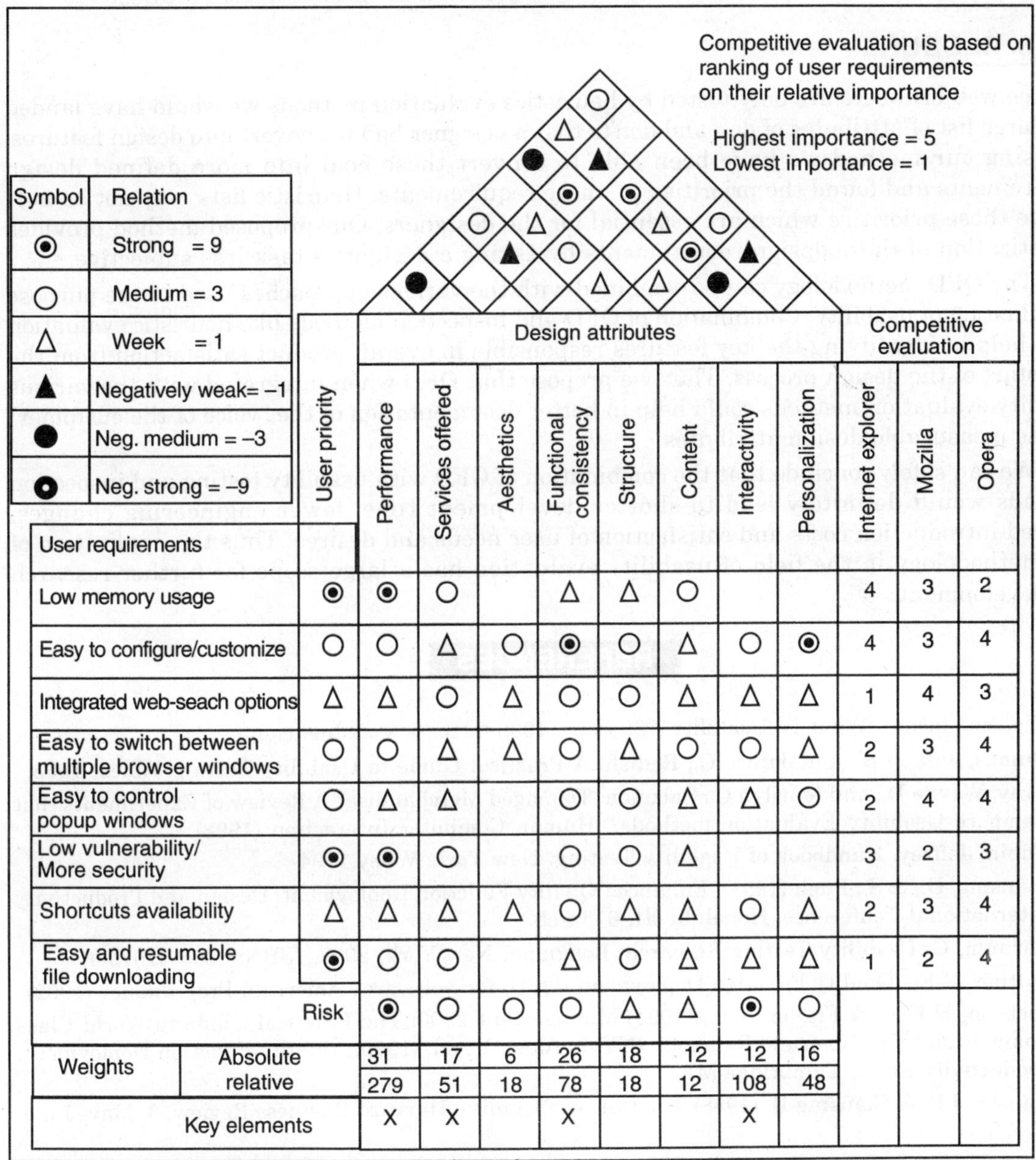

Fig. 33.4: A comparative usability analysis of web browsers using QFD

33.15. OPERATING THE PROPOSED MODEL

The user ratings from the conducted experiment were fed into the QFD model shown in Fig. 33.4. The calculation was done to identify key design attributes of each browser.

33.16. CLOSURE

If three web-browsers are only tested by heuristics evaluation methods we would have landed on a large list of attributes of do's and don'ts that a designer had to convert into design features. By using our method we have been able to convert these goal into more defined design requirements and found the priorities of these requirements. Heuristic lists may not be able to give these priorities which are essential for the designers. Our proposed method provides prioritization of these design requirements by asking a designer's task less subjective.

The QFD methodology can be combined with the various approaches used in the purpose of evaluation of usability. Combination of QFD and inspection methods like heuristic evaluation could help in identifying the key features responsible in overall product satisfaction from the very start of the design process. Thus we propose that QFD when integrated with the various usability evaluation methods could help in better documentation of the "voice of the customer" and its measurable design attributes.

We can safely conclude that the combination of QFD with usability testing and inspection methods would definitely lead to shorter development time, fewer engineering changes, reduced introduction costs and satisfaction of user needs and desires. Thus the application of this methodology in the field of usability evaluation has a large scope for further research and development.

REFERENCES

1. Nielsen, Jacob- "What is 'Usability'?"29 Sept. 1998<http://www.zdnet.com>.
2. Dumas, Joseph S., and Janice C., Redish. A Practical Guide to Usability Testing. Ablex, 1993.
3. Gray, Wayne D., and Marilyn C., Salzman "Damaged Merchandise? A Review of Experiments that Compare Usability Evaluation methods." Human Computer Interaction (1998).
4. Rubin, Jeffrey. Handbook of Usability Testing. New York: Wiley,1994.
5. Clausing, D., & S., Pugh (1991), Enhanced Quality Function Deployment. Design and Productivity International Conference, Honolulu HI, 6–8 Feb.
6. Barnum, C. Usability Testing Research, Longman, New York, 2002. ISBN:0-205-31519-4.
7. Zultner, R.E., 'Quality Function Deployment (QFD) for Software', American Programmer (1992).
8. Jackson, H.K.J., & Frigon, N.L., (1994) Management 2000: The Practical Guide to World Class Competition, Van Nostrand Reinhold, (NY. 9) Akao, Y., ed. (1990). Quality Function Deployment, Productivity Press, Cambridge MA.
9. Hauser, J.R. & Clausing D. (1988) The House of Quality. Harvard Business Review, 3, May–June, 63–73.
10. King, B. (1989). Better Designs in Half the Time: Implementing Quality Function Deployment in America, GOAL/QPC, Methuen MA.
11. Clausing, D. & S. Pugh (1991) Enhanced Quality Function Deployment. Design and Productivity International Conference, Honolulu HI, 6-8 Feb.
12. C. Wharton et. al. "The cognitive walkthrough method: a practitioner's guide" in J. Nielsen & R. Mack "Usability Inspection Methods" pp. 105-140.

SECTION - IV

PRODUCT DESIGN FOR X

DESIGN FOR SIX SIGMA (DFSS)

34.1. INTRODUCTION

Many organizations today follow Six Sigma methodologies to enhance quality in their products and services, to achieve total customer satisfaction. These organizations over their course of progress find that to move from the level of four sigma to five sigma, they have not just alone improvise their existing processes but need to design the systems from *ab initio* imbibing the error freeness and failure less performance throughout the life cycle of the product. This has in fact given birth to Design for Six Sigma (DFSS).

DFSS is the expansion of the Six Sigma improvement process to tackle critical issues in the design of new products and services. DFSS is an approach to prevent problems in the first place, and to incorporate positive customer satisfaction into the initial design intent from the beginning.

DFSS uses event, pattern and structural tools in its methodology and the extent of the creativity and usage of creativity tools is still grossly underutilized and undermined. Further improvement requires use of pattern thinking in the customer, functional and physical domains, as well as in the process domain. This article brings in forth the creativity tools that are available in various contemporary change management methodologies and sums up to a checklist, which has to be used/can be used during the design phase of DFSS.

According to Smith [1], Quality tools and methods have evolved utilizing three stages of thinking (event, pattern and structure) across various domains associated with product development. The transition from event thinking to pattern thinking is also the transition from Six Sigma to Design for Six Sigma (DFSS). Six Sigma addresses event level concerns that occur in product development after the fact. DFSS is a rigorous approach to designing products and services from the very beginning to ensure that they meet customer expectations DFSS represents a higher evolution of the methodology, utilizing pattern level thinking and tools to build quality into the product or service.

DFSS is the expansion of the Six Sigma improvement process to tackle critical issues in the design of new products and services. While traditional Six Sigma focuses on improving existing products and processes, DFSS is an approach to prevent problems in the first place, and to incorporate positive customer satisfaction into the initial design intent from the beginning.

DFSS is an integration of all the quality prevention tools across the pattern level domains. Further improvement requires implementation of structural thinking tools. Design for Six Sigma (DFSS) is a methodology for determining the values of the design parameters, which maximize the performance of some system without tightening the material, manufacturing or environmental tolerances. According to Stahl [2], similar to DMAIC (a principle recommending application of a methodology consisting of Design, Measure, Analyze, Improve and Control for improvement of existing business methodology; we will discuss that in detail in a future chapter), DFSS is a five-step process represented by the acronym DMADV (Define, Measure, Analyze, Design and Validate). The steps of *Define* and *Measure* of a DMADV project may be summarized as a process of CTQ (Critical-To-Quality) flow-down. The analyze phase can be summarized as a process of capability flow-up. This is where DMAIC and DMADV differ significantly.

DFSS is an example of offensive quality. Offensive quality approaches produce customer value by creating what customers do like, in contrast with defensive quality approaches that eliminate what customers do not like (variation from target). While defensive quality makes existing processes and products error free, offensive quality involves creativity and innovation to introduce new products and services, and bundles of new product features, therefore market researchers engaged in CI must probe and explore beyond traditional customer preference inquiries to develop approaches for creating offensive quality [3].

According to Smith [1], an effective DFSS program must utilize tools that make a difference in each design domain. For example, a DFSS program that does not interface with customers or does not utilize powerful methods like taguchi's parameter design (to hit the design with noise or uncontrolled variation and adjust controllable factors in the design to make the design robust against noise) is a program that is guaranteed to leave significant portions of the work of design incomplete.

34.2. CREATIVITY

Plsek [4] thinks Creativity is not a magical ability. It simply requires us to approach situations from a fresh perspective. Plsek [5] advises anyone who can think can learn to think creatively. Learning to think creatively is essential to the success of any business that wants to be competitive. Finster [3], the creativity literature identifies two approaches to creativity. One approach identifies those who are most likely to be creative and then employs those candidates in organizational positions requiring creative output, positions like research and development, product conceptualization, and advertising. The second approach to creativity discusses processes, environment, mechanisms, and tools for raising the creative capacity of everyone in an organization. Smith [1], thinking at a level of fundamental structure offers even higher leveraged opportunities to create products and services that not only function as intended, but also deliver unprecedented customer satisfaction. When the foundational structure of design is properly established, the methods at the pattern level are much more effective. When pattern level methods work well, the event outcomes become world-class.

Successful creators of innovation mindsets combine efficiency and effectiveness and stimuli. It is seen that creativity occurs primarily at the early stages of innovation processes with innovation implementation later.

Rank et al. [6] distinguish between creativity and innovation. Creativity typically refers to the production of new and useful ideas by an individual or a small group of individuals working together. Innovation is defined as the intentional introduction and application within a role, group or organization of ideas, processes or procedures, new to the relevant unit of adoption, designed to significantly benefit the individual, the group, organization or wider society. Hence, creativity refers to idea generation, whereas innovation refers to idea implementation. Voice may function as a mediator between creativity and innovation. If new ideas are not articulated, they can hardly be implemented.

Classification of Creativities

There are three major principles that underlie all tools of creativity. Plsek [7] says cole and others has examined the concepts of Continuous improvement, Continuous innovation, and Discontinuous innovation, Incrementalism, Exploitation and Exploration. And concluded that Focusing on alternative ways to creatively build quality improvement through continuous innovation into the development process. Miller [8] presented Innovation styles model as: Visioning, modifying, experimenting and exploring styles, team participation. McFadzean [9] offers such a classification, which categorises the creative problem solving techniques into paradigm preserving, paradigm stretching and paradigm breaking.

Tools: West et al. [10] describe how, using an input-process-output model, we can identify the key elements necessary for developing team innovation. They propose that it is the implementation of ideas rather than their development that is crucial for enabling organizational change.

Vansuch, [11] suggests that creativity generates new ideas, while innovation applies those ideas. Both activities can improve customer/supplier relationships. In an environment of creativity and innovation, there should be a balance of convergent and divergent thinking. The best creative and innovative people are adventurous, persistent, spontaneous, and open to possibilities and opportunities. He suggests that the following tools are useful for divergent thinking:

- Fluency tools, like brainstorming;
- Excursion tools, like random words;
- Pattern breaker tools, like metaphors; and
- Shake-up tools, like humor.

There are a lot of idea generation tools available now. Greetings cards, Heuristic ideation technique (HIT), Cherry Split, Backward Forward Planning, Slice and Dice, Component detailing, Morphological analysis, Morphological Forced Connections, Listing, Collective notebook, Attribute Listing and variants. The following tools also help in idea generation. Cartoon storyboard, Collective notebook (CNB), Component detailing, Crawford slip writing, Drawing, Visual brainstorming, Essay writing, Story writing, Sculptures etc.

According to Plsek [7], Creative thinking means moving away from the mental valleys of De Bono's model of mental mechanics. To do so requires the three principles of directed creativity:

attention, escape, and movement. The concept of customer needs has evolved from thinking about specifications and dimensions of quality to addressing the three categories of product characteristics: must be, more is better, and surprising quality.

Directed creativity provides many tools for customer needs analysis, of which five follow. First, identify non-obvious customers. Second, look for customers who might benefit from part of an existing product or service. Third, create alternatives by studying customer complaints and comments. Fourth, work on poorly developed dimensions of quality. Fifth, transform must be quality characteristics into ideas for surprising quality and more-is-better.

Plsek [4] opines that by using customer surveys and focus groups, organizations are asking customers to be creative on their behalf, but customer research alone does not result in the kind of innovation that goes beyond generating customer satisfaction to encourage customer retention. Internal brainstorming sessions are often used to product innovative ideas, but brainstorming is a weak tool. Organizations need to innovative and deliver surprising quality through the use of directed creativity.

Directed creativity requires individuals and teams to use the basic mental processes of attention, escape, and movement to develop new ideas. The approach goes beyond customer needs analysis.

While it has been said that brainstorming is a weak creativity tool, with the following galore of many variants, it cannot be taken lightly. Brainstorming, Classic Brainstorming, Imaginary Brainstorming, Negative (Reverse) brainstorming, Rawlinson Brainstorming, Role storming, Value brainstorming, Visual brainstorming, Trigger method, Trigger Sessions, Brain Writing, Force-fit game, Brain writing pool, Brain writing game, Constrained brain writing, Brain writing 6-3-5, Idea card method, Pin cards, Varying level of constraint, Crawford Slip Writing. Also belong to this family are Nominal Group Technique and the Improvised Nominal Group Techniques, Nominal-interacting technique and the Free association.

Plsek [12] adds that Directed creativity is creativity based on theory and principles. These include de Bono's mental mechanics model of perception and memory; the principles of attention, escape, and movement; and the use of tools described in the literature. Tools for enhancing creative thinking include:

- Random word;
- Break the rules;
- Element or scene modification;
- Be someone else; and
- The concept fan.

The tools that would be of help here are:

- Relational words;
- Paraphrasing key words;
- Brute think;
- Listing;
- Listing pros and cons;
- Morphological analysis;
- Paired comparison;
- Comparison tables;
- Attribute listing and variants;
- Listing pros and cons;
- PMI (Plus, Minus, Interaction);
- Personal balance-sheet;
- Progressive hurdles;
- Delphi;
- Q-sort;
- Paired comparison;

- Interpretive structural modeling;
- Plusses;
- Advantages;
- Synectics.
- Receptivity to ideas;
- Potentials and concerns;
- Limitations and unique qualities; and

Plsek [13] feels that we all use a flexible, pattern-matching system as we do routine and repetitive tasks; however, pattern matching is not a practical approach when we need fresh, creative ideas. Tools that enhance mental creativity are based upon three principles: attention, escape, and movement. Some techniques that apply these principles include:

- Selecting random words from a book or dictionary, then finding connections between the concepts associated with the random words;
- Modifying the "givens (or substituting the elements) in situations;
- Approaching issues as if we were someone else; and
- Using a concept fan to discover multiple creative possibilities.

The best way to go about improvising the design is the usage of Osborn's checklist, and the subsequent improvisation techniques like SCAMPER, SCAMMPERR, Reversals, Sequential-attributes matrix, Exaggeration (magnify or minify), Implementation checklists, Excursions, Problem Reversal, False Faces, which formed later the mainstream value engineering.

Norausky [14] adds that the creative process and the innovative process require an environment of freedom for exploration, openness in communication and trust. He suggests the following tools for the creative process.

- Analogy;
- Free association; and
- Scenario writing;
- Technology fiction.

Wycoff and Snead [15] recommend Collaboration rooms, innovation centers, creativity labs are easy to set up/design and have enormous pay-off. In the example of Franklin Covey's Innovation theme rooms, Robinson [1] suggests the use of illustrations and metaphor as most valuable in demonstration.

According to McFadzean [9] there are a number of different techniques that can enhance creative thinking. Some of these techniques require less imagination and produce less novel results than others. One method of encouraging novel ideas is to undertake creative problem solving (CPS).

The following creative thinking tools can do wonders. Creative thinking, Creative problem solving (CPS), Phases of Integrated Problem Solving (PIPS), Lateral Thinking, Bunches of bananas, Random stimuli of various kinds, Stimulus analysis, TILMAG, DO IT, Force-field analysis and Simplex.

Sometimes it is much useful to think the other way around. Wondering what might happen If.... The tools that can help in this kind of situation are: Help hinder, Bullet proofing, Negative (Reverse) brainstorming, Potential-problem analysis, (PPA), Stake holder Analysis, Strategic assumption testing, Kepner and Tregoe's method, Bug listing etc.

Mann and Dewulf [16] suggest that some people may want just a few tools or strategies to help them, others may be looking for a higher level start-to-finish process, and still others are looking for a higher level creativity philosophy from which they hope everything else might emerge.

To reach innovative performance creative people need to take the initiative in promoting their ideas, with the possible corresponding price of low performance quality. Multi-level methods such as hierarchical linear modeling may also be applied to analyse the influence of cultural factors on creativity and innovation. Mapping tools also help in generating creative ideas. Radical tree diagrams, Hierarchical tree diagrams, Clustering methods, Mind mapping, Causal mapping, Consensus mapping, Mind mapping (Spider diagram), Card storyboards, Highlighting, KJ-method, Snowball technique, Fishbone diagram and SODA.

Boedker et al. [17] warns that it is necessary to understand more about where these tools come from and how they might be used in a systematic and purposeful design activity, and not just at random. They propose to look for things and notions that in similar ways help us see, or rather do, something in a new and different ways in the context of use. Reminding us of the need for cooperation and hands-on experience, such tools need to support action as well as reflection.

Fritz and McManus [18] think that the Innovation and the creative process may not be found in the ways we might think: brainstorming and free association. The nature of invention is about focusing the mind and looking at something in a structural versus free-associative way.

Preliminary questioning, asking repeatable questions 'Why?' 'Why?' etc., use of Five W's and H and Dimensional Analysis will go a long way in improving the designs in a creative and innovative way.

TRIZ: Smith [1], says DFSS is used to prevent problems by building quality into the design process across domains at the pattern level of thinking. The fact that problems exist in the event realm means that the original process of design had serious flaws that is pattern or structural work that should have been done is either missing or incomplete. The answer is simply that higher-level thinking can always be used as a methodology to solve lower level problems.

According to Smith [1], both Six Sigma and DFSS can be made more effective by incorporating structural tools such as axiomatic design and TRIZ (a Russian acronym for "theory of inventive problem solving") into the methodology. Because these tools address design foundation flaws, they will enhance every aspect of Six Sigma and DFSS, making the process of problem solving and problem prevention much more insightful, productive and efficient than programs that do not utilize these methods. Use of new structural tools such as TRIZ and axiomatic design provide a foundation for future enhancement of Six Sigma methodologies. Companies at the pattern level interact with customers and find out what customers believe is important today. No customer can tell the designer what will be important tomorrow. At the structural level, TRIZ directed evolution predicts what will be important to customers tomorrow. "TRIZ" means "Theory of Inventive Problem Solving." It is the analytical, "left-brain" creativity method that works exceedingly well in the analytical world of DFSS and other methods of quality improvement. It is a method that helps everyone be creative, but bases that creativity on data and analysis.

Smith [1], says suppose the designer cannot find a set of design parameters that keep all the functional requirements independent. In this situation, improving one function typically degrades another. Altshuller called the situation where functions oppose each other contradictions and developed a methodology in which design teams could systematically innovate and find design parameters that resolved contradictions, creating win-win functional situations. He created a matrix of contradictions and resolution principles and used this information to guide design teams so that they could brainstorm in areas that are likely to lead to win-win solutions. Altshuller also believed in minimizing information content; he called this the principle of ideality. Later, TRIZ was expanded to include an entire algorithm of innovation techniques, including the study of system evolution.

Design for Six Sigma (DFSS) and the Ideation/TRIZ methodology (I-TRIZ) have been integrated to ensure that new products are designed to meet Six Sigma levels and marketing plans. Use of this single, coordinated process synergizes efforts to increase productivity in the product design process. It allows organizations to stimulate innovative thinking in solving difficult problems related to designing new products, services and processes; and to create the likely evolutionary successors of these products, services and processes. What is more difficult is to integrate TRIZ thinking with other psychologically based creativity and assessment tools. Users and trainers for these various tools tend to be very protective about each process and do not spend sufficient time thinking about ways to integrate the best of all tools. Hipple [19] reviews suggested ways to effectively integrate TRIZ innovation and problem-solving principles with these other tools.

Mann and Dewulf [16] A systematic programme of research to compare and contrast different creativity tools, methods and philosophies in terms of their relevance to primarily scientific, engineering and business applications has concluded that TRIZ currently offers the most useful foundation for a higher order systematic creativity model and that given this foundation, the other available methods that are best able to complement and help deliver the higher order model.

TRIZ, ARIZ, Contradiction Analysis and Ideal Final Result, are the most acknowledged creativity tools rampantly used in DFSS. In New Product Development and Designs the following tools find good application. Super Group, Role Storming, Super heroes, CATWOE, Criteria for idea-finding potential, Bug Listing, Focus Group, Delphi, and Problem inventory analysis (PIA).

Hyatt [20] introduces Synoptics, an approach to creativity, in the context of six challenges to creativity. Synoptics was developed from a desire to understand the nature of existing creativity methods, asking: why do we have the current methods? How are they related? What other potential methods exist? He says Synoptic is offered as an additional block in the foundation of serious creativity that will help in its acceptance and use. Systematic creative thinking is an extraordinary lever that can move difficult obstacles. He proposes metaphoric models of thought and tree structures.

34.3. CLOSURE

Mann and Dewulf [16] have speculated here that TRIZ is but one component of a higher level creativity capability we have chosen to label 'systematic creativity' and believe that it is fundamental to the evolution of such a 'systematic creativity' model is that it will emerge – initially at least–from the integration of the different tools, methods and philosophies that currently exist. They opine that there are several emerging creativity models that have not so far been explored in the context of their place in a bigger 'systematic creativity' picture. These include game theory, chaos theory, spiral dynamics and general periodicity.

The future of Six Sigma and DFSS involves incorporating structural thinking methods such as axiomatic design and TRIZ. Use of these methods will make Six Sigma and DFSS more effective and more productive with less effort.

There is currently no single 'creativity' entity that will satisfy every individual desire. If there ever is, one thing it will have to encompass is due recognition of individual difference, and be self-adapting to accommodate those differences.

The following checklist can /should be followed during the design stages to make the products more creative.

Idea Generation Tools	**Brain Storming Tools**	**Creative Thinking Tools**
Cartoon story board, *Collective notebook (CNB),* *Component detailing,* *Crawford slip writing, Drawing,* *Visual brainstorming* Essay writing, Story writing, Sculptures, Brain sketching, Pin cards, Gallery method, Component detailing, Brute think, Random word Greetings cards Heuristic ideation technique (HIT) Cherry Split Backward forward planning Slice and Dice Component detailing Morphological analysis Morphological Forced Connections Listing Collective note book Attribute Listing and variants Relational words, Paraphrasing key words,	Brainstorming, Classic Brainstorming, Imaginary Brainstorming, Negative (Reverse) brainstorming, Rawlinson Brainstorming, Role storming, Value brainstorming, Visual brainstorming, Trigger method, Trigger Sessions, Brain Writing, Force-fit game, Brain writing pool, Brain writing game, Constrained brain writing, Brain writing 6-3-5, Idea card method, Pin cards, Varying level of constraint, Crawford Slip Writing. Nominal Group Technique a Improvised Nominal Group Techniques, Nominal-Interacting Technique Free association.	Creative thinking, Creative problem solving (CPS), Phases of integrated problem solving (PIPS) Lateral thinking, Bunches of bananas, Random stimuli of various kinds, Stimulus analysis TILMAG, DO IT, Force-field analysis, Simplex. **Thinking Other Way Tools** What might happen If… Help hinder, Bullet proofing Negative (Reverse) brainstorming, Potential-problem analysis (PPA), Stake holder Analysis, Strategic assumption testing, Kepner and Tregoe's method, Bug listing
Listing	**Value Engineering Tools**	**Mapping Tools**
Listing pros and cons, Morphological analysis, Paired comparison, Comparison tables, PMI (Plus, Minus, Interaction), Personal balance-sheet,	Osborn's checklist, SCAMMPERR, SCAMPER Reversals, Sequential-attributes matrix, Exaggeration (magnify or minify), Implementation checklists	Radical tree diagrams Hierarchical tree diagrams Clustering methods Causal mapping Consensus mapping Mind mapping (Spider diagram),

Progressive hurdles, Delphi Q-sort, Paired comparison Interpretive structural modeling Receptivity to ideas Plusses, potentials and concerns Advantages, limitations and unique Qualities *Synectics*	Excursions, Problem Reversal False Faces, Value engineering	Card storyboards, Snowball technique, K.J. Method Contradiction Analysis Progressive hurdles, Delphi, Q-sort, Paired comparison, Interpretive structural modeling Highlighting Fishbone diagram SODA
	Questioning Tools	
	Preliminary questions, 'Why?' 'Why?' etc. - repeatable questions Five W's and H, Dimensional analysis	
TRIZ Family Tools		Design/ NPD
TRIZ, ARIZ, Contradiction Analysis, Ideal Final Result	NLP Observer and merged viewpoints Other people's definitions Other people's viewpoints	Super Group, Role Storming, Super heroes, CATWOE, Criteria for idea-finding potential, Bug Listing, Focus Group, Delphi, Problem inventory analysis (PIA)

REVIEW QUESTIONS

1. What is meant by the principle of *six sigma*? Explain with an example.
2. How the design of a product can ensure achievement of six sigma? Explain with an example and justify your answer.
3. What is the importance of CTQ? How do you identify the CTQs?
4. What are the various stages of DMADV? Apply DMADV to a simple manufacturing problem and exemplify the use of this technique.
5. What is creativity? Why it is not a magical ability?
6. What is the difference between creativity and innovation?
7. What are the three major principles underlying the tools of creativity?
8. What is TRIZ? Why do you think it is novel?

REFERENCES

1. Smith L. R., (2001), "*Six Sigma and the Evolution of Quality in Product Development*", November 2001, Volume 1, Number 1.
2. Stahl R., Schultz B. and Pexton C., (2003), GE Medical Systems Healthcare's Horizon From incremental improvement to designing the future. February 2003, Volume 2, Number 2.

3. Finster M.P, (2001), "*Creating Continuous Innovation*", Quality Management Journal, October 2001, Vol. 8, No. 4.
4. Plsek P.E., (2000), "*Creative Thinking for Surprising Quality Quality Progress*", Vol. 33, No. 5, May 2000, pp. 67–73.
5. Plsek, P.E., (2003), "*Creativity, Innovation and Quality*", Prentice Hall of India , New Delhi.
6. Rank J., Pace V.L., Frese M., (2004), "*Three Avenues for Future Research on Creativity*", Innovation, and Initiative Applied Psychology: an International Review, Vol. 53, Issue 4, Page 518 —October 2004.
7. Plsek, P., (1998a), "*Surprising Quality: Creative Customer Needs Analysis*", Annual Quality Congress, Philadelphia, PA, Vol. 52, No. 0, May 1998, pp. 716–724.
8. Miller W.C., (1995), Four distinct strategies for quality improvement: Is innovation built into your improvement processes? "*Journal For Quality and Participation*", Vol. 18, No. 1, January 1995, 46–48.
9. McFadzean E., (1998), The Creativity Continuum: Towards a Classification of Creative Problem Solving Techniques Creativity and Innovation Management, Vol. 7, Number 3, September 1998, pp. 131–139.
10. West M. A., Hirst G., Richter A., Shipton H., (2004) Twelve steps to heaven: Successfully managing change through developing innovative teams, "*European Journal of Work and Organizational Psychology*", Volume 13, Number 2, June 2004, pages 269–299.
11. Vansuch, G. M., (1998) CIA for CS: Creativity and Innovation Applications for Customer/ Supplier Relationships Annual Quality Congress, Philadelphia, PA, Vol. 52, No. 0, May 1998, pp. 447–451.
12. Plsek P.E., (1998b), Incorporating the Tools of Creativity into Quality Management, "*Quality Progress*", Vol. 31, No. 3, March 1998, pp. 21–28.
13. Plsek, P., (1996, Bringing Creativity to the Pursuit of Quality, Annual Quality Congress, Chicago, IL, Vol. 50, No. 0, May 1996, pp. 99–105.
14. Norausky, P. H., (1998), A Competitive Advantage: Customer and Supplier Innovation Teams, Annual Quality Congress, Philadelphia, PA, Vol. 52, No. 0, MAY 1998, pp. 436–446.
15. Wycoff, Joyce; Snead, Lynne; 1999, Column: Tools@Work: Stimulating Innovation with Collaboration Rooms, "*Journal for Quality and Participation*", Vol. 22, No. 2, March 1999, pp. 55–57.
16. Mann D., and Dewulf S., (2001) Evolving The World's Systematic Creativity Methods, 7th European Association of Creativity and innovation conference, University of Twente, Netherlands, December 2001.
17. Boedker S., Nielsen C., Petersen M.G., (2000), Creativity, cooperation and interactive design, Proceedings of the conference on Designing interactive systems: processes, practices, methods, and techniques, New York City, New York, United States, pp. 252–261.
18. Fritz R. and McManus K. (1998) How Creativity Nurtures the Nature of Invention: Journal for Quality and Participation, Vol. 21, No. 6, November 1998, pp. 8–15.
19. Jack Hipple, The Integration of TRIZ with Other Ideation Tools and Processes as well as with Psychological Assessment Tools, Creativity and Innovation Management, Vol. 14, Issue 1, 2005, p. 22.
20. Hyatt M., (2002), An Overview of Synoptics and the Six Challenges of Creativity, downloaded from http://www.triz-journal.com/archives/2002/10/f/ , downloaded on Jan 22, 2006.

ANALYSIS OF MEANS (ANOM)

35.1. INTRODUCTION

Output variations in production processes are due to operator errors, improperly adjusted machines and process capability limitations. Life cycle cost is the analysis of cost issues over the entire life cycle of a given product, rather than a specific fiscal period. It includes factors based on stages (introduction, growth, maturity, decline) and all research and development, design, marketing, production and logistics costs. The key element of Quality Cost is the internal failure cost which when contained, will end up with low cost of manufacture besides quality Improvement. The analysis of mean (ANOM) developed by Ott [1] is a graphical tool resulting from designed experiments. It gives a graphical view of the out of control situation and whether it is due to treatment effects and contrasts. The technique combines the advantages of control charts and designed experiments. This chapter explains in detail as to how life cycle cost economics was achieved in needle manufacturing by a continuous improvement project with Quality Cost as driver and by using ANOM as a problem identification tool for feedback adjustment of process to reduce process scrap.

Goldratt states that a firm's goal is to make money now and in the future. A company will not exist if it is not making money. Any activity that does not help make money is a waste of time and resources. His Theory of Constraints is implemented through three measures: throughput, operating expenses, and inventory. Cost is a major driver in many, if not all, of the decisions that a firm must make in managing its operations. Thus, in determining a level of quality improvement (as measured by process capability), cost is a major consideration. The integration of product and process design through improved business practices, management philosophies and technology tools will result in a more producible product to better meet customer needs, quicker and smoother transition to manufacturing, and a lower total program/life cycle cost. Life cycle cost is the analysis of cost issues over the entire life cycle of a given product, rather than a specific fiscal period. It includes factors based on stages (introduction, growth, maturity, decline) and all research and development, design, marketing, production and logistics costs.

For more comprehensive analysis of the total cost of a product, entire life-cycle cost considerations have to be taken into account. To achieve this it will be necessary to develop ways to reduce all forms of waste in manufacturing. This requires the use of a robust tool to reduce variations in manufacturing. Statistical Process Control (SPC), which give a snap shot of statistical control of production process is a time-tested and more-handy toolbox for this purpose.

A systematic approach used to understand that the process is producing without much variability is known as the Statistical Process Control. The control charts are important tools of the SPC, which is the running record of the job. The primary function of the control chart is to detect causes of variation that are due to some assignable reasons. The random or common cause variability is inherent to the capability of the process. Random variation cannot be eliminated without modifying the process. The system itself must be changed to improve its capability.

Assignable cause variability represents a change in the process in a systematic way. This change can be attributed to some identifiable cause or causes that are not an inherent part of the process and can, therefore, be eliminated. It is worth pointing out that SPC, as a technique is actually process evaluation, rather than control, since it does not directly control the process. It helps determine whether a process is in statistical control and flags out-of-control conditions when they occur.

Sewing machine needles are an important accessory to sewing machines and a key commodity for the haberdasher. In 1750 BC first needles with eye were used. The eye was drilled in bones, horn or ivory splinter. The history of modern needle dates back to Sung Dynasty (1127–1279) when Chinese needle maker Liu of Tsheilam made his needle. In modern times the hub of sewing machine needle manufacture is in Germany.

The objective of the chapter is to detail a continuous improvement project undertaken in a precision light engineering needle factory to reduce process scrap with quality cost as driver. The aim was to economize on product life cycle cost by reducing the internal failure cost. To fulfill this objective ANOM was used to pin point the treatment effects for the purpose of feedback adjustments. This was in the place of elaborate designed experiments. Thus an effort has been made to achieve the long standing objective of integrating engineering process control and statistical process control.

35.2. BACKGROUND

The geometry of a sewing machine needle is shown in Fig. 35.1. The manufacturing of sewing machine needle involves machining, metal forming and metal finishing processes. The raw material used is high carbon steel wire. There are mainly eleven stages of manufacturing operations. The key parts of the sewing machine needle are Blade, Shank, Point, Eye and Grooves.

The significance of addressing the life cycle economics of the product in design and production stage need not be overemphasized in precision engineered products like needles. Throughout production process, life cycle cost objectives are accomplished by identifying high cost contributors.

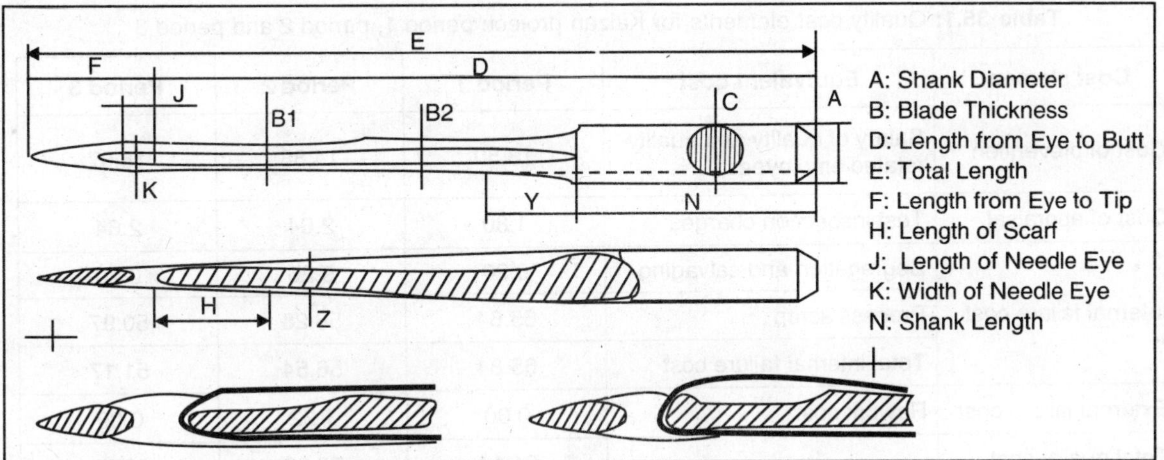

Fig. 35.1: A sketch of needle

35.3. QUALITY COST

As money is the most common language for all stakeholders like customers, employees, management in the organization and as measurement is easier, "Quality Cost Analysis" is being considered as a driver for improving quality and reducing cost of manufacturing. The American Society of Quality classifies Quality Cost into 93 elements. Juran the quality guru pointed out that by analyzing and reducing the *internal failure cost*, we will be digging a gold mine as the internal failure costs give major opportunity for reduction.

Kaizen project was undertaken to reduce process scrap % and in turn quality cost due to internal failure. By definition all costs due to defective products identified before shipment from factory form the Internal Failure Cost. The rework done as a result of rejections in Sampling stages and the resultant segregation cost and cost of defective products segregated and rejected represent part of this cost. Another important facet of this cost is the process scrap cost. The process scrap percentage represents the defective products identified in process and at the final inspection stages. All these form part of internal failure cost, which is a significant factor in our objective to achieve product life cycle economics in Production.

35.4. PROBLEM DISCUSSION

The following table shows an overview of the project period 1, where in the proof of need for improvement was identified. Period 2 is a transition phase and period 3 is a phase when improvements have been made.

In precision grinding of shank of needles, the control charts were used as the monitoring tool. A set up adjustment was made when there is an action signal showing out of control situation as set out by the General Electric run rules as follows:

- A single point lying beyond the three-sigma limits;
- Two out of three consecutive points lying beyond the two sigma limits;
- Four out of five consecutive points lying beyond the one sigma limits; and
- Eight consecutive points lying on one side of the target value.

Table 35.1: Quality cost elements for Kaizan project: period 1, period 2 and period 3

Cost element	Equivalent cost	Period 1	Period 2	Period 3
Cost of prevention	Salary of quality and quality related employees	13.50	14.88	15.07
Cost of appraisal	Test inspection charges	1.80	2.64	2.64
Internal failure cost	Segregation and salvaging	0.20	0.26	0.20
	Process scrap	65.64	56.28	50.97
	Total internal failure cost	65.84	56.54	51.17
External failure cost	Rework cost	0.00	1.92	0.00
Total quality cost		81.14	75.98	68.88

A control chart depicting the process situation in Period 1 is shown in Table 35.2.

Table 35.2: Control chart data–period

		1	2	3	4	5	6	7	8	9	10
Sample Measurements	1	1.57	1.59	1.59	1.59	1.57	1.58	1.59	1.57	1.55	1.56
	2	1.58	1.59	1.60	1.58	1.56	1.58	1.58	1.55	1.54	1.57
	3	1.58	1.60	1.57	1.59	1.56	1.58	1.58	1.59	1.55	1.55
	4	1.60	1.59	1.60	1.60	1.58	1.58	1.58	1.56	1.56	1.56
	5	1.59	1.60	1.58	1.59	1.58	1.59	1.60	1.55	1.56	1.55
Sum		7.92	7.97	7.94	7.95	7.85	7.91	7.93	7.82	7.76	7.79
Average X		1.58	1.59	1.59	1.59	1.57	1.58	1.59	1.56	1.55	1.56
Range R		0.03	0.01	0.03	0.02	0.02	0.01	0.02	0.04	0.02	0.02
		11	12	13	14	15	16	17	18	19	20
Sample Measurements	1	1.58	1.57	1.57	1.55	1.55	1.57	1.56	1.56	1.57	1.56
	2	1.57	1.60	1.57	1.57	1.56	1.55	1.58	1.57	1.55	1.58
	3	1.56	1.57	1.57	1.56	1.55	1.60	1.57	1.57	1.57	1.57
	4	1.57	1.56	1.56	1.56	1.55	1.58	1.57	1.59	1.57	1.60
	5	1.55	1.59	1.55	1.55	1.58	1.57	1.55	1.57	1.58	1.58
Sum		7.83	7.89	7.82	7.79	7.79	7.87	7.83	7.86	7.84	7.89
Average X		1.57	1.58	1.56	1.56	1.56	1.57	1.57	1.57	1.57	1.58
Range R		0.03	0.04	0.02	0.02	0.03	0.05	0.03	0.03	0.03	0.04

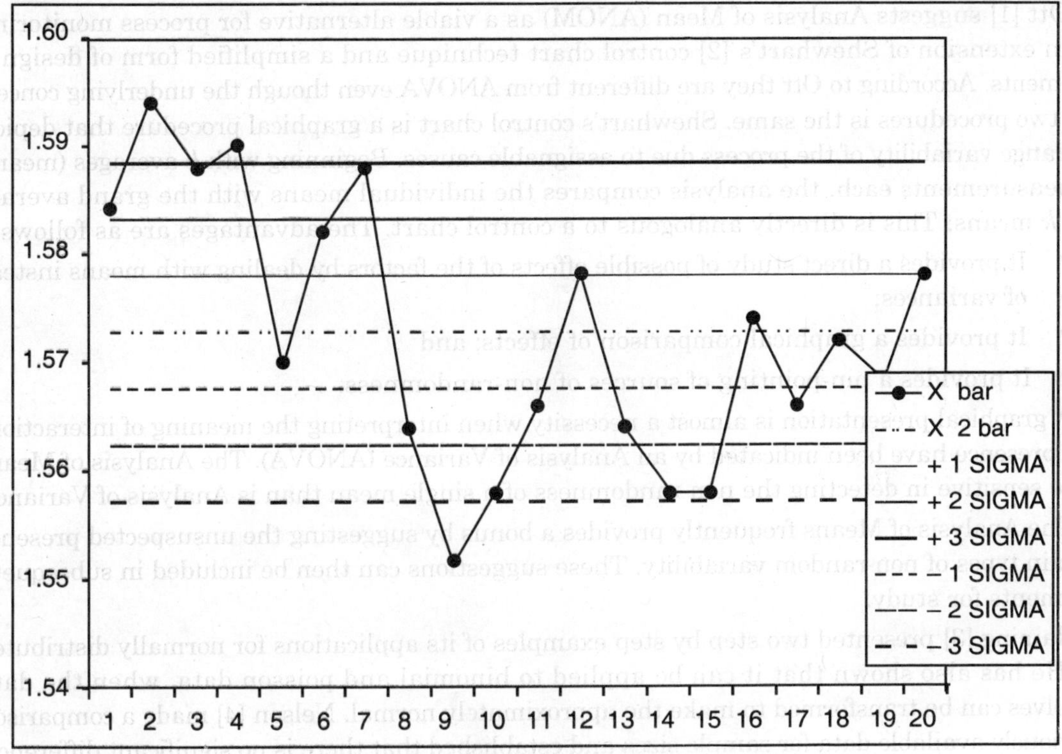

Fig. 35.2: Control chart for averages – period 1

With the aim of reduction of internal failure costs through reduction of process scarp, a kaizen project was conducted. X bar chart has been the traditional tool for process monitoring in this process (Fig. 35.2). The theory behind control charts is based upon the *central limit theorem.* It states that the averages of sub-groups of samples taken from a production process will form a normal distribution regardless of the distribution of the individual readings.

Shewhart lays down the practical requirements of criteria of control as follows:

- Criteria for control should indicate the presence of assignable causes of variation;
- It should not only indicate the presence of assignable causes but also should do this in a way to facilitate the discovery of these causes;
- It should be as simple as possible and adaptable in a continuing and self-corrective operation of control; and
- It should be such that the chance of looking for assignable causes when they are not present does not exceed some prescribed value.

35.5. ANALYSIS OF MEANS (ANOM)

As the control charts only depict the process situation, the need to go in for the tools for finding out the reasons or causes as a basis for feedback adjustment was felt. ANOM is an integral tool which depicts both statistical significance and engineering significance and this aids identifying the root cause of the problem due to treatment effects.

Ott [1] suggests Analysis of Mean (ANOM) as a viable alternative for process monitoring. It is an extension of Shewhart's [2] control chart technique and a simplified form of design of experiments. According to Ott they are different from ANOVA even though the underlying concept of the two procedures is the same. Shewhart's control chart is a graphical procedure that depicts short range variability of the process due to assignable causes. Beginning with k averages (means) of n measurements each, the analysis compares the individual means with the grand average of the k means. This is directly analogous to a control chart. The advantages are as follows:

- It provides a direct study of possible effects of the factors by dealing with means instead of variances;
- It provides a graphical comparison of effects; and
- It provides a pin-pointing of sources of non-randomness.

A graphical presentation is almost a necessity when interpreting the meaning of interactions whose presence have been indicated by an Analysis of Variance (ANOVA). The Analysis of Means is more sensitive in detecting the non-randomness of a single mean than is Analysis of Variance.

The Analysis of Means frequently provides a bonus by suggesting the unsuspected presence of certain types of non-random variability. These suggestions can then be included in subsequent experiments for study.

Raming [3] presented two step by step examples of its applications for normally distributed data. He has also shown that it can be applied to binomial and poisson data when the data themselves can be transformed to make the approximately normal. Nelson [4] made a comparison of previously available data for sample sizes and established that there is no significant difference. He constructed tables with substantial numerical data.

He gave the step by step procedure for carry out ANOM:

- Compute the group means, $X'_i = (i = 1,2,\ldots,k)$;
- Compute the grand mean, X'', using equation 1;
- Compute s, an estimate of the standard deviation of an individual observation. This is the square root of s^2, where s^2 using equations 2 and 3;
- Obtain the value h_α from the table for type-I risk level α, number of means k, and degrees of freedom $(n-1)k$;
- Compute the upper and lower decision lines, UDL and LDL where,
  UDL = $X'' + h_\alpha s\sqrt{[(k-1)/(k_n)]}$
  LDL = $X'' - h_\alpha s\sqrt{[(k-1)/(k_n)]}$; and
- Plot the means against the decision lines. If any mean falls outside the decision lines, conclude there is a statistically significant difference among the means.

The grand mean X'' of the group mean (X'_i) is simply the average of these mean values, which is written as:

$$X'' = \Sigma(X'_i)/n$$

The pooled estimate for the standard deviation is the square root of the average of s^2_i for the individual observations.

The lower and upper decision lines (LDL and UDL) are:

$$LDL = X'' - h_\alpha s\sqrt{[(k-1)/(k_n)]}$$

$$UDL = X'' + h_\alpha s\sqrt{[(k-1)/(k_n)]}$$

Where h_α is from table I (Appendix-D) for risk level α, number of means k, and degrees of freedom $[(n-1)k]$. The means are then plotted against the decision lines. If any mean falls outside the decision lines, there is a statistically significant difference among the means.

Lloyd (1974) gave the factors for means. His chapter presented an extensive table of factors of significance levels 0.10, 0.05, 0.01 and 0.001 for carrying out the analysis of means. Examples are shown for analysis involving unequal and equal samples.

An ANOM chart was drawn for the data in period 1 as detailed in Fig. 35.3.

Table 35.3: ANOM chart data and calculation for period 1

Sl. No.	**1**	**2**	**3**	**4**	**5**	**6**	**7**	**8**	**9**	**10**
X′	1.584	1.594	1.594	1.594	1.570	1.582	1.592	1.564	1.548	1.558
X″	1.57	1.57	1.57	1.57	1.57	1.57	1.57	1.57	1.57	1.57
UDL	1.59	1.59	1.59	1.59	1.59	1.59	1.59	1.59	1.59	1.59
LDL	1.55	1.55	1.55	1.55	1.55	1.55	1.55	1.55	1.55	1.55
Sl. No	**11**	**12**	**13**	**14**	**15**	**16**	**17**	**18**	**19**	**20**
X′	1.566	1.578	1.564	1.558	1.558	1.574	1.566	1.572	1.568	1.578
X″	1.57	1.57	1.57	1.57	1.57	1.57	1.57	1.57	1.57	1.57
UDL	1.59	1.59	1.59	1.59	1.59	1.59	1.59	1.59	1.59	1.59
LDL	1.55	1.55	1.55	1.55	1.55	1.55	1.55	1.55	1.55	1.55

The processes parameters have been recorded for period 1 and based on the same counter measures were installed. After installing counter measures, a confirmation control chart and ANOM chart was made and the details of the same for period 3 is detailed in Table 35.4.

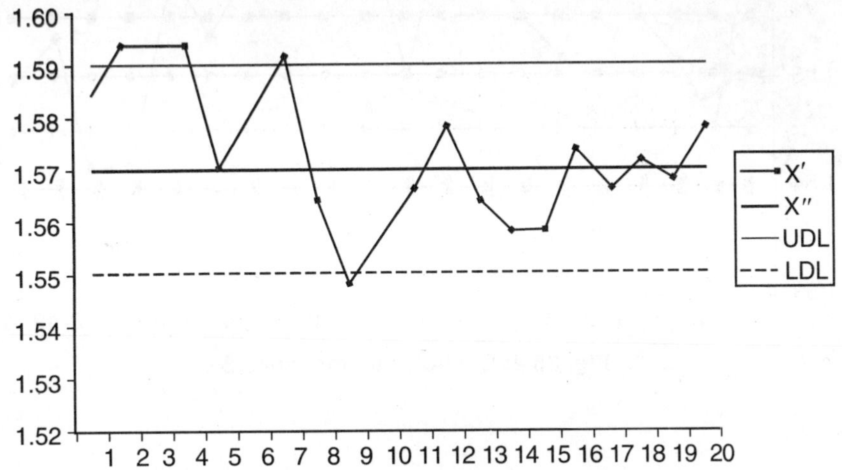

Fig. 35.3: ANOM for period 1

Table 35.4: Control chart data for period 3

		1	2	3	4	5	6	7	8	9	10
Sample Measurements	1	1.56	1.54	1.55	1.55	1.55	1.56	1.55	1.55	1.55	1.55
	2	1.60	1.55	1.55	1.56	1.57	1.55	1.57	1.59	1.56	1.55
	3	1.55	1.55	1.56	1.55	1.55	1.54	1.55	1.55	1.58	1.55
	4	1.57	1.55	1.56	1.56	1.58	1.53	1.55	1.55	1.56	1.55
	5	1.54	1.54	1.54	1.56	1.57	1.58	1.56	1.57	1.57	1.55
Sum		7.82	7.73	7.76	7.78	7.82	7.76	7.78	7.81	7.82	7.75
Average X		1.56	1.55	1.55	1.56	1.56	1.55	1.56	1.56	1.56	1.55
Range R		0.06	0.01	0.02	0.01	0.03	0.05	0.02	0.04	0.03	0.00
		11	12	13	14	15	16	17	18	19	20
Sample Measurements	1	1.58	1.58	1.57	1.55	1.55	1.56	1.54	1.55	1.55	1.55
	2	1.58	1.58	1.55	1.56	1.55	1.58	1.54	1.54	1.56	1.54
	3	1.56	1.56	1.54	1.56	1.57	1.58	1.54	1.53	1.55	1.54
	4	1.55	1.55	1.58	1.57	1.55	1.55	1.54	1.55	1.54	1.56
	5	1.56	1.56	1.55	1.56	1.55	1.53	1.55	1.54	1.57	1.56
Sum		7.83	7.83	7.79	7.80	7.77	7.80	7.71	7.71	7.77	7.75
Average X		1.57	1.57	1.56	1.56	1.55	1.56	1.54	1.54	1.55	1.55
Range R		0.03	0.03	0.04	0.02	0.02	0.05	0.01	0.02	0.03	0.02

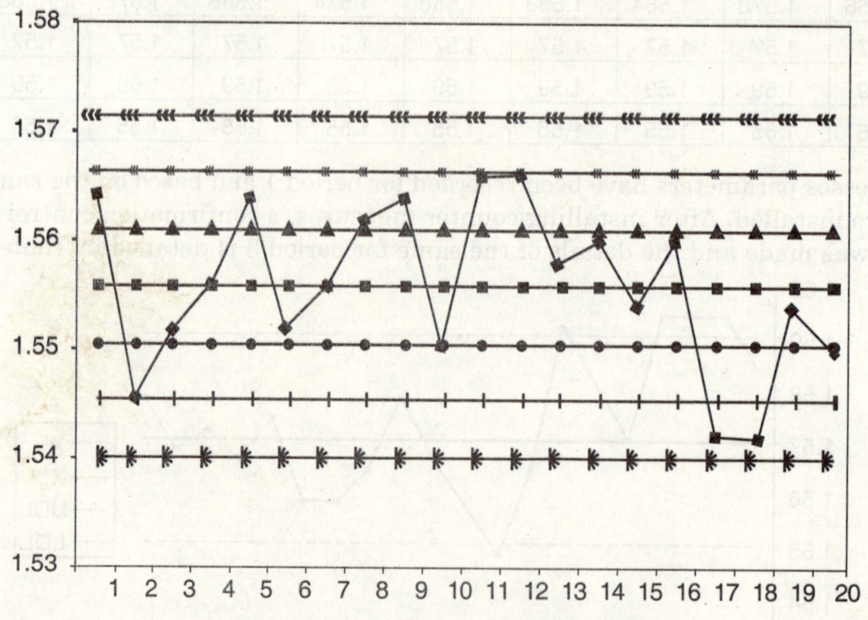

Fig. 35.4: Control chart for period 3

Table 35.5: Anom data and calculations for period 3

Sl. No.	**1**	**2**	**3**	**4**	**5**	**6**	**7**	**8**	**9**	**10**
X′	1.564	1.546	1.552	1.556	1.564	1.552	1.556	1.562	1.564	1.550
X″	1.56	1.56	1.56	1.56	1.56	1.56	1.56	1.56	1.56	1.56
UDL	1.58	1.58	1.58	1.58	1.58	1.58	1.58	1.58	1.58	1.58
LDL	1.54	1.54	1.54	1.54	1.54	1.54	1.54	1.54	1.54	1.54
Sl. No	**11**	**12**	**13**	**14**	**15**	**16**	**17**	**18**	**19**	**20**
X′	1.566	1.566	1.558	1.560	1.554	1.560	1.542	1.542	1.554	1.550
X″	1.56	1.56	1.56	1.56	1.56	1.56	1.56	1.56	1.56	1.56
UDL	1.58	1.58	1.58	1.58	1.58	1.58	1.58	1.58	1.58	1.58
LDL	1.54	1.54	1.54	1.54	1.54	1.54	1.54	1.54	1.54	1.54

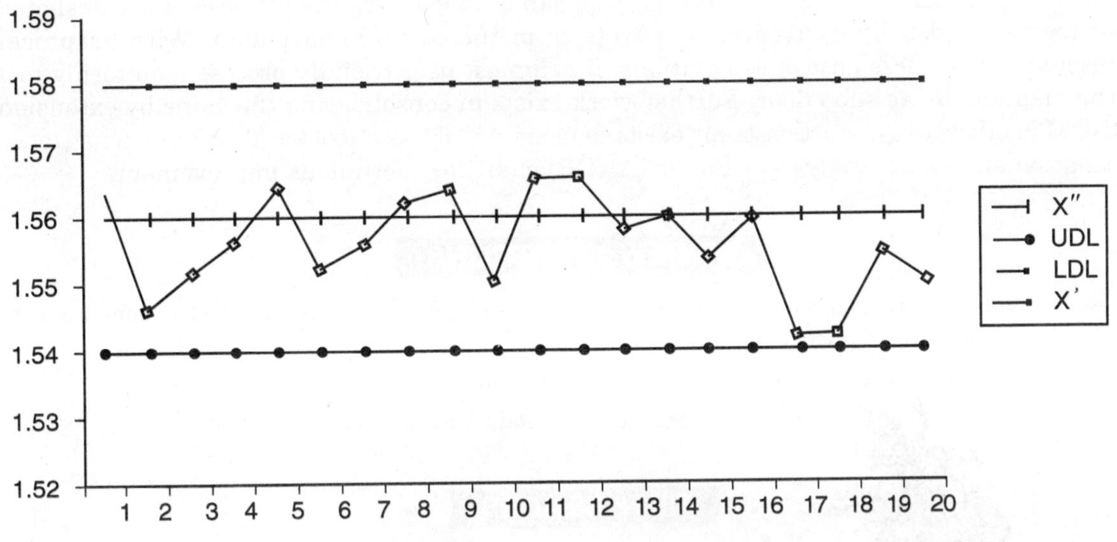

Fig. 35.5: Anom chart for period 3

35.6 CLOSURE

In practice process monitoring tools like Shewhart's control charts is used to monitor and adjust the process with the aim of reducing process scrap. These control charts depict short range variations due to assignable reasons. The ANOM combines the hypothesis testing features of ANOVA with the control chart's power of graphical presentation of the experiment means. This chapter details as to how ANOM is used in a precision manufacturing process to reduce variations and in turn process scrap and manufacturing cost.

The ANOM chart in period 1 shows the statistical and engineering significance of variations. A cause and effect analysis was made and the operating conditions for treatments were recorded.

When counter measures are taken in the area of treatments where there is an effect, improvements were seen.

The table shows in period 1 when significant improvements have been called for and period 2 which is a transition stage and period 3 where reduction in scrap and reduction in internal failure cost. This will result in overall product life cycle economics.

	Period 1	Period 3
1. Turnover for the year	Rs. 300 Lacs	Rs. 320 Lacs
2. Total quality cost	Rs. 81.14 Lacs	Rs. 68.88 Lacs
Increase in profit	= Rs. 12.26 Lacs	
Percentage to turnover	= 3.83%	

Thus there will be a swing of 3 % in cost which improve the overall lifecycle cost. The usage of ANOM has lessened the need for doing extensive design and experiment studies as the area for improvement could be identified from control chart data itself. It is a tool that can integrate statistical process control and engineering process control. It lessens the need to conduct elaborate experimental design work and at the same time serves the purpose of the designed experiments in identifying treatment effects from the control chart data. With graphical representation and less dose of calculations, it is a most user friendly process monitoring tool for the manufacturing shop floor. Further work exists in consolidating the gains by extension of ANOM in other areas. Further scope exists in using ANOM as a tool for identifying consistency in variation and using Analysis of Range (ANOR) chart for continuous improvement.

REVIEW QUESTIONS

1. Do you agree with the statement, "Analyzing and reducing internal failure cost is equivalent to digging a gold mine"? Justify your answer.
2. How ANOM differs from ANOVA, in both procedure and performance?
3. How a control chart for means is constructed? Explain with an example.

REFERENCES

1. Ott, E.R., "Analysis of Means–A Graphical Procedure", Journal of Quality Technology, Vol. 15, No.1, (1983), pp. 10–18.
2. Shewhart, W.A., "Statistical Methods from View Point of Quality Control," Dover Publications Inc., New York (1986).
3. Raming, P.F., "Applications of the Analysis of Means," Journal of Quality Technology, Vol.15, No.1, (1983), pp. 19–25.
4. Nelson, L.S., "Factors for the Analysis of Means, "Journal of Quality Technology", Vol. 6, No. 4, (1974), pp. 175–181.

Risk Management, The DMAIC Way

36.1. INTRODUCTION

A design risk assessment is the act of determining potential risk in a design process, either in a concept design or a detailed design. It provides a broader evaluation of design beyond just CTQs (critical to quality issues), and will enable to eliminate possible failures and reduce the impact of potential failures. This ensures a rigorous and systematic examination of the reliability of the design and allows capturing of the system-level risk. The art of managing risk is more challenging than ever. Designers face a wide range of demands, from working with multiple variables to finding technology solutions that generate comprehensive risk analysis.

The six sigma methodology seeks to uncover the root causes of defects and treat them at the source. Developing a systematic means of classifying and responding to risk, however, can significantly reduce the number of surprises that threaten success and help cope with those that do surface at a later date in usage. The acronym DMAIC stands for **D**efine, **M**easure, **A**nalyze, **I**mprove, and **C**ontrol and it is a rigorous and both quantitative and qualitative analytical process. The DMAIC process is not so different from the classical broad risk management framework. This paper examines the risk management in design in a different way, through the six sigma way, in identifying, assessing, prioritizing, monitoring, controlling and mitigating risk in design and sums up to a checklist to be followed at the design stage.

Uncertainty is a fact of life! It exists in any design environment. While it may not be possible to predict project outcomes with 100% certainty, we can influence the outcome, avoid many potential risks and be ready to calmly and efficiently respond to unavoidable challenges instead of reacting to them, possibly making things worse.

36.2. RISK

Risk is not a problem; rather the problem is a risk whose time has come. That is a situation, condition, or an action, that has the potential of causing a loss (i.e., impact) in the future. Risk is the cumulative effect of the chances of uncertain occurrences adversely affecting design objectives (e.g., cost, schedule, functionality, quality). Risk is an uncertain event or condition that, if it occurs, has a positive or negative effect on a project objective. A risk has a cause and, if it occurs, a consequence. Risk has two elements. First, that occurrence of the risk condition is probabilistic and secondly if it does occur, it can damage to the design/project. The risk is of direct or indirect loss resulting from inadequate or failed internal processes, people, and systems or from external events.

Design risk can result in increased write offs, additional expenditure or loss of revenue. Finn [1] says studies have shown that schedule risk is derived primarily from several common factors, including unintentional iterations, intentional iterations, completeness of activities and information, activity duration variances, among others. It is the category of "unintentional iterations" that is of most concern here, because the inability to control the engineering product quality is a major cause of unnecessary and unplanned iteration. Specifically, design mistakes (errors and omissions) cause not only delays due to time required for correction, but also consequential re-work in coupled and related activities.

36.3. RISK MANAGEMENT

If the design is robust, then six-sigma levels of quality and high levels of reliability are possible if not probable. If the design is not robust, then it is very difficult and labor intensive to achieve those levels of quality. Robust products are not developed by happenstance; it takes a well-defined process, effective tools, and professional team members to create robust products. Glaccum and Hirschauer [2] say that the discipline of risk management is hardly new, but the threats that confront designs are constantly changing and increasing in complexity. Risk management aims to identify these risks and then take actions such that their effects on the design is minimized.

Risks faced by designers from results are sometimes magnified by the actions of other designers. Risk management systems, which neglect this feature, may give a seriously misleading picture of the true risks. The hazards arising from this potential blind spot are at their most dangerous when the prevailing conventional wisdom lulls designers into a false sense of security on the attractiveness of a designed position. The efforts of one designer to reverse his design make more acute the need to follow suit on the part of others.

Thornton et al. [3] feel that to understand why companies still fail to systematically address variation, propose an ideal model of variation management is, entitled Variation Risk Management (VRM). It was found that many problems with industry implementation are due to a lack of quantitative models that enable a design team to make quick and accurate decisions. They suggest that the Risk Management via the DMAIC way may be is one of the ways. Jones (2003) opines that some recommend use of a broad risk management framework such as six-sigma to ride herd over a project's many potential risk factors. Six-sigma is an approach, structure or framework that allows you to look at a project upfront and go through a methodical process to define, to measure and to control risks. There hasn't been any of this kind in the literature, hence this paper.

36.4. SIX-SIGMA

Six-sigma means data-driven decision-making. Beyond the simple short-term goals of cost savings through quality or productivity improvement, most companies would tell you that it's to move them from a "fire fighting" culture to one of "fire prevention." The Six-sigma methodology seeks to uncover the root causes of defects and treat them at the source. Similarly, we can make Risk management a routine and systematic part of our design planning process. No design manager can predict every conceivable obstacle that will throw itself into the path of design completion. Developing a systematic means of classifying and responding to risk, however, can significantly reduce the number of surprises that threaten success and help us cope with those that do surface. Six-sigma is divided into five distinct phases, Define, Measure, Analyze, Improve, and Control (DMAIC). A successful Six Sigma practitioner employs a variety of analytical tools, mixing and matching the appropriate tools during each of the different phases. Because many traditional product and process engineering tools fail to quantify the value of improved design, they are less useful for Six Sigma. The implementation of a framework for effective management and supervision of risk is defined in a DMAIC format as "define, measurement, analysis, improving and control of risk. Six-sigma, the DMAIC way, provides three powerful methodologies for designing robust processes. New processes have high capability (satisfy customers and produce low defects) right from the start; new processes are designed to minimize the risk of failure; Fixing broken processes. Uses powerful statistical (and non-statistical) tools to locate and eliminate the root causes of otherwise intractable problems, focuses on removal and prevention of defects, Reduces process variability, Controlling processes, so that their behaviour is predictable (within limits); Controlling processes on an ongoing basis (i.e., keeping them from decaying and producing errors). For measuring and judging process variation, for detecting special causes, to providing early warning of process changes and the ability to calculate Process Sigma, an index of process performance.

It is argued that the numeric interpretation of Six Sigma is less important than process rigour, which can be applied to engineering design by using combinations of statistical and computer aided analytical methods. In this paper the author outlines how Risk can be identified, assessed, planned for, and controlled. He arms the designer with a simple, common sense approach through the Six Sigma way, to addressing uncertainty in designs.

36.5. DEFINE

Glaccum and Hirschauer [2] say that this seemingly innocuous step is the most important. Without a clear definition of what the product is supposed to do, the rest of the process will not be as effective. The mission statement must be written clearly enough to be tested consistently so that results from multiple rounds of testing are possible. Once the right resources are identified and a team charter is documented, the definition of the project objectives and outcomes can be formalized. A design need is always defined in terms of something the client considers to be Critical to Quality, a CTQ. In the design context the CTQ could be less or no uncertainty, reliability and lesser risk.

With the objectives and outcomes defined, a project plan can be documented. This plan expressly articulates the activities necessary to complete the project and details the overall schedule including milestones and resource requirements [4].

The next step is to identify what essential operations of the device are necessary to fulfill the mission. This definition is from an operational perspective rather than from a technological perspective. The definition should include major subsystems that comprise the system as it executes in reality (function) to fulfill its mission instead of the conventional technological subsystems/subassemblies [2].

36.6. VOICE OF THE CUSTOMER

Taking a proactive approach by capturing the *voice of customers* to enhance the quality and design process and putting in a measurement system to measure the effectiveness of the internal controls system could lead to more satisfied customers. At the earliest possible stage, it is important to classify those CTQ performances, product or service characteristics that most significantly affect customer satisfaction, safety (serious danger to human life), environment and business performance. Ideally, this would use a cross-functional team approach that encompassed the full range of product, process and service knowledge and utilized methods such as quality function deployment (QFD), which is a useful methodology for identifying and understanding customer and technical requirements. Capability assessments carried out on CTQ processes during the pre-introduction stages provide management with their initial quality indicators of how well the product/service will perform in the customer's hand [5].

These include self-assessment, risk mapping, key risk indicators or scorecards. Whilst all have their merits and some may be better suited to certain designs, they all take time to implement and have a price tag that can be substantial. For example, in today's designs environment the pace of change is such that risk mapping can become rapidly obsolete and therefore requires constant adjustment as processes are modified. To ensure a complete and common understanding of a process and to establish a base line from which to generate new ideas, process mapping is usually essential. At the highest level the process map captures SIPOC (The Suppliers, their Inputs, the Process steps, the Outputs and the Clients). Detailed process maps can be quite complex. The next step is to create a statistically sound plan for data collection and graphic representation of the data collected, taking into consideration what data already exists. One of the most interesting aspects of this phase is graphic representation, because the manner in which the information is displayed is critical to understanding the Design being evaluated. Gantt charts. Run charts, Pareto charts, scatter displays, histograms and fish bone diagrams, used in project management each of which offers a different perspective on the design [4].

Kwak [5] thinks that the biggest issue is discovering the risks. Brainstorming has been the method most used but this is sometimes difficult if the team members are not committed to make the RM process work. Use of historical records of risk associated with past products has been minimal.

Define individual risk events and overall design risk using qualitative and quantitative approaches. Both qualitative and quantitative risk assessments help understand potential threats to designs, and propose controls to helps us mitigate risk and proactively prevent failures. Not all disruptions can be prevented, but we can apply time-tested methodologies for accepting, reducing or transferring risk. The ultimate goal of Risk Management is to help us take immediate actions that shield our design from risks that can halt our operations.

36.7. QUALITATIVE RISK ASSESSMENT

Qualitative risk assessment is a subjective approach that assigns a low, medium or high impact to the potential threats to our designs. We examine five asset groups: facilities, network, technology, information and human resources. This risk assessment identifies potential threats to designs, assesses controls in place to eliminate or reduce the threats, identifies remaining exposures and recommends mitigation techniques to eliminate or reduce risks.

Identification practices are performed well for "high risk/high probability of occurrence" areas and for "low risk/low probability of occurrence" areas. If the risk and probability of occurrence are high, then mitigation occurs; if they are low, nothing is done. In some cases, the "high risk/low probability of occurrence" and "low risk/high probability of occurrence" areas are not completely assessed because of schedule pressure and a lack of understanding how systematic variation propagates [6]. Following are some of those practices.

- Quantitative risk assessment
- Inputs and outputs
- Qualitative risk scoring methods
- Prioritizing risk events – probability × impact
- Failure mode and effect analysis (FMEA)
- Monte carlo analysis
- PERT as a risk management technique
- Expected value analysis

36.8. MEASURING RISK

A design risk assessment is the act of determining potential risk in a design process, either in a concept design or a detailed design. It provides a broader evaluation of design beyond just CTQs, and will enable to eliminate possible failures and reduce the impact of potential failures. This ensures a rigorous, systematic examination in the reliability of the design and allows capturing system-level risk. Kelly [4] warns that the significance of the measurement phase of the process cannot be overemphasized. The fundamental principle of Six Sigma is that if something can't be measured then there is no point in discussing it, as we will never be able to assess improvement. Implicit in this principle, however, is the belief that every process can be quantified and measured in some way.

The key to capturing and measuring data is to have the capability to analyze the design model at all phases of this process. Once these data have been captured, they can be analyzed with respect to the number of parts-per-million that are out of specification, or deviate from the defined standards. Using this Six Sigma analysis, a process measurement can be made regarding the effectiveness of the engineering design/modeling process. A hierarchical decomposition of the models within the statistical population of the models can be analyzed, with respect to the granularity of the data required. For example, at an assembly-level analysis, it may be meaningful to provide a basic part-level six sigma analysis [1]. Kelly [4] suggests the following measurement phase:

- Did we view the process holistically: activities, information, people, systems, etc.?
- Was the process viewed at the appropriate level?
- Were the client needs identified and prioritized?
- Were measurable critical to quality requirements identified? Do these correlate to the important client needs?
- Was the current sigma level of performance determined for the critical metrics?
- Were targets established based on client needs?
- Did we refine the problem and goal statements [4]?
- Identify all specific lists of risks it will have to address.
- Have a framework in place for systematically identifying and addressing risks?
- Understand the project's endpoint or deliverables.
- Identify the single points of failure for our design function or facility.
- Identify underlying procedural problems.

In the last several years the number of statistical add-ins available for use with spreadsheets has continued to grow. For routine procedures and tests of hypotheses, basic graphics and even regression modeling, as well as for most introductory statistics courses, a statistical add-in product for a spreadsheet is likely adequate. The functionality of products for use with spreadsheets is growing, particularly for risk analysis and Monte Carlo sampling. In order to improve design quality, or optimize for design quality, design quality must first be measured. Two measures of design quality, coming from two separate communities, are reliability and robustness. One drawback of reliability-based optimization methods is that the objective is evaluated at the nominal or mean value design point. Ertan [6] identified three practices in the Measurement and Feedback category. During assessment, the interrelationships and dependencies of the expected variation from the feature level to the system level are evaluated against the acceptable variation.

Crevling et al. [7] suggest that any high-risk requirements are fully discussed and strategies put in place to deal with worst-case scenarios that are likely to surface later. These risk assessments focus on technical difficulties in terms of form, fit and function. They also assess cost risks due to system complexity, assemblability and serviceability. The risk assessment is underwritten by a formal FMEA and system noise mapping.

36.9. PROCESS CAPABILITY AND PROCESS CAPABILITY UNCERTAINTY

The one consistency between the measure of reliability and the measure of robustness is the measure of variation: sigma. The term "sigma" refers to standard deviation, σ. Standard deviation or variance, σ^2, is a measure of dispersion of a set of data around the mean value (μ) of this data. This property can be used both to describe the known variability of factors that influence a system (product or process), and as a measure of performance variability, and thus reliability, robustness, or simply quality. Performance variation can be characterized as a number of standard deviations from the mean performance. The areas under the normal distribution associated with each σ-level relate directly to the probability of performance falling in that particular range. These probabilities are presented as percent variation (equivalent to reliabilities) and number of defective parts per million parts [8]. Ertan [6] says that the management of this uncertainty by taking a system view to variation and deploying effective control plans beginning during design and continuing through production. Some process uncertainty exists and special attention is needed to qualify the process capability.

36.10. FMEA/DFMEA

Ertan [6] says Failure Mode Effect Analysis (FMEA) considers the risk factor being the components of potential occurrence, severity, and non-detection; however, it is based on qualitative previous history and predictions for future behavior. A DFMEA would be a quantified estimate of the criticality of CTQ internal failures and CTC performance factors using the classic FMEA form but for design reviews and technology transfers. Performance to customer needs uses CTC's, design parameters to help meet those requirements and also the manufacturability requirements might be called CTQ's. Both would be appraised in the DFMEA document and then updated as the product moves into production, based on real failure frequencies and severities, becoming a living risk-management and design feedback document [9]. Ertan [6] adds Quantitative analysis including Variation System Analysis (VSA) helps prioritize from where the major sources of problems are coming.

Variance reduction sampling techniques have been developed to reduce the sample size (number of simulations) without sacrificing the quality of the statistical description of the behavior of the system. The methods and tools developed for structural reliability analysis and robust design facilitate this estimation. Three classes of methods that have been described by Koch and Cofer [8]:

- Monte Carlo Simulation;
- Simple Random Sampling;
- Descriptive Sampling;
- Design of Experiments;
- Sensitivity Based Estimation;
- First Order Taylor's Expansion; and
- Second Order Taylor's Expansion.

36.11. DESIGN OF EXPERIMENTS (DOE)

Koch and Cofer [8] explain an approach for estimating performance variability due to uncertain design parameters is through more structured sampling, using a designed experiment, a common tool for robust design analysis. In DOE, a design matrix is constructed in a systematic fashion that specifies the values for the design parameters (uncertain parameters in this context) for each sampled point, or experiment. A number of experimental design techniques exist for efficiently sampling values of design parameters. For the most computationally expensive analyses, requiring days for a single analysis, a set of sampled design points can be executed in parallel and used to construct surrogate models, or approximation models, to replace the high fidelity code during probabilistic optimization. Polynomial response surfaces and more recently kriging surrogate models have been employed for this purpose.

According to Kelly [4] the objective of the analysis phase is to study the data collected in the measurement phase, to determine why the process performs as it does today, identifying and quantifying cause and effect relationships. The reality is that each has its own unique risk factor, which requires the knowledge of its relative importance. It is also a prudent activity to carry out risk analysis on these [5].

Kwak [10] says Qualitative risk analysis is a driving force behind our ability to stay ahead of the competition. He says that we can still apply the issue to a risk matrix and assign a value to the issue, or prioritize the risk. As a briefing tool or snapshot assessment of risk issues, qualitative methods can be very effective. The key benefit is in its ability to represent complex information in simple terms. Qualitative risk analysis is an excellent communications tool. Most people do not deal with probabilities, Monte Carlo analysis results and other black box approaches. But they certainly understand low, medium and high. Nothing communicates more easily and effectively than qualitative risk methods. Examples of qualitative analysis could be expanded to include risk impact analysis. Qualitative risk analysis can be used to change the level of urgency of a management response and so it does have business value.

An uncertain situation often involves risk, which is defined as the possibility of loss, damage, or any other undesirable event. When using spreadsheets, to capture uncertainty, analysts can perform simple "what-if analysis" or "scenario analysis" by manually changing model variables and analyzing their effect on the key outputs. One established solution to the limitations of spreadsheet risk analysis is Monte Carlo simulation.

Monte Carlo methods have long been recognized as the most exact method for all calculations that require knowledge of the probability distribution of responses of uncertain systems to uncertain inputs. To implement a Monte Carlo simulation, a defined number of system simulations to be analyzed are generated by sampling values of random variables (uncertain inputs), each with defined probabilistic distributions and associated properties. Several sampling techniques exist for simulating a population; two techniques are well known – *simple random sampling*, the traditional approach to Monte Carlo, and *descriptive sampling*, a more efficient variance reduction technique [8]. His simulation-based probabilistic design analysis and optimization methods, including a combination of those from structural reliability-based design and robust design communities. They also presented the quality concepts of six-sigma and strategies for combining deterministic and probabilistic design methods to effectively trade-off design performance and performance variation, and avoid the pitfalls of deterministic optimization.

There are enough similarities between the aims and methods of Design for Six Sigma (DFSS) and reliability and robustness engineering methods to propose an approach that combines them. The engineering effort required to achieve high integrity design needs to be applied to the areas of design where the greatest reliability risks to the customer is evident. The DFSS tools will help with achieving the level of detail required to deliver high reliability for complex products. However, if applied where unnecessary, the additional resource required applying. The in-depth analysis tool could be wasted. Efficient design planning and analysis tools need to be utilized to position each technology or system within the strategy map and apply the appropriate level of design effort. Employing tools such as Quality Function Deployment, Failure Mode and Effect Analysis, Fault Tree Analysis, Reliability and Robustness Specification processes, Combined Design Functional Specifications, Function Structure Trees, could generate means of quantification to achieve this level of definition. If fully quantifiable, then the TF can make design optimization straightforward. However, in practice it is most likely that the TF will be only partly quantifiable, therefore for other factors, Risk Management methods (Design Failure Mode and Effects Analysis (DFMEA) must be applied. DFMEAs help to ensure that similar designs are capable from current data and documented evidence and data is available to prove that new design can only move in a positive direction.

Kwak [10], Understanding project risk tolerance is a crucial part of any Risk Management plan. While it is at the heart of decision-making, it is all too often overlooked. Risk tolerance should be clearly understood and communicated among firm, project managers, and stakeholders to have a meaningful qualitative risk analysis. The levels and perspectives of risk tolerance are dynamic throughout the life of the project.

Infosys believes that risks have to be systematically identified and managed in a proactive manner to maximize shareholder value. Infosys used quality models such as the Software Engineering Institute's Capability Maturity Model (SEI-CMM) to ensure that risks were identified and measures taken to mitigate them at the project planning stage. (Vedpuriswar et al. [11].

36.12. IMPROVE

The objective of the subsequent improvement phase is to identify improvement breakthroughs, identify high gain alternatives, select the preferred approach, design the future state, determine the new sigma level, perform cost benefit analyses, design dashboards and scorecards and create a preliminary implementation plan [4]. Improvement starts with knowing what quality goals are important to the customer. Having established what they are, demonstrated the capability to achieve them economically (without any waste) and lain down the basis of process reliability, focus can now change to continual improvement on the CTQ during the 'life cycle'. This would consist of reviewing internal and supplier process performance, reduction in variation, and feedback from the field of product/service performance and customer satisfaction [5].

It is the use of an improvement-oriented process to actively pursue design flaws, understand the nature of the flaw, and then implement a design solution to eliminate the flaw. By eliminating design flaws during all phases of development, the product is less susceptible to unforeseen uses, process variation, and has a higher level of reliability than products developed with the traditional demonstration testing philosophy. This is what makes the products more robust [8]. This practice is defined as how design modifications are performed when there is an inability to achieve the function of the system at a reasonable cost. The design process needs to be iterative with both design and manufacturing making changes concurrently. If the product has a producibility problem during production ramp-up, a reactive change is made before the product is at rate [6].

Kwak [5] advises that oftentimes creativity greatly complements systematic Risk Management process. Recent literatures also discuss the importance of balancing between structural and analytical approach with creative and out of ordinary thinking that could have great impact in better implementing Risk Management. He cites the example of Sprint conference rooms in Dallas; there are models that have been placed in the rooms to encourage out-of-the-box creative thinking. These models help the folks in these meetings not just with Risk Management, but with all kinds of other creative thinking problems. One can think of everything that can go wrong with an airplane or mountain in a brainstorming session for risk identification.

In this direction, AFD is a good creative tool from the TRIZ/ARIZ communities. Anticipatory Failure Determination (AFD) is a failure analysis method. Like FMEA, it has the objective of identifying and mitigating failures. Rather than asking developers to look for a cause of a failure mode, it reverses the problem by asking developers to view the failure of interest as the intended consequence and try to devise ways to assure that the failure always happens reliably. At the heart of the improvement phase are many tools and techniques to find flaws. These tools consist of accelerated testing tools, risk analysis tools, statistical analysis tools, modeling tools, and simulation tools.

An important success factor in designing robust products is the proper selection of reliability tools for detecting design flaws. Since there are so many tools at the disposal of the quality engineer, the issue is the knowing of what tools are appropriate to use at a particular point [2].

Kwak [5] informed that Motorola has a process called M Gates that has been implemented across the corporation. It is a process that has broken down new concepts or business ideas into 15 gates that need to be "cleared" prior to moving forward. Each gate has its own unique purpose and at each gate there is various risk mitigation tools used where participants in the process identify and assist in brainstorming up possible solutions.

Highly Accelerated Life Testing (HALT) was used with the early prototype units to detect design flaws with the electronics and non-time (cycle) related mechanical interconnects of the device. Applying stressors that are above normal operating parameters are a way to quickly "precipitate" failures [2].

Stress calculations are the first line of defense in creating a mechanically robust device. Cycle testing has been recognized as an effective tool when applied to systems with failures that occur during the wear out period of the bath tub curve for which there are no other means of accelerating or inducing failure precipitation. Verification phase serves two purposes: First, it confirms that the corrective action was effective. It then takes the product to higher stress levels in order to identify additional failure modes. However, the resultant increase in product reliability is huge! [2].

Following a current paradigm shift from design optimization to comprehensive design exploration and quality assessment and improvement, additional numerical design tools have been incorporated into this simulation-based design environment to support and augment the optimization capabilities. These tools include design of experiments (DOE), approximation methods, and a suite of probabilistic quality engineering methods including: Monte Carlo simulation, reliability analysis and optimization, Taguchi's robust design, six sigma based robust optimization [8]. Design of experiments is conducted prior to the first prototype build and several rounds of experimentation were needed to determine the optimum settings [2].

The next stage of the quality planning is to flow the CTQ characteristics into control plans, where the detail of the high level of control, which might be mistake proofing, SPC (control charts) or duplicate inspection, is necessary to ensure a Six Sigma process is achieved [5]. Variation risk mitigation is the final step. Risk mitigation utilizes the quantitative and qualitative analysis to determine if process control, process change, or design change is necessary. A process control or process change can reduce the variation while a design modification changes the latitude of the allowable variation. The trade-offs are whether it is more costly to control the process, it is more risky to change the process, or it is worth spending the time and human resources to iterate through another design [6].

Statistical Process Control (SPC) takes data from a process, translates it to information via analysis using statistics and communicates the information to an earlier stage in the process where the effect of change is more efficient. Engineering management need to achieve the same for the design process. A last "risk" is the one that actions are done with little or no value added. Properly documenting all the processes and making sure that all actions are documented, gives a chance to think about the tasks people do in the organisations. Double work, documenting for the sake of documenting, different ways to do a similar task are all things that should be avoided in efficient operations.

In other words internal controls system could be referred to as a set of policies and procedures aimed at reducing the risk of errors or omissions resulting in improved quality of design produced by an entity. In order to enhance internal controls it is important to have a sound Risk Management process in place. In Six Sigma terms it can be referred to as reducing the error rate to improve the qualities of a product i.e., design. One can only manage what can be measured. If improving internal controls system is dependent on better Risk Management process then one needs a measurement system to better manage the risk. Therefore we cannot continue to rely on subjective opinions.

36.13. CLOSURE

The implementation of the above simple tools effectively across the design will go a long way to reducing risks substantially. All said and done the objective is to improve controls and reduce risks going forward, not just create another reporting system that shows what has happened in the past.

The framework should provide a firm-wide definition of design risk and lay down the principles of how design risk is to be identified, assessed, monitored, and controlled/mitigated. The designers should ensure that the design Risk Management framework is subject to effective and comprehensive internal audit by independent, appropriately trained and competent staff. The internal audit function should not be directly responsible for design Risk Management. Designers should identify and assess the design risk inherent in all material products, activities, processes and systems. Designers should also ensure that before new products, activities, processes and systems are introduced or undertaken, the design risk inherent in them is subject to adequate assessment procedures. Designers should implement a process to regularly monitor design risk profiles and material exposure to losses. There should be regular reporting of pertinent information to senior management that supports the proactive management of design risk. Risk Management helps us to identify designs, establish objective value, assess potential vulnerabilities and impacts, and propose safeguards and mitigation tactics.

Kwak [10] listed the challenges and realities in applying effective Risk Management processes are difficult, especially, integrating the Risk Management processes into the organizations. However, the benefits of implementing effective Risk Management tools and techniques in the organization are equally great. Implementing effective Risk Management process will succeed by changing the organizational culture (Kwak & Stoddard, [5]. Conclude that senior management holds the key to establishing an organization that encourages Risk Management infrastructure, culture, and behavior. A documented process does not guarantee the process will be followed. As the size and complexity of the project increases, the efforts for Risk Management increase exponentially. Here is a Risk Management plan template describing the non-exhaustive tools and techniques that can/have to be used during the design.

Risk Management - Tools/Techniques

Define	Measure	Analysis	Improve	Control
Gantt charts	Probability distributions	Spread sheet		SPC
Run charts	Monte carlo simulation	Monte carlo simulation	Finite element analysis (FEA)	Mistake proofing
Pareto charts	Simple random sampling		Fault tree analysis	Duplicate inspection
Scatter displays	Descriptive sampling		Free body diagrams	Dashboards
Histograms	Design of Experiments (DOE)		DOE	
Fish bone diagrams		Combined design functional specifications	Bell core reliability prediction	
SIPOC	Data analysis empirically from historical data	SEI-CMM	Safety factor	
VOA	Sensitivity based estimation; first order Taylor's expansion second order Taylor's expansion	Reliability and robustness specification processes	Modeling/Simulati on	
QFD		Quality function deployment		
Risk scoring methods			Tolerance stack-up	
FMEA	DFMEA	Failure mode and effect analysis	FMEA	
PERT	PFMEA	Design failure mode and effects analysis	Circuit analysis	
Monte Carlo analysis			Component stress analysis (Safety margin)	
Value analysis	Assume theoretical p.d.f. and use data to get parameters		Sub-system/ component stress testing	
Probability x impact matrix	Regression over variables that affect risk	Risk matrix	Modeling/Simulati on	
Self-assessment	Parallel processing and Approximation methods			

Risk mapping	Surrogate model		Highly Accelerated Life Testing (HALT)	
Key risk indicators	Probabilistic optimization		Environmental Stress Screening (ESS)	
Scorecards	Polynomial response surfaces		Cycle Testing w/Stressors	
FTA	Kriging surrogate model	Fault tree analysis	Highly Accelerated Stress Screening (HASS)	
Event tree analysis	Anticipatory failure determination (AFD)	Function structure trees		
Event sequence diagrams				
Master logic diagrams				
Reliability block diagrams				

REVIEW QUESTIONS

1. What is a design risk? What can be its liabilities?
2. How the six-sigma method can enable achievement of risk management? Explain with an example.
3. What DMAIC stands for? Explain how DMAIC can be applied to a problem with is a simple example.
4. What are CTQs? How do we identify the CTQs?
5. What is a process map? What is SIPOC? How the two are related to each other?
6. What are variation reduction sampling techniques? How they ensure design robustness?

REFERENCES

1. Finn G.A., (1999), "*Six Sigma in the Engineering Design Process*", (1999) ASA Quality & Productivity Research Conference, 1999.
2. Glaccum J.F., and Hirschauer M.A., (2002), "*Proactive Reliability*". Using the Right Improvement Tools to Grow Reliability in Product Development, Annual Quality Congress, Denver, CO, Vol. 56, May 2002, pp. 151–164.
3. Thornton A.C., Donnelly S., Ertan B., (2000), More than Just Robust Design: Why Product Development Organizations Still Contend with Variation and its Impact on Quality, Research in Engineering Design Volume 12, Number 3 Pages: 127–143.
4. Kelly W.J., (2000), Getting from feelings to facts through Six Sigma, Marriott Hotel, Toronto, Ontario, Canada February 16, 2000 ORIMS Annual Douglas Barlow Professional Development Day (www.rims.org/ifrima).
5. Kwak Y.H. and Stoddard J. (2003), "*Project Risk Management*": lessons learned from software development environment, article in press, Technovation XX (2003).

6. Ertan B., (1998), Analysis of Key Characteristic Methods and Enablers Used in Variation Risk Management, Submitted to the Department of Mechanical Engineering on May 8, 1998, in partial fulfillment of the requirements for the Degree of Master of Science at the Massachusetts Institute of Technology June, 1998.
7. Crevling C.M., Slutsky J.L., &. Antis. D., (2003), "*Design for Six Sigma*", Pearson Education (Singapore) Pte. Ltd., India.
8. Koch P.N., and Cofer J.I., (2002), IV, Simulation-Based Design In The Face Of Uncertainty, http://www.sightnacorp.com/download/iSIGHTSolutionCD/Turbo_apps/Koch_JANNAF_2002.pdf.
9. Clayton M., (2003), "*Design Risk Assessment*", http://www.isixsigma.com/dictionary/Design_Risk_Assessment-223.htm.
10. Kwak Y.H. (2003), "*Perceptions and Practice of Project Risk Management*", Aggregating 300 Project management years, Proceedings of the Project Management Institute Global Congress North America, September 18-25, 2003, Baltimore Maryland USA.
11. Vedpuriswar A.V., and Singh R.K., Enterprise Risk Management at Infosys, "*Enterprise Risk Management – Concepts and Cases*" – Vol. I, ICFAI Press, Hyderabad, India.
12. Brunson D., Hauam B., Mistry S., and Campean F., (2002), Designing for six sigma reliability.

SIX SIGMA IN DESIGN

37.1. INTRODUCTION

Designing for products are normally carried out with in fixed specification. This design, when released is operated in a tolerance band for almost all process variables. In reality, the product performance over the process tolerance limits are only assumed based on experience and not tested. This leads to initial trial provided from the lab being okay and then the successive trials from shop floor (which are supposed to be manufactured in an identical way), not being okay. Another aspect of the design failure is that the work piece material variations by the user not being captured at the design stage. This paper proposes specific practical techniques for capturing many of the failures during design and based on these, a ready reckon for further design is proposed.

It is shown that using the above concepts, First Time Right (FTR) designs can be maximized. This results in enhanced reliability in the performance that can be achieved at the customer end.

There are many metrics for product development [1–3]. Most of them measure only the revenue generated through new products. As a process, the best measure for a designer will be First Time Right (FTR). This shows how many times he is able to strike at the right design at the first iteration itself. Naturally, if the first trial itself works, the reputation of the person and the organization improves and automatically ensures the revenue growth. This can be achieved by a set of specific activities as listed below.

37.2. STEP 1—PROVIDING A PRODUCT THAT WORKS

Start with a reference product. Any development request comes from the marketing team, through specified formats for product development (3). Mostly, the details given do not completely cover all the implied needs.

One good way to get out of this is to find out what product is now acceptable to the user and personally understanding that product. It may be one of our own existing products being used by the user, who wants an improvement over the present performance. In this case, we can derive the needs from the present product capabilities.

Another case may be that he is using a competitor product and then it is important to grab a sample of the competitor product, do a breakdown analysis and then get to know the implied needs. It is always safer to start with a product reference.

Every year, all R and D persons work on a large number of thumb rules and these thumb rules stay back as individual possessions, some are valid and some are only beliefs. Also, such of those ideas which have not been really validated, can be verified by proper experimentation in the lab.

It is risky to try unproven concepts at the customer end [4]. The following method can be attempted to develop a list of proven concepts.

We propose a simple tool as shown in Fig. 37.1 for validation of such thumb rules. This is termed as Design Failure Analysis (DFA).

Present design	Present problem	Concept— What to change?
Changed design	Measure of problem stated	Result of concept
Conclusion	Concept works Add to design guidelines	Doc.ref no: DG/05/01
	Concept does not work Add to "Do Not Do" list	DND/05/1

Fig. 37.1: Format for design failure analysis

For a case study, let us assume that we have a grinding wheel that works and the customer wants a 20% improvement in life of the grinding wheel.

The present design is filled in the first column. The problem is stated as "Life is less by 20%". In the column for change, suppose I want to improve bond % by 2 units, I fill it as "increase bond by 2%". Then the new design with increased bond % is written in the second row first column. The lab trial is done and the life of the wheel with new design is evaluated. If the wheel shows twenty or more % improvement, then the result is filled as "OKAY". Otherwise, it is filled as "NOT OKAY".

If it is "OKAY", we allot a serial number for the concept and add to design guidelines. If not okay, we allot a serial number and add to the Do-Not-Do list.

A preferred numbering system could be:

- DG/year/serial number for successful concepts (DG—Design guideline)
- DND/year/serial number for failed concepts (DND—Do not do)

At the end of the year, when many lab trials have been done, all these can be compiled into two separate lists, arranged product wise. Of course, if done in a simple excel sheet, this data can become searchable on many parameters also.

Table 37.1: Design guidelines

Product	Required improvement	Concept to be used
Snagging wheel	Life	Raw material type

Table 37.2: Do-Not-Do list

Product	Concept	Failure observed
Precision wheel	Brown grain instead of White alumina	Wheel acts hard

Using the design guidelines, for any product, for a given performance improvement, concept to be used can be located. If such a concept is not available, then a set of experiments must be planned to verify several concepts and only those that are successful and find their way into this list must be taken to the field testing. When we did this for three years, the following were observed.

- Almost everyone uses only the same two or three concepts, passionately for every development, as their Universal solution;
- When shared, each one was richer by the concepts used by others;
- FTR improved from 35% to 65%; and
- The biggest challenge was that we found it difficult to add any new concept as years passed by—showing the need to be more innovative and not allowing us to live on the past.

37.3. STEP 2—MAKING YOUR DESIGN SALEABLE

A general mistake a designer does is that he looks at only performance and so always chooses a high performing design, without considering the price performance aspect. Evaluating the successful designs for performance per unit of cost would be a simple tool to choose the design that will sell for it (Fig. 37.2). Select the best design by using performance versus contribution graph. Conduct the lab level experiments. Present the data (not your conclusions) of all the experiments to HOD. All the designs should be plotted for performance versus contribution.

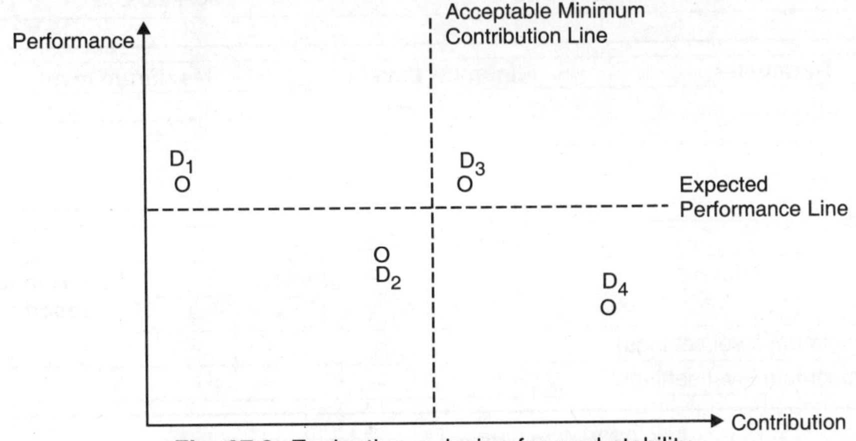

Fig. 37.2: Evaluating a design for marketability

In Fig. 37.2, the *x*-axis indicates contribution percentage and the *y*-axis indicates performance parameter (life, or number of parts machined etc). It can be seen the design D_1 meets performance but contribution is poor, design D_2 is very close to expected performance and acceptable contribution, design D_4 is very low in performance but high in contribution and D_3 is just of acceptable performance and acceptable contribution. The conclusion can be that the design D_4 is to be selected for scaling up whereas design D_2 can be worked upon for further development. On *y*-axis, any selected parameter of performance of the product can be mapped. For example in metal cutting, it could be life of the tool, number of components that can be machined before the tool wears, or surface finish achieved etc. Mostly, the performance expected by the customer is taken in this graph. For example, if the customer wanted the development only to improve the finish obtained on the component, then finish of the component will be the main parameter for comparison.

37.4. STEP 3—RETAINING THE PERFORMANCE DURING SCALING UP

Know the inconsistency band before launching the product. A basic mistake every R and D team does is to have equipment in R and D that is better than that in shop floor. As a result, products that come out of pilot production of R and D are always better in quality. A general failure everyone faces is the threat of losing the performance when scaled up. Many researchers are also branded as test tube researchers, by the manufacturing team. This can be avoided by the simple step of validating the chosen design over all the parameters and their tolerances that exist in shop floor.

Redo the selected design with the shop floor tolerance on various parameters, like density, viscosity, temperature (one set at lower tolerance level and another set at higher end of the tolerance level), as shown in Table 37.3. For example, for a grinding wheel, if density is specified as 33.5 ± 0.5 grams/inch^3 and temperature specified as 250 ± 20 °C, then the tolerance verifying experiments will be as shown in Table 37.3.

Table 37.3: Tolerance level of the specification

Parameter	Minimum level	Maximum level
Density	33.5 – 0.5 grams/inch^3	33.5 + 0.5 grams/inch^3
Temperature	250 – 20°C	250 + 20°C

Parameter	Minimum Level	Maximum level

Experimental Plan					
Trial no.	Parameter				Performance result
	1	2	3	4	
1. (minimum level settings)					
2. (maximum level settings)					

Fig. 37.3: Consistency clearance matrices

Trial 1 will be done with a density of 33.0 and a temperature of 230°C and Trial 2 will be done with a density of 34.0 and a temperature of 270°C. The results of these two products will be considered as the acceptable performance range for the existing process tolerances. The results show the performance variation possible with the permitted process tolerance levels that currently exist in the shop floor. This data is used to decide whether this variation in the new product performance can be allowed or design to be reexamined at. If the performance is allowable, then send for field trial. If not allowable, redo the design. This can be termed as the Consistency Clearance matrices Fig. 37.3.

Any one of the following conclusions may be chosen.

- The maximum performance range is acceptable. Proceed to next stage; or
- The performance band is not acceptable. Fine tune the design.

If the variation is too high, it is always better to look for a better design rather than trying to change the tolerance specification exclusively for this product alone. It is common knowledge that exclusive specifications, away from the normal practice will either lead to delays or mistakes.

37.5. STEP 4—GETTING GEARED TO HANDLE COMPLAINTS

Keep Track of the indicators. Loss has a direct link to complaints. Any defect, if detected inside the factory, is termed loss and whatever passes our gate and is detected by the customer, becomes a complaint.

So, any abnormal process loss happening during the manufacturing of the new product is a possible indicator of what kind of complaint may come up later. So, it is important to monitor the loss nature and loss level of the new product.

Hold on to the design that worked initially. In the unlikely case of a new product ending up in a performance complaint, the first accusation by most others will be on the design– because it is natural human tendency that the manufacturing team, by virtue of their hands on and long lasting experience of using materials and processes, start evaluating the way in which a design could have been made.

The most important aspect of a researcher is to have the guts to hold on to the design, which worked initially, without hastily getting into design corrections. It should be almost important to "Ban design changes for addressing complaints". Looking back, there are many process steps that convert a design recipe into the final product. Anything in between could have gone wrong. So, one must have the patience to analyze every aspect and to locate the process defect, before even thinking of changing the design.

Following a format called Process Failure Analysis (PFA), as shown in Fig. 37.4 can systematically do this.

First trial process details	Complaint batch process details	What could have gone wrong?
Change in process	Verification of complained parameter	Conclusion

Fig. 37.4: Process failure analysis

For example, let us say the third batch of grinding wheels got a complaint that they are very hard and not cutting at all. Then we start comparing various process parameters (from shop floor records) like, density (from QA report), dry house temperature (from oven charts), firing temperature (from kiln charts) etc., for both the first and third batch. The variations will then be analyzed to understand the cause of the complaint and then the necessary process corrections will be done, to simulate, the first batch process conditions, as closely as possible.

By following this method, we have been able to reduce the need for design changes for handling complaints.

For example, the defect could be due to one of the kilns overshooting in temperature on and off. Without correcting this, if the base product design is changed, the corrected design may also be liable to be affected by the kiln temperature overshooting. This may lead to another kind of complaint.

It is important to look beyond the manufacturing person's statement "I am doing only as usual. I have not done any change". Maybe he is right. But the tolerances in each parameter and the actual number of process parameters make it very complex in the possible number of ways in which something can go wrong.

Knowing the possible abuse of your product is very important. A customer may use your product in more than one way. We have had many pleasant experiences where the customer knows more than us in how a grinding wheel can be used. So, get to know how many ways are possible to abuse your product and evaluate those conditions in the lab.

For example, if a product is designed to grind mild steel, there are chances that the user may suddenly use it on stainless steel also. So, it is better to have evaluated in all possible materials used by the customer and offer that design which works quite okay in both the work piece materials.

Table 37.4: Improving an existing design

Steps	Sub Steps
Providing a product that works	Start with a reference product Use proven concepts only
Making your design saleable	Select the best design by performance versus contribution graph
Retaining the performance during scaling up	Know the consistency band before launching the product
Getting geared to handle complaints	Keep track of the indicators Hold on to the design that worked initially Know the possible abuse of your product Be ready with the next design.

Be ready with the next design. As everyone knows, competition also has some research person like you. As soon as you launch the product, your competitor also gives a product that works better than what you recently gave. So, within two months, through our marketing person the same requirement comes back for further development. No point starting all over again, because, it will be a loss of time. So, the best and simplest way is to plan an "all in one experiment" by using Taguchi methods [5–6]. We can use the possible design parameters, the possible process tolerance parameters etc. — all together in a Taguchi matrix and do a set of experiments. From the result of these experiments, it is possible to improve upon an existing design, without every time having to redo the experiments from scratch. This will help in releasing the next development very quickly. Table (37.4) exemplifies these concepts.

REVIEW QUESTIONS

1. Why often the specified formats for product development coming from marketing teams cannot cover all the implied needs of the customers?
2. What is DFA?
3. How to evaluate a design for marketability? Explain the tools available for it.
4. What is meant by performance scaling up and how it can be boosted up?
5. Describe the steps and sub-steps that may be followed for improving the existing design?

REFERENCES

1. James B. Ayers, Apply Management tools to development activities, "*Chemical Engineering Progress*", 1999, pp. 31–38.
2. S.D., Natu , New Product Strategy, "*Technology Trends*", 2001, pp. 64–65.
3. George J., Kidd Jr, The place of research and development in industry, Quality at Work in Research and Development, "*Quality Resources*", New York, 1992, pp. 27–41.
4. Radhika Dhawan, How to develop the best new products, "*Business Today*", 1999, pp. 74–85.
5. Lewis P. Yasenchak, "FMEA can head off potential problems before they occur", http://www.modplas.com, 2000, downloaded in Jan 2006.
6. M.S., Phadke, "*Quality Engineering Using Robust design*" Prentice Hall, Englewood Cliffs, N.J., Association, 1985, pp. 49–53.